Science, Law, and Hudson River Power Plants

A Case Study in Environmental Impact Assessment

Cover art and design by Angelika Peck, Carol Watson, and Daryl George
Johns Hopkins University Applied Physics Laboratory

Publication of the Monograph was made possible by grants from

Central Hudson Gas and Electric Corporation
Consolidated Edison Company of New York, Inc.
New York Power Authority
Niagara Mohawk Power Corporation
Orange and Rockland Utilites, Inc.

Hudson River Foundation for Science and Environmental Research

Electric Power Research Institute

U.S. Department of Energy

The views expressed in this Monograph do not necessarily reflect the views of the grantors, nor do the grantors assume any liability for the monograph's contents or for the use of the information contained herein.

Science, Law, and Hudson River Power Plants

A Case Study in Environmental Impact Assessment

Edited by

Lawrence W. Barnthouse
Ronald J. Klauda
Douglas S. Vaughan
Robert L. Kendall

American Fisheries Society Monograph 4

Bethesda, Maryland
1988

The American Fisheries Society Monograph series is a registered serial. Suggested citation formats follow.

Entire book

Barnthouse, L. W., R. J. Klauda, D. S. Vaughan, and R. L. Kendall, editors. 1988. Science, law, and Hudson River power plants: a case study in environmental impact assessment. American Fisheries Society Monograph 4.

Article within the book

Christensen, S. W., and T. L. Englert. 1988. Historical development of entrainment models for Hudson River striped bass. American Fisheries Society Monograph 4:133–142.

Publication 3020 of the Environmental Sciences Division
Oak Ridge National Laboratory

Library of Congress Catalog Card Number: 88-72144

ISSN 0362-1715   ISBN 0-913235-51-2

Address orders to

American Fisheries Society
5410 Grosvenor Lane, Suite 110
Bethesda, Maryland 20814, USA

CONTENTS

SECTION 5: CLOSING PERSPECTIVES

Preface

This retrospective on the Hudson River power plant case evolved from a symposium held at the 1982 annual meeting of the American Fisheries Society on Hilton Head Island, South Carolina. John Boreman stimulated the organization of that symposium. Its program represented another important demolition of the barriers between scientists for the Hudson River utilities and for the U.S. federal agencies. These barriers had arisen during several years of adjudicatory hearings over mitigation of power plant impacts on the Hudson River fish fauna. They began to erode during the negotiations over settlement of the case in 1979–1980.

Preparation of this monograph encouraged even more discussion and collaboration among key scientists who had been on opposing sides of the case. The cooperative spirit, persistence, and patience exhibited by authors during many years of monograph preparation are admirable, and appreciated by the editors. Publication of this monograph attests to our collective conviction that lessons learned from this power plant case must be told in the open and peer-reviewed literature.

Several monograph authors contributed reviews of other chapters, and we appreciate their efforts. We are especially grateful to the following external reviewers, whose comments led to substantial improvements in the book: Dean W. Ahrenholz, Herbert M. Austin, Philip T. Briggs, Richard C. Browne, Glen F. Cada, Alexander J. Chester, Ian Chisholm, David R. Colby, Dennis J. Dunning, Joseph H. Elrod, Michael C. Healey,

Joseph E. Hightower, John G. Holsapple, Wayne E. Hubert, Gene R. Huntsman, Peter E. Ihssen, A. L. Jensen, Robert L. Kellogg, Ron J. Kernehan, Boyd E. Kynard, Karin E. Limburg, William J. Madden, Jr., Barton C. Marcy, Jr., F. Joseph Margraf, John E. Matuszek, Jack S. Mattice, Ishwar Murarka, Russell S. Nelson, Stephen J. Nepszy, Dan B. Odenweller, James J. Orsi, William J. Overholtz, R. H. Peterson, Allyn B. Powell, Kent S. Price, Paul J. Rago, William A. Richkus, Saul B. Saila, Roger W. Schneidervin, George R. Sedberry, Eileen M. Setzler-Hamilton, William L. Shelton, Brian J. Shuter, C. Lavett Smith, Albert V. Tyler, Michael J. Van Den Avyle, David Welch, David R. Wolfert, Charles M. Wooley, and Byron H. Young.

This monograph could not have reached fruition without the financial support of several organizations: Central Hudson Gas and Electric Corporation, Consolidated Edison Company of New York, U.S. Department of Energy, Electric Power Research Institute, Hudson River Foundation for Science and Environmental Research, New York Power Authority, Niagara Mohawk Power Corporation, and Orange and Rockland Utilities. Their generosity was extended without constraints on the book's editorial and technical content, and we are grateful for it.

LAWRENCE W. BARNTHOUSE
RONALD J. KLAUDA
DOUGLAS S. VAUGHAN
ROBERT L. KENDALL
Editors

List of Fish Species

The colloquial names of most United States and Canadian fish species have been standardized in *A List of Common and Scientific Names of Fishes from the United States and Canada*, Fourth Edition, 1980, American Fisheries Society Special Publication 12. Throughout this monograph, species covered by the *List* are cited by their common names only. Their respective scientific names follow.

Alewife *Alosa pseudoharengus*
American eel *Anguilla rostrata*
American plaice *Hippoglossoides platessoides*
American sand lance *Ammodytes americanus*
American shad *Alosa sapidissima*
Atlantic cod *Gadus morhua*
Atlantic croaker *Micropogonias undulatus*
Atlantic herring *Clupea harengus harengus*
Atlantic mackerel *Scomber scombrus*
Atlantic menhaden *Brevoortia tyrannus*
Atlantic moonfish *Selene setapinnis*
Atlantic needlefish *Strongylura marina*
Atlantic salmon *Salmo salar*
Atlantic silverside *Menidia menidia*
Atlantic sturgeon *Acipenser oxyrhynchus*
Atlantic tomcod *Microgadus tomcod*
Banded killifish *Fundulus diaphanus*
Barndoor skate *Raja laevis*
Bay anchovy *Anchoa mitchilli*
Black crappie *Pomoxis nigromaculatus*
Black sea bass *Centropristis striata*
Blacknose dace *Rhinichthys atratulus*
Blueback herring *Alosa aestivalis*
Bluefish *Pomatomus saltatrix*
Bluegill *Lepomis macrochirus*
Bluespotted cornetfish *Fistularia tabacaria*
Bluespotted sunfish *Enneacanthus gloriosus*
Bluntnose minnow *Pimephales notatus*
Bridle shiner *Notropis bifrenatus*
Brook stickleback *Culaea inconstans*
Brook trout *Salvelinus fontinalis*
Brown bullhead *Ictalurus nebulosus*
Brown trout *Salmo trutta*
Butterfish *Peprilus triacanthus*
Central mudminnow *Umbra limi*
Chain pickerel *Esox niger*
Channel catfish *Ictalurus punctatus*
Chinook salmon *Oncorhynchus tshawytscha*
Chum salmon *Oncorhynchus keta*
Cobia *Rachycentron canadum*
Coho salmon *Oncorhynchus kisutch*
Comely shiner *Notropis amoenus*
Common carp *Cyprinus carpio*

Common shiner *Notropis cornutus*
Conger eel *Conger oceanicus*
Creek chub *Semotilus atromaculatus*
Crevalle jack *Caranx hippos*
Cunner *Tautogolabrus adspersus*
Cutlips minnow *Exoglossum maxillingua*
Eastern mudminnow *Umbra pygmaea*
Eastern silvery minnow *Hybognathus regius*
Emerald shiner *Notropis atherinoides*
Fallfish *Semotilus corporalis*
Fat sleeper *Dormitator maculatus*
Fathead minnow *Pimephales promelas*
Flying gurnard *Dactylopterus volitans*
Fourbeard rockling *Enchelyopus cimbrius*
Fourspine stickleback *Apeltes quadracus*
Fourspot flounder *Paralichthys oblongus*
Gizzard shad *Dorosoma cepedianum*
Golden shiner *Notemigonus crysoleucas*
Goldfish *Carassius auratus*
Gray snapper *Lutjanus griseus*
Green sunfish *Lepomis cyanellus*
Grubby *Myoxocephalus aenaeus*
Haddock *Melanogrammus aeglefinus*
Hickory shad *Alosa mediocris*
Hogchoker *Trinectes maculatus*
Inshore lizardfish *Synodus foetens*
Ladyfish *Elops saurus*
Largemouth bass *Micropterus salmoides*
Lined seahorse *Hippocampus erectus*
Logperch *Percina caprodes*
Longhorn sculpin *Myoxocephalus octodecemspinosus*
Longnose dace *Rhinichthys cataractae*
Lookdown *Selene vomer*
Mackerel scad *Decapterus macarellus*
Mummichog *Fundulus heteroclitus*
Naked goby *Gobiosoma bosci*
Ninespine stickleback *Pungitius pungitius*
Northern hog sucker *Hypentelium nigricans*
Northern kingfish *Menticirrhus saxatilis*
Northern pike *Esox lucius*
Northern pipefish *Syngnathus fuscus*
Northern puffer *Sphoeroides maculatus*

Northern searobin *Prionotus carolinus*
Northern stargazer *Astroscopus guttatus*
Pacific halibut *Hippoglossus stenolepis*
Pacific sardine *Sardinops sagax*
Petrale sole *Eopsetta jordani*
Pinfish *Lagodon rhomboides*
Pink salmon *Oncorhynchus gorbuscha*
Pollack *Pollachius virens*
Pumpkinseed *Lepomis gibbosus*
Rainbow smelt *Osmerus mordax*
Rainbow trout *Salmo gairdneri*[1]
Red hake *Urophycis chuss*
Redbrest sunfish *Lepomis auritus*
Redfin pickerel *Esox americanus americanus*
Redfish *Sebastes marinus*
Rock bass *Ambloplites rupestris*
Rock gunnel *Pholis gunnellus*
Rosyface shiner *Notropis rubellus*
Rough silverside *Membras martinica*
Round herring *Etrumeus teres*
Satinfin shiner *Notropis analostanus*
Scup *Stenotomus chrysops*
Sea lamprey *Petromyzon marinus*
Sea raven *Hemitripterus americanus*
Seaboard goby *Gobiosoma ginsburgi*
Sharksucker *Echeneis naucrates*
Sheepshead minnow *Cyprinodon variegatus*
Shield darter *Percina peltata*
Shortnose sturgeon *Acipenser brevirostrum*

---

[1]Recent taxonomic evidence indicates that *Salmo* species of Pacific Ocean drainages have close affinity with Pacific salmon *Oncorhynchus* spp., and should be classified therewith. Further, *S. gairdneri* now is considered to be conspecific with Kamchatkan trout, the name for which, *S. mykiss*, has taxonomic priority. Accordingly, the American Fisheries Society's Names of Fishes Committee has endorsed *Oncorhynchus mykiss* as the scientific name for rainbow trout.

Silver hake *Merluccius bilinearis*
Silver lamprey *Ichthyomyzon unicuspis*
Silver perch *Bairdiella chrysoura*
Smallmouth bass *Micropterus dolomieui*
Smallmouth flounder *Etropus microstomus*
Sockeye salmon *Oncorhynchus nerka*
South American anchoveta *Engraulis ringens*
Spiny dogfish *Squalus acanthias*
Spot *Leiostomus xanthurus*
Spotfin shiner *Notropis spilopterus*
Spottail shiner *Notropis hudsonius*
Spotted hake *Urophycis regia*
Striped anchovy *Anchoa hepsetus*
Striped bass *Morone saxatilis*
Striped cusk-eel *Ophidion marginatum*
Striped killifish *Fundulus majalis*
Striped mullet *Mugil cephalus*
Striped searobin *Prionotus evolans*
Summer flounder *Paralichthys dentatus*
Tautog *Tautoga onitis*
Tessellated darter *Etheostoma olmstedi*
Threespine stickleback *Gasterosteus aculeatus*
Tidewater silverside *Menidia peninsulae*
Trout-perch *Percopsis omiscomaycus*
Walleye *Stizostedion vitreum*
Warmouth *Lepomis gulosus*
Weakfish *Cynoscion regalis*
White bass *Morone chrysops*
White catfish *Ictalurus catus*
White crappie *Pomoxis annularis*
White mullet *Mugil curema*
White perch *Morone americana*
White sucker *Catostomus commersoni*
Windowpane *Scophthalmus aquosus*
Winter flounder *Pseudopleuronectes americanus*
Yellow bullhead *Ictalurus natalis*
Yellowfin tuna *Thunnus albacares*
Yellow perch *Perca flavescens*
Yellowtail flounder *Limanda ferruginea*

American Fisheries Society Monograph 4:1–8, 1988

SECTION 1: GENERAL INTRODUCTION

Introduction to the Monograph

LAWRENCE W. BARNTHOUSE

Environmental Sciences Division, Oak Ridge National Laboratory
Post Office Box 2008, Oak Ridge, Tennessee 37831-6036, USA

RONALD J. KLAUDA[1]

Texas Instruments Incorporated
Ecological Services Group, Buchanan, New York 10511, USA

DOUGLAS S. VAUGHAN[2]

Environmental Sciences Division, Oak Ridge National Laboratory

Between 1963 and 1980, the Hudson River estuary (Figure 1) was the focus of one of the most ambitious environmental research and assessment programs ever performed. The studies supported a series of U.S. federal proceedings involving licenses and discharge permits for two controversial electric power generating facilities: the Cornwall pumped storage facility, and units 2 and 3 of the Indian Point nuclear generating station. Both facilities were to draw large volumes of water from a region of the Hudson used as spawning and nursery habitat by several fish species, including the striped bass. Fishermen and conservationists feared that a major fraction of the striped bass eggs and larvae in the Hudson would be entrained with the pumped water and killed. Additional fish would be killed (and had been killed at Indian Point unit 1) when they were impinged (trapped) on trash screens at the intakes. Scientists were asked to aid the utility companies and regulatory agencies in determining the biological importance of entrainment and impingement.

Scientists spent more than 15 years studying the physical and chemical characteristics and biological productivity of the estuary and documenting the abundance, distribution, and life histories of the major fish species. New sampling equipment and analytical methods were developed. Sophisticated models of the striped bass and other important Hudson River fish populations were devel-

oped. Utility and agency scientists attempted to predict, on the basis of these studies, the likely impacts of major electric power generating facilities—especially Cornwall and Indian Point—on the biological resources of the Hudson. They defended their predictions in roughly a half dozen major adjudicatory hearings and assisted the negotiators, who ultimately arranged an out-of-court settlement.

Most of the results of these studies are buried in contractor reports, magnetic tape libraries, and hearings transcripts. Our objective in producing this monograph is to make this information accessible to the scientific community. Our intent is to (1) summarize the scientific issues and approaches, (2) present the significant results of the Hudson River biological studies, (3) describe the role of the studies in the decision-making process, (4) evaluate the successes and failures of the studies, and (5) present recommendations for future estuarine impact assessments.

This monograph contains both technical papers that present research results and synthesis papers that summarize and interpret the results. It is no secret that major disagreements arose among scientists over the validity of some of the data and the correctness of some of the analyses. We have sought to include all points of view in this book, and have not attempted to suppress legitimate scientific disagreements. Here, we present background information to establish the context for the papers that follow.

Study Areas and Major Research Groups

The Hudson River is unimpounded and tidal for 243 km between its confluence with upper New

---

[1]Present address: The Johns Hopkins University, Applied Physics Laboratory, Environmental Sciences Group, Shady Side, Maryland 20764, USA.

[2]Present address: Southeast Fisheries Center, National Marine Fisheries Service, Beaufort, North Carolina 28516, USA.

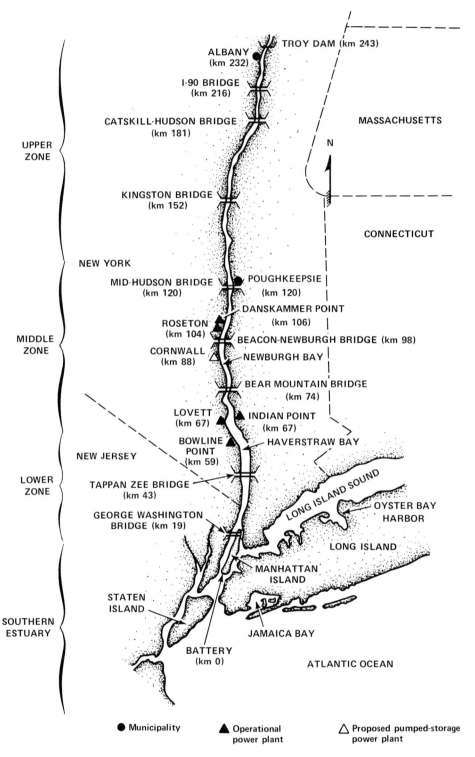

FIGURE 1.—Map of the Hudson River estuary with key geographic features and locations of power plants. Throughout this monograph, river distances are calculated from the Battery on southern Manhattan Island, km 0.

York Bay at the southern tip of Manhattan Island (the Battery, km 0) and the federal lock and dam at Troy, New York (km 243). This tidal portion of the Hudson River and adjacent coastal streams and bay (Figure 1) were the focus of intensive sampling in the late 1960s and throughout the 1970s.

Based on physiographic differences, the study area can be divided into three distinct zones: (1) a shallow, wide, and brackish lower zone extending from the George Washington Bridge to the upper end of Haverstraw Bay; (2) a deep, narrow, and usually freshwater middle zone from Haverstraw Bay through the Hudson Highlands to Newburgh Bay; and (3) a shallow, wide freshwater zone from Newburgh Bay to Troy Dam. Most sampling occurred in the 224-km segment of the river between the George Washington Bridge and Troy Dam, and was most concentrated in the vicinities of the six proposed and existing power plants (Figure 1) that were the primary focus of regulatory attention.

The cooperative involvement of many research groups was required to intensively study such a large region for so long a period. When the power-plant-related studies began on a relatively small scale in the mid-1960s, only New York University (NYU) and Northeastern Biologists were active in data collection. When the scope of studies and budgets expanded in response to the growing controversy concerning Cornwall and Indian Point, many research groups became involved as contractors for the Hudson River utility companies. The major contractors included Ecological Analysts (EA), Lawler, Matusky & Skelly Engineers (LMS; originally Quirk, Lawler & Matusky [QLM]), NYU, Raytheon Company, Stone and Webster Engineering Company, and Texas Instruments Incorporated (TI). James T. McFadden of the University of Michigan also played a key role in designing and reviewing the utilities' ecological programs. Many other groups participated on a lesser scale at times between 1963 and 1980. These include Boyce Thompson Institute, Edenton National Fish Hatchery, Hazelton Environmental Corporation, Marine Protein Corporation, Normandeau Associates, Incorporated, Oklahoma Department of Wildlife Conservation, Pennsylvania State University, Southern Illinois University, Johns Hopkins University, Oceanic Society, University of Rhode Island, and Welaka National Fish Hatchery.

Although the field and laboratory studies were funded and managed entirely by the utility companies, state and federal agencies funded a variety of modeling studies and data analyses in support of the regulatory proceedings. The primary agencies involved were the Federal Power Commission (FPC) and its successor, the Federal Energy Regulatory Commission (FERC); the U.S. Atomic Energy Commission (AEC) and its successors, the U.S. Nuclear Regulatory Commission (NRC), the Energy Research and Development Agency (ERDA), and the U.S. Department of Energy (DOE); the U.S. Army Corps of Engineers (COE); the U.S. Environmental Protection Agency (EPA); and the New York State Department of Environmental Conservation (NYSDEC). Research groups from the Oak Ridge National Laboratory (ORNL) and the U.S. Fish and Wildlife Service's National Power Plant Team (NPPT) performed the bulk of the agency-funded work. In addition, a large number of university-based individual consultants were employed by several of the agencies, most notably by EPA.

Historical Background

Most of the papers in this monograph emphasize analyses performed for the 1977–1980 EPA hearings and the 1979–1980 settlement negotiations. So that readers may understand the evolution of the scientific issues and methodological approaches, we provide a brief historical outline of the pre-1977 studies and the legal proceedings that motivated them. Readers interested in further details are referred to papers by Christensen et al. (1981) and Barnthouse et al. (1984) and to reports by Sandler and Schoenbrod (1981) and by Limburg et al. (1986).

The origins of the Hudson River biological studies can be traced to a river survey program known as the Hudson River fisheries investigations (HRFI), which was administered between 1965 and 1968 by the Hudson River Policy Committee. The HRFI surveys were intended to address an issue raised by two citizens' groups, the Scenic Hudson Preservation Council and the Hudson River Fishermen's Association. These groups contended that entrainment of striped bass eggs and larvae by the proposed Cornwall pumped storage facility might have a devastating impact on the Hudson River striped bass population. The objective of the HRFI program was to define the spatial and temporal distribution of striped bass eggs, larvae, and juveniles in relation to the Cornwall intake so that the impact of the water withdrawals could be estimated.

The summary report produced by the Hudson River Policy Committee (1968) concluded that the impact of Cornwall would be negligible. The report was used to support the FPC's 1970 decision to license the Cornwall facility. The HRFI findings were later challenged on the grounds that an erroneous method had been used to estimate striped bass entrainment (see Christensen and Englert 1988, this volume). Construction of Cornwall was halted in 1974 pending a resolution of this issue.

During and after the HRFI studies, NYU and later Raytheon conducted modest fish sampling programs in the vicinity of the Indian Point nuclear generating station. Indian Point unit 1 was then in operation, unit 2 was nearing completion, and unit 3 was under construction. The sampling at Indian Point assumed major importance when, in 1971, the U.S. Court of Appeals for the District of Columbia held that the National Environmental Policy Act required the AEC to explicitly consider nonradiological environmental impacts in its licensing decisions. The AEC was required to prepare an environmental impact statement before issuing an operating license for Indian Point unit 2.

As in the controversy over the Cornwall facility, the major point of contention concerning Indian Point unit 2 was mortality of striped bass eggs, larvae, and juveniles due to entrainment through the plant's once-through cooling system. The Consolidated Edison Company of New York (ConEd), the owners of Indian Point unit 2, submitted an environmental report concluding that entrainment would be insignificant. The AEC staff, in its environmental impact statement (USAEC 1972), agreed that entrainment of zooplankton and phytoplankton would have a negligible impact. However, the impact statement concluded that entrainment of striped bass by Indian Point unit 2 could substantially deplete the Hudson River striped bass population and recommended that a closed-cycle cooling system (i.e., a cooling tower) be built to reduce water withdrawals.

The most notable scientific aspect of the hearings that followed was the prominent role of mathematical models in the impact calculations performed by both agency and utility scientists. The AEC's prediction of a significant depletion was derived from an entrainment model conceived by C. P. Goodyear of ORNL (an early version of the model was described by Hall 1977). This model was the first attempt to simulate the movement of ichthyoplankton in response to the tidal circulation of the Hudson. Later, J. P. Law-

ler of QLM (consultants to ConEd) employed a one-dimensional advection–dispersion model for the same purpose and concluded that entrainment of striped bass would be inconsequential (Lawler 1972). Lawler was also the first to use a model of the striped bass life cycle to project reductions in the abundance of adult striped bass as a result of entrainment of the young-of-the-year fish.

The assumptions and parameters of these models and the validity (or lack thereof) of the projections derived from them were considered at great length in decisions issued by the Atomic Safety and Licensing Board (1973) and the Atomic Safety and Licensing Appeals Board (1974). The net result of these two decisions was to allow Indian Point unit 2 to operate with once-through cooling, but only until May 1, 1979. A cooling tower would have to be in place by that date unless new studies funded by ConEd showed that the impact of once-through cooling was negligible.

In late 1974, the FPC held hearings to reconsider the potential impact of the Cornwall facility in light of information gathered since the HRFI study. The data and models developed for the Indian Point unit 2 hearings (as well as some new ones) were presented at the FPC hearings. The FPC was unable to reconcile the conflicting assessments; the hearings were adjourned with no resolution and no licensing decision. In 1975, the NRC (successor to the AEC) agreed to allow Indian Point unit 3 to operate with once-through cooling, subject to the same restrictions it had applied to unit 2: cessation of once-through cooling by May 1, 1979, unless new evidence showed closed-cycle cooling to be unnecessary.

From 1973 onward, ConEd and other utility companies (Orange and Rockland Utilities, Incorporated, Central Hudson Gas and Electric Company, and the Power Authority of the State of New York) operating power plants on the lower and middle zones of the Hudson funded a research effort of unprecedented magnitude to obtain the new evidence needed to resolve the entrainment question. The largest component of the utility research program was a river-wide ecological study performed by TI. Ichthyoplankton surveys conducted by TI provided weekly estimates of ichthyoplankton distribution and abundance from the George Washington Bridge to Troy Dam, from April through mid-August, for the years 1973 through 1980. Although the primary objective of the ichthyoplankton survey was to obtain accurate information on striped bass, the data gathered provided a basis for estimating the impact of entrainment on other species, notably white perch, Atlantic tomcod, and American shad. A fisheries survey intended to characterize the spatial and temporal distributions of juvenile and older fish vulnerable to impingement at power plants was also conducted. In addition to the river-wide surveys, TI conducted a variety of special studies, including mark–recapture programs, stock assessments, feeding studies, a striped bass rearing and stocking program, and an evaluation of the contribution of the Hudson River stock to the Atlantic coastal striped bass population.

Complementing the river-wide ecological survey program, near-field and in-plant sampling programs were conducted at the Bowline Point, Lovett, Indian Point, Roseton, and Danskammer power plants. The primary objectives of these programs were to estimate the abundance and survival of fish entrained and impinged at the plants.

Because the Atomic Safety and Licensing Appeals Board, in 1974, had endorsed the advection–dispersion models and population simulation models developed by QLM over ORNL's simpler model, both the utilities and the federal agencies funded ambitious research programs to develop state-of-the-art striped bass models. Many of the field and laboratory studies discussed above were designed partly or primarily to provide data for calibrating and verifying these models.

The ultimate objective of both the utility and agency modeling efforts was to obtain the most accurate possible estimates of the long-term effects of water withdrawals by Cornwall, Indian Point, and other power plants on the abundance and persistence of the Hudson River striped bass population. Within a few years, however, many scientists on both sides of the licensing issue became convinced that this objective would never be achieved. By 1974 TI's ichthyoplankton surveys had demonstrated that transport models relying solely on hydrodynamic principles significantly overestimated rates of downstream movement of all life stages of striped bass (Christensen et al. 1981). Although attempts were made to fix up the models by incorporating empirically fitted "transport-avoidance factors," purely empirical models that estimated entrainment directly from weekly distributional data were soon developed by both utility and agency scientists (TI 1975; Boreman et al. 1981). There were other serious problems with both the QLM–LMS and the ORNL striped bass models. Swartzman et al.

(1978) performed detailed comparisons of several versions of the LMS and ORNL models. They concluded that (1) differences in predictions obtained from the models were due primarily to differences in parameter estimates and assumptions concerning the importance of biological compensation rather than to differences in the underlying model structures, and (2) none of the models were useful for predicting entrainment impacts. Given the inevitable uncertainties in the data and the large numbers of functions and parameters in the models, plausible cases for either very small or catastrophic impacts could readily be based on any of the models.

Subsequent assessment efforts performed in connection with EPA proceedings de-emphasized simulation modeling. Following the enactment of the 1972 amendments to the Federal Water Pollution Control Act, the EPA took steps to issue discharge permits for all the Hudson River power plants, both nuclear and fossil-fuel. The permits issued in 1975 for Indian Point units 2 and 3, Bowline Point units 1 and 2, and Roseton units 1 and 2 specified the construction of cooling towers. The four utility companies involved contested the permits and asked for adjudicatory hearings.

In preparing for the EPA hearings, which began in 1977 and ended with the 1980 settlement agreement, all parties devoted their major effort to analyzing the rapidly accumulating data base (Vaughan et al. 1988, this volume).

The most controversial of the issues raised at these hearings was the potential importance of density-dependent regulatory mechanisms in offsetting direct mortality caused by the power plants. Attempts to identify density-dependent mortality or growth in Hudson River fish populations (McFadden 1977) and to quantify them with stock–recruitment models (McFadden and Lawler 1977) were vigorously contested.

In contrast to previous proceedings, considerable attention was paid in the EPA hearings to species in addition to striped bass. Substantial efforts were devoted to estimating the impacts of entrainment on individual year classes based on near-field and river-wide data on the distribution and abundance of vulnerable life stages, and on estimates of through-plant mortality obtained from laboratory studies and in-plant sampling. Analogous methods and data were used to estimate the impacts of impingement on single year classes. In addition to estimating the actual historical impacts of the Hudson River power plants, scientists working for both sides, utility and gov-

TABLE 1.—Sources of technical reports and legal documents relevant to the Hudson River power plant case.

Data or document set	Sponsor	Source
	Utility-sponsored biological studies	
Station-specific and near-field surveys		
Bowline Point	Orange and Rockland Utilities (ORU)	1
Lovett	ORU	1
Indian Point		
Unit 2	Consolidated Edison (ConEd)	2
Unit 3	New York Power Authority (NYPA)	3
Cornwall	ConEd	2
Roseton	Central Hudson Gas and Electric (CHG&E)	4
Danskammer Point	CHG&E	4
Far-field surveys	ORU, ConEd, NYPA, CHG&E	2
	Agency-sponsored analyses (Technical reports)	
All	Federal Power Commission (FPC), Federal Energy Regulatory Commission, Atomic Energy Commission (AEC), Nuclear Regulatory Commission (NRC), Energy Research and Development Agency, Department of Energy, Army Corps of Engineers, Environmental Protection Agency (EPA), New York State Department of Environmental Conservation	5, 6
	Hearing dockets (Environmental impact statements, testimonies, decisions)	
Indian Point units 2 and 3	AEC, NRC	7
Cornwall	FPC	8
Hearings and settlement	EPA	9

Addresses of sources

1 Manager, Environmental Services
   Orange and Rockland Utilities, Incorporated
   One Blue Hill Plaza
   Pearl River, New York 10965, USA

2 Director, Biological Studies
   Consolidated Edison Company of New York, Incorporated
   4 Irving Place
   New York, New York 10003, USA

3 Manager, Hudson River Studies
   New York Power Authority
   123 Main Street
   White Plains, New York 10601, USA

4 Director, Environmental Affairs
   Central Hudson Gas and Electric Corporation
   284 South Avenue
   Poughkeepsie, New York 12602, USA

5 National Technical Information Service
   5285 Port Royal Road
   Springfield, Virginia 22151, USA

6 (Oak Ridge National Laboratory reports)
   U.S. Department of Energy Technical Information Center
   Post Office Box 62
   Oak Ridge, Tennessee 37831, USA

7 (On-site inspection and copying only)
   U.S. Nuclear Regulatory Commission Public Document Room
   1717 H Street, NW
   Washington, D.C. 20555, USA

8 Office of Public Reference
   Federal Energy Regulatory Commission
   U.S. Department of Energy
   825 North Capitol Street, NE
   Washington, D.C. 20426, USA

9 (On-site inspection and copying only)
   Environmental Sciences Division
   Oak Ridge National Laboratory
   Post Office Box 2008
   Oak Ridge, Tennessee 37831

ernment, attempted to estimate the magnitudes of impacts that would have occurred if various mitigating measures had been employed. At first, these efforts were independent. During the settlement negotiations, however, a cooperative effort involving both agency and utility scientists provided the scientific foundation for the settlement described in Section 4 of this monograph.

Monograph Organization

The monograph is organized into five sections. Section 1 consists of this introduction. Section 2 contains concise summaries of essential descriptive material on the Hudson River estuary, including discussions of river hydrography, physico-chemistry, carbon and nutrient budgets, trophic structure, and fish community composition. The life histories and spatiotemporal distributions of the three most intensively studied fish species (striped bass, white perch, and Atlantic tomcod) are described, and the major power plants on the Hudson are discussed. Section 3 presents findings related to estimating the short-term and long-term effects of entrainment and impingement on fish populations. Both the techniques used to estimate entrainment and impingement impacts and the results of assessment studies are presented. Section 3 also contains four papers on the controversial topics of population regulation and stock–recruitment modeling as well as some mutual commentary by the respective authors. Section 4 summarizes the major settlement conditions and features discussions of the three key measures agreed on for mitigating entrainment and impingement: flow reductions and scheduled outages, a barrier net at the Bowline Point generating station, and a striped bass hatchery. Section 5 presents five summary and synthesis papers. Three of these discuss the roles of science and scientists in the adjudicatory process, as seen by scientists, by a lawyer, and by an administrative law judge. The remaining two papers present our view of the successes and failures of the Hudson River research and assessment efforts.

This monograph contains concise, peer-reviewed presentations and interpretations of most of the major data sets compiled during the major study years and not already published elsewhere. Detailed descriptions of sampling procedures, quality assurance programs, analytical methods, and results are contained in the technical reports cited throughout the monograph. Many of these reports may be requested from the individuals listed in Table 1 or from the authors who cited the reports.

Acknowledgments

We thank the Hudson River Foundation, the Consolidated Edison Company of New York, Orange and Rockland Utilities, Central Hudson Gas and Electric, the Power Authority of the State of New York, the Electric Power Research Institute, and the U.S. Department of Energy for their financial support. Research was sponsored by the Office of Health and Environmental Research, U.S. Department of Energy, under contract DE-ACO5–84OR21400 with Martin Marietta Energy Systems, Incorporated. This is publication 2891, Environmental Sciences Division, ORNL.

References[3]

Atomic Safety and Licensing Appeals Board (of the U.S. Atomic Energy Commission). 1974. Decision in the matter of Consolidated Edison Company of New York, Inc. (Indian Point station, unit no. 2). Docket 50-247.

Atomic Safety and Licensing Board (of the U.S. Atomic Energy Commission). 1973. Initial decision in the matter of Consolidated Edison Company of New York, Inc. (Indian Point station, unit no. 2). Docket 50-247.

Barnthouse, L. W., J. G. Boreman, S. W. Christensen, C. P. Goodyear, W. Van Winkle, and D. S. Vaughan. 1984. Population biology in the courtroom: the Hudson River controversy. BioScience 34:14–19.

Boreman, J., C. P. Goodyear, and S. W. Christensen. 1981. An empirical methodology for estimating entrainment losses at power plant sites on estuaries. Transactions of the American Fisheries Society 110:253–260.

Christensen, S. W., and T. L. Englert. 1988. Historical development of entrainment models for Hudson River striped bass. American Fisheries Society Monograph 4:133–142.

Christensen, S. W., W. Van Winkle, L. W. Barnthouse, and D. S. Vaughan. 1981. Science and the law: confluence and conflict on the Hudson River. Environmental Impact Assessment Review 2:63-88.

Hall, C. A. S. 1977. Models and the decision making process: the Hudson River power plant case. Pages 345–364 in C. A. S. Hall and J. W. Day, editors. Ecosystem modeling in theory and practice. Wiley, New York.

Hudson River Policy Committee. 1968. Hudson River fisheries investigations (1965–1968). Report to Consolidated Edison Company of New York.

---

[3]See Table 1 for sources of legal documents and unpublished reports pertaining to the Hudson River.

Lawler, J. P. 1972. Effect of entrainment and impingement at Indian Point on the population of the Hudson River striped bass. Modifications and additions to testimony of April 5, 1972. Testimony (October 30, 1972) before the U.S. Atomic Energy Commission in the matter of Consolidated Edison Company of New York, Inc. (Indian Point station, unit no. 2). Docket 50-247.

Limburg, K. E., M. A. Moran, and W. H. McDowell. 1986. The Hudson River ecosystem. Springer-Verlag, New York.

McFadden, J. T., editor. 1977. Influence of Indian Point unit 2 and other steam electric generating plants on the Hudson River estuary, with emphasis on striped bass and other fish populations. Report to Consolidated Edison Company of New York.

McFadden, J. T., and J. P. Lawler, editors. 1977. Influence of Indian Point unit 2 and other steam electric generating plants on the Hudson River estuary with emphasis on striped bass and other fish populations, supplement 1. Report to Consolidated Edison Company of New York.

Sandler, R., and D. Schoenbrod, editors. 1981. The Hudson River power plant settlement. (Materials prepared for a conference sponsored by New York University School of Law and the Natural Resources Defense Council.) New York University School of Law, New York.

Swartzman, G. L., R. B. DeRiso, and C. Cowan. 1978. Comparison of simulation models used in assessing the effects of power-plant-induced mortality on fish populations. University of Washington, College of Fisheries, Center for Quantitative Science, UW-NRL-10, Seattle.

TI (Texas Instruments). 1975. First annual report for the multiplant impact study of the Hudson River estuary. Report to Consolidated Edison Company of New York.

USAEC (U.S. Atomic Energy Commission). 1972. Final environmental statement related to operation of Indian Point nuclear generating plant, unit no. 2, volumes 1 and 2. Docket 50-247.

Vaughan, D. S. 1988. Introduction [to entrainment and impingement impacts]. American Fisheries Society Monograph 4:121–123.

American Fisheries Society Monograph 4:9–10, 1988

SECTION 2: STUDY AREA, FOCAL FISH SPECIES, AND POWER PLANTS

Introduction

RONALD J. KLAUDA[1]

Texas Instruments Incorporated
Ecological Services Group, Buchanan, New York 10511, USA

The Hudson River power plant case grew out of clashes, part real and part potential, among many parties because of interactions between a few power plants and several fish species. In the end, as this monograph attempts to explain, the perceived significance of these clashes to each party depended heavily upon which fish species, which power plants, and which impact projections were of most interest and could be most comfortably accepted. The papers in this section describe the mute and disinterested parties to the complex drama (mostly legal and political, but partially ecological) that was played out (and to a lesser extent continues) by the large and diverse cast of utility executives, regulatory agency staffs, consultant scientists, resource managers, commercial and sport fishers, conservationists, preservationists, concerned citizens, lawyers, judges, and interested observers from many professions during the 1960s and 1970s.

The purpose of this section is to describe the portion of the Hudson River estuary, the fish species, and the power plants that were foci in the controversy that absorbed over a decade of intensive and expensive study. The papers in this section summarize a vast amount of basic descriptive information. This background information is essential for an understanding of the range of power plant impact estimates debated during the case, and also contributes importantly to the available data base for Hudson River fishes.

The power plants and fish species involved in the case make use of a 243-km segment of the estuary that is unimpounded and tidal. Far from pristine, the study area embraces navigable water that has become cleaner from one perspective (less raw sewage) but contaminated from another

(more PCBs). Boyle (1969) summed up his view, shared by others, when he wrote, "To those who know it, the Hudson River is the most beautiful, messed up, productive, ignored, and surprising piece of water on the face of the earth." In this section of the monograph, Cooper et al. provide a physicochemical overview of the estuary and describe the biologically relevant characteristics (e.g., flow, salinity) that influence the distribution of fishes and determine their relative vulnerabilities to each power plant.

In spite of its problems, those who have studied the lower Hudson River know it supports a large assemblage of resident, diadromous, and marine fish species (Smith 1985). The paper by Beebe and Savidge discusses this assemblage and reveals that species diversity has remained high in the face of decades of intensive human activity. Gladden et al. delineate the pathways of energy flow to the fish populations and provide a basis for understanding observed variations in species abundance.

Among the river's 140 finfish species, three were of most concern during the power plant studies. Striped bass was always considered the most important species because it spawns near several power plants, migrates to coastal waters, and is extremely popular with recreational and commercial fishers. Boreman and Klauda describe the distributions of early life stages, and Hoff et al. discuss adult population characteristics of the striped bass in the Hudson River. White perch was selected as a focal species because it is a ubiquitous and abundant resident that is frequently impinged on cooling water intake screens at several power plants. The life history and population dynamics of white perch in the Hudson River are summarized by Klauda et al. Atlantic tomcod was the third focal species. As McLaren et al. describe, the Atlantic tomcod population in the Hudson River is at the southern end of the

[1]Present address: The Johns Hopkins University, Applied Physics Laboratory, Environmental Sciences Group, Shady Side, Maryland 20764, USA.

species' range, may be thermally stressed much of the time, and depends heavily on a single age-group for reproduction. This anadromous species was viewed by study participants as a potentially sensitive indicator of environmental stress that might be induced by power plant operations.

To assess the effects of power plants on fishes, it is essential that population sizes be estimated accurately. This is difficult enough in a 10-hectare freshwater impoundment; in a large open-ended system like the Hudson River estuary, the task presents special problems, especially when two focal species, striped bass and Atlantic tomcod, are anadromous. Young et al. outline the methods by which fish population sizes were gauged and the extent to which sampling and analytical problems were solved during the power plant studies.

The section's final paper by Hutchison describes the power plants. These facilities include six fossil-fueled stations, the Indian Point nuclear plant, and a pumped storage plant proposed for construction near Cornwall, New York. Plans to construct the Cornwall facility were abandoned in late 1980 (Christensen et al. 1981; Barnthouse et al. 1988, this volume), but the proposal itself spawned much of the controversy and strongly influenced environmental study designs in the Hudson River during the 1960s and 1970s. Location, generating capacity, cooling water intake and discharge design, and operation mode vary among the river's seven existing and one proposed power plants. These design and operational differences contribute to an understanding of which fish species were entrained or impinged at each power plant and why the mortality rates of entrained and impinged fish varied among plants, topics that are discussed in Sections 3 and 4 of this monograph. A major characteristic held in common by all the power plants, however, was that they were designed to withdraw large quantities of cooling water from the river. These water withdrawals were the central issue of the Hudson River power plant case.

References

Barnthouse, L. W., J. Boreman, T. L. Englert, W. L. Kirk, and E. G. Horn. 1988. Hudson River settlement agreement: technical rationale and cost considerations. American Fisheries Society Monograph 4:267–273.

Boyle, R. H. 1969. The Hudson River, a natural and unnatural history. Norton, New York.

Christensen, S. W., W. Van Winkle, L. W. Barnthouse, and D. S. Vaughan. 1981. Science and the law: confluence and conflict on the Hudson River. Environmental Impact Assessment Review 2:63–88.

Smith, C. L. 1985. Inland fishes of New York state. New York State Department of Environmental Conservation, Albany.

American Fisheries Society Monograph 4:11-24, 1988
© Copyright by the American Fisheries Society 1988

Overview of the Hudson River Estuary

JON C. COOPER,[1] FRANK R. CANTELMO,[2] AND CHARLES E. NEWTON[3]

Texas Instruments Incorporated
Ecological Services Group, Buchanan, New York 10511, USA

Abstract.—This paper presents an overview of the geography, hydrology, and physicochemistry of the Hudson River estuary as a basis for understanding the life histories of fishes and other aquatic organisms. Salinity distribution in the estuary is mainly influenced by freshwater flow and tidal currents. Two-thirds of the freshwater inflow is controlled by a dam near Troy, New York, 243 river kilometers above the Battery in New York City; this dam defines the upper limit of the estuary. South of Troy Dam, flow increases linearly through the remaining third of the watershed. Even in times of high freshwater flow, however, net flow is dominated by tidal flow. Freshwater discharge influences seasonal water temperatures, hence dissolved oxygen concentrations, as well as turbidity; pH, however, shows no temporal or spatial patterns. The estuary is generally limnetic (<0.3‰ salinity) above km 80, oligohaline (0.3–5‰) between km 80 and 40, and mesohaline (5–18‰) below km 40. The salt front (0.1‰) is generally found near km 80, but can move rapidly as freshwater flow changes. Salinity increases regularly between the salt front and the sea, but irregular variations in the estuary's width, depth, and configuration cause local, unpredictable distortions in the overall gradient, especially when flows are high. The estuary is stressed by sewage and industrial discharges, polychlorinated biphenyls, pesticides, and agricultural chemicals, though loadings of some pollutants have abated in recent years. Major power plants are sited in the mesohaline zone, where tidal energy and biological productivity are large. By their withdrawal of large volumes of river water, which they return at elevated temperatures, power plants have potentially large effects on the organisms that are found in these waters. The prevailing opinion among many researchers is that fluctuations in salinity, temperature, and freshwater input are dominant influences on the biota of the Hudson River. For example, salinity patterns are important to the distribution of fish and benthos; temperature patterns determine species distributions, and reproductive strategies. Because of their size and placement, power plants have the potential to affect all of these factors.

The Hudson River arises at Lake Tear of the Clouds in the Adirondack Mountains of northern New York State and flows south 507 km to the Atlantic Ocean (Figure 2). North of Albany, it receives a major tributary, the Mohawk River; south of there, the Hudson watershed is contained by the Catskill Mountains and the Helderberg escarpment to the west and the eastern highlands along the New York–New England border. The lower Hudson River receives several modest tributaries from these mountains and highlands (Figure 3), but this region of the river itself is estuarine, being under tidal influence and, in its lower reaches, saline. The discussions here relate mostly to the Hudson estuary.

---
[1]Present address: International Scholars for Environmental Studies, 107 Canner Street, New Haven, Connecticut 06511, USA.
[2]Present address: Department of Biological Sciences, St. John's University, Grand Central and Utopia Parkways, Jamaica, New York 11439, USA.
[3]Present address: WAPPORA, Incorporated, 6709 Joy Drive, The Colony, Texas 75056, USA.

The tidal portion extends 243 km northward from the Battery on Manhattan Island (km 0) to Troy Dam (Lock) just north of Albany (Figure 3). The Hudson Valley has been scoured by glaciers. Where mountains imposed lateral constrictions on ice flow, the glaciers gouged the valley deeper, and the modern river bottom still has not been graded to base level. As the river passes by the Catskill and through the Appalachian mountains, from km 170 to 65, its bottom is deeper than it is farther downstream (Figure 4); the maximum depth is 66 m near West Point. These deep holes aside, the river gradient is small—only 1.5 m—over the 224 km from Albany to New York, and the river bottom at Albany is at sea level (Fenneman 1938). In general, river width and cross-sectional area vary inversely with depth (Figure 4). The average width of the Hudson River at the mouth is about 1,000 m; above km 50, it reaches its maximum width, nearly 6 km, and then narrows.

Hydrology

The upper Hudson and Mohawk watersheds (25,927 km^2) contribute approximately 80% of

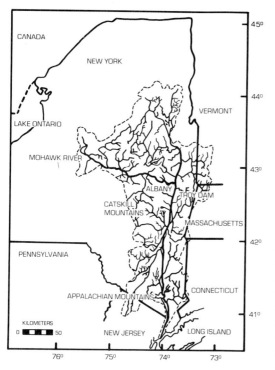

FIGURE 2.—Hudson River watershed.

The maximum tidal excursion in the Hudson is approximately 21 km. The morphometry of the river channel is such that simple upriver and downriver currents do not exist during flood and ebb, respectively. Flow in the deepest part of the channel reverses with the tides, but large eddies form at turns and along shoreline projections. The major currents in the vicinity of sharp bends in the channel may be on opposite sides of the channel during the ebb and flood tides. The mean ebb current velocity is 0.4 m/s, and the mean tidal flood current velocity is 0.36 m/s (LMS 1978a, 1978b).

The river's greatest tidal amplitude (the difference between the mean of high and the mean of low tides within a 24-h period) occurs at Albany (mean, 1.56 m), although it is only slightly higher there than at the mouth of the river (1.37 m: Geise and Barr 1967). During the period 1890–1951, the mean tidal amplitude at Albany increased from approximately 0.8 to 1.6 m principally because the navigation channel in the upper estuary was deepened and widened, increasing the river's cross-sectional area (Geise and Barr 1967). Mean tidal amplitude is least at West Point (0.8 m).

Water Chemistry

The water chemistry of the Hudson River estuary is influenced by temporal and localized spatial variations. Although descriptive equations may provide good general measures of longitudinal trends for chemical variables, they may not always describe local conditions accurately.

The most intensive need for power plant impact assessment was near the Indian Point facility at km 61. Thus, in the temperature and dissolved oxygen models (described below) km 63 is taken as the reference location in the estuary. Although this area is only one-quarter of the way from Manhattan to Troy Dam, it is within a zone of transition from fresh to salt water. Above this zone is a large reach (often over 100 km long) of fresh water where the chemistry is relatively constant. Downriver of this zone, salinity increases fairly steadily. The estuary's largest and most dramatic changes in water chemistry are expected in the transitional zone, where the fresh and saline water masses mix.

The main sources of data in the following sections are the surveys conducted on the Hudson River by Texas Instruments Incorporated from 1972 through 1977 (TI 1976b, 1977a, 1977b, 1978a, 1978b, 1979a, 1981). Most samples were collected as part of an ongoing study on fish populations.

annual freshwater flow through the Hudson estuary. The drainage basin below Troy Dam (18,753 km²) contributes the remainder. The annual flows entering the estuary from tributaries are directly related to the watershed area of its tributaries ($r = 0.987$; $P < 0.01$), and there is a linear increase in watershed area from Albany southward ($r = 0.986$; $P < 0.01$: TI 1976a). The annual discharge at Green Island, where the U.S. Geological Survey maintains a gauging station just downstream of Troy Dam, averaged 13.1×10^9 m³ over the years 1966–1983 (Table 2); the net flow (total outflow minus tidal incursions) at New York City has been 18.2×10^9 m³ over the same period. Monthly discharges are greatest in spring and fall, associated with snow melt and rains (Table 2).

Even high freshwater flows are small compared to the magnitude of tidal flows (Darmer 1969), which are 10–100 times greater; they range from about 5,679 to 8,500 m³/s, but may exceed 14,000 m³/s (Busby 1966). The tidal regime of the Hudson River estuary essentially reverses the current pattern twice each day, although strong winds from the south and north can push water into or out of the estuary, obscuring the true tidal regime (Busby 1966).

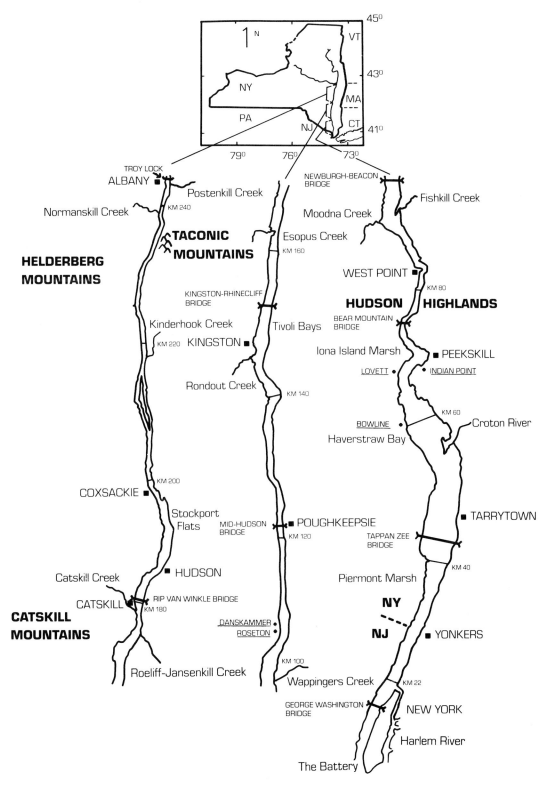

FIGURE 3.—Geographical features of the estuarine lower Hudson River.

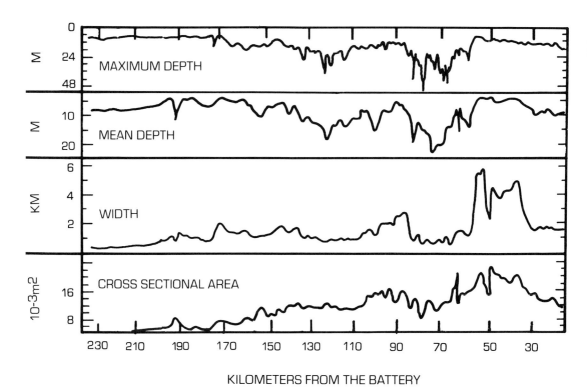

FIGURE 4.—Depths, widths and cross-sectional areas of the estuarine lower Hudson River.

Water temperature and dissolved oxygen were measure in situ; conductivity, turbidity, and pH were measured in the field or in samples returned to the laboratory for analysis by standard methods.

Freshwater discharge data for the Hudson River (at Green Island, New York) were taken from USGS (1966–1984; Table 2). Tidal amplitudes for the Hudson estuary were derived from

TABLE 2.—Monthly and annual freshwater discharge of the Hudson River at Green Island, below Troy Dam, 1966–1983 (USGS 1966–1984).

Year	Jan	Feb	Mar	Apr	May	Jun	Jul	Aug	Sep	Oct	Nov	Dec	Annual discharge ($10^9 m^3$)
1966	230.2	329.3	653.8	442.5	521.2	234.2	104.0	119.8	159.4	165.6	199.4	258.2	8.874
1967	272.3	216.1	321.8	876.0	483.1	175.5	143.7	162.8	139.7	197.5	332.5	467.5	9.836
1968	251.1	269.4	704.0	518.2	523.5	444.8	277.4	125.7	126.4	146.5	407.8	441.7	10.999
1969	330.8	361.4	494.6	1,153.4	592.2	283.0	153.8	172.8	117.0	137.5	404.1	334.2	11.774
1970	232.4	434.3	426.4	1,116.7	411.9	180.9	169.8	111.1	174.6	231.8	264.3	322.4	10.584
1971	254.9	342.9	572.5	1,055.5	997.9	207.1	176.5	252.8	263.8	221.2	206.5	481.3	13.067
1972	379.8	309.4	760.6	1,075.0	1,147.5	839.0	520.4	215.7	178.7	206.5	740.5	764.8	18.532
1973	742.3	579.5	832.9	876.6	781.6	369.6	294.2	158.3	135.7	160.0	234.5	748.1	15.351
1974	623.3	527.8	587.1	854.2	650.3	248.9	333.7	180.1	294.2	256.2	486.4	548.8	14.516
1975	539.9	548.5	670.7	724.4	566.3	367.4	211.4	253.9	482.2	662.6	637.0	530.5	16.083
1976	417.4	885.0	897.3	1,040.8	900.5	431.1	432.6	417.1	271.0	657.9	507.7	398.6	18.841
1977	225.3	227.4	1,232.9	1,148.6	453.7	207.4	162.4	154.0	408.0	853.5	663.8	721.8	16.769
1978	744.9	401.2	606.2	950.2	530.3	281.8	131.5	169.2	175.4	243.7	227.1	303.0	12.369
1979	570.8	335.5	1,252.6	1,079.4	554.1	233.5	131.5	148.6	220.8	313.7	464.9	430.1	20.084
1980	256.1	119.7	633.7	748.4	273.9	192.4	143.8	130.1	118.1	158.2	241.9	272.7	8.540
1981	148.3	851.1	349.4	348.5	327.9	169.1	140.0	133.6	232.9	456.4	395.0	321.4	10.151
1982	320.8	361.0	619.8	1,084.5	354.2	431.5	181.8	122.5	122.0	124.4	196.1	233.4	10.780
1983	259.3	352.2	580.4	1,062.9	1,036.9	357.9	126.8	155.1	132.6	153.8	338.9	799.0	8.712
Mean 1966–1983	357.9	550.1	642.0	852.0	584.5	297.6	201.8	167.4	197.5	290.4	365.7	441.0	13.103

yearly tide tables for the Battery in New York City (NOAA 1947–1984). Time and height of tides at points upstream were calculated with adjustments published by National Oceanic and Atmospheric Administration.

Salinity

Salinity distributions in the lower Hudson River, as in other estuaries, are primarily influenced by tidal currents, freshwater discharge and associated currents, and basin morphometry. The intrusion of denser and cooler saline waters influences energy and mass transport in the estuary and creates important habitats for such species as Atlantic tomcod during the summer (McLaren et al. 1988, this volume). This intrusion of water leads to density-induced circulation which, after tidal mixing, is the most important factor in the movements of nutrients, plankton, silt, and pollutants (Lauff 1967). Large gradients in the contour of the river basin, especially the presence of sill structures (e.g., between km 58 and 75), tend to increase mixing (as does tidal activity) by intensifying turbulence. Secondary influences are imposed by meteorological conditions (wind, relative humidity, and air temperature) interacting with water temperatures. Depending on tidal amplitude, freshwater flow, and occasionally local winds, the Hudson estuary may be defined as mixed, partially mixed, or stratified with respect to salinity (Pritchard 1952, 1955, 1967).

Vertical gradients of salinity show that salt water in the Hudson is generally well mixed with fresh water during low flow conditions. Only a 10% increase is found, on the average, from the top of the water column to the bottom layers. Under high-flow conditions, fresh water overrides the salt water layer, and salinity differences of up to 20% can be established (Busby and Darmer 1970).

The Venice System classification (Symposium on the Classification of Brackish Waters 1958) has become a standard method of describing salinity distribution of estuaries. Based on salinity, zones within estuaries can be described as limnetic (<0.3‰), oligohaline (0.3–5‰), mesohaline (5–18‰), or polyhaline (18–30‰). The locations of these zones in the Hudson River are variable (Figure 5). The Hudson River estuary is limnetic above km 80 and this zone may extend to km 40. The oligohaline zone generally occurs from km 80 to km 40 but may extend to km 19. The mesohaline zone generally occurs from km 45 to the lower end of the Hudson River estuary in New York harbor. Salinity distribution patterns for the Hud-

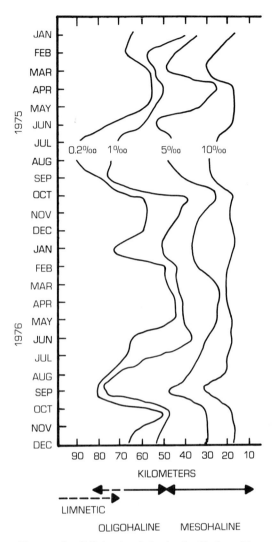

FIGURE 5.—Salinity isopleths in the Hudson River estuary, 1975–1976.

son are similar to those of the York River system (which Boesch 1977 took as representative of the Chesapeake Bay) in that isohalines are quite evenly spaced and the spacing does not change much during a tidal cycle or seasonally (Figure 5). The San Francisco Bay system (Kelley 1966) and the Pamlico Sound system in North Carolina (Tenore 1972) show markedly different salinity patterns.

An index of saltwater distribution is the location of the leading edge of the mass of intruding seawater, called the salt front (Figure 6). Conductivity of the water is a measure of salinity. As the intruding seawater mixes with fresh water and is

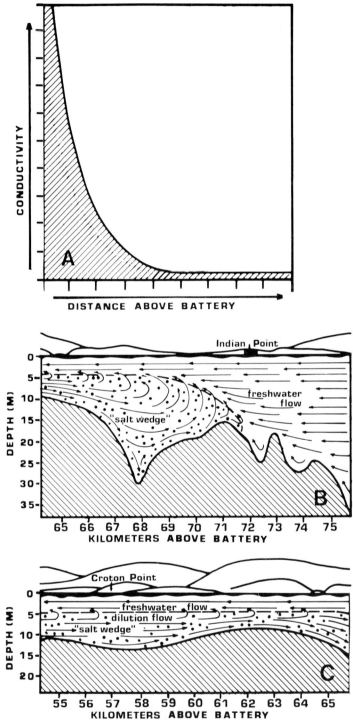

FIGURE 6.—Salinity patterns in the Hudson River estuary and the influence of geographical structures on them, in a perspective from the river's eastern shore. In a schematic representation (A), saline oceanic waters advance into the Hudson River and establish a salt wedge under the lighter fresh water; at the wedge's leading edge, a salt front is operationally defined at a salinity of 0.1‰. Sill structures (B and C) have important effects on the mixing of saline and fresh water in the Indian Point and Croton Point areas of the river.

diluted, conductivity declines. Conductivity declines nearly exponentially with upstream distance until it reaches values characteristic of the incoming fresh water (Figure 6A). In reverse perspective, the point at which conductivity begins to increase exponentially from the freshwater value is referred to as the salt front. For the Hudson estuary, the salt front is identified as the point associated with a conductivity of 0.3 mS/cm, equivalent to a salinity of approximately 0.1‰. This operationally distinguishes the northern limnetic zone from the saline southern region. The salt front's position fluctuates upstream and downstream. Twenty-four surveys that have yielded salinity data since 1929 (TI 1976a; Abood 1977) have documented the complexity of salinity distributions in the Hudson estuary.

Despite the difficulties with predicting salinity distributions at a particular time and place, a relatively simple deterministic model has been developed (TI 1976a) to predict the general location of the salt front in the main channel of the Hudson. The model is based on a regression of freshwater flow and tidal amplitude. The predicted salt front position has correlated well ($r = 0.94$; $P \leq 0.01$) with the empirical data set (TI 1976a, 1977a, 1977b, 1978a, 1978b). The distance (K_s, km) of the salt front above the Battery is estimated by

$$K_s = -27.88 \, (\log_e D_5) + 12.55/A_4 + 26.39; \quad (1)$$

D_5 is freshwater discharge ($10^3 m^3/s$) at Green Island 5 d previously, and A_4 is the tidal amplitude (m) at Indian Point (km 61) 4 d previously. The tidal amplitude at Indian Point was used because research was focused there, but amplitudes at other localities can be substituted with different coefficients. The 5-d lag for Green Island discharges and the 4-d lag for tidal amplitudes gave the best model fits of several alternative lag times tried, but why they work well is not understood, because transit times of water vary with the magnitude of the flow.

The location of the salt front, K_s, may be used as a reference point from which to describe the longitudinal distribution of salt isopleths around the front:

$$\frac{K_s}{K_c} = 0.150C + 0.966; \quad (2)$$

K_c is the position (kilometers above Battery Park) of a point having a conductivity C (mS/cm, on the bottom at midchannel). Substituting equation (1) for K_s and solving for K_c gives

$$K_c = \frac{-27.88 \, (\log_e D_5) + 12.55/A_4 + 26.39}{0.150C + 0.966}. \quad (3)$$

This model estimates longitudinal salinity profiles in the river (Figures 7, 8).

The validity of the salt front model is based on its ability to estimate salt front locations in the main channel. The calculated positions (from equation 1) and the data from field studies (in 1974, for example: Figure 8) were not significantly different (Wilcoxon matched-pair signed rank test; $Z = 0.204$; $P > 0.10$). When hydrological conditions do not change quickly, the predictions are accurate to within 3 km of the actual salt front location. However, the model is sensitive to the time parameters; on occasions when the salt front is moving rapidly (due to freshwater flow, for instance), the model is less accurate. Because events that induce rapid salt front movement usually occur at times when the salt front moves across the sills near river km 64, the front's passage may be delayed by several days. In such cases, the model predicts a salt position that is no farther than 8 km away from the actual position. The problem of accurately predicting salt front position when the front moves rapidly or across sill structures has been tackled by other, often more complex, models. In general they yield the same degree of accuracy as equations (1) or (3).

A series of sill-like structures across the channel between km 58 and km 75 promote mixing of fresh and salt water as the salt front oscillates with tidal forces. The channel depth decreases abruptly in several places, forcing the intruding saltwater mass upward into the path of opposing freshwater flows (Figure 6). In the region downriver from these sills, flow tends to be stratified (but marked stratification only occurs in times of high freshwater flow). As the salt wedge moves through the sill area, the salt front steepens, greatly reducing stratification. When saline water reaches the deeper area north of km 77, stratification apparently resumes.

Less mixing of fresh and salt water by tidal action occurs in shallow areas that border the deeper channel. Channel water attains higher velocities and moves earlier than that of shoal areas; further, dense saline water tends to follow this deeper channel. The reduced circulation and delay in mixing result in less variable salinity in

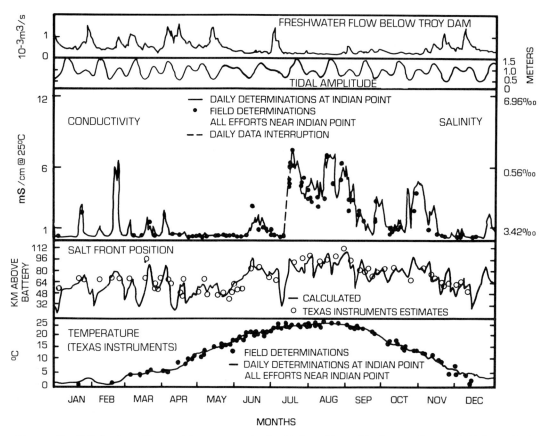

FIGURE 7.—Salt front position and temperature in the Hudson River. Freshwater flow and tidal amplitude primarily determine the salt front position, as indicated by conductivity. The calculated salt front position, from equation (1), correlates well with positions actually observed or estimated by interpolation from measured conductivities near the salt front.

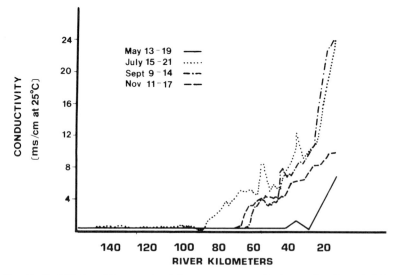

FIGURE 8.—Longitudinal Hudson River conductivity measurements for late spring through fall, 1974. (Data are from TI 1977b; the figure is adapted from Limburg et al. 1986.)

TABLE 3.—Mean monthly temperature (°C) of Hudson River water, recorded at the Hudson River Water Works, Green Island (below Troy Dam), New York, 1970–1977.

Year	Jan	Feb	Mar	Apr	May	Jun	Jul	Aug	Sep	Oct	Nov	Dec
1970	1.0	1.1	2.0	7.1	15.3	21.1	23.5	26.1	22.7	18.0	11.3	3.5
1971	0.7	0.8	1.9	6.8	11.7	20.0	24.7	25.0	22.4	17.8	11.8	3.5
1972	1.0	0.6	1.7	6.0	12.9	19.3	22.4	24.3	22.8	16.2	7.2	1.6
1973	0.8	1.7	3.6	8.6	13.9	20.1	29.4	24.7	23.7	18.0	10.5	4.1
1974	0.9	0.8	3.2	7.8	14.4	19.0	22.9	24.4	21.1	13.9	8.5	1.8
1975	1.3	1.2	3.5	6.9	14.5	21.2	24.9	25.0	20.8	14.6	10.5	3.4
1976	0.6	1.2	3.7	8.6	13.2	19.7	24.4	23.2	21.7	13.8	5.5	1.3
1977	0.6	0.7	2.6	8.7	14.4	20.3	24.5	25.1	22.1	12.7	9.4	2.1

shallow areas and less abrupt flushing of backwater bays.

Although salinity isopleths in the Hudson have a relatively stable distribution with respect to one another, their spatial positions can change quickly due to variations in freshwater flow. A crude index of stability—the resistance to changes in salinity distribution—is the ratio of the volume of an estuary (in km^3) to the mean flow of fresh water (in m^3/s) into the system (Simpson et al. 1973). The Hudson estuary has a stability ratio of 0.4 years, lower than those of the Delaware Bay and estuary (0.6 years), the Potomac River (0.7 years), and the Chesapeake Bay (1.2 years). After hurricane Agnes in 1972, for example, the salt front moved 27 km in only 5 d (TI 1976a).

Temperature

Water temperatures in the estuary range from average annual lows in January of 0.6–2°C to average annual highs in July or August of 22–29°C (Table 3). During winter, ice may completely cover the estuary as far south as Peekskill, although it is broken up by ship traffic and tidal action. Maximum summer water temperatures may exceed 30°C in shallow areas.

Freshwater discharge and ocean waters both influence temperature patterns in the Hudson. High freshwater discharges keep temperatures low downstream. During spring and early summer, when the major annual discharges occur, temperature increases are retarded and, in the lower estuary, temperatures may even be reduced. In the late fall, freshwater discharges again increase and accelerate cooling downstream. In spring and summer, ocean water that enters the river with the tides is cooler than fresh water, so temperature is lower towards New York City. By the late fall and winter, the horizontal temperature gradient has reversed, and the temperature increases from the north to the south, because oceanic waters cool to a lesser extent than shallow

sources of fresh water (McFadden et al. 1977). The magnitude of temperature differences at any one time between the upper and lower reaches of the channel can approach 11°C (Abood et al. 1976).

The river's annual temperatures cycle at km 63 can be described with a simple model (TI 1976a) that relates observed temperatures in the channel (T_c) to discharge 5 d earlier at Green Island (D_5) and to the days (d_i) from August 6 (arbitrary point):

$$T_c = 11.89971 \cos(2\pi d_i/365.25) - 1.6333 \ D_5 + 13.29842. \quad (4)$$

Temperatures measured at km 63 during 1972–1973 correlated strongly with T_c ($r = 0.998$; $P < 0.01$).

Dissolved Oxygen

Dissolved oxygen (DO) in the Hudson estuary is generally undersaturated, although supersaturation occurs in turbulent areas such as beaches (which may also be areas of major algae blooms). Freshwater discharges influence temperature (equation 4) which, in turn, influences DO concentrations. Dissolved oxygen may be locally depleted due to high detrital loads but only a few such instances have been observed above km 32 since 1972 (TI 1976a). Low DO was reported between km 80 and km 96 during 1974 (Klauda et al. 1980; Dey 1981). McFadden et al. (1977) reported a general drop in DO levels below Albany, a recovery and peak near Saugerties (km 161), relatively high levels south to Croton Point (km 56), and a decline again in the New York City region. There is considerable fluctuation with the seasons (Figure 9). In general, the highest DO levels occur during late winter and early spring when the river is coolest and least saline. As the growing season progresses, oxygen is depleted in some areas (Storm and Heffner 1976). Typical dissolved oxygen levels are between 5 and 14 mg/L.

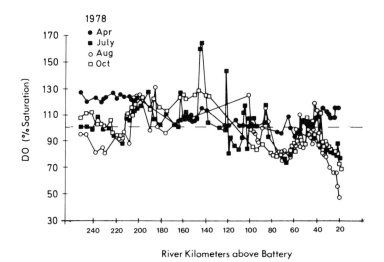

FIGURE 9.—Hudson River dissolved oxygen (DO) profiles for 1978, expressed as percent of saturation. Data are from midchannel surface waters. (Adapted from Limburg et al. 1986, with permission, and based on data from USGS 1969; Abood et al. 1976; Fleming et al. 1976; and TI 1979a.)

Temporal variation in dissolved oxygen concentration at km 63 can be described by the model

$$DO = 2.469 \cos(2\pi d_j/365.25)$$
$$- 0.101 \, T_c - 0.600 \, D_5 + 11.025; \quad (5)$$

DO is dissolved oxygen concentration (mg/L) at d_j days from February 12, an arbitrary date, in a given year; T_c is the temperature (°C) at time d_j; and D_5 is 5 d prior to d_j. The DO values predicted by this model are strongly correlated ($r = 0.97$; $P < 0.01$) with the empirical data set (TI 1976a).

Turbidity, pH, and Alkalinity

The main component of turbidity in the Hudson estuary is silt transported by terrestrial runoff. Turbidity normally varies between 10 and 50 Formazin turbidity units (FTU), but is higher during periods of heaviest runoff (over 100 FTU), particularly those associated with seasonally reduced watershed ground cover (TI 1976a). Additional silt is suspended through erosional processes within the estuary itself. No longitudinal variations in turbidity are clearly discernible (TI 1976a). However, algal blooms in various river regions may increase local turbidity in summer (mid-June to mid-September). If summer turbidity readings are excluded to eliminate the effect of these blooms, a significant positive correlation ($r = 0.891$; $P = 0.001$) between turbidity and freshwater flow is obtained. Temporal variation in turbidity at km 63 can be described by the model

$$FTU_b = 13.87 \, D_b + 0.689; \quad (6)$$

FTU_b is the mean biweekly (2-week) turbidity (FTU) and D_b is mean biweekly discharge (m³/s) at Green Island during the previous 2 weeks (TI 1976a). High turbidity in the estuary attenuates light and limits primary productivity (Sirois and Fredrick 1978).

Data on pH collected in the Hudson River estuary since 1966 have shown no predictable temporal or spatial patterns (TI 1976a). Mean weekly pH varies between 6.4 and 8.2. Most measurements were above 7.0; lower exceptions occurred during late spring and summer. A New York State water quality survey performed in the mid-1960s for the entire river showed average pH values to range from 6.8 to 7.8 (NYSDEC 1967). The lower Hudson River is well buffered. Alkalinity at the head of the estuary (Green Island) is 29–71 mg/L and hardness is 49–93 mg/L, both as $CaCO_3$ (USGS 1966–1984), and both measures increase down river into the saline zone (McFadden et al. 1978).

Pollutants and Nutrients[4]

Releases of municipal wastewater into the Hudson River have increased steadily since records began in 1952. During the mid-1970s, these releases approached 4×10^6 m³/d and included some 200 tonnes of biological oxygen demand

---

[4]Much of the information summarized in this section is contained in Limburg et al. (1986); that work should be consulted for source literature.

(BOD) daily, two-thirds of which originated in the New York metropolitan area. Despite increasing municipal releases, however, the river's water quality has been improving because of more and better treatment facilities; for example, the municipal BOD loadings mentioned above were a third less than their 1965 peak, and phosphorus loadings were down 40% to some 20 tonnes/d. Municipal nitrogen loadings have remained steady at around 200 tonnes total N/d because treatment plants were not designed to remove this nutrient.

Industrial wastewater discharges, some of which go through municipal systems, have been about 10 times greater than municipal releases, but 95% of this has been cooling water, most of it cycled through power plants. Important industrial pollutants in the Hudson River include cadmium, nickel, and polychlorinated biphenyls (PCBs); the latter, originating near Schenectady on the upper Hudson, were taken up by striped bass, causing the commercial fishery for that species in the lower estuary to be closed since 1971. Agricultural runoff brings pesticides and additional nutrients to the river, but the magnitudes of these nonpoint effluents are not well documented.

Some 70–75% of the phosphorus reaching the river (chiefly as phosphate $PO_4^=$) does so from sewer outfalls along the lower 40 km of the estuary. This reach of the estuary is saline; the phosphate is mixed by tidal action and moves with the saline water mass. Phosphate concentrations in the lower estuary approach 6 μM in summer, when freshwater flows are low and the salt front is well upriver. They are typically less than 2 μM in spring, when high freshwater discharges flush the lower river. In contrast to other eutrophic estuaries, algae remove only a small fraction of imported phosphate from the Hudson system, leaving a strong correlation between phosphate concentration and salinity.

Sewage also is the major source of nitrogen in the estuary. Typical concentrations of total nitrogen (organic and inorganic forms) are 100–130 μM near the urban areas of Albany and New York City; between those endpoints, values are half these amounts. The predominate forms of nitrogen are ammonia (NH_4, NH_3^+) near sewage outfalls and nitrate (NO_3^-) elsewhere; nitrite (NO_2^-) contributes little to the nitrogen total.

Organic carbon concentrations in the river typically are 3–8 mg/L, supported by inputs of 246×10^6 kg C/year to the estuary. Of the total input, 44% is of terrestrial origin (chiefly from leaf litter), 24% comes from sewage, 15% is of marine origin,

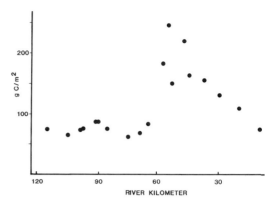

FIGURE 10.—Estimated 1972 semiannual primary productivity in the lower Hudson River estuary, May–October. Semiannual productivity for this period is approximately 90% of annual productivity (Adapted, with permission, from Limburg et al. 1986, with data from Sirois and Fredrick 1978.)

15% is fixed by phytoplankton (Figure 10), and 2% is fixed by aquatic macrophytes. The balance among these sources changes seasonally. Terrestrial contributions are greatest during the spring runoff; primary productivity accounts for the larger fraction of carbon inputs during the warm, low-flow periods of later summer. Marine incursions bring substantial amounts of carbon into the estuary, but this is the most poorly documented source.

Biological Implications for Power Plant Placement

As with all estuaries, the transition zone between fresh and salt water in the Hudson River is of great biological importance. The Indian Point and other nearby power plants were a major focus of the Hudson River studies because of their proximity to the transition zone. The estuary's largest and most dramatic changes in water chemistry would be expected in this zone, and it is an area where enormous physical (tidal) energies are converted to high biological productivity. Anthropogenic events there must be carefully examined and monitored.

The complex mosaic of riverine, tidal, and density currents of the transition zone often allows phytoplankton, zooplankton, and fish eggs and larvae to hold consistent positions in the river for extended periods, where they are particularly vulnerable to localized stresses from human activities. Freshwater and marine faunas and floras converge at the transition zone; the fish community of the Hudson transition, for example, is quite diverse (Table 4), and different components

TABLE 4.—Trophic strategies and salinity preferences of Hudson River fish (adapted from Limburg et al. 1986 with permission).[a]

Salinity regime	Food preference			
	Plankton	Benthos, detritus	Epibenthos	Fish, macroinvertebrates
Fresh water	Emerald shiner Spottail shiner Tesselated darter[b]	Goldfish Common carp Eastern silvery minnow Golden shiner White catfish Brown bullhead Bluegill White sucker	Yellow perch Pumpkinseed Redbreast sunfish	Largemouth bass
Euryhaline	Fourspine stickleback Banded killifish[b] Tidewater silverside[b] Alewife[b,c] American shad[b] Bay anchovy[b,c] Blueback herring[b,c] Rainbow smelt[b,c]	Mummichog	White perch Shortnose surgeon Hogchoker Atlantic tomcod[c]	White perch Shortnose sturgeon Striped bass[c] American eel[c]
Marine	Atlantic menhaden[c] Atlantic silverside[b,c] Winter flounder Northern pipefish[b,c]	Longhorn sculpin Winter flounder	Longhorn sculpin Fourbeard rockling[c] Winter flounder	Longhorn sculpin Atlantic sturgeon[c] Bluefish[c] Weakfish[c]

[a] From Weinstein (1977), TI (1981), and Gladden et al. (1988, this volume).
[b] Includes zooplankton, eggs, or larvae.
[c] Migratory species.

of it are locally abundant as the zone moves up- or downriver. Both adults of anadromous species and their progeny pass through (and may linger in) the transition zone.

Each transition zone is uniquely influenced by its basin's hydrology and topography. Just above Indian Point, for instance, the Hudson River makes its sharpest turn, affecting current patterns in the area. Just below Indian Point, the Hudson changes from a deep, narrow river to a wide, shallow one; in the shallow areas bordering the channel, water circulation is reduced, temperatures are higher, salinities are more stable, and primary productivity is greater (Figures 3, 10) than elsewhere in the river. Such features of the local ecology must be considered when power plants and other facilities are sited.

It is not enough to focus only on transition zones, however. Hudson River studies by Cooper and H. K. Baker (unpublished) indicate that local relative abundances of fish species are closely correlated with the species' estuary-wide abundances. Changes in year-class strength appear to be reflected throughout the estuary. It is possible that what affects species in one place may affect them in others, even in areas that are widely separated. This, it seems prudent to consider an entire river system when even local anthropogenic impacts are assessed.

Acknowledgments

Much of this work was supported by a contract between Consolidated Edison Company of New York, Incorporated, and Texas Instruments Incorporated. Funding was also provided by the Power Authority of the State of New York, Orange and Rockland Utilities, Incorporated, Central Hudson Gas and Electric Corporation, and the Hudson River Foundation. We thank John Gladden for use of his unpublished analysis; Karen Kachura and Chris Rossi for the illustrations; Larry Wehr, Ronald Klauda, and other members of the technical staff at Ecological Services (of TI) for review; and Springer-Verlag, Karen Limburg, and Mary Ann Moran for permission to use some previously published materials. Special thanks go to Elizabeth Dreyer for support above and beyond the call of duty.

References[5]

Abood, K. A. 1977. Evaluation of circulation in partially stratified estuaries as typified by the Hudson River. Doctoral dissertation. Rutgers University, New Brunswick, New Jersey.

Abood, K. A., E. A. Maikish, and R. R. Kimmel. 1976. Field and analytical investigations of ambient tem-

[5]See Table 1 for sources of legal documents and unpublished reports pertaining to the Hudson River.

perature distribution in the Hudson River. *In* Hudson River ecology, 4th symposium. Hudson River Environmental Society, New Paltz, New York.

Boesch, D. F. 1977. A new look at the zonation of benthos along the estuarine gradient. Pages 245–266 *in* B. Coull, editor. Ecology of marine benthos. University of South Carolina Press, Columbia.

Busby, M. W., and K. I. Darmer. 1970. A look at the Hudson estuary. Water Resources Bulletin 6:802–812.

Busby, W. 1966. Flow, quality, and salinity in the Hudson River estuary. Pages 135–145 *in* Hudson River ecology. Hudson River Valley Commission of New York, Albany.

Darmer, K. I. 1969. Hydrologic characteristics of the Hudson River estuary. Pages 40–55 *in* G. P. Howell and G. J. Lauer, editors. Second symposium on Hudson River ecology. New York State Department of Environmental Conservation, Albany.

Dey, W. P. 1981. Mortality and growth of young-of-the-year striped bass in the Hudson River estuary. Transactions of the American Fisheries Society 110:151–157.

Fenneman, N. M. 1938. Physiography of eastern United States, volume 14. McGraw-Hill, New York. (Not seen; cited in Howells 1972.)

Fleming, A. K., and six coauthors. 1976. The environmental impact of PL 92-500 on the Hudson River estuary. *In* Hudson River ecology, 4th symposium, Hudson River Environmental Society, New Paltz, New York.

Giese, G. L., and J. W. Barr. 1967. The Hudson River estuary, a preliminary investigation of flow and water characteristics. New York State Water Commission Bulletin 61, Albany.

Gladden, J. B., F. R. Cantelmo, J. M. Croom, and R. Shapot. 1988. Evaluation of the Hudson River ecosystem in relation to the dynamics of fish populations. American Fisheries Society Monograph 4:37–52.

Howells, G. P. 1972. The estuary of the Hudson River, United States of America. Proceedings of the Royal Society of London B, Biological Sciences 180:521–534.

Kelley, D. W. 1966. Ecological studies of the Sacramento–San Joaquin estuary. Description of the Sacramento–San Joaquin estuary. California Department of Fish and Game, Fish Bulletin 133:8–17.

Klauda, R., W. P. Dey, T. P. Hoff, J. B. McLaren, and Q. E. Ross. 1980. Biology of juvenile Hudson River striped bass. Marine Recreational Fisheries 5:101–123.

Lauff, G., editor. 1967. Estuaries. American Association for the Advancement of Science Special Publication 83, Washington, D.C.

Limburg, K. E., M. A. Moran, and W. H. McDowell. 1986. The Hudson River ecosystem. Springer-Verlag, New York.

LMS (Lawler, Matusky & Skelly Engineers). 1978a. Annual progress report for 1974. Report to Central Hudson Gas and Electric Corporation.

LMS (Lawler, Matusky & Skelly Engineers). 1978b.

Roseton and Danskammer Point generating stations. Hydrothermal analysis. Report to Central Hudson Gas and Electric Corporation.

McFadden, J. T., Texas Instruments, and Lawler, Matusky & Skelly Engineers. 1977. Influence of Indian Point unit 2 and other steam electric generating plants on the Hudson River estuary, with emphasis on striped bass and other fish populations. Report to Consolidated Edison Company of New York.

McFadden, J. T., Texas Instruments, and Lawler, Matusky & Skelly Engineers. 1978. Influence of the proposed Cornwall pumped storage project and steam electric generating plants on the Hudson River estuary, with emphasis on striped bass and other fish populations, revised. Report to Consolidated Edison Company of New York.

McLaren, J. B., T. H. Peck, W. P. Dey, and M. Gardinier. 1988. Biology of Atlantic tomcod in the Hudson River estuary. American Fisheries Society Monograph 4:102–112.

NOAA (National Oceanographic and Atmospheric Administration). 1947–1984. Tide tables for the east coast of North America including Greenland (annual predictions). U.S. Department of Commerce, National Ocean Survey, Washington, D.C.

NOAA (National Oceanographic and Atmospheric Administration). 1982. Monthly normals of temperature, precipitation, and heating and cooling degree days 1951–1980. New York. (Climatography of the United States 81.) U.S. Department of Commerce, National Climatic Center, Asheville, North Carolina.

NYSDEC (New York State Department of Environmental Conservation). 1967. Periodic report of the water quality surveillance network. 1965 through 1967 water years. NYSDEC, Albany, New York.

Pritchard, D. W. 1952. Salinity distribution and circulation in the Chesapeake Bay estuarine system. Journal of Marine Research 11:106–123.

Pritchard, D. W. 1955. Estuarine circulation patterns. Proceedings of the American Society of Civil Engineers. 81:717.

Pritchard, D. W. 1967. What is an estuary? Physical viewpoint. Pages 3–5 *in* G. Lauff, editor. Estuaries. American Academy for the Advancement of Science Special Publication 83, Washington, D.C.

Simpson, H. J., R. Bopp, and D. Thurber, 1973. Salt movement patterns in the Hudson. *In* Hudson River ecology, 3rd symposium. Hudson River Environmental Society, New York.

Sirois, D. L., and S. W. Fredrick. 1978. Phytoplankton and primary production in the lower Hudson River estuary. Estuarine and Coastal Marine Science 7:413–423.

Storm, P. C., and R. L. Heffner. 1976. A comparison of phytoplankton abundance, chlorophyll *a* and water quality facts in the Hudson River and its tributaries. *In* Hudson River ecology, 4th symposium. Hudson River Environmental Society, New Paltz, New York.

Symposium on the Classification of Brackish Waters. 1958. The Venetian System for the classification of marine waters according to salinity. Oikos 9:311–312.

Tenore, K. R. 1972. Macrobenthos of the Pamlico River estuary, North Carolina. Ecological Monographs 42:51–69.

TI (Texas Instruments). 1976a. A synthesis of available data pertaining to major physiocochemical variables within the Hudson River estuary emphasizing the period from 1972 through 1975. Report to Consolidated Edison Company of New York.

TI (Texas Instruments). 1976b. Hudson River ecological study in the area of Indian Point. Thermal effects report to Consolidated Edison Company of New York.

TI (Texas Instruments). 1977a. 1976 fisheries data display. Report to Consolidated Edison Company of New York.

TI (Texas Instruments). 1977b. 1976 ichthyoplankton data display. Report to Consolidated Edison Company of New York.

TI (Texas Instruments). 1978a. 1977 fisheries data display. Report to Consolidated Edison Company of New York.

TI (Texas Instruments). 1978b. 1977 ichthyoplankton data display. Report to Consolidated Edison Company of New York.

TI (Texas Instruments). 1979a. 1978 water quality data display. Report to Consolidated Edison Company of New York.

TI (Texas Instruments). 1979b. 1976 year class report for the multi-plant impact study of the Hudson River estuary. Report to Consolidated Edison of New York.

TI (Texas Instruments). 1981. 1979 year-class report for the multiplant impact study of the Hudson River estuary. Report to Consolidated Edison Company of New York.

USGS (U.S. Geological Survey). 1966–1984. Water resources data for New York. Water quality records. USGS, Albany, New York.

Weinstein, L. H., editor. 1977. An atlas of the biologic resources of the Hudson estuary. Boyce Thompson Institute for Plant Research, New York.

American Fisheries Society Monograph 4:25–36, 1988

Historical Perspective on Fish Species Composition and Distribution in the Hudson River Estuary

C. Allen Beebe[1] and Irvin R. Savidge[2]

Texas Instruments Incorporated
Ecological Services Group, Buchanan, New York 10511, USA

Abstract.—Research on the fish fauna of the Hudson River estuary has been conducted for over 160 years, but most of our knowledge of species composition has been obtained since 1974. One hundred forty species have been reported for the estuary south of Albany. Species composition has been influenced by both geological and human activities. Species composition is relatively distinctive, although variable, in each of three ecological zones (marine, brackish, freshwater); however, species often overlap these zones. On an annual basis, species composition is influenced by salinity and season. The highest numbers of species occur in the brackish zone and during summer.

The Hudson River estuary, stretching from the river mouth to the upstream limit of tidal influence near Albany, offers a variety of fish habitats. Large aquatic meadows, deep channels, rocky shorelines, and gravel beaches occur along a continuum of salinity regimes (Cooper et al. 1988, this volume). These features create distinctive distribution patterns of species (Smith 1985; Gladden et al. 1988, this volume) and variations in species composition. Periods of high freshwater flow or drought can also alter the distribution of many fish species.

Although the fish fauna of the Hudson River estuary is considered typical of the fauna expected in a temperate estuary (Greeley 1937; Boyce Thompson Institute 1977; McFadden et al. 1978), the 140 recorded species form one of the most diverse fish groups in Atlantic coastal rivers. Research on Hudson River fishes has been conducted since the 1800s, but the studies were limited in scope and duration until the mid-1960s, when the large-scale power plant impact studies began. Species composition was summarized for the period 1936–1975 during the course of the power plant studies conducted during the 1960s and 1970s (McFadden et al. 1978; Barnthouse et al. 1988, this volume). This paper updates that summary with information on fishes collected between 1976 and 1980, includes additional information from historical records prior to 1936, and summarizes the occurrence and distribution of species within the estuary from the George Washington Bridge to the Troy Dam to provide a historical perspective on the present composition of the finfish fauna. Recent reports of studies downstream of the George Washington Bridge include LMS (1980a, 1980b), Malcolm Pirnie (1982), USACE (1984).

Historical Studies

The lower Hudson River from New York City to Albany played a historical role in the development of ichthyology in the United States. Some of the earliest studies of fishes were done in the lower Hudson River or its tributaries. The studies prior to 1900 were primarily systematic classifications, descriptions of new or unusual species, or parts of larger faunal surveys (Jordan 1905). In 1815, Samuel Mitchill prepared a detailed systematic account of fishes found on Long Island and near New York City, mentioning species found in the lower Hudson (Mitchill 1815). Later, he also described new species found in a lower Hudson tributary, the Wallkill River (Mitchill 1817), and revised his original species list (Mitchill 1818).

James DeKay's (1842) extensive listing of New York fishes, published as part of New York's first biological survey, included many species collected from the lower Hudson River. The series of annual reports prepared by the Regents of the University of the State of New York (1855–1900) listed species of fish collected in the lower Hudson River, of which specimens were deposited in the New York State Museum. Another series of annual reports (U.S. Fish Commission 1882–1900) provides historical data on commercial catches of

[1]Present address: NEC Information Systems, Incorporated, 289 Great Road, Acton, Massachusetts 01720, USA.

[2]Present address: Department of Chemistry and Biochemistry, University of Colorado, Boulder, Colorado 80309, USA.

American shad and striped bass. Smith (1897) prepared a descriptive faunal list of fish taken near New York City adjacent to portions of the lower Hudson River and Mearns (1898) prepared a similar publication dealing with the Hudson Highlands, north of the city. The last extensive publication that included treatment of Hudson River fishes during this period was Bean's (1903) descriptive catalogue of New York fishes.

Several publications focusing on individual species are also available from this period. Cheney (1887) and Mather (1887) discussed the unsuccessful attempts to establish a breeding population of Atlantic salmon in the headwaters of the estuary. Webster (1982) provided a detailed review of these efforts. Phillips (1883) described capture of Atlantic cod. Clinton (1824) provided the taxonomic description and name for the spottail shiner taken from the lower Hudson River and Fisher (1890) reported the capture of cobia, a marine species, near Ossining. Fish collections during the nineteenth century were made primarily with seines, gill nets, or hook and line. The researchers often visited fish markets and often relied upon local fishermen for information and specimens.

The New York State Conservation Department conducted a series of 1-year biological surveys of the aquatic flora and fauna of each of the major watersheds throughout the state from the 1920s through the 1930s. Two of these surveys (1934 and 1936) covered the Hudson River from Albany to the Tappan Zee. Fish were collected with seines, gill nets, and a small trawl in numerous tributaries, lakes, and ponds as well as in the Hudson River. Specimens collected in the surveys were deposited in collections of the New York State Museum, Cornell University, University of Vermont, or the University of Michigan Museum of Zoology. Commercial fishermen also provided specimens and information on several species such as American shad, Atlantic sturgeon, and striped bass (Greeley 1935, 1937; Curran and Ries 1937).

Fisheries research on the Hudson during the 1940s focused on the commercial fishery for American shad (New York State Conservation Department 1943). This research continued into the 1950s and was expanded to include work on striped bass. Although other species were taken in these seine and gill-net collections, the publications during this period dealt only with the biology of American shad (Burdick 1954; Talbot 1954; Fischler 1959) and striped bass (Raney 1952; Raney et al. 1954; Rathjen and Miller 1957).

After 1965, fisheries research on the Hudson River increased in response to concerns about the impacts of power plants on aquatic life (e.g., Perlmutter et al. 1966, 1967; Carlson and McCann 1969; Dovel 1977; McFadden et al. 1978; TI 1980c). These studies were the first in which data were collected for several successive years over the entire length of the estuary to examine distribution and abundance of all life stages of selected species. Basic research also increased during this period (e.g., Howells 1972; Medeiros 1974; Holsapple and Foster 1975; Wilk 1977; Grabe 1978, 1980; Kendall and Walford 1979; Reider 1979a, 1979b). During the 1970s, standardized sampling designs and techniques improved the usefulness of the data for describing the fish community.

Data Sources

Information on the occurrence and distribution of various species reported from the Hudson River estuary was drawn from all the sources known to the authors. In addition to the literature cited in the previous section, museum collections of the American Museum of Natural History, Cornell University, U.S. National Museum, Museum of the Hudson Highlands, New York State Museum, and the University of Michigan Museum of Zoology were examined for specimens collected from the Hudson River. The historical zoogeographical effects of the construction of the various canals connecting the Hudson River with other drainage systems was evaluated from the relevant literature.

Information on patterns of seasonal abundance of common species in the nearshore areas of the estuary was obtained from beach-seine collections by Texas Instruments from 1974 to 1979. During these years, over 19,000 standardized beach-seine samples were taken between the George Washington Bridge and the Troy Dam (Table 5).

Gear deployment procedures and the sampling design for the beach-seine survey are presented by Young et al. (1988, this volume). In order to weight the 12 sampling regions equally within three ecological zones—marine, brackish, and freshwater—and the sampling periods equally within the seasons, catch per tow was first calculated for each half-month by region and year. These data subsets were then averaged to provide the average species abundance by ecological zone and season.

TABLE 5.—Seasonal and zonal distributions of beach seine samples taken by Texas Instruments during the 1974–1979 surveys.

Season	Zone[a]			Total
	Marine (km 19–53)	Brackish (km 54–88)	Freshwater (km 89–243)	
Spring	1,817	3,019	2,036	6,872
Summer	1,591	2,774	2,234	6,599
Fall	1,446	2,625	1,539	5,610
All	4,854	8,418	5,809	19,081

[a] River kilometers are measured from the Battery at km 0; George Washington Bridge is at km 19, and Troy Dam is at km 243 near Albany.

Species Occurrence and Distribution

Altogether, 140 species have been reported from the Hudson River estuary (Table 6). The spatial and temporal occurrences of a species reflect both the intensity of the sampling effort and the abundance of the species. Uncommon species tend to be most frequently captured in the regions of most intense sampling, such as the Indian Point region surrounding the Indian Point and Lovett power plants (Hutchison 1988, this volume). Common species were captured throughout the estuary even though their regional abundance varied by several orders of magnitude.

Except for a few species that entered the estuary through direct introductions or through canals connecting other watersheds, the species composition of the Hudson River estuary has probably remained similar to what it was at the time the area was settled by Europeans. All but five species (barndoor skate, Atlantic salmon, cobia, ninespine stickleback, and sharksucker) have been collected within the last 20 years. The spatial and seasonal patterns of the species composition are correlated with freshwater flow, which controls the salinity patterns. These factors and water temperature are correlated with season of the year. Together, these variables affect distribution patterns of the various species. During low-flow periods or drought (often during late summer), more marine species enter the estuary and some move upstream into fresh water; during high-flow periods (usually in spring), more freshwater species extend further south into normally brackish or marine sections of the estuary. Cooper et al. (1988) describe physical and chemical patterns of the estuary that undoubtedly influence the temporal and spatial distribution patterns of the fish species.

The increasing intensity of fish faunal studies since the 1800s has resulted in continuing increase in the number of species reported from the Hudson River estuary. The 140 species from the Hudson River form one of the most diverse groupings found in Atlantic coast rivers. In addition to native species resulting from postglacial repopulation, various species have gained access through the canals connecting the Hudson to the Great Lakes, Lake Champlain, and the Delaware River. The net result of human activities on the Hudson has been an increase in fish species diversity.

The estuary's physical features provide a range of habitat conditions conducive to a diverse fauna. The extensive surveys from 1965 through 1980 demonstrated that, despite a long period during which the Hudson's environmental quality was neglected, the estuary is capable of sustaining a diverse and abundant first fauna.

Freshwater Fauna

The composition of the freshwater fauna has resulted from geological and human actions (Gilbert 1980b; Smith 1983, 1985). Most of New York State was covered by the Pleistocene glaciers, which pushed the freshwater fauna southward and probably eliminated some species (Greeley 1937; Smith 1983). Beginning about 16,000 years ago, postglacial repopulation of the Hudson system began. Coastal lowland species (e.g., eastern mudminnow and bluespotted sunfish) entered from the south and inland species (e.g., spotfin shiner and satinfin shiner) entered through streams flowing east from the Great Lakes region (Greeley 1937; Gilbert 1980b; Smith 1983, 1985). The Erie Canal, completed in 1825 between the Hudson's Mohawk River tributary and Lake Erie, and the Barge Canal, a later modification and partial replacement of the Erie Canal, gave fish from the upper Great Lakes more modern access to the Hudson River. Because no substantial collections had been made in the Mohawk River before the canal was built, the time and route by which several species colonized the Hudson River remains ambiguous. Gilbert (1980a) noted that the number of native (not purposely introduced) freshwater species (66) occurring in the Hudson is higher than in neighboring systems such as the Delaware and Connecticut rivers because of the greater opportunity for access to the Hudson by midwestern species.

White bass were found in Oneida Lake during the New York state biological survey of 1934, but not in the Mohawk River, the Erie–Barge Canal, or the Hudson River (Greeley 1935, 1937). This

TABLE 6.—Fish species collected from the Hudson River,[a] by ecological zone and region.[b]

Family and species	Marine		Brackish			Freshwater						
	YK	TZ	CH	IP	WP	CW	PK	HP	KG	SG	CS	AL
Petromyzontidae												
Silver lamprey			×	×								
Sea lamprey				×				×				
Rajidae												
Barndoor skate												×
Acipenseridae												
Shortnose sturgeon	×	×	×	×	×	×	×	×	×	×	×	×
Atlantic sturgeon	×	×	×	×	×	×	×	×	×	×	×	×
Elopidae												
Ladyfish			×									
Anguillidae												
American eel	×	×	×	×	×	×	×	×	×	×	×	×
Congridae												
Conger eel	×	×		×								
Clupeidae												
Blueback herring	×	×	×	×	×	×	×	×	×	×	×	×
Alewife	×	×	×	×	×	×	×	×	×	×	×	×
Hickory shad	×	×		×								
American shad	×	×	×	×	×	×	×	×	×	×	×	×
Atlantic menhaden	×	×	×	×	×	×	×	×	×	×	×	×
Atlantic herring	×		×									
Gizzard shad	×	×	×	×	×	×	×	×	×	×	×	×
Round herring	×											
Engraulidae												
Bay anchovy	×	×	×	×	×	×	×	×	×	×	×	×
Striped anchovy	×			×								
Salmonidae												
Sockeye salmon							×					
Rainbow trout				×						×		
Atlantic salmon												×
Brown trout	×	×	×	×	×	×						
Brook trout				×								
Osmeridae												
Rainbow smelt	×	×	×	×	×	×	×	×	×	×	×	×
Umbridae												
Central mudminnow			×	×								×
Eastern mudminnow				×		×						
Esocidae												
Redfin pickerel				×						×	×	×
Northern pike				×								×
Chain pickerel		×	×	×	×	×						
Synodontidae												
Inshore lizardfish	×	×	×	×								
Cyprinidae												
Goldfish	×	×	×	×	×	×	×	×	×	×	×	×
Common carp	×	×	×	×	×	×	×	×	×	×	×	×
Cutlips minnow		×	×	×	×	×	×	×	×	×	×	×
Eastern silvery minnow		×				×				×	×	×
Golden shiner	×	×	×	×	×	×	×	×	×	×	×	×
Comely shiner										×		
Satinfin shiner		×	×			×			×		×	×
Emerald shiner		×	×		×	×	×	×	×	×	×	×
Bridle shiner					×				×	×		
Common shiner		×	×				×	×		×	×	×
Spottail shiner		×	×	×	×	×	×	×	×	×	×	×
Rosyface shiner												×
Spotfin shiner	×	×	×	×	×	×	×	×			×	×
Fathead minnow	×	×	×			×	×	×	×		×	
Bluntnose minnow		×						×			×	×
Blacknose dace		×	×	×	×	×	×	×				×
Longnose dace				×								
Creek chub		×	×	×	×	×	×		×	×	×	
Fallfish				×		×		×		×	×	×
Catostomidae												
White sucker	×	×	×	×	×	×	×	×	×	×	×	×
Northern hog sucker										×	×	×
Ictaluridae												
White catfish	×	×	×	×	×	×	×	×	×	×	×	×
Yellow bullhead							×					
Brown bullhead	×	×	×	×	×	×	×	×	×	×	×	×
Channel catfish				×	×		×					

TABLE 6.—*Continued.*

Family and species	Marine		Brackish			Freshwater						
	YK	TZ	CH	IP	WP	CW	PK	HP	KG	SG	CS	AL
Percopsidae												
Trout-perch				×								×
Gadidae												
Fourbeard rockling				×								
Atlantic cod			×		×							
Silver hake	×		×	×	×							
Atlantic tomcod	×	×	×	×	×	×	×	×	×	×	×	×
Pollock	×			×								
Red hake		×		×		×						
Spotted hake	×	×	×	×								
Ophidiidae												
Striped cusk-eel				×								
Belonidae												
Atlantic needlefish		×	×	×			×				×	
Cyprinodontidae												
Sheepshead minnow							×					
Banded killifish	×	×	×	×	×	×	×	×	×	×	×	×
Mummichog	×	×	×	×	×	×	×	×	×	×	×	×
Striped killifish	×											
Atherinidae												
Rough silverside	×	×	×	×	×	×	×	×	×	×	×	
Tidewater silverside	×	×	×	×	×				×	×		×
Atlantic silverside	×	×	×	×	×		×	×	×	×	×	
Gasterosteidae												
Fourspine stickleback	×	×	×	×	×	×	×	×	×	×	×	×
Brook stickleback				×							×	
Threespine stickleback		×	×	×	×	×	×	×			×	
Ninespine stickleback			×									
Fistulariidae												
Bluespotted cornetfish		×										
Syngnathidae												
Lined seahorse				×								
Northern pipefish	×	×	×	×	×	×	×	×	×	×	×	×
Percichthyidae												
White perch	×	×	×	×	×	×	×	×	×	×	×	×
White bass				×			×					
Striped bass	×	×	×	×	×	×	×	×	×	×	×	×
Serranidae												
Black sea bass				×								
Centrarchidae												
Rock bass		×	×	×	×	×		×		×	×	×
Bluespotted sunfish					×							
Redbreast sunfish	×	×	×	×	×	×	×	×	×	×	×	×
Green sunfish			×							×		
Pumpkinseed	×	×	×	×	×	×	×	×	×	×	×	×
Warmouth						×	×					
Bluegill	×	×	×	×	×	×	×	×	×	×	×	×
Smallmouth bass				×		×	×	×	×	×	×	×
Largemouth bass	×	×	×	×	×	×	×	×	×	×	×	×
White crappie				×	×		×		×	×		×
Black crappie			×	×	×	×	×	×	×	×	×	×
Percidae												
Tessellated darter	×	×	×	×	×	×	×	×	×	×	×	×
Yellow perch	×	×	×	×	×	×	×	×	×	×	×	×
Logperch			×	×	×					×		
Shield darter									×			
Walleye			×						×		×	
Pomatomidae												
Bluefish	×	×	×	×	×	×	×	×	×	×	×	×
Rachycentridae												
Cobia			×									
Echeneidae												
Sharksucker			×									
Carangidae												
Crevalle jack	×	×	×	×	×	×						
Mackerel scad				×								
Lookdown	×	×		×								
Atlantic moonfish				×		×						

TABLE 6.—*Continued.*

Family and species	Marine		Brackish			Freshwater						
	YK	TZ	CH	IP	WP	CW	PK	HP	KG	SG	CS	AL
Lutjanidae												
Gray snapper		×	×	×								
Sparidae												
Pinfish				×								
Scup		×		×								
Sciaenidae												
Silver perch				×								
Weakfish	×	×	×	×	×	×	×	×	×	×	×	×
Spot	×		×	×								
Northern kingfish	×		×									
Atlantic croaker	×	×	×									
Labridae												
Tautog				×								
Cunner				×								
Mugilidae												
Striped mullet		×	×	×								
White mullet	×		×	×								
Uranoscopidae												
Northern stargazer		×		×								
Pholidae												
Rock gunnel				×								
Ammodytidae												
American sand lance		×	×	×	×	×	×	×	×	×		
Eleotridae												
Fat sleeper					×							
Gobiidae												
Naked goby				×								
Seaboard goby			×									
Scombridae												
Atlantic mackerel				×								
Stromateidae												
Butterfish	×	×		×	×							
Triglidae												
Northern searobin	×		×	×								
Striped searobin	×		×	×	×	×			×	×		
Cottidae												
Grubby		×		×								
Longhorn sculpin				×								
Sea raven					×							
Dactylopteridae												
Flying gurnard				×								
Bothidae												
Smallmouth flounder	×		×	×								×
Summer flounder	×	×	×	×	×	×	×	×	×	×	×	×
Fourspot flounder		×										
Windowpane	×				×				×			
Pleuronectidae												
Winter flounder	×	×	×	×	×	×		×	×	×	×	×
Yellowtail flounder	×											
Soleidae												
Hogchoker	×	×	×	×	×	×	×	×	×	×	×	×
Tetraodontidae												
Northern puffer			×	×								×

[a] Specimens of all species reported from the Hudson River, except cobia, Atlantic cod, Atlantic salmon, and sea raven, are located in one or more of the following museum collections: American Museum of Natural History, Cornell University Museum, Museum of the Hudson Highlands, New York State Museum, University of Michigan Museum of Zoology.

[b] Locations of the 12 regions (see Figure 17), measured from the Battery at river kilometer 0 (George Washington Bridge is at km 19 and Troy Dam is at km 243):

AL	Albany	199–243	CW	Cornwall	89–98
CS	Catskill	171–198	WP	West Point	75–88
SG	Saugertes	150–170	IP	Indian Point	62–74
KG	Kingston	137–149	CH	Croton–Haverstraw	54–61
HP	Hyde Park	123–136	TZ	Tappan Zee	38–53
PK	Poughkeepsie	99–122	YK	Yonkers	19–37

species has been collected in the Hudson River since 1975, and probably entered from Oneida Lake through the canal. Smith (1985) noted that white bass may also have gained access from Lake Champlain to the north through the Champlain–Hudson Canal.

The occurrence of emerald shiner in the Hudson River estuary may also be due to access through the Erie–Barge Canal, although it could have entered the Mohawk River earlier during the period of postglacial repopulation (Snelson 1968; Smith 1985). Access through the canal appears more likely, because this species was collected in the Mohawk River and upper portion of the Hudson estuary (above Catskill) during the 1934 survey (Greeley 1935), but it was not collected below Catskill in the 1936 survey (Greeley 1937). Emerald shiners were common throughout the estuary in later surveys. If the emerald shiner had entered the Mohawk during the postglacial period, it probably would have moved into the lower Hudson estuary by the time of the 1936 survey and detected in the hundreds of samples taken at that time.

The central mudminnow is also believed to have entered the estuary through the Erie–Barge Canal (Smith 1985). This species has been collected at several locations within the estuary since 1967. These occurrences indicate sympatry with the eastern mudminnow in the Hudson River. Previous researchers considered the respective distributions of these two species to be separated by the Appalachian Mountains (Scott and Crossman 1973; Gilbert 1980a).

The distributions of several other species have also been affected by the Erie–Barge Canal. Gizzard shad have a wide native distribution south and west of the Hudson (Megrey 1980). From 1973 through 1980, this species was collected throughout the estuary. The northern limit of this species along the Atlantic coast previously was considered to be New York Harbor (Breder 1938; Scott and Crossman 1973). No specimens were collected in the estuary during the 1936 survey (Greeley 1937), although specimens have been reported from New York Harbor (Breder 1938), indicating a potential route of access. Because this species is common in the Great Lakes, entrance to the Hudson through the Erie–Barge Canal is another possibility. This canal has also allowed white perch, a species common in the Hudson, to gain access to Lake Ontario (Scott and Christie 1963) and eventually to Lake Erie (Busch et al. 1977).

The construction of the Champlain–Hudson Canal has permitted the silver lamprey to enter the Hudson from Lake Champlain (Reider 1979b). A third canal linking the Hudson and the Delaware River has allowed the shield darter to become established in Rondout Creek, the tributary used for the canal route. This species was taken in the estuary near the confluence of Rondout Creek with the Hudson at Kingston.

Other freshwater species are occasionally washed into the estuary from tributaries during high flow periods. These include the comely shiner, rosyface shiner, yellow bullhead, and brook stickleback.

Freshwater species collected on a regular basis throughout the estuary include species in several families. Examples include the spottail shiner, tessellated darter, pumpkinseed, redbreast sunfish, and brown bullhead (Table 6).

Several exotic freshwater species have also been introduced into the Hudson River and its tributaries. Some, including common carp, goldfish, brown trout, and some of the Centrarchidae, are established in the estuary and its tributaries and were collected in most surveys.

Marine and Brackish-Water Faunas

The extensive marine and brackish-water fish faunas found in the Hudson River estuary are a result of the mid-Atlantic location of the estuary and its proximity to the Gulf Stream. Many marine species ride the Gulf Stream into coastal nurseries, including the Hudson River estuary. In addition, several tropical or pelagic species stray into the estuary during the summer (Smith 1985), and deep-water species occasionally enter the estuary during winter. Individuals of these species have been collected frequently in the lower, more saline, portion of the estuary. Shallow-water marine species often move upstream into the estuary, sometimes beyond the salt front. Euryhaline species may be found throughout the estuary, often in large numbers in their preferred habitats. The estuary also supports several anadromous species of commercial and recreational importance and one catadromous species.

Examples of marine species in the Hudson whose young disperse along the Gulf Stream are bluefish, Atlantic menhaden, weakfish, and spot. Juvenile bluefish were commonly collected from Yonkers to just below Poughkeepsie and as far upstream as Castleton (river kilometer 220) in the Albany region during the drought of 1979 and 1980. Adult bluefish are frequently taken in the brackish portion of the estuary by commercial fishermen. Juvenile Atlantic menhaden were collected between Yonkers and Poughkeepsie and as far north as Coxsackie (km 200). Weakfish and spot were collected together in channel areas upstream of Indian Point. Weakfish were collected as far upstream as the Albany region.

Deepwater species occasionally collected in the lower portions of the estuary include the flying gurnard, striped cusk-eel, Atlantic cod, and longhorn sculpin. Adult specimens of Atlantic cod have been reported by commercial fishermen (Phillips 1883; Boyle 1969) and early life stages were collected by Texas Instruments during 1980. Longhorn sculpin have been taken in commercial nets (Boyle 1969) and in impingement collections at power plants (TI 1980b). Both the flying gurnard and striped cusk-eel have been taken in impingement collections.

Nearshore marine species taken in the Hudson River estuary include inshore lizardfish (Tabery et al. 1978; TI 1980a, 1980b), bluespotted cornetfish (Young et al. 1982), three species of silverside (rough silverside, tidewater silverside, Atlantic silverside: Greeley 1937; Boyle 1969; McFadden et al. 1978; TI 1980a, 1980b), and northern pipefish (Greeley 1937; Boyle 1969; McFadden et al. 1978). These species were usually collected in the lower portion of the estuary, although the silversides and northern pipefish were also found further upstream during low-flow periods. The silversides and northern pipefish were collected regularly during the various surveys of the estuary, but the inshore lizardfish and bluespotted cornetfish were taken only within the last 15 years.

Other marine species such as the fat sleeper, Atlantic mackerel, gray snapper, and black sea bass were collected infrequently in the lower portion of the estuary (Table 6). Although some were taken in beach seines (Boyle 1969), most specimens of these infrequently collected species came from the intake screens at the Indian Point power plant (TI 1979, 1980a, 1980b).

The abundant euryhaline species occurring in the Hudson River estuary include white perch, white catfish, banded killifish, mummichog, and others. These common species were collected during most surveys of the estuary along its entire length. The two killifish species were abundant in shore-zone areas, white catfish were common in offshore areas, and white perch were collected from both habitats in large numbers (TI 1980a, 1980b). The distribution and abundance data on white perch suggest two large concentrations, one in the area of Tappan Zee through Haverstraw Bay and the other upstream from Kingston (TI 1980c; Klauda et al. 1988, this volume).

Anadromous species using the Hudson River estuary include striped bass, Atlantic tomcod, Atlantic sturgeon, shortnose sturgeon, rainbow smelt, blueback herring, alewife, and American shad. Although the shortnose sturgeon is listed as an endangered species, it was not uncommon in gill-net catches. The other species were common throughout much of the estuary.

The American eel, a catadromous species, occurs throughout the estuary and its watershed. Elvers, spawned in the Sargasso Sea, enter the estuary and move upstream. Males remain in brackish water while females move into freshwater areas (Boyce Thompson Institute 1977).

Seasonal and Spatial Nearshore Zone Abundance

The species composition of beach-seine samples varied among ecological zones and seasons (Table 7). Bay anchovy was the dominant species in the marine zone during all seasons, using that zone as a spawning ground and nursery (Boyce Thompson Institute 1977). The spottail shiner was the most abundant species in the brackish and freshwater zones during the spring. Young-of-the-year blueback herring dominated collections from these zones during both the summer and fall, although relative abundances of this species in the two zones reversed between the two seasons. During the fall, these young fish begin a downstream movement into the brackish zone, then leave the estuary. Prior to leaving in the fall, the young-of-the-year clupeids tend to remain in the brackish zone rather than the marine zone.

The combined catch per effort in beach seines for all species increased during the summer, reflecting both recruitment and the movement of marine species into the river as the salt front extended further upstream. McFadden et al. (1978) reported that the numbers of species also increased through the spring months, reaching a peak in July, and then decreased through December in both the shore zone and deepwater areas of the Hudson River. This trend has also been reported for other Atlantic coast estuarine systems including those in Georgia (Dahlberg and Odum 1970), Chesapeake Bay (McErlean et al. 1973; Massman 1962), Delaware Bay (Derickson and Price 1973), and estuaries in New England (Oviatt and Nixon 1973; Haedrich and Haedrich 1974).

TABLE 7.—Seasonal and spatial abundances of common fish species in the near-shore areas of the Hudson River as reflected in the average catch per tow of 30.5-m beach seines, 1974–1979. An empty cell means the species was not caught; a + means <0.05 fish/tow.

| | Season[a] and zone[b] | | | | | | | | |
| | Spring | | | Summer | | | Fall | | |
Species	M	B	F	M	B	F	M	B	F
Blueback herring	1.8	1.7	3.1	9.3	52.5	200.8	6.1	72.2	30.8
Bay anchovy	68.6	3.4	0.1	59.6	11.8	0.3	11.4	4.3	+
White perch	8.2	3.4	0.1	14.2	28.0	17.2	3.6	3.8	2.8
Spottail shiner	0.2	8.7	23.3	0.2	9.8	25.8	0.1	8.4	15.9
American shad	0.1	2.9	3.1	10.2	18.7	28.0	3.6	4.5	2.3
Banded killifish	0.2	3.9	3.2	1.1	15.7	12.8	1.5	7.1	5.5
Striped bass	3.2	0.8	0.3	11.1	9.2	3.1	5.7	1.0	0.4
Alewife	1.1	1.4	6.5	3.8	6.9	7.0	0.8	2.0	1.1
Tessalated darter	0.1	1.7	1.9	0.1	3.7	5.8	0.1	2.2	2.4
Pumpkinseed	0.1	1.3	1.5	0.1	2.0	2.5	0.2	2.3	1.6
Atlantic tomcod	8.1	0.3	+	1.8	0.1		0.2	0.1	
Golden shiner	+	1.2	1.8	+	0.7	2.3	+	0.7	1.0
Bluefish	0.8	0.7	+	2.5	2.3	+	+	0.1	
Atlantic menhaden	0.6	0.4	+	1.5	1.7	+	0.4	0.2	
American eel	0.3	0.3	0.4	0.4	0.4	1.0	0.2	0.2	0.2
Emerald shiner	+	0.1	1.1		+	0.6			1.2
Goldfish	+	0.2	0.5	+	0.6	0.8		0.2	0.3
Redbreast sunfish		0.2	0.5	+	0.4	1.0		0.1	0.2
Atlantic silverside	0.1	+		1.3	0.1		0.6	0.1	
Fourspine stickleback	0.1	0.4	0.3	0.1	0.4	0.3	0.1	0.4	0.3
Rainbow smelt	1.6	0.6	0.1				+		
Mummichog	0.1	0.4	0.1	0.1	0.7	0.2	0.1	0.5	+
Bluegill	+	0.1	0.2	+	0.3	0.3	+	0.3	0.4
Hogchoker	0.3	+	+	0.7	0.2	0.1	+	+	+
Largemouth bass		+	0.1		0.3	0.4	+	0.1	0.1
Northern pipefish	0.2			0.4	0.1		0.1	0.1	
Spot	+			0.5	0.2	+	0.1	+	
Tidewater silverside	+	+		0.2	0.2		0.2	0.2	
Brown bullhead		0.1	0.1	+	0.5	0.1		+	+
Gizzard shad		+				+	+	0.4	0.2
Eastern silvery minnow			+			0.6			0.1
Common carp	+	0.1	0.1	+	0.1	0.1	+	+	+
Yellow perch		0.1	0.1		0.1	0.2		0.1	+
White catfish	0.1	+	+	0.1	0.1	+	+	+	
Alosa unidentified		0.5	9.2		13.2	52.8			
Lepomis unidentified	+	0.1	0.1	0.6	1.4	3.6	0.3	0.9	1.7

[a] Spring = Apr 1–Jun 30;
Summer = Jul 1–Sep 30;
Fall = Oct 1–Dec 30.

[b] M = marine zone (km 19–53);
B = brackish zone (km 54–88);
F = freshwater zone (km 89–243).

The spring and summer increases in both total catch per effort and number of species collected in beach seines also reflected movements of resident brackish and freshwater species into the shore zone from deepwater areas (McFadden et al. 1978). The decrease in seine catches in the fall reflected both emigration from the estuary and movement of fish into deeper water. Gladden et al. (1988) discuss the fish faunal relationships with biotic and abiotic components of the Hudson River estuary.

Species compositions, as reflected in the occurrence records and abundance data, are approximate because of differences among gear in selectivity and intensity of sampling and, to a lesser extent, because of differences in the efficiency of the same gear in collecting different species (TI 1978). Bottom-dwelling species such as the sturgeons, white catfish, hogchoker, and Atlantic tomcod, as well as the pelagic clupeids, silversides, and anchovies, are underrepresented in the nearshore samples and represent a higher proportion of the total estuarine fish fauna than indicated by their relative abundances in beach seine catches.

Recent interest in and concern for the Hudson estuary should allow this diverse fish fauna to persist. Knowledge of the fauna gained from the power plant impact studies will benefit future management of the resource.

References[3]

Barnthouse, L. W., R. J. Klauda, and D. S. Vaughn 1988. Introduction to the monograph. American Fisheries Society Monograph 4:1–8.

Bean, T. H. 1903. Catalogue of the fishes of New York. New York State Museum Bulletin 60.

Boyce Thompson Institute. 1977. An atlas of the biologic resources of the Hudson estuary. Boyce Thompson Institute of Plant Research, Estuarine Study Group, Yonkers, New York.

Boyle, R. H. 1969. The Hudson River. A natural and unnatural history. Norton, New York.

Breder, C. M., Jr. 1938. The species of fish in New York harbor. New York Zoological Society Bulletin 41:23–29.

Burdick, G. E. 1954. An analysis of factors, including pollution, having possible influence on the abundance of shad in the Hudson River. New York Fish and Game Journal 1:188–205.

Busch, W. D. N., D. H. Davies, and S. J. Nepszy. 1977. Establishment of white perch, *Morone americana*, in Lake Erie. Journal of the Fisheries Research Board of Canada 34:1039–1041.

Carlson, F. T., and J. A. McCann. 1969. Hudson River fisheries investigations. 1965–1968. Evaluations of a proposed pumped storage project at Cornwall, New York in relation to fish in the Hudson River. Hudson River Policy Committee, New York.

Cheney, A. N. 1887. Salmon in the Hudson River. U.S. Fish Commission Bulletin 6:351–352.

Clinton, D. 1824. Description of a new species from Hudson River (*Clupea hudsonia*). Annals of the Lyceum of Natural History 1:49–50.

Cooper, J. C., F. R. Cantelmo, and C. E. Newton. 1988. Overview of the Hudson River estuary. American Fisheries Society Monograph 4:11–24.

Curran, H. W., and D. T. Ries. 1937. Fisheries investigations in the lower Hudson River. Pages 124–145 *in* A biological survey of the lower Hudson watershed. New York State Conservation Department, Supplement to 26th Annual Report for 1936, Albany, New York.

Dahlberg, M. D., and E. P. Odum. 1970. Annual cycles of species occurrence, abundance, and diversity in Georgia estuarine fish populations. American Midland Naturalist 83:382–392.

DeKay, J. E. 1842. Natural history of New York. Part I. Zoology. Reptiles and fishes. Part 4—fishes. Appleton, and Wilby and Putnam, Albany, New York.

Derickson, W. K., and S. K. Price, Jr. 1973. The fishes of the shore zone of Rehoboth and Indian River bays, Delaware. Transactions of the American Fisheries Society 102:552–562.

Dovel, W. L. 1977. Performance report for the biology and management of shortnose and Atlantic sturgeons of the Hudson River. Report to New York Department of Environmental Conservation, Project AFS-9-R-2, Albany, New York.

Fisher, A. K. 1890. Notes on the occurrence of a young crab eater (*Elacate canadum*) from the lower Hudson Valley, New York. Proceedings of the U.S. National Museum 13:195.

Fishler, K. G. 1959. Contributions of the Hudson and Connecticut rivers to New York–New Jersey shad catch of 1956. U.S. Fish and Wildlife Service Fishery Bulletin 60:161–174.

Gilbert, C. R. 1980a. Genus *Umbra*, mudminnows. Pages 129–130 *in* D. S. Lee, C. R. Gilbert, C. H. Hocutt, R. E. Jenkins, D. E. McAllister, and J. R. Stauffer Jr., editors. Atlas of North American freshwater fishes. North Carolina State Museum of Natural History, Raleigh.

Gilbert, C. R. 1980b. Zoogeographic factors in relation to biological monitoring of fish. Pages 309–355 *in* C. H. Hocutt and J. R. Stauffer, Jr., editors. Biological monitoring of fish. Lexington Books, Lexington, Massachusetts.

Gladden, J. B., F. R. Cantelmo, J. M. Croom, and R. Shapot. 1988. Evaluation of the Hudson River ecosystem in relation to the dynamics of fish populations. American Fisheries Society Monograph 4:37–52.

Grabe, S. A. 1978. Food and feeding habits of juvenile Atlantic tomcod, *Microgadus tomcod,* from Haverstraw Bay, Hudson River, New York. U.S. National Marine Fisheries Service Fishery Bulletin 76:89–94.

Grabe, S. A. 1980. Food of age 1 and 2 Atlantic tomcod, *Microgadus tomcod,* from Haverstraw Bay, Hudson River, New York. U.S. National Marine Fisheries Service Fishery Bulletin 77:1003–1006.

Greeley, J. R. 1935. Annotated list of fishes occurring in the watershed. Pages 88–101 *in* A biological survey of the Mohawk–Hudson watershed. New York Conservation Department, Supplement to 24th Annual Report for 1934, Albany.

Greeley, J. R. 1937. Fishes of the area with annotated list. Pages 45–103 *in* A biological survey of the lower Hudson watershed. New York State Conservation Department, Supplement to 26th Annual Report for 1936, Albany, New York.

Haedrich, R. L., and S. O. Haedrich. 1974. A seasonal survey of the fishes in the Mystic River, a polluted estuary in downtown Boston, Massachusetts. Estuarine and Coastal Marine Science 2:59–73.

Holsapple, J. G., and L. E. Foster. 1975. Reproduction of white perch in the lower Hudson River. New York Fish and Game Journal 22:122–127.

Howells, G. P. 1972. The estuary of the Hudson River, U.S.A. Proceedings of the Royal Society of London, B, Biological Sciences 180:521–534.

Hutchison, J. B., Jr. 1988. Technical descriptions of Hudson River electricity generating stations. American Fisheries Society Monograph 4:113–120.

Jordan, D. S. 1905. A guide to the study of fishes, volume 1. Henry Holt, New York.

Kendall, A. W., Jr., and L. A. Walford. 1979. Sources and distribution of bluefish, *Pomatomus saltatrix,* larvae and juveniles off the east coast of the United

---

[3]See Table 1 for sources of legal documents and unpublished reports pertaining to the Hudson River.

States. National Marine Fisheries Service Fishery Bulletin 77:213–227.

Klauda, R. J., J. B. McLaren, R. E. Schmidt, and W. P. Dey. 1988. Life history of white perch in the Hudson River estuary. American Fisheries Society Monograph 4:69–88.

LMS (Lawler, Matusky, & Skelly Engineers). 1980a. Biological and water quality data collected in the Hudson River near the proposed Westway project during 1979–1980, volume 1. Reports LMSE-80/9031 and 386/002 to New York Department of Transportation, Albany.

LMS (Lawler, Matusky, & Skelly Engineers). 1980b. Biological and water quality data collected in the Hudson River near the proposed Westway project during 1979–1980, volume 2. Reports LMSE-80/9051 and 232/009 to New York Department of Transportation, Albany.

Malcolm Pirnie. 1982. Hudson River estuary fish habitat study. Report to U.S. Army Corps of Engineers, New York District, New York.

Massman, W. H. 1962. Water temperatures, salinities, and fishes collected during trawl surveys of the Chesapeake Bay and York and Pamunky Rivers 1956–1959. Virginia Institute of Marine Science Special Science Report 27.

Mather, F. 1887. Report of operations at Cold Springs Harbor, New York, during the season of 1885. Pages 110–112 in Report of the commissioner, part XIII. U.S. Commission of Fish and Fisheries, Washington, D.C.

McErlean, A. J., S. B. O'Connor, J. A. Mihursky, and I. C. Gibson. 1973. Abundance, diversity and seasonal patterns of estuarine fish populations. Estuarine and Coastal Marine Science 1:19–36.

McFadden, J. T., Texas Instruments, and Lawler, Matusky & Skelly Engineers. 1978. Influence of the proposed Cornwall pumped storage project and steam electric generating plants on the Hudson River estuary with emphasis on striped bass and other fish populations, revised. Report to Consolidated Edison Company of New York.

Mearns, E. A. 1898. A study of the vertebrate fauna of the Hudson highlands, with observations on Mollusca, Crustacea, Lepidoptera and flora of the region. Part II. Vertebrates of the Hudson Highlands. Fisheries Bulletin of the American Museum of Natural History 10:311–322.

Medeiros, W. H. 1974. The Hudson River shad fishery: background, management problems, and recommendations. New York State Sea Grant College Program, NYSSGP-RS-75–011, SUNY (State University of New York), Stony Brook.

Megrey, B. A. 1980. Dorosoma cepedianum (Lesueur), gizzard shad. Page 69 in D. S. Lee, C. R. Gilbert, C. H. Hocutt, R. E. Jenkins, D. E. McAllister, and J. R. Stauffer Jr., editors. Atlas of North American freshwater fishes. North Carolina State Museum of Natural History, Raleigh.

Mitchill, S. L. 1815. The fishes of New York, described and arranged. Transactions of the Literary and Philosophical Society of New York 1:355–492.

Mitchill, S. L. 1817. Report on the ichthyology of the Wallkill, from specimens of fishes presented to the society (Lyceum of Natural History) by Dr. B. Akerly. American Monthly Magazine Critical Review 1:289–290.

Mitchill, S. L. 1818. Memoir on ichthyology. The fishes of New York, described and arranged. American Monthly Magazine Critical Review 2:241–248, 321–328.

New York State Conservation Department. 1943. Report by the Bureau of Inland Fisheries on Hudson River (south of Troy Dam). Pages 175–178 in 32nd Annual Report for 1942, Albany.

Oviatt, C. A., and S. W. Nixon. 1973. The demersal fish of Narragansett Bay: an analysis of community structure, distribution, and abundance. Estuarine and Coastal Marine Science 1:361–378.

Perlmutter, A., E. Leff, E. E. Schmidt, R. Heller, and M. Sicilano. 1966. Distribution and abundance of fishes along the shores of the lower Hudson River during the summer of 1966. Pages 147–200 in Hudson River ecology. Hudson River Valley Commission of New York, Albany.

Perlmutter, A., E. E. Schmidt, and E. Leff. 1967. Distribution and abundance of fishes along the shores of the lower Hudson River during the summer of 1965. New York Fish and Game Journal 14:47–75.

Phillips, B. 1883. A stray cod up the Hudson. U.S. Fish Commission Bulletin 3:416.

Raney, E. C. 1952. The life history of the striped bass. Bulletin of the Bingham Oceanographic Collection, Yale University 14(1):5–97.

Raney, E. C., W. S. Woolcott, and A. G. Mehring. 1954. Migratory patterns and racial structure of Atlantic coast striped bass. Transactions of the North American Wildlife Conference 19:376–396.

Rathjen, W. F., and L. C. Miller. 1957. Aspect of the early life history of the striped bass (Roccus saxatilis) in the Hudson River. New York Fish and Game Journal 4:43–60.

Regents of the University of the State of New York. 1855–1900. On the condition of the state cabinet of natural history. In Annual reports, Albany.

Reider, R. H. 1979a. Occurrence of a kokanee in the Hudson River. New York Fish and Game Journal 26:94.

Reider, R. H. 1979b. Occurrence of the silver lamprey in the Hudson River. New York Fish and Game Journal 26:93.

Scott, W. B., and W. J. Christie. 1963. The invasion of the lower Great Lakes by the white perch, Roccus americanus (Gmelin). Journal of the Fisheries Research Board of Canada 20:1189–1195.

Scott, W. B., and E. J. Crossman. 1973. Freshwater fishes of Canada. Fisheries Research Board of Canada Bulletin 184.

Smith, C. L. 1983. Going with the flow. Natural History 92:49–56.

Smith, C. L. 1985. Inland fishes of New York State. New York State Department of Environmental Conservation, Albany.

Smith, E. 1897. The fishes of the fresh and brackish

waters in the vicinity of New York City. Proceedings of the Linnean Society of New York 9:1–51. (Abstract.)

Snelson, F. F. 1968. Systematics of the cyprinid fish *Notropis amoenus,* with comments on the subgenus *Notropis.* Copeia 1968:776–802.

Tabery, M. A., A. P. Ricciardi, and T. J. Chambers. 1978. Occurrences of larval inshore lizardfish in the Hudson River estuary. New York Fish and Game Journal 25:87–88.

Talbot, G. B. 1954. Factors associated with fluctuations in abundance of Hudson River shad. U.S. Fish and Wildlife Service Fishery Bulletin 56:373–413.

TI (Texas Instruments). 1978. Catch efficiency of 100-ft (30–m) beach seines for estimating density of young-of-the-year striped bass and white perch in the shore zone of the Hudson River estuary. Report to Consolidated Edison Company of New York.

TI (Texas Instruments). 1979. Hudson River ecological study in the area of Indian Point. 1977 annual report to Consolidated Edison Company of New York.

TI (Texas Instruments). 1980a. Hudson River ecological study in the area of Indian Point. 1978 annual report to Consolidated Edison Company of New York.

TI (Texas Instruments). 1980b. Hudson River ecological study in the area of Indian Point. 1979 annual report to Consolidated Edison Company of New York.

TI (Texas Instruments). 1980c. 1978 year class report for the multiplant impact study of the Hudson River estuary. Report to Consolidated Edison Company of New York.

USACE (U.S. Army Corps of Engineers). 1984. Final supplemental environmental impact statement, West Side highway project, volume 2. Fisheries portion. USACE, New York District, New York.

U.S. Fish Commission. 1882–1900. Bulletins of the United States Fish Commission 1–20.

Webster, D. A. 1982. Early history of the Atlantic salmon in New York. New York Fish and Game Journal 29:26–44.

Wilk, S. J. 1977. Biological and fisheries data on bluefish, *Pomatomus saltatrix* (Linnaeus). U.S. National Oceanic and Atmospheric Administration Northeast Fisheries Center Sandy Hook Laboratory Technical Series Report 21.

Young, B. H., I. H. Morrow, and S. R. Wanner. 1982. First record of the bluespotted cornetfish in the Hudson River. New York Fish and Game Journal 29:106.

Young, J. R., R. J. Klauda, and W. P. Dey. 1988. Population estimates for juvenile striped bass and white perch in the Hudson River Estuary. American Fisheries Society Monograph 4:89–101.

American Fisheries Society Monograph 4:37–52, 1988

Evaluation of the Hudson River Ecosystem in Relation to the Dynamics of Fish Populations

JOHN B. GLADDEN[1]

Savannah River Ecology Laboratory
Post Office Drawer E, Aiken, South Carolina 29801, USA

FRANK R. CANTELMO

Department of Biological Sciences
St. John's University, Jamaica, New York 11439, USA

J. M. CROOM

Quantitative Analysis
3898 Westwood Path, Stone Mountain, Georgia 30083, USA

R. SHAPOT[2]

Envirosphere Corporation
160 Chubb Avenue, Lyndhurst, New Jersey 07071, USA

Abstract.—Components of the Hudson River estuary ecosystem were evaluated to elucidate aspects of trophic function that could contribute to the observed variability in fish population abundance. Over half (55%) of the organic carbon available in the estuary was imported from the surrounding watershed, while phytoplankton (17%) and benthic algae (14%) provided most of the remaining food base. Allochthonous carbon inputs were highest in spring as a result of high freshwater flow rates, and algal production was highest in summer during peak insolation. Invertebrate populations provide food for most fish species; benthic invertebrate communities were found to provide a more temporally stable food base than either epibenthic invertebrates or zooplankton, both of which showed large seasonal changes in abundance. Dominant resident fish species primarily used shallower areas of the estuary, feeding omnivorously or on the benthic and epibenthic invertebrates. Migrant species occurred in all major estuarine habitants, but dominated the fish community in deepwater portions of the estuary and depended on zooplankton and other fish for food. Migrant fish species appeared to have larger year-to-year variations in abundance than did resident species, possibly as a result of less stable (less predictable) food resources for the migrant fish than for the resident fish. Overall, freshwater flow rates may be the single largest factor influencing trophic functioning in the estuary. Such flows strongly influence organic carbon inputs from the watershed, phytoplankton standing crops, and the spatiotemporal distributions of salinity and temperature in the estuary.

The objective of the Hudson River ecological studies was to assess the impact of operating several electricity generating stations on the function and persistence of the Hudson River biota. The sources of potential impact included nuclear power stations, coal-fired stations, and a proposed pumped-storage facility. Entrainment and impingement mortality at the power plant intakes was the impact of greatest concern. All these facilities are or were to have been situated in the middle reaches of the estuary, where they could affect not only resident freshwater and euryhaline species but also migratory species that used the estuary for only a part of their life cycles.

Although most aspects of the Hudson ecosystem were examined during the impact assessment, the major emphasis was on evaluating changes in fish species populations. Numerous approaches toward assessing the long-term fate of affected fish populations were attempted by scientists working for the U.S. Environmental Protection Agency and the utilities. In addition to field studies sponsored by the utilities, extensive population modeling efforts were undertaken to integrate data and predict long-term changes. Two major problems complicated these modeling efforts: the nature and magnitude of density-dependent processes in the populations considered, and the

---

[1]Present address: Savannah River Laboratory, Building 773-42A, Aiken, South Carolina 29808, USA.

[2]Present address: Roy F. Weston, Inc., Weston Way, West Chester, Pennsylvania 19380, USA.

TABLE 8.—Selected physiographic characteristics of the Hudson River estuary. (Adapted from TI 1975b, 1981.)

River region	Distance[a] (km)	Surface area (km^2)	Volume (10^6 m^3)	Surface:volume ratio	Area <3 m deep (%)
Yonkers	19–37	27.55	229.4	0.120	24.7
Tappan Zee	38–53	62.75	321.8	0.195	32.6
Croton–Haverstraw	54–61	30.74	147.7	0.208	39.4
Indian Point	62–74	16.87	208.3	0.081	24.6
West Point	75–88	10.03	207.5	0.048	11.8
Cornwall	89–98	17.59	139.8	0.126	27.2
Poughkeepsie	99–122	24.55	298.1	0.082	13.0
Hyde Park	123–136	11.67	165.5	0.071	4.8
Kingston	137–149	17.63	141.5	0.125	22.0
Saugerties	150–170	25.38	167.3	0.144	31.1
Catskill	171–198	28.58	160.7	0.178	31.0
Albany	199–225	13.00	71.1	0.183	32.7

[a]River kilometers upstream from the Battery on Manhattan Island (Figure 17).

spatial and temporal variability in abundances of the populations. The magnitude of the second problem caused substantial difficulties for the empirical resolution of the first. Toward the end of the controversy, both government and utility scientists used stochastic population modeling to address these problems (TI 1980, 1981; O'Neill et al. 1981; Christensen and Englert 1988 and other papers in Section 3 of this volume). With few exceptions (Levins 1979), these models dealt with populations of single fish species in isolation from their biotic and abiotic surroundings.

The high year-to-year variability observed in fish cohort abundances could be caused by changes in spawning stock abundance, cannibalism, predation, and food supply. Spatiotemporal distributions of temperature and salinity can influence the location and timing of a species' peak abundance and spawning activity and the survival of its eggs and larvae. The unpredictability of temperature and salinity distributions from year to year in the Hudson River estuary (Cooper et al. 1988, this volume) could result in substantial variations in habitat suitability for both fish and prey species.

In spite of the importance of population variability in analyses of fish population data, little effort was made to determine the underlying causes of that variability. Two approaches to understanding the dynamics of fish populations are to more clearly define the environment in which the fish function and to determine which biotic and abiotic factors are most likely to limit the production and persistence of fish species. The objective of this study was to determine whether spatiotemporal variations in abiotic or biotic characteristics of the estuary could be related to variations in the abundance of fish species or to the structure of the estuarine fish community.

Physiography and Physicochemical Nature of the Estuary

Texas Instruments (TI 1976a) and Cooper et al. (1988) have discussed the physical structure of the Hudson River estuary and the dynamics of numerous physicochemical variables. It is clear from those analyses that freshwater input to the estuary strongly influences such variables as salinity, water temperature, dissolved oxygen, and turbidity (Cooper et al. 1988). From the perspective of an organism using the estuary, the spatiotemporal distributions of temperature and salinity in the middle and lower positions of the estuary are highly variable and have only limited predictability over distances of several kilometers.

Substantial differences in physiographic characteristics exist through the estuary (Table 8). Downstream areas of the estuary (km 38–61) are broad and shallow with high surface:volume ratios and a high percentage of the water area in shallows. Midregions of the estuary (km 62–170) are generally deeper with lower proportions of shallow-water area. The uppermost segment of the estuary (km 171–225) is also relatively shallow.

Food Chain Bases in the Hudson River

Maintenance of the trophic structure in estuaries has been considered to depend primarily on the input of allochthonous organic matter from surrounding terrestrial or marsh ecosystems (Odum and de la Cruz 1967). However, Correll (1978) has suggested that phytoplankton production may also contribute importantly to estuarine food chains, and Welsh et al. (1982) found that benthic primary production can be substantial in shallow New England estuaries. Additional organic carbon may enter streams and estuaries in sewage effluents from urbanized regions. The objective of this section is to outline the pattern and magnitude of organic carbon availability in the Hudson River estuary.

Phytoplankton and Benthic Plant Productivity

Phytoplankton primary production was measured in 1972, by means of carbon-14 assimilation, in the lower half of the Hudson River estuary (from 3.5 km below the Battery to km 122: Sirois and Frederick 1978). Phytoplankton productivity was greatest in June and July, peaking at over 5.5 g $C \cdot m^{-2} \cdot d^{-1}$ in 1972 (Figure 11). Primary production in the most productive sites monitored decreased sharply in August, but most sites maintained productivity above spring levels through September or October 1972. Primary production for the period May through October (semiannual) increased steadily from the Battery upstream to the Tappan Zee and Croton–Haverstraw regions,

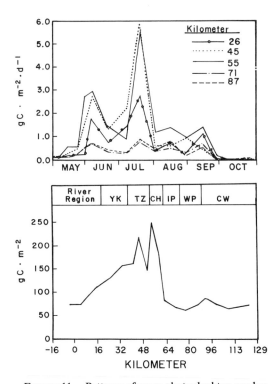

but declined farther upriver (Figure 11). The peak semiannual productivity during 1972 in the Tappan Zee and Croton–Haverstraw regions (Sirois and Frederick 1978) was less than reported in lower New York Bay (O'Reilly et al. 1976) but higher than reported for other areas of the Hudson River (Garside et al. 1976; Malone 1977) and several other Atlantic coast estuaries (Smayda 1973).

The reasons for these large spatial and temporal variations in primary production are believed to be quite complex. Seasonal variations in light intensity are a major influence on primary production rates because Hudson River phytoplankton populations are light-limited rather than nutrient-limited (Heffner 1973; O'Reilly et al. 1976; Storm and Heffner 1976). High turbidities result in a shallow euphotic zone (NYU Medical Center 1977, 1978). Nitrogen and phosphorus concentrations are usually adequate to high (Hetling 1976). The summer season is also marked by a shift of phytoplankton community structure from diatoms to a green algae community composed largely of nannoplankton (O'Reilly et al. 1976; NYU Medical Center 1977, 1978). O'Reilly et al. (1976) found that nannoplankton accounted for 67% of the annual primary production in New York Bay, and Malone (1977) concluded that nannoplankton accounted for productivity increases above 0.25 g $C \cdot m^{-2} \cdot d^{-1}$ in the lower estuary.

Variations in river morphometry, flow patterns, and flow rates also contribute to spatial differences in phytoplankton productivity. River regions having the highest primary production (Figure 11) are relatively broad and shallow (Table 8; Cooper et al. 1988). The lower flushing rates in these shoal areas facilitate the rapid increase in photosynthetic capacity (i.e., in phytoplankton cell numbers and chlorophyll), particularly when the community is dominated by plankton with short generation times. Malone (1977) also concluded that flushing rates and algal community composition are important influences on phytoplankton productivity in the New York Bay area. The occurrence of Hurricane Agnes in late June 1972 may have contributed to the reduced primary production rates in the lower estuary (Figure 11) by flushing algae from these shoal-dominated regions. However, a comparable pattern occurred for chlorophyll-a concentrations in 1975 (NYU Medical Center 1977). Few long-term studies of phytoplankton productivity are available, but Boynton et al. (1982) reported that annual productivity varied over twofold from 1972 to 1977 in the Chesapeake Bay estuary.

FIGURE 11.—Patterns of gross phytoplankton productivity in the Hudson River estuary in 1972. Above: May–October productivity trends at five locations. Below: longitudinal profile of semiannual productivity in the lower estuary. Kilometers are river distances from the Battery, Manhattan Island. Regions: YK, Yonkers; TZ, Tappan Zee; CH, Croton–Haverstraw; IP, Indian Point; WP, West Point; CW, Cornwall to Poughkeepsie. (Data are from Sirois and Frederick 1978.)

Welsh et al. (1982) found that benthic primary production can be substantial in New England estuaries and that the magnitude of productivity is closely related to estuarine surface-area:volume ratios. Surface:volume ratios ranged from 0.05 to 0.21 in the Hudson River regions used for calculation of phytoplankton production (Table 8). These values were in the lower range of estuarine area:volume ratios considered by Welsh et al. (1982). Benthic primary production calculated for the Hudson River estuary (km 22–244) was approximately 28,845 tonnes of organic carbon per year. However, it should be noted that phytoplankton production calculated by this technique (by difference from total production) gave estimates that were usually higher than our estimates calculated from Sirois and Frederick (1978). The dependence of benthic primary production on light and temperature is likely to result in concurrent maximum production by benthic and pelagic primary producers.

Marsh Productivity

Macrophytes growing submerged and in tidal marshes are believed to represent a major food source in estuarine ecosystems (Odum and de la Cruz 1967; Heinle et al. 1976). Although Kiviat (1973) and Quirk, Lawler & Matusky (QLM 1974) provided descriptions of the rooted aquatic vegetation in the Hudson River estuary, productivity values for these communities are not available. Primary production, however, has been estimated for several marshes in the mid-Atlantic states having comparable species composition (Jervis 1969; Udell et al. 1969; Whigham and Simpson 1975). Net macrophyte production in these marshes averaged about 0.5 kg C·m^{-2}·year^{-1} in these studies. Thus, the 12.04 km^2 of Hudson River estuary marshland from km 22 to km 244 (A. J. Bonavist, New York State Department of Environmental Conservation, personal communication) should produce about 6.02 × 10^6 kilograms of organic carbon annually. Moran and Limburg (1985) estimated that 5.4 × 10^6 kg of organic carbon are produced annually by macrophytes in the Hudson River estuary. It is not known how much of the organic matter produced in these marshes enters river food chains; Woodwell et al. (1977) suggested that estuarine marshes may be net consumers of organic carbon rather than net exporters.

Watershed Inputs of Carbon

Total organic carbon (TOC) concentrations in the Hudson River and its tributaries ranged from less than 2 to 25 mg·L^{-1} during 1974 to 1977, but the majority of values were between 3 and 8 mg·L^{-1} (USGS 1975a, 1975b, 1976, 1977). Values above 10 mg·L^{-1} occurred in both the Hudson and Mohawk rivers but were largely restricted to February through May in each year with occasional peaks in collections from the fall. However, the total movement of organic matter from the terrestrial watershed into the Hudson River is determined by both the TOC concentration in water and the magnitude of stream flow. Carbon flux rates based on carbon concentrations and water flow rates indicated two peaks in TOC export from the surrounding watershed. The late-fall maximum was probably a result of the coincident leaf loss from trees, increased precipitation, and freshwater runoff. Northeastern watersheds are normally covered by snow in winter; in spring, snowmelt and heavy rains result in another large influx of organic carbon into the streams and rivers. The fall and spring peaks of organic matter output from Hudson River watersheds correspond to the patterns observed by Fisher and Likens (1973) in New Hampshire watersheds.

The total organic carbon output was calculated for four sites within the Hudson River basin over a 9-month period from August 1974 to May 1975. Comparable and relatively complete TOC concentration and instantaneous discharge data were available for these sites during this period. The average daily carbon output for these four sites was 8.6 kg C·km^{-2} of watershed area (Table 9), and the values were surprisingly similar among the sites.

We tested the similarity of watershed organic carbon export more rigorously by examining USGS surface water records for sites within the

TABLE 9.—Average daily organic carbon output from four watersheds in the Hudson River basin in 1974–1975. Organic carbon concentrations and flow data are from USGS (1975a,1975b) and organic carbon output rates are time-weighted averages.

Watershed	Watershed area (km^2)	Sampling Begin (1974)	Sampling End (1975)	Average daily carbon output (kg C·km^{-2}·d^{-1})
Sacandaga River (East Branch)	295	Aug 8	May 15	7.2
Wappinger Creek	469	Aug 14	May 7	9.1
Esopus Creek (Saugerties)	1,101	Aug 12	May 13	9.1
Mohawk River (Crescent Dam)	8,943	Aug 12	May 12	9.2
Mean				8.6

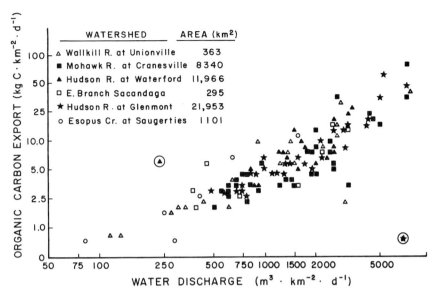

FIGURE 12.—Relationship between standardized organic carbon export and standardized surface discharge from watersheds of various sizes within the Hudson River drainage. Circled values were not used in statistical analysis of these data.

Hudson River basin that had at least 10 concurrent discharge and TOC measurements during the period 1974–1977 (USGS 1975a, 1975b, 1976, 1977). Six such sites had drainage areas ranging from 363 to 21,953 km^2 and the data were standardized to a unit-area basis for both carbon export (kg $C \cdot km^{-2} \cdot d^{-1}$) and water discharge ($m^3 \cdot km^{-2} \cdot d^{-1}$). Analysis of covariance within the log-transformed data (Figure 12) indicated that water movement was extremely important for predicting carbon export ($r^2 = 0.78$; $P = 0.001$) and that there was no difference ($P = 0.41$) in unit-area, flow-adjusted carbon export among the drainages examined. The two points that were excluded from the analysis are assumed to be associated with storm events, which can produce exceptional short-term patterns of carbon movement in streams (Fisher and Likens 1973; Bilby and Likens 1979). This finding may have considerable importance for current developments in stream ecology (Minshall et al. 1983), but its primary importance for this analysis is to show that organic carbon input to the estuary appears to be largely controlled by the short-term (daily) rates of freshwater input. Furthermore, the analysis indicates that allochthonous organic carbon inputs to the estuary may vary substantially from year to year because of the large annual variation in freshwater flow (Cooper et al. 1988).

Sewage Inputs

Additional organic carbon inputs to the Hudson River occur from anthropogenic sources such as sewage outfalls. Analysis of Hetling's (1976) data indicates that although sewage input may be locally important, its overall effect on the organic carbon budget of the river is small. The total organic carbon input as sewage was about 57.6 kg $C \cdot d^{-1}$ in the early 1970s; inputs in the Albany (km 201–244) and Yonkers (km 22–39) regions constituted about two-thirds of this total. Sewage inputs downstream from the George Washington Bridge (km 19) were not included in Hetling's data but may influence river organic carbon levels as far north as Indian Point (Schaffer 1978). Moran and Limburg (1985) obtained similar estimates of sewage carbon inputs, but included areas downstream from the George Washington Bridge. However, the current magnitude of organic carbon inputs from sewage treatment plants should be much lower because treatment facilities have been upgraded all along the estuary.

Relative Magnitudes of Organic Carbon Inputs

Inputs of organic carbon from the surrounding watershed dominate the carbon flux in the Hudson estuary (Table 10). The upper watershed, upriver of the estuary, and the lower watershed

TABLE 10.—Summary of organic carbon sources for Hudson River estuary (km 22–244).

Source	Organic carbon tonnes·year^{-1}	Percent of total
Phytoplankton[a]	34,632	17
Benthic algae	28,845	14
Macrophyte[b]	6,020	3
Sewage effluents	21,026	10
Upper watershed (above km 244)[c]	65,771	32
Lower watershed (km 22–244)[d]	46,790	23
Total	203,084	

[a] Estimated from May to October 1972; likely underestimated by 5–10% for November–April productivity.
[b] Assumes 0.5 g C·g^{-1} dry weight biomass (Odum 1971).
[c] 8.6 kg C·km^{-2}·d^{-1} × timas 365 d·year^{-1} × 20,953 km^2/1,000 kg·tonne^{-1}.
[d] 8.6 kg C·km^{-2}·d^{-1} × 365 d·year^{-1} × 14,906 km^2/1,000 kg·tonne^{-1}.

together contribute over half of the total organic carbon input. Phytoplankton and benthic primary production within the estuary make nearly equal contributions of organic carbon to the estuary and the combined input from these sources in approximately one-third of the total.

The data in Table 10 do not reflect the temporal patterns of carbon inputs to the estuary, however. Allochthonous carbon inputs should be highest in March through May because of the strong relationship between carbon input and freshwater flow (Figure 12; Cooper et al. 1988), and little input of allochthonous carbon should occur during summer and early fall. The mean residence time for water entering the upper estuary (0.4 year: Simpson et al. 1973) suggests that allochthonous carbon entering during high spring flows should be available for processing within the estuary through most of the summer. As allochthonous inputs are declining, phytoplankton and benthic production within the estuary is increasing, particularly in the shoal-dominated areas downstream from Indian Point. Thus, high levels of organic carbon availability are maintained in the estuary from early spring through summer if allochthonous inputs and autochthonous primary production are combined.

The trophic pathways and the efficiency with which organic carbon is used are not clear, however. Particulate components may be grazed in the water column or may settle and become incorporated into sediments. Both these pathways lead food resources directly to invertebrates. However, Schaffer (1978) found that more organic carbon in water of the Indian Point region was in dissolved than in particulate form and thus unuseable by zooplankton (Stephens 1967). Most of this carbon must be first incorporated into microbial biomass, with resultant respiratory energy losses, before becoming available to invertebrates. It is likely that organic carbon that reaches the estuary from the surrounding watershed is highly refractory, further reducing its nutritional value relative to the phytoplankton. Thus, autochthonous production in the estuary may actually provide closer to half of the energy available to higher trophic levels and the magnitude of annual production is likely to vary substantially from year to year (Boynton et al. 1982).

Invertebrate Communities

Invertebrates in the Hudson River estuary make up a heterogenous group of organisms that forms essential links in the food web between organic carbon sources and fish populations. Population fluctuations, dispersal patterns, and generation times of invertebrates may be very important influences on the life history strategies and year-class strengths of Hudson River fishes.

The role of invertebrates in estuarine processes is often related to body size and habitat preference (Levinton 1972; Rhoads et al. 1978), so it is appropriate to categorize Hudson River invertebrates into three major groups: pelagic invertebrates, epibenthic invertebrates, and benthic invertebrates.

Pelagic Invertebrates

Pelagic invertebrates are dominated by microzooplankton (0.08–2.0 mm) and dominant taxa include the calanoid copepods, pelagic cladocerans, and rotifers. Their feeding is restricted to small particles such as bacteria-laden detritus, phytoplankton, and small microzooplankton (Saunders 1969). In addition, they have limited mobility and their distributions are largely determined by water movements in the estuary. Large population pulses can result as individuals drift into and out of favorable conditions for growth and reproduction. Population dynamics of microzooplankton are strongly affected by temperature, salinity, and availability of appropriate-sized food (Hulsizer 1976). Heinle (1974) and Chervin (1977) have shown that population pulses of microzooplankton often occur soon after blooms of phytoplankton.

Data from NYU Medical Center (1978) reveal that population pulses of pelagic invertebrates occurred at Indian Point during the spring and summer of 1976 (Figure 13). Total densities were greatest from late June (180,400·m^{-3}) to late August (220,000·m^{-3}) when they varied almost 5-fold on alternate biweekly sampling periods and over 10-fold from May to November. These pulses of pelagic invertebrates represent a series of increases in several taxa including rotifers, cladocerans, and copepods. Generation times for Hudson River microzooplankton are temperature-dependent and can vary from 4 d for amictic rotifers to a few months for calanoid copepods. Most species exhibit short generation times (1–3 weeks) and can, therefore, respond rapidly to favorable abiotic conditions.

Epibenthic Invertebrates

Epibenthic invertebrates live in close association with surficial sediments but also move regularly into the overlying water column. This group includes both microzooplankton-sized organisms (e.g., cyclopoid copepods, cladocerans, and ostracods) and macrozooplankton-sized organisms (e.g., mysids, isopods, amphipods, and dipterans). NYU Medical Center (1978) found that the mysid *Neomysis americana* and the amphipods *Gammarus* spp. and *Monoculodes edwardsii* were often numerically dominant and may collectively account for greater than 75% of the total epibenthos in some regions. Species composition changed seasonally and generally reflected changes in physicochemical conditions. For ex-

ample, the halophilic *Neomysis americana* usually replaced *Gammarus* spp. when freshwater flow decreased and the salt front moved upriver of Indian Point in the summer (NYU Medical Center 1978). Additionally, species of epibenthic invertebrates vary widely in their ability to respond to favorable environmental conditions because generation times range from weeks for the microzooplankton species to several months for the larger species.

Epibenthic macrozooplankton and microzooplankton consume a wide variety of food items ranging from particulate matter and bacteria to diatoms and zooplankton; some species may switch from one group of food organisms to another during their various life stages (Cummins et al. 1969). Because many epibenthic organisms share both benthic and pelagic habitats, they are able to exploit a wider range of food resources than either the more sedentary benthos or less mobile pelagic invertebrates.

Hudson River epibenthic invertebrates exhibit population maxima in late spring to early summer and again in late fall (Figure 13) as is common in other Atlantic coast estuaries (Sanders et al. 1965; Boesch and Diaz 1974; Watling et al. 1974). NYU Medical Center (1978) found that epibenthic populations near Indian Point were highest from late May (8,100·m^{-3}) to early August (10,700·m^{-3}) and peaked again during early November (15,600·m^{-3}). Populations near Indian Point var-

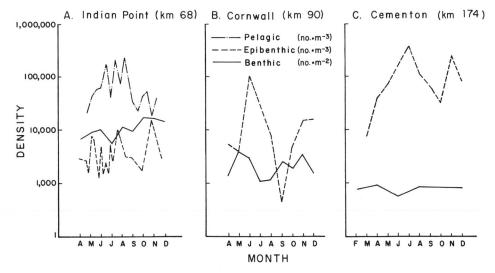

FIGURE 13.—Mean daytime abundances of invertebrates in pelagic, epibenthic, and benthic habitats at three locations in the Hudson River estuary. Data were derived from various sources (TI 1975a; NYU Medical Center 1978; PASNY 1986) and each curve represents a single annual survey (February or April through December) between 1973 and 1976.

ied over 10-fold from early June to November; increased densities occurred primarily in amphipod, mysid, and dipteran populations (NYU Medical Center 1978). Epibenthic populations in the vicinities of Cornwall (TI 1975a) and Cementon (PASNY 1976) exhibited even larger variations in abundance than communities near Indian Point (Figure 17).

Benthic Invertebrates

Benthic invertebrates live predominantly in the sediments and populations in the Hudson River estuary are numerically dominated by polychaetes, oligochaetes, molluscs, and harpacticoid copepods. These taxa account for more than 70% of the benthos in many regions (TI 1975a, 1976b, 1976c; PASNY 1976). Benthic invertebrates are generally large (5–10 mm) and consume a wide variety of living and nonliving food items. Suspension-feeding organisms such as *Streblospio benedicti* and *Scolecolepides viridis* filter food particles from the overlying water, whereas deposit feeders such as *Amnicola* sp. and *Limnodrilus* sp. derive nutrition from the organisms and organic detritus in the sediment.

Benthic invertebrates in the Hudson River estuary exhibited much smaller population variations than either pelagic or epibenthic communities (Figure 13). For example, benthic infaunal densities near Indian Point varied only threefold throughout the year (TI 1976b); greatest abundances occurred during October (17,000·m^{-2}) and November (16,000·m^{-2}). Benthic communities in upstream areas (Cornwall: TI 1975a; Cementon: PASNY 1976) also exhibited more stable abundances than epibenthic invertebrate communities.

Benthic infauna have long generation times and, due to their limited mobility, must tolerate a broader range of physicochemical conditions than pelagic invertebrates. Benthic invertebrates also have a more constant food supply because most species consume a variety of food particle sizes, and organic carbon concentrations are temporally more stable in sediments (Cantelmo 1978) than in the water column (Schaffer 1978).

Pelagic, epibenthic, and benthic invertebrates have differences in life history characteristics that may have profound effects on carnivores that depend on invertebrates as food sources. Pelagic and epibenthic invertebrates (micro- and macrozooplankton) exhibit large seasonal density pulses in response to spring inputs of allochthonous detritus and phytoplankton blooms. Benthic invertebrates, with longer generation times and broader food bases, exhibit smaller seasonal changes in abundance than either the macrozooplankton or microzooplankton. Large seasonal fluctuations in pelagic invertebrate populations and smaller changes in the infaunal benthic community are common in estuaries (Cronin et al. 1962; Painter 1966; Herman et al. 1968; Peters 1968; McGrath 1973; Hulsizer 1976; Crumb 1977). Tenore (1972) found relatively few changes in the densities of benthic macrofauna in the lower-salinity region of the Pamlico River estuary, while zooplankton densities in the same region exhibited very large seasonal changes (Peters 1968). Seasonal 50- to 80-fold changes in zooplankton abundance also have been observed in the Sacramento–San Joaquin estuary, California (Painter 1966). Thus, the relative numerical fluctuations of pelagic and benthic invertebrate populations in the Hudson River are similar to those in other estuaries.

Fish Community Structure

Analysis of fish community structure requires recognition of the seasonal cycles of occurrence and relative abundances of Hudson River fishes. The estuary is an open system subject to ingress and egress of fish species in generally predictable life cycle rhythms that are modified in time-phase by river temperatures and flows. The temperature cycle is chiefly responsible for seasonal movements and other cyclic activities of fishes, and salinity can influence the degree of penetration into the estuary by anadromous and marine species. Many species are year-round residents of the river, but anadromous species enter the river as adults to spawn and return to the ocean afterwards, leaving their progeny to occupy the river as a nursery area for the remainder of the year or longer (Beebe and Savidge 1988, this volume). Still other species enter the river periodically in feeding migrations to prey on plankton or aggregations of young euryhaline and anadramous fishes.

Adult and juvenile fisheries data reported in this section were collected by Texas Instruments from the lower and middle reaches of the Hudson River estuary. Data for the period 1974–1977 were used because the types and deployment of sampling gear and the spatial and temporal allocation of sampling effort were consistent during this interval.

Results from three sampling efforts were used to evaluate adult and juvenile fish species abundances in the shore zone (sampled with beach seines), channel bottom (bottom trawls), and pelagic zone (midwater trawls). Beach-seine surveys were conducted weekly from April to December with a 30.5-m-long seine. Approximately 100 seining sites were sampled from km 19 to km 122; seven of these sites were near Indian Point (km 63–69). A 13.5-m-long otter trawl was used to sample channel bottom habitats biweekly from April to December. Thirty-eight stations (7 near Indian Point) were sampled from km 45 to km 121. Seven stations near Indian Point were sampled biweekly from July to December with a 15-m-long midwater trawl; comparable pelagic-zone sampling data were not available for areas elsewhere. Further details of sampling methodologies and distributions of sampling efforts are contained in TI (1980).

Monthly changes in fish species richness were determined from 1975–1977 collections pooled over all sampling efforts throughout the estuary (km 19–244); data were summarized by TI (1979). The maximum fish species richness occurred in June–September, when 55–59 species were caught each month during 1975–1977 (Figure 14). Minimum observed species richness occurred in December but data were not available for the

winter months. However, 31 species were captured almost every month throughout the 3-year period. This suggests that many species use the estuary only for short periods; 100 fish species were collected during this 3-year period and 140 species are known from the estuary (Beebe and Savidge 1988). Of the 31 species consistently observed in the estuary, 22 were considered residents (i.e., normally completing their life cycle within the estuary) and 9 were considered migrants, all but the American eel being anadromous.

Spatial Segregation

The spatial heterogeneity of the Hudson River should be reflected in spatial differences in fish distributions. For example, the Tappan Zee and Croton–Haverstraw regions (km 39–61: Table 8) are predominantly shoal areas with large fluctuations in salinity through the year, whereas the West Point and Cornwall regions (km 76–98) are deep, narrow, and have much lower salinity. Indian Point (km 69) is near the midpoint of salt front movement within the estuary (Cooper et al.

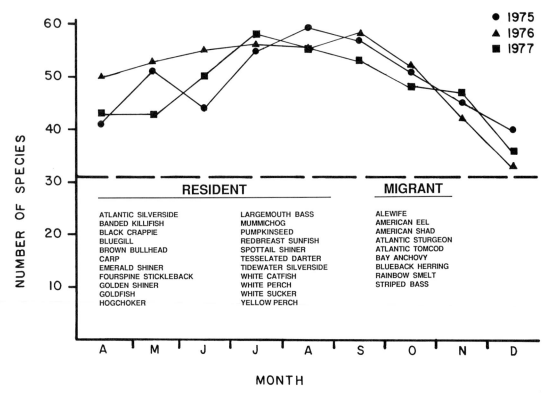

FIGURE 14.—Species richness of fish collected in all gears throughout the Hudson River estuary, April–December 1975–1977, and 31 species that were collected in nearly all year–month combinations.

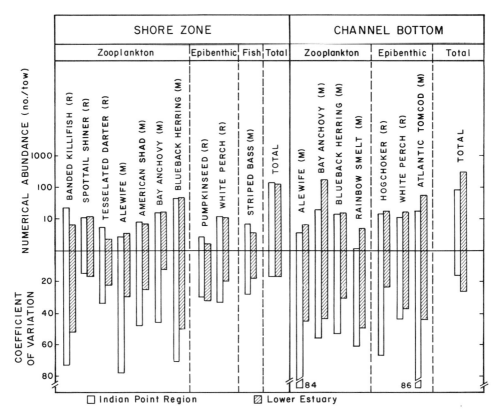

FIGURE 15.—Means and coefficients of variation of annual mean numerical abundance for dominant resident (R) and migrant (M) fish species collected in the vicinity of Indian Point and throughout the lower Hudson estuary, 1974–1977. Species are segregated by primary feeding habit (zooplankton, epibenthic, or fish prey) within the shore-zone and channel-bottom assemblages, which were collected by beach seines and trawls, respectively. Means and coefficients of variation were calculated from untransformed annual means (numbers/tow); means are presented on a logarithmic scale.

1988) and would be expected to have one of the least stable salinity regimes in the estuary.

The spatial segregation of fish species within the estuary was evaluated by comparing numerical abundances near Indian Point (km 62–69) with abundances throughout the lower estuary (km 19–98 for beach seining; km 38–98 for bottom trawling). Annual mean catches (number/tow) were available for several species from 1974 through 1977 (TI 1980) and means ($N = 4$) and coefficients of variation (CV = 100 × standard deviation/mean) were calculated for the annual mean catches from the shore zone and channel bottom. Annual mean species catch/effort was calculated from monthly mean catches/effort after the monthly means were adjusted for any differences in sampling effort (TI 1980). If species mean abundances are approximately equal, the CV provides a measure of the relative numerical stability of the population among years. Species were

further divided according to their predominant adult feeding preferences: benthos, epibenthic crustaceans and insects, pelagic zooplankton, or fish and macroinvertebrates. Finally, fish species were designated either as residents (R) that complete their entire life cycles in fresh or euryhaline waters or as migrants (M) that normally complete a portion of the life cycle in the ocean.

The results (Figure 15) suggest that fish species shift their regions of peak abundance in the lower estuary among years. The CVs for nearly all species in both habitats were higher for the 7-km-long Indian Point region that for the lower estuary as a whole. This reflects larger year-to-year variations in mean abundance at Indian Point than throughout the lower estuary. The fish probably were responding to local variations in physico-chemical variables such as salinity distributions.

In contrast to the pattern for individual species, the CVs for total shore zone catches were identical for Indian Point and the lower estuary (Figure 15), indicating that total fish abundance at Indian Point was stable among years. Individuals of different species replaced each other on a nearly one-for-one basis as varying physicochemical or feeding conditions favored different species. The pattern of CVs for total channel bottom catches was opposite that for individual species, being greater riverwide than locally. This probably is related to the pattern of mean abundances discussed below.

The relationship among mean abundances (all species) near Indian Point and throughout the river differed conspicuously between shore zone and river channel (Figure 15). Generally, mean shore-zone catches were nearly equal at Indian Point and elsewhere; where they differed, Indian Point catches usually were higher. In conjunction with the higher CVs, this indicates that the Indian Point area (km 62–69) may contain better shore-zone habitat, but is unstable from year-to-year in relation to the optimum conditions for individual species. In contrast, river channel catches were consistently higher riverwide than in the Indian Point region, indicating that the latter region was neither optimal nor temporally stable for bottom-dwelling species.

Species Biomass Relationships

Fish species vary considerably in size and, consequently, patterns of numerical abundance may distort perceptions of individual species importance in trophic relationships within an ecosystem. For example, most species in Figure 15 reach maximum individual weights of one to a few kilograms, but striped bass can exceed 25 kg. Comparisons involving trophic considerations should be based on species biomass, rather than on numerical relationships. Fish weights were measured in collections from the seven beach-seine and trawl stations in the Indian Point region and annual mean wet-weight species biomass (g/tow) was calculated from 1974 to 1977 by the same methods used for numerical catch/effort. Overall (4-year) means and CVs were calculated from annual means for shore zone (beach seine), channel bottom (bottom trawl), and pelagic zone (midwater trawl). Individual species were considered if their mean annual biomass exceeded 1 g/tow in a gear type. Species were categorized by trophic and life history status as before.

Fish in the benthic omnivore feeding group were too rare to be included in the numerical analyses (Figure 15) but they were well repre-

sented in the biomass calculations (Figure 16), indicating that they have high individual weights. All of these species are residents and were largely restricted to the shore-zone habitat. Species in the epibenthic feeding group are also predominantly resident species. Three of these four species occurred primarily in the offshore benthic areas and only the pumpkinseed occurred primarily in the shore zone.

In contrast, the zooplankton-feeding fish species are predominantly migratory and occurred in all habitats sampled (Figure 16). Among this group of nine species, only the banded killifish, spottail shiner, and tesselated darter are residents, and the residents were generally less abundant or less spatially ubiquitous than the migrant species. Five of the migrant species exceeded 1 g/tow in all three habitats for the 4-year period, and catches for most of these species–habitat combinations exceeded 10 g/tow. Thus, most of the migrant fish species in the estuary fed predominantly on zooplankton and occurred in all available habitats.

The fish–macroinvertebrate feeding group was also dominated by migrant species. The resident largemouth bass was restricted to the shore zone, whereas all migrant species of this group foraged in the channel habitat. Only striped bass and bluefish were abundant in all areas sampled.

Benthic omnivores formed the most temporally stable group examined in the Indian Point region, as indicated by their CVs (Figure 16). Although white perch biomass was quite variable, epibenthic feeders made up the second most stable group in the region. It is notable that these two groups were dominated by resident species.

The relatively low CVs for fish species biomass in the shore zone indicates that fish population biomass was more stable temporally there than in other habitats (Figure 16). Biomass of most species collected in channel bottom habitats was highly variable, as was the biomass for several species (e.g., alewife, bluefish) in the pelagic zone. Differences in the variability of fish biomass among years in the three habitats is most evident if total collections are considered; the CVs for channel-bottom and pelagic-zone trawl catches are over four times greater than for beach-seine catches.

Discussion

The overall pattern that emerges from these analyses is one of spatial, temporal, and trophic partitioning among the dominant Hudson River fish species. Most resident species occur predom-

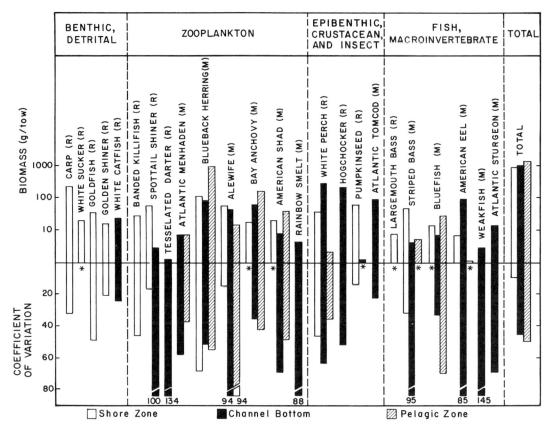

FIGURE 16.—Means and coefficients of variation of annual mean biomass for dominant resident (R) and migrant (M) fish species collected in the vicinity of Indian Point, 1974–1977. Taxa are segregated by primary feeding habit (top labels). Means and coefficients of variation were calculated from untransformed annual wet-weight means (g/tow); means are presented on a logarithmic scale. Asterisks (*) denote species that were not collected in all years and for which coefficients of variation were not calculated.

inantly in shore-zone and channel-bottom habitats. Resident adults feed at a relatively low trophic position, and the omnivorous species among them achieve large body sizes (but not large numerical abundances). Thus, those species that remain within the estuary year-round tend to have large individual and population biomasses, feed low in the trophic structure, and are capable of consuming a wide variety of food items of predominantly benthic origin. Such an approach provides not only flexibility in feeding but also considerable trophic buffering against the large variations in the type and magnitude of food available in pelagic habitats of the estuary.

The dominant migrant species in the estuary feed higher in the trophic structure, consuming zooplankton, macroinvertebrates, and fish. These resources are seasonally abundant and undoubtedly tied, directly or indirectly, to the spring input of allochthonous organic matter and spring–summer phytoplankton production; both input and production cycles have large seasonal, and probably year-to-year, variations in amplitude. The migrant species also appear to be much more spatially ubiquitous than resident species. Alewives, American shad, bay anchovy, and blueback herring among the zooplankton feeders, and bluefish and striped bass among the piscivores, were important components of community biomass in all three habitats; only white perch was so widely dispersed among the resident species. Furthermore, it appears that the pelagic habitat is exploited almost exclusively by migrant species.

Although numerous (probably interrelated) factors are involved in the habitat and foraging partitioning among fish species, trophic factors undoubtedly play an important role. Three dynamically distinct trophic bases account for most of the organic carbon availability in the estuary: allochthonous carbon input in dissolved and particulate forms, phytoplankton production, and benthic algal production (Table 10). Of the three, benthic production is probably the most predictable and stable food base from year to year. Benthic primary production is concentrated in shoal and shore-zone areas where shallow water permits light to reach the river bottom.

Allochthonous carbon and phytoplankton production are relatively less stable and predictable energy sources. Allochthonous carbon inputs are determined by highly variable freshwater flows (Figure 12), and phytoplankton production is strongly influenced by flushing rates (O'Reilly et al. 1976; Malone 1977) that are also controlled by freshwater flows. The influences of freshwater flow on turbidity and temperature (Cooper et al. 1988) also should contribute to variability in pelagic phytoplankton production. However, a substantial portion of the allochthonous carbon and pelagic production may ultimately contribute to benthic production through sedimentation in low-velocity shoals areas and consumption by filter-feeding benthic invertebrates. Thus, benthic consumers in shore and shoal zones likely experience not only greater total carbon availability, but also more constant carbon availability than either pelagic or channel-bottom consumer populations (Cantelmo 1978; Schaffer 1978). The areas of the Hudson having well-developed shore-zone and shoals habitats support the most extensive resident fish populations and serve as nursery areas for juveniles of both resident and migratory species (Beebe and Savidge 1988).

The resource partitioning by resident and migrant fish species represent low-risk–persistence and high-risk–exploitation strategies, respectively. Resident fish species predominantly exploit the benthic habitat in shore and shoal areas. Benthic organic carbon levels are more stable throughout the year than water column organic carbon levels (Cantelmo 1978; Schaffer 1978), and benthic invertebrate abundance is more stable than zooplankton abundance (Figure 13). Furthermore, resident species exploit a wide variety of food sources from benthic invertebrates to organic detritus, although with a resultant decrease in energy yield per mass consumed (Cummins and Wuycheck 1971).

Migrant species are predominantly carnivorous, feeding on zooplankton and other fish species. Food resources for these species are thus somewhat unpredictable due to the temporal vari-

ability of both the epibenthic and zooplankton communities (Figure 13) and the fish production that is dependent on zooplankton production. Thus, the higher variability of the migratory fish species biomass and numerical abundance can be explained in terms of the relative variability of the different estuarine trophic bases.

The analysis of fish trophic dynamics presented herein may be somewhat deceptive in that it has emphasized adult feeding relationships. It is probably the foraging success of larvae and early juveniles that is most important in determining year-class success. The predominant food for most larval fish is zooplankton. It is during the late-spring–early-summer peak spawning period that food limitations are most likely to occur. For example, Thurow (1974) has shown that extremely weak year classes of some demersal fish may be due to reduced availability of food during the larval feeding period. Cushing (1975) also referred to the availability of appropriate food in his "match or mismatch" hypothesis, which stresses that the growth of invertebrate food organisms in temperate ecosystems is triggered at critical times by abiotic environmental conditions. Cushing (1975) proposed that these critical times of increased food demands by larval fish must coincide with increased populations of invertebrates to insure a strong year class. The maximum zooplankton development should coincide with the beginning of larval feeding (May 1974), which is the end point of a sequence of events that begins with appropriate temperatures for spawning and development of eggs and larvae. Reproduction and growth responses to a given set of temperature and feeding conditions undoubtedly vary among species of both fish and zooplankton. By this mechanism, varying spring river flow and temperature conditions will favor the success of different species in different years and the variability should be reflected most strongly in those species that exploit the open-water habitats, as do the carnivorous migrant species. Prediction of fish species abundances, if that is desired, clearly requires a better understanding of events during this period.

The scenario developed above applies primarily to migratory species and those residents (e.g., white perch) that spawn in open water. Spawning success of resident species may be controlled by factors other than those emphasized above. It is probable that temperature and salinity, combined with availability of suitable spawning substrates, are important. If such is the case, then river flow conditions could also strongly influence success of these species by altering the temperature and salinity distributions in space and time. It is also possible that the first winter, rather than the spring–early-summer period, may be critical for residents. Organic carbon inputs and primary production are minimal in winter and invertebrate production is suppressed (although fish food requirements should also be reduced). Size-related overwintering mortality is not uncommon in freshwater systems (Shelton et al. 1979; Toneys and Coble 1979) and could reduce a marginally successful cohort to a poor one.

River flow is undoubtedly the largest single factor affecting most fish populations in the Hudson. Ulanowicz and Polgar (1980) and Polgar (1982) have clearly demonstrated the statistical relationships between striped bass success and estuarine flow and temperature conditions in the Potomac River. They have also proposed several mechanisms through which these factors are related. Comparable statistical relationships have been shown for Hudson River striped bass (TI 1979) and white perch (Klauda et al. 1988, this volume). The degree to which river flow affects organic carbon availability and temperature (growth factors) and salinity distributions (habitat suitability) strongly suggests that flow affects the success of other species in the estuary also. Such relationships have not been examined for most Hudson River species because abundance records are not sufficiently long for reliable testing. However, such tests are a place to start to remove "environmental noise" from abundance data so that trends potentially related to power plant or other anthropogenic impacts may be assessed.

Acknowledgments

The ideas and data for this manuscript were developed with the support of Texas Instruments Incorporated, through a contract with Consolidated Edison Company of New York, the Power Authority of the State of New York, Central Hudson Gas and Electric Company, and Orange and Rockland Utilities. Ronald Klauda and Quentin Ross provided numerous helpful early discussions and the comments of Larry Barnthouse, Robert Kendall, and several reviewers have substantially improved the manuscript. The final preparation of this manuscript was supported by the U.S. Department of Energy and the University of Georgia's Institute of Ecology through contract DE-AC09-76SR00819.

References[3]

Beebe, A., and I. R. Savidge. 1988. Historical perspective on fish species composition and distribution in the Hudson River estuary. American Fisheries Society Monograph 4:25–36.

Bilby, R. E., and G. E. Likens. 1979. Effect of hydrologic fluctuations on the transport of fine particulate organic carbon in a small stream. Limnology and Oceanography 24:69–75.

Boesch, D. F., and D. J. Diaz. 1974. New records of peracarid crustaceans from oligohaline waters of the Chesapeake Bay. Chesapeake Science 15:56–59.

Boynton, W. R., W. M. Kemp, and C. W. Keefe. 1982. A comparative analysis of nutrients and other factors influencing estuarine phytoplankton production. Pages 69–90 in V. S. Kennedy, editor. Estuarine comparisons. Academic Press, New York.

Cantelmo, F. R. 1978. The ecology of sublittoral meiofauna in a shallow marine embayment. Doctoral dissertation. City University of New York, New York.

Chervin, M. B. 1977. The assimilation of particulate organic carbon by estuarine and coastal copepods. Doctoral dissertation. City University of New York, New York.

Christensen, S. W., and T. L. Englert. 1988. Historical development of entrainment models for Hudson River striped bass. American Fisheries Society Monograph 4:133–142.

Cooper, J. C., F. R. Cantelmo, and C. E. Newton. 1988. Overview of the Hudson River estuary. American Fisheries Society Monograph 4:11–24.

Correll, D. L. 1978. Estuarine productivity. BioScience 28:646–650.

Cronin, L. E., J. C. Daiber, and E. M. Hulburt. 1962. Quantitative seasonal aspects of zooplankton in the Delaware River estuary. Chesapeake Science 3:63–93.

Crumb, S. E. 1977. Macrobenthos of the tidal Delaware River between Trenton and Burlington, New Jersey. Chesapeake Science 18:253–265.

Cummins, K. W., and J. C. Wuycheck. 1971. Caloric equivalents for investigations in ecological energetics. Mitteilungen Internationale Vereinigung fuer Theoretische und Angewandte Limnologie 18:1–158.

Cummins, K. W., R. R. Costa, R. E. Rowe, G. A. Moshiri, R. M. Scanlon, and R. K. Zajdel. 1969. Ecological energetics of a natural population of the predaceous zooplankton Leptodora kindtii Focke (Cladocera). Oikos 20:189–223.

Cushing, D. H. 1975. Marine ecology and fisheries. Cambridge University Press, Cambridge.

Fisher, S. G., and G. E. Likens. 1973. Energy flow in Bear Brook, New Hampshire: an integrative approach to stream ecosystem metabolism. Ecological Monographs 43:421–439.

Garside, C., T. C. Malone, O. A. Roels, and B. A. Sharfstein. 1976. An evaluation of sewage-derived

---

[3]See Table 1 for sources of legal documents and unpublished reports pertaining to the Hudson River.

nutrients and their influence on the Hudson Estuary and New York Bight. Estuarine and Coastal Marine Science 4:281–289.

Heffner, R. L. 1973. Phytoplankton community dynamics in the Hudson River estuary between mile points 39 and 77. *In* Hudson River ecology, 3rd symposium. Hudson River Environmental Society, New Paltz, New York.

Heinle, D. R. 1974. An alternative grazing hypothesis of the Patuxent estuary. Chesapeake Science 15:146–150.

Heinle, D. R., D. A. Flemer, and J. F. Ustach. 1976. Contribution of tidal marshlands to mid-Atlantic estuarine food chains. Pages 309–320 *in* M. L. Wiley, editor. Estuarine processes, volume 2. Circulation, sediments and transfer of material in the estuary. Academic Press, New York.

Herman, S. S., J. A. Mihursky, and A. J. McErlean. 1968. Zooplankton and environmental characteristics of the Patuxent River estuary 1963–1965. Chesapeake Science 9:67–82.

Hetling, L. J. 1976. An analysis of past, present and future Hudson River wastewater loadings. *In* Hudson River ecology, 4th symposium. Hudson River Environmental Society, New Paltz, New York.

Hulsizer, E. E. 1976. Zooplankton of lower Narragansett Bay. Chesapeake Science 17:260–270.

Jervis, R. A. 1969. Primary production in the freshwater marsh ecosystem of Troy Meadows, New Jersey. Bulletin of the Torrey Botanical Club 96:209–231.

Kiviat, E. 1973. A freshwater tidal marsh on the Hudson Tivoli North Bay. *In* Hudson River ecology, 3rd symposium. Hudson River Environmental Society, New Paltz, New York.

Klauda, R. J., J. B. McLaren, R. E. Schmidt, and W. P. Dey. 1988. Life history of white perch in the Hudson River estuary. American Fisheries Society Monograph 4:69–88.

Levins, R. 1979. Community structure and population change. Testimony (April 24, 1979) before the U.S. Environmental Protection Agency in the matter of the Hudson River Power Case, Docket C/II-WP-77-01.

Levinton, J. 1972. Stability and trophic structure in deposit-feeding and suspension-feeding communities. American Naturalist 106:472–486.

Malone, T. C. 1977. Environmental regulation of phytoplankton productivity in the lower Hudson estuary. Estuarine and Coastal Marine Science 5:157–171.

May, R. C. 1974. Larval mortality in marine fishes and the critical period concept. Pages 3–19 *in* J. H. S. Baxter, editor. The early life history of fish. Springer-Verlag, New York.

McGrath, R. A. 1973. Benthic macrofaunal census of Raritan Bay. *In* Hudson River ecology, 3rd symposium. Hudson River Environmental Society, New Paltz, New York.

Minshall, G. W., and six coauthors. 1983. Interbiome comparison of stream ecosystem dynamics. Ecological Monographs 53:1–25.

Moran, M. A., and K. E. Limburg. 1985. The Hudson River ecosystem. Pages 8–55 *in* K. E. Limburg,

M. A. Moran, and W. H. McDowell, editors. Environmental impact assessment of the Hudson River ecosystem, multiple case study and data base review. Cornell University, Ecosystems Research Center, Ithaca, New York.

NYU (New York University) Medical Center. 1977. Effects of entrainment by the Indian Point power plant on Hudson River biota: a progress report for 1975. Report to Consolidated Edison Company of New York.

NYU (New York University) Medical Center. 1978. Effects of entrainment by the Indian Point power plant on biota in the Hudson River estuary: a progress report for 1976. Report to Consolidated Edison Company of New York.

Odum, E. P. 1971. Fundamentals of ecology, 3rd edition. Saunders, Philadelphia.

Odum, E. P., and A. A. de la Cruz. 1967. Particulate organic detritus in a Georgia salt marsh–estuarine ecosystem. Pages 383–388 *in* G. Lauff, editor. Estuaries. American Association for the Advancement of Science Special Publication 83, Washington, D.C.

O'Neill, R. V., R. H. Gardner, S. W. Christensen, W. Van Winkle, J. H. Carney, and J. B. Mankin. 1981. Some effects of parameter uncertainty in density-independent and density-dependent Leslie models for fish populations. Canadian Journal of Fisheries and Aquatic Sciences 38:91–100.

O'Reilly, J. E., D. P. Thomas, and C. Evans. 1976. Annual primary production (nannoplankton, netplankton, dissolved organic matter) in the lower New York Bay. *In* Hudson River ecology, 4th symposium. Hudson River Environmental Society, New Paltz, New York.

QLM (Quirk, Lawler & Matusky). 1974. Hudson River study: aquatic factors governing the siting of power plants. Report to Empire State Energy Research Corporation.

Painter, R. E. 1966. Ecological studies of the Sacramento–San Joaquin estuary: zooplankton of San Pablo and Suisun bays. California Department of Fish and Game, Fish Bulletin 133:18–56.

PASNY (Power Authority of the State of New York). 1976. Application to the N.Y. state board on electric generating siting and the environment 1980, 700 MW nuclear field unit Greene County nuclear power plant.

Peters, D. 1968. A study of relationships between zooplankton abundance and selected environmental variables in the Pamlico River estuary of eastern North Carolina. Master's thesis. North Carolina State University, Raleigh.

Polgar, T. T. 1982. Factors affecting recruitment of Potomac River striped bass and resulting implications for management. Pages 427–442 *in* V. S. Kennedy, editor. Estuarine comparisons. Academic Press, New York.

Rhoads, D. C., P. L. McCall, and J. Y. Yongst. 1978. Disturbance and protection on the estuarine seafloor. American Scientist 66:577–586.

Sanders, H. L., P. C. Mangelsdorf, Jr., and G. R. Hampson. 1965. Salinity and faunal distribution in the Pocasset River, Massachusetts. Limnology and Oceanography 10 (supplement):R216–R229.

Saunders, G. W. 1969. Some aspects of feeding in zooplankton. Pages 556–573 in Eutrophication: causes, consequences, correctives. National Academy of Science, Washington, D.C.

Schaffer, S. A. 1978. Concentrations of organic carbon and protein in the Hudson estuary near Indian Point. Master's thesis. New York University, New York.

Shelton, W. L., W. D. Davies, T. A. King, and T. J. Timmons. 1979. Variation on the growth of the initial year class of largemouth bass on West Point Reservoir, Alabama and Georgia. Transactions of the American Fisheries Society 108:142–149.

Simpson, H. J., R. Bopp, and D. Thurber. 1973. Salt movement patterns in the lower Hudson. In Hudson River ecology, 3rd symposium. Hudson River Environmental Society, Bronx, New York.

Sirois, D. L., and S. W. Fredrick. 1978. Phytoplankton and primary production on the Hudson River estuary. Estuarine and Coastal Marine Science 7:413–423.

Smayda, T. J. 1973. A survey of phytoplankton dynamics in the coastal waters from Cape Hatteras to Nantucket. University of Rhode Island Marine Publication Series 2:3-1-3-100.

Stephens, G. C. 1967. Dissolved organic material as a nutritional source for marine and estuarine invertebrates. American Association for the Advancement of Science Special Publication 83:367–373.

Storm, P. A., and R. L. Heffner. 1976. A comparison of phytoplankton abundance, chlorophyll 'a' and water quality factors in the Hudson River and its tributaries. In Hudson River ecology, 4th symposium. Hudson River Environmental Society, New York.

Tenore, K. R. 1972. Macrobenthos of the Pamlico River estuary, North Carolina. Ecological Monographs 42:51–69.

Thurow, F. 1974. Zur Starke des Dorschjahrganges 1972 in der Westlichen Ostsee. Berichte der Deutschen Wissenshaftlichen Kommission fuer Meeresforschung 23:129–136.

TI (Texas Instruments). 1975a. Benthic landfill studies, Cornwall, final report. Report to Consolidated Edison Company of New York.

TI (Texas Instruments). 1975b. First annual report for the multiplant impact study of the Hudson River estuary. Report to Consolidated Edison Company of New York.

TI (Texas Instruments). 1976a. A synthesis of available data pertaining to major physicochemical variables within the Hudson River estuary emphasizing the period from 1972 through 1975. Report to Consolidated Edison Company of New York.

TI (Texas Instruments). 1976b. Hudson River ecological study in the area of Indian Point. Thermal effects report. Report to Consolidated Edison Company of New York.

TI (Texas Instruments). 1976c. Liberty State Park eco-

logical study final report. Report to the Port Authority of New York and New Jersey.

TI (Texas Instruments). 1979. Hudson River ecological study in the area of Indian Point. Annual report (1977) to Consolidated Edison Company of New York and Power Authority of the State of New York.

TI (Texas Instruments). 1980. Hudson River ecological study in the area of Indian Point. Annual report (1978) to Consolidated Edison Company of New York and Power Authority of the State of New York.

TI (Texas Instruments). 1981. 1979 year class report for the multiplant impact study of the Hudson River estuary. Report to Consolidated Edison Company of New York.

Toneys, M. L., and D. W. Coble. 1979. Size-related, first winter mortality of freshwater fishes. Transactions of the American Fisheries Society 108:415–419.

Udell, H. F., J. Zarvdsky, and T. E. Doheny. 1969. Productivity and nutrient values of plants growing in the salt marshes of the town of Hempstead, Long Island. Bulletin of the Torrey Botanical Club 96:42 51.

Ulanowicz, R. E., and T. T. Polgar. 1980. Influences of anadromous spawning behavior and optimal environmental conditions upon striped bass (Morone saxatilis) year-class success. Canadian Journal of Fisheries and Aquatic Sciences 37:143–154.

USGS (U.S. Geological Survey). 1975a. 1974 Water resources data for New York, part I. Surface water records. USGS, Albany, New York.

USGS (U.S. Geological Survey). 1975b. 1974 Water resources data for New York, part II. Water quality records. USGS, Albany, New York.

USGS (U.S. Geological Survey). 1976. Water resources data for New York, water year 1975. USGS, Report NY-75-1, Albany, New York.

USGS (U.S. Geological Survey). 1977. Water resources data for New York, water year 1976, volume 1. New York excluding Long Island. USGS Report NY-76-1, Albany, New York.

Watling, L., J. Lindsay, R. Smith, and D. Maurer. 1974. The distribution of isopoda in the Delaware Bay region. Internationale Revue der gesamten Hydrobiologie 59:343–351.

Welsh, B. L., R. B. Whitlatch, and W. F. Bohlen. 1982. Relationship between physical characteristics and organic carbon sources as a basis for comparing estuaries in southern New England. Pages 53–68 in V. S. Kennedy, editor. Estuarine comparison. Academic Press, New York.

Whigham, D. F., and R. L. Simpson. 1975. Ecological studies of the Hamilton marshes. Progress report, June 1974 to January 1975. Hamilton Township Environmental Commission, Hamilton, New Jersey.

Woodwell, G. M., D. E. Whitney, C. A. S. Hass, and R. A. Houghton. 1977. The Flax Pond ecosystem study: exchanges of carbon in water between a salt marsh and Long Island Sound. Limnology and Oceanography 22:833–838.

American Fisheries Society Monograph 4:53–58, 1988

Distributions of Early Life Stages of Striped Bass in the Hudson River Estuary, 1974–1979

JOHN BOREMAN

National Marine Fisheries Services
Northeast Fisheries Center, Woods Hole, Massachusetts 02543, USA

RONALD J. KLAUDA[1]

Texas Instruments Incorporated
Ecological Services Group, Buchanan, New York 10511, USA

Abstract.—Spatial and temporal distribution patterns for early life stages of striped bass in the Hudson River estuary during 1974–1979 are described. For these years, the early life stages exhibited a low annual variability in the patterns. Eggs were collected in all sampling regions of the estuary, usually from the first week of May until early June. Yolk-sac larvae were found during the same period and in the same regions as were eggs, but peak larval densities were consistently upriver from the peak egg densities. Post–yolk-sac larvae were collected from late May to late July and generally in the same regions as yolk-sac larvae. Juveniles were found from late June into August distinctly downriver from the larval life stages. Estimates of year-class strength varied by less than twofold during 1974–1979, which makes it difficult to relate distribution patterns for the early life stages to year-class success.

Estimates of entrainment mortality for striped bass, needed to settle the Hudson River power plant case, relied on the spatial and temporal distributions of the entrainable life stages: eggs, yolk-sac and post–yolk-sac larvae, and juveniles less than 50 mm in total length. During presettlement testimony, the 1974 and 1975 distributions of these life stages were used to generate both historical and projected entrainment estimates (Boreman and Goodyear 1988, this volume). Estimates for the settlement negotiations relied on averages of the 1974 through 1978 distributions (Englert et al. 1988, this volume).

In this paper, we present the spatial and temporal distributions of the entrainable life stages of striped bass in the Hudson River estuary during 1974–1979. We examine the year-to-year variability in the distributions and compare the distribution patterns for the Hudson River population to patterns for striped bass populations in other major spawning rivers. We also discuss the sampling protocol with respect to its ability to generate relevant data for early life histories of fish species in large estuarine systems.

Methods

Distribution patterns for entrainable life stages of striped bass in the Hudson River estuary were obtained from riverwide surveys conducted by

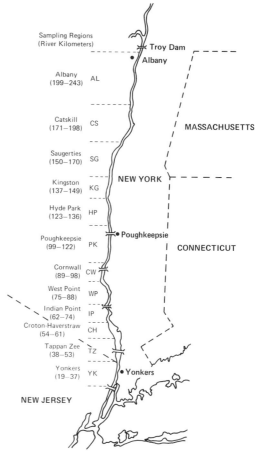

Sampling Regions
(River Kilometers)

Troy Dam
Albany

Albany
(199–243) AL

Catskill
(171–198) CS

MASSACHUSETTS

Saugerties
(150–170) SG

NEW YORK

Kingston
(137–149) KG

Hyde Park
(123–136) HP

Poughkeepsie
(99–122) PK — Poughkeepsie

CONNECTICUT

Cornwall
(89–98) CW

West Point
(75–88) WP

Indian Point
(62–74) IP

Croton-Haverstraw
(54–61) CH

Tappan Zee
(38–53) TZ

Yonkers
(19–37) YK — Yonkers

NEW JERSEY

FIGURE 17.—Locations of 12 sampling regions in the Hudson River estuary (modified from Dey 1981).

[1]Present address: The Johns Hopkins University, Applied Physics Laboratory, Environmental Sciences Group, Shady Side, Maryland 20764, USA.

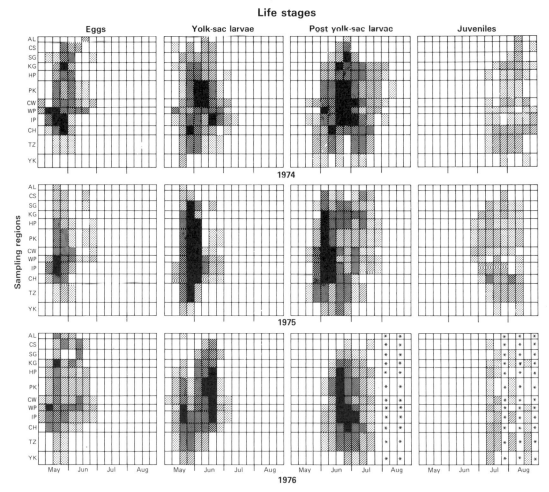

FIGURE 18.—Distribution patterns of early life stages of striped bass in the Hudson River estuary during 1974–1979. Sampling regions are those in Figure 17; asterisks (*) mark weeks when there was no sampling.

Texas Instruments Incorporated (TI 1977, 1978, 1979, 1980a, 1980b, 1981). In 1974 and 1975, the surveys were done weekly from mid-April to mid-August; during 1976–1979, the surveys were weekly from mid-April to mid-July and biweekly for the following month. Approximately 200 samples were obtained during each survey from three depth strata within 12 geographic regions (Figure 17). At least three samples were allocated to each depth stratum per region; the remaining samples were allocated to regions and depth strata based on distributions of entrainable life stages of striped bass in previous years. A physicochemical description of the survey regions is provided by Cooper et al. (1988, this volume).

Surveys were conducted during daylight through early June and at night thereafter to reduce possible gear avoidance by the developing fish. The 12 regions were sampled sequentially over a 3- or 4-d period each week. Stratified random samples were taken in each region with a 1.0-m^2 epibenthic sled in the shoal strata (6 m deep or less) and channel-bottom strata (bottom 3 m), and with a 1.0-m^2 Tucker trawl in the surface strata overlying the channel-bottom strata. Both gears were equipped with a 505-μm-mesh net and were towed horizontally. Tow speeds were 1.0 and 0.9 m/s for the sled and trawl, respectively. Tow duration was 5 min. Gear mouth opening, net surface area and porosity, tow speed, and duration yielded theoretical filtration efficiencies greater than 85% (TI 1977). A calibrated digital flow meter was suspended inside the net to record sample volume; a calibrated electronic flowmeter

Life stages

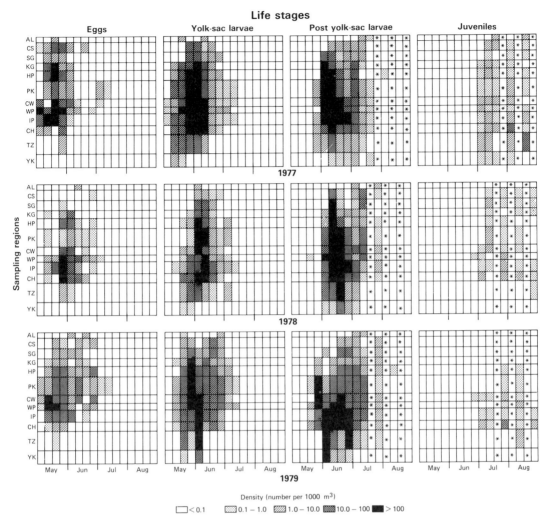

FIGURE 18.—Continued.

was mounted on the towing cable to record tow speed. Flowmeters were calibrated after each survey by the Chesapeake Bay Institute of The Johns Hopkins University. Double-trip release mechanisms opened and closed the nets at specified depths.

Yolk-sac larvae were those in the transitional life stage from hatching through development of a complete and functional digestive system, regardless of the degree of yolk absorption and oil retention. Post–yolk-sac larvae had a complete and functional digestive system, but had not yet acquired the full complement of adult fin rays. Regional densities for these life stages, as well as those for eggs and juveniles up to 50 mm total length, were derived from an average of the individual stratum densities weighted by the volumes of the strata within the region.

Results

Eggs

Striped bass eggs were collected from the Hudson River estuary over 7–10-week periods during 1974–1979, usually from the first week of May until early June (Figure 18). Maximum sample densities in a region (>100/1,000 m³) were encountered during May of each year, when water temperatures were 12–22°C (Figure 19). Earlier reports of striped bass spawning activity in the Hudson River (Raney 1952) indicate that no overt shift in the timing of egg deposition has occurred.

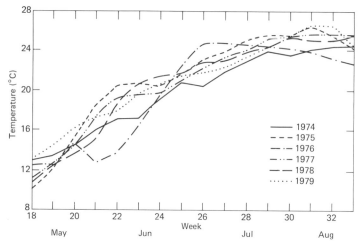

FIGURE 19.—Average weekly water temperatures in the Hudson River estuary, 1974–1979, measured at the Poughkeepsie Water Works (km 120).

Eggs were collected in all regions of the estuary in 1976, and in all regions except the lowermost (Yonkers, km 19–37) in the other years. Above-average freshwater runoff during May 1976 (TI 1980b) may have accounted for the deposition of eggs further downriver than usual, because the downriver limit of spawning of striped bass in estuaries is usually associated with position of the salt front (Hardy 1978).

A spatial peak in egg density occurred within the regions bounded by km 54 and km 98 each year. A concurrent but secondary peak was evident upriver within km 123–149 in all years except 1975 and 1979. Peaks in weekly egg densities shifted upriver as the season progressed, presumably a result of an upriver shift in spawning in response to an upriver movement of the salt front as freshwater runoff diminished (McFadden 1977).

Yolk-Sac Larvae

Within river regions, densities of yolk-sac larvae peaked either the same week as did egg densities or the week after (Figure 18). These observations are consistent with laboratory studies by Rogers et al. (1977), in which striped bass eggs hatched in 2–6 d in water temperatures similar to those of the Hudson River during May (Figure 19).

Yolk-sac larvae were collected in all 12 regions each year. Peak densities of yolk-sac larvae were consistently upriver from the peak egg densities, suggesting a higher hatchability of eggs or survival of yolk-sac larvae in the upper river regions. Similar observations were made in the Potomac

River in 1974 and 1976 (Setzler-Hamilton et al. 1981). Polgar et al. (1976) and Ulanowicz and Polgar (1980) attributed the 1974 Potomac River observation to either a continual upriver movement of active spawners or a higher survival of eggs and larvae further upriver.

An apparent reduction in yolk-sac larva abundance during the third week in May 1976 coincided with a sudden drop in water temperature (Figure 19). Dey (1981) and Boreman (1983) concluded that the sudden drop in temperature caused almost complete mortality of striped bass spawned in the Hudson River estuary before that week. Failure of the peak density of yolk-sac larvae in the previous week to appear as a peak in the later life stages (Figure 18) supports this conclusion. An identical temperature-related mortality of yolk-sac larvae occurred in the area of upper Chesapeake Bay and the Chesapeake and Delaware Canal in 1976, when water temperature decreased from 17 to 11.3°C within 36 h after the peak spawning period (Kernehan et al. 1981).

Post–Yolk-Sac Larvae

Post–yolk-sac larvae appeared in the survey collections from late May until late July each year except 1976, when they appeared only during June and early July. Their late appearance in 1976 can be attributed to the previously mentioned temperature-related mortality of the earlier-spawned fish. The period of peak sample densities of post–yolk-sac larvae occurred approximately 1 week following those of yolk-sac larvae, corresponding with the expected 4- to 10-d duration of the yolk-sac-larva life stage (Hardy 1978).

A distinctive downriver shift in distribution between the yolk-sac and post–yolk-sac-larva stages was apparent only in 1979 (Figure 18). In other years, peak sample densities of yolk-sac and post–yolk-sac larvae both occurred within km 54–149, implying a relatively stable spatial expanse of nursery area for striped bass larvae in the Hudson River estuary. Striped bass larvae in the Potomac River also showed little spatial shift during development in 1974–1977 (Setzler-Hamilton et al. 1981).

Juveniles

Juvenile striped bass were collected in the riverwide trawl and sled surveys from late June through the end of the survey period in mid-August. Peak regional densities of juveniles appeared approximately 4–6 weeks after the respective peaks of post–yolk-sac larvae, indicating a longer life stage duration of post–yolk-sac larvae than reported for the Potomac River population (20–30 d: Setzler et al. 1980). In the laboratory, the post–yolk-sac larva stage lasted 68 d at 15°C and 24 d at 21°C (Rogers et al. 1977). Hudson River temperatures were usually within this range when post–yolk-sac larvae were present (Figure 19); corresponding Potomac River temperatures are somewhat higher.

Peak sample densities of juvenile striped bass principally occurred from km 38 to km 74, indicating a downriver shift in distribution compared to the larval stages. The area of peak juvenile occurrence was considerably more restricted than the areas for the larvae. The highest regional density of juveniles occurred between km 54 and 61 every year, a region that is characterized by extensive shallow water (Cooper et al. 1988). However, the shallow water may have reduced gear avoidance by juveniles, resulting in a relatively higher catchability than in the other river regions.

Discussion

For the period 1974 to 1979, variability in spatial and temporal distribution patterns of early life stages of striped bass in the Hudson River estuary was low. Abundance estimates for the fall juvenile life stage varied by less than twofold during the same period (TI 1981). Year-class success and survival conditions for striped bass in the Hudson River did not vary as radically as they did for striped bass in the Potomac River, where a 35-fold variation in year-class success was observed for 1974–1976 (Ulanowicz and Polgar 1980).

The power plants of concern in the Hudson River power case (Bowline, Lovett, Indian Point, Danskammer, Roseton) are located in the stretch of estuary bounded by the Croton–Haverstraw and Poughkeepsie sampling regions (Figure 17; Hutchison 1988, this volume). Early life stages of striped bass present in these regions are vulnerable to entrainment in the cooling water withdrawn by the power plants (Boreman and Goodyear 1988). The highest sample densities of the early life stages were consistently present in the power plant regions during each of the sample years (Figure 18), indicating that the power plants are located in the prime spawning and nursery habitat for Hudson River striped bass.

The observed distribution patterns for early life stages of striped bass are influenced by confounding variables, including mortality, movement, and gear efficiency. Attempts to separate these variables to determine the optimal location and time for survival of the early life stages of striped bass have met meager success (Ulanowicz and Polgar 1980), primarily because of the assumptions necessary to accept transport modeling results. The limited number of sampling years (six) and the lack of wide variability in indices of year-class success during those years (less than twofold) prohibit any definitive conclusions about the relationship between early life stage distribution patterns and year-class success of striped bass in the Hudson River estuary.

Aside from separating distribution patterns into mortality and movement components, other difficulties arise in using weekly survey data to assess timing and locations of the early life stages of striped bass. For those life stages, egg and yolk-sac larva, that have durations of less than 1 week, peaks in abundance may be missed both spatially and temporally, especially if the spawning peak is sharply defined in time and location. However, the consistent distribution patterns of these life stages among the six sample years suggest that this was not a serious problem in studies of the Hudson River population.

Acknowledgments

We thank Larry Barnthouse and Ron Kernehan for providing useful suggestions. We also acknowledge the contribution of the Texas Instruments Incorporated employees who participated in the collection and tabulation of the data used in this paper.

References[2]

Boreman, J. 1983. Simulation of striped bass egg and larva development based on temperature. Transactions of the American Fisheries Society 112:286–292.

Boreman, J., and C. P. Goodyear. 1988. Estimates of entrainment mortality for striped bass and other fish species inhabiting the Hudson River estuary. American Fisheries Society Monograph 4:152–160.

Cooper, J. C., F. R. Cantelmo, and C. E. Newton. 1988. Overview of the Hudson River estuary. American Fisheries Society Monograph 4:11–24.

Dey, W. P. 1981. Mortality and growth of young-of-the-year striped bass in the Hudson River estuary. Transactions of the American Fisheries Society 110:151–157.

Englert, T. L., J. Boreman, and H. Y. Chen. 1988. Plant flow reductions and outages as mitigative measures. American Fisheries Society Monograph 4:274–279.

Hardy, J. D., Jr. 1978. Development of fishes of the mid-Atlantic Bight, volume 3. U.S. Fish and Wildlife Service FWS/OBS-78-12.

Hutchison, J. B., Jr. 1988. Technical descriptions of Hudson River electricity generating stations. American Fisheries Society Monograph 4:113–120.

Kernehan, R. J., M. R. Headrick, and R. E. Smith. 1981. Early life history of striped bass in the Chesapeake and Delaware Canal and vicinity. Transactions of the American Fisheries Society 110:137–150.

McFadden, J. T., editor. 1977. Influence of Indian Point unit 2 and other steam electric generating plants on the Hudson River estuary with emphasis on striped bass and other fish populations. Report to Consolidated Edison Company of New York.

Polgar, T. T., J. A. Mihursky, R. E. Ulanowicz, R. P. Morgan, and J. S. Wilson. 1976. An analysis of 1974 striped bass spawning success in the Potomac estuary. Pages 151–165 *in* M. L. Wiley, editor. Estuarine processes, volume 1. Academic Press, New York.

Raney, E. C. 1952. The life history of the striped bass, *Roccus saxatilis* (Walbaum). Pages 5–97 *in* E. C.

Raney, E. F. Tresselt, E. H. Hollis, V. D. Vladykov, and D. H. Wallace. The striped bass, *Roccus saxatilis*. Bulletin of the Bingham Oceanographic Collection Yale University 14(1).

Rogers, B. A., D. T. Westin, and S. B. Saila. 1977. Life stage duration studies on Hudson River striped bass. University of Rhode Island, Sea Grant Program, Marine Technical Report 31, Kingston.

Setzler, E. M., and eight coauthors. 1980. Synopsis of biological data on striped bass, *Morone saxatilis* (Walbaum). NOAA (National Oceanic and Atmospheric Administration) Technical Report NMFS (National Marine Fisheries Service) Circular 433.

Setzler-Hamilton, E. M., W. R. Boynton, J. A. Mihursky, T. T. Polgar, and K. V. Wood. 1981. Spatial and temporal distribution of striped bass eggs, larvae, and juveniles in the Potomac estuary. Transactions of the American Fisheries Society 110:121–136.

TI (Texas Instruments). 1977. 1974 year class report for the multiplant impact study of the Hudson River estuary. Report to Consolidated Edison Company of New York.

TI (Texas Instruments). 1978. 1975 year class report for the multiplant impact study of the Hudson River estuary. Report to Consolidated Edison Company of New York.

TI (Texas Instruments). 1979. 1976 year class report for the multiplant impact study of the Hudson River estuary. Report to Consolidated Edison Company of New York.

TI (Texas Instruments). 1980a. 1977 year class report for the multiplant impact study of the Hudson River estuary. Report to Consolidated Edison Company of New York.

TI (Texas Instruments). 1980b. 1978 year class report for the multiplant impact study of the Hudson River estuary. Report to Consolidated Edison Company of New York.

TI (Texas Instruments). 1981. 1979 year class report for the multiplant impact study of the Hudson River estuary. Report to Consolidated Edison Company of New York.

Ulanowicz, R. E., and T. T. Polgar. 1980. Influence of anadromous spawning behavior and optimal environmental conditions upon striped bass (*Morone saxatilis*) year-class success. Canadian Journal of Fisheries and Aquatic Sciences 37:143–154.

---

[2]See Table 1 for sources of legal documents and unpublished reports pertaining to the Hudson River.

American Fisheries Society Monograph 4:59–68, 1988
© Copyright by the American Fisheries Society 1988

Stock Characteristics of Hudson River Striped Bass

Thomas B. Hoff,[1] James B. McLaren,[2] and Jon C. Cooper[3]

Texas Instruments Incorporated
Ecological Services Group, Buchanan, New York 10511, USA

Abstract.—Striped bass, because of their tremendous popularity both commercially and recreationally, were a principal focus of the Hudson River power plant case. Between 1976 and 1979, over 23,000 age-II and older striped bass were studied as one facet of an extensive research program on the spring population in the Hudson River. Samples were collected from the overwintering as well as the spawning portion of the striped bass population, and included immature as well as mature fish. At least 12 age-groups contributed to spawning each year (some fish live to 18 years of age). Of these 12, age-groups III, IV, and V usually were most abundant, but the percentage of the population represented by any single age-group varied as the result of fluctuations in year-class strength. The 1973 year class was the strongest in recent years. Males first became sexually mature at age II and females at age IV. Fast-growing individuals within a year class tended to mature earlier. Fecundity increased with the size of fish, reaching an observed maximum of about 3 million eggs per female. The Schumacher–Eschmeyer population estimate for the 1979 population (slightly over 250,000 fish) was the largest during the interval 1976–1979. Although significant annual variations in maturity and growth were detected for Hudson River striped bass, there was no evidence of a consistent change in either variable that might be associated with increasing power plant operations (over 3,700 MW of electrical generating capacity came on-line during 1974–1976) and a reduction in striped bass abundance. Age at maturity and age structure are the two life history components that differ the most between the Hudson River population and other striped bass populations.

Striped bass, because of their commercial and recreational importance, were a principal focus of the Hudson River power plant case. The Hudson River produces 5–30% of the striped bass that enter the Atlantic coastal fishery (Van Winkle et al. 1988, this volume); only the Chesapeake Bay system contributes more. Commercial catches of striped bass reached 6,700 tonnes in 1973, of which 790 tonnes were taken in New York State, though harvests declined nearly 90% through 1983 (MAFMC 1984). Since the Marine Recreational Fishery Statistics Survey began in 1979 (NMFS 1984), Atlantic coast marine anglers have caught a further average of 263 tonnes through 1985 (MAFMC 1988).

Since the 1930s, when Merriman (1941) began his landmark study, the coastal stocks of striped bass have shown large-amplitude cycles of abundance due to irregular recruitment of strong year classes, excessive mortality within other year classes, or both. The last strong year class in the Atlantic fishery was produced in 1970. Reasons for the subsequent decline of the stocks are not fully understood, but because striped bass are anadromous and rely on the upper reaches of estuaries, including the Hudson estuary, for their reproduction, human perturbations of the environment have been cited frequently as contributing factors. Conservationists were particularly alarmed by the planned development of nuclear and conventional power plants along the Hudson River because of perceived threats to young striped bass and other species that might be entrained and killed in the large volumes of cooling water the plants would withdraw from the river. It was their concern, reflected by federal regulatory agencies, that stimulated extensive research on Hudson River fish during the 1970s.

Others have presented direct estimates of entrainment and other mortality imposed on larval and young juvenile striped bass (Muessig et al. 1988; Boreman and Goodyear 1988, both this volume). Here, we focus on the population of age-II and older striped bass in the Hudson River prior to and during the spawning period. Because of its role in producing subsequent generations, the adult population of fish can offer indirect evidence of the long-term effects of power plants

[1]Present address: Mid-Atlantic Fishery Management Council, Federal Building, Room 2115, Dover, Delaware 19901, USA.

[2]Present address: Beak Consultants Incorporated, 12072 Main Road, Akron, New York 14001, USA.

[3]Present address: International Scholars for Environmental Studies, 107 Canner Street, New Haven, Connecticut 06511, USA.

and other environmental stresses on the species' well-being.

Most of what is now known about adult Hudson River striped bass (MMES 1987) was learned during the power plant studies that began in the late 1960s. Some of this information has been published, including the relative contribution of the Hudson stock to the coastal fisheries (Berggren and Lieberman 1978), the biology of juvenile striped bass (Klauda et al. 1980), movements of adults (McLaren et al. 1981), diet (Gardinier and Hoff 1982), reproductive effort (Young and Hoff 1988), the commercial fishery within the Hudson River (McLaren et al. 1988), and some stock characteristics from a limited sampling of the commercial fishery (Dew 1981). In this chapter, we address the age structure, size, maturity, and spawning potential of adult striped bass in the Hudson River.

Methods

Field collection.—Striped bass were collected annually by gill nets and haul seines as soon as the Hudson River became free of ice (approximately mid-March), and collecting continued until catches became greatly reduced (late June). Gill nets were tended around the clock for 4–6 d/week. The location of gill nets changed weekly to maximize the catch and to follow the bulk of the population as it moved. Sampling was concentrated in the vicinity of the Tappan Zee Bridge and Haverstraw Bay in the spawning season (March and April), upriver into and above the Indian Point area as the season progressed (May), and downriver (June) at the end of the spawning season. Overall, gill-net sampling was restricted to the 57-km reach from approximately the Tappan Zee Bridge to the New-burgh–Beacon bridge (see Figure 1). The spawn-ing migration of striped bass generally begins in the Hudson River estuary around the third week in April (McLaren et al. 1981). Peak spawning usually occurs in mid-May when water tempera-tures are approximately 14°C (Klauda et al. 1980). Spawning activity ranges from Croton Point to Coxsackie but appears to be concentrated just upriver of West Point (Kahnle and Brandt 1985). Following spawning, most adults leave the estu-ary (McLaren et al. 1981) and some apparently join the coastal migratory stock. —Anchored 91-m gill nets were set in two clusters, each containing at least four nets of different standard stretched-mesh sizes (10.2, 11.4, 12.7, and 15.2 cm in 1976; 10.2, 12.7, 15.2, and 17.8 cm in 1977–1979). Two to four additional nets were

usually fished per cluster. A net cluster usually spanned 2–10 km of the river, and mesh sizes were placed randomly within the clusters.

To alleviate and characterize the biases in the information collected by gill-net sampling and to gain a greater understanding of the characteristics of the striped bass, a haul seine also was used. Haul (beach) seines are less size-selective in the fish they catch than gill nets. (This gear is impor-tant in the work of several current striped bass researchers [Kahnle and Brandt 1985; Young 1986] and the cornerstone of the coastwide adult stock monitoring program begun by the Atlantic States Marine Fishery Commission in 1987 [B. H. Young, New York Department of Environmental Conservation, personal communication].) A 274-m haul seine was used to sample beaches primarily within Haverstraw Bay; on occasion, a 61-m haul seine was used to sample areas inaccessible to the larger net.

The catch of commercial fishermen, contracted to fish 2 d/week with their own fishing gear and techniques, was used to supplement each year's data on body length and weight, maturity, and fecundity. Each fisherman was accompanied by study personnel when he tended nets. The com-mercial gear consisted of 23–439-m-long staked, anchored, or drift gill nets of 11.4–35.6-cm stretched mesh. Four fishermen with relatively constant fishing locations were employed per year (McLaren et al. 1988).

All striped bass were measured to the nearest millimeter total length and scale samples were removed for age analysis. Annuli on the scales, which are the basis for aging the fish, are laid down in late spring or early summer (MMES 1987); however, fish were conventionally pro-moted to the next age-group on 1 January. Size-stratified subsamples of fish were further analyzed for determination of weight, sex, maturation state, and fecundity. The remaining fish that were active and in good physical condition were marked with nylon internal anchor tags (Floy D-67c) and released at least 100 m from the capture location.

Age composition.—The migratory nature of striped bass and the size selectivity of sampling gear, particularly gill nets, presented a challenge for accurate estimation of the stock's age compo-sition. Our analytical approach was to assess the age structure of the population during the 6 weeks from approximately mid-April through late May, a period just prior to and including the time of spawning (Boreman and Klauda 1988, this volume). We relied on both the haul-seine and noncommercial gill-net data obtained from a region where both types of gear were fished concurrently, the Tappan Zee and Croton and Haverstraw bays. To minimize the bias introduced by the size selectivity of gill nets (Hamley 1975), only the catch from the 274-m haul seines was used to describe size composition. Gill-net data, however, provided information on the age and sex of fish of specified sizes once the catch was partitioned into 20-mm total length (TL) intervals. With this approach we assumed that the bias in age or sex that might be related to size selectivity within each 20-mm interval would be low. —Age and sex proportions for striped bass larger than 200 mm TL were calculated by:

$$P_{jk} = \frac{\sum_{i=1}^{m} C_i \left(\dfrac{N_{ijk}}{T_i} \right)}{\sum_{i=1}^{m} C_i};$$

P_{jk} = proportion of population for fish age j (j = 1, 2, . . ., l) and sex k;

C_i = total number collected in length interval i (i = 1, 2, . . ., m);

N_{ijk} = number of fish of length i, age j, and sex k in the combined haul-seine–gill-net catch;

T_i = total number of fish subsampled for age and sex in length interval i.

The proportions (P_{jk}) were considered the most reliable representations of the sex and age composition of the population falling within each 20-mm interval, because they included fish caught both in the shore zone (less than 3-m depths) by haul seines and in the shoals (3–6-m depths) by gill nets. Each of these proportions was weighted by the fraction of the 274-m haul-seine catch (C_i) that represented the least size-selective estimate of relative abundance of that 20-mm length grouping in the river.

Maturity.—Maturity was determined from striped bass collected in haul seines and gill nets (including commercial gear) over the entire sampling region during mid-March through June of each year. Catches were sampled on a stratified basis, biweekly sampling period, length, and sex being the strata.

All fish were classified by inspection into four groups: obviously mature (eggs developed in females, milt running in males); obviously immature (gonads undeveloped); indeterminable maturity; and spent (most of eggs or milt gone). Obviously mature and immature fish were then used to calculate a total-body-weight:gonad-weight ratio that could be used as a criterion for separating mature from immature fish; spent fish were not used in this calculation. Fish in the indeterminable category and those visually classified as mature and immature then were reclassified on the basis of their individual total-body-weight:gonad-weight ratios. All spent fish were added to fish classified as mature by the weight-ratio method, and the overall percentage of mature fish in each age-group was calculated.

Fecundity.—Fecundity was estimated for ripe females by counting the number of eggs in a sample aliquot of ovaries. The aliquot was removed from the center of one ovary as a triangular section 1–2 mm thick and constituting approximately one-eighth of the cross section of the ovary. A single aliquot per fish was chosen because Lewis and Bonner (1966) found no significant differences in the number of mature ova found in the anterior, mid, or posterior sections of striped bass ovaries or between right and left ovaries. The ratio of the total weight of both ovaries to the weight of the aliquot was multiplied by the number of eggs in the aliquot to determine the total number of eggs per female. Fecundity analyses were performed on several ripe females incidentally collected during the 1973–1975 spawning season as well as during the large-scale directed efforts from 1976 through 1979.

Length and weight.—Mean total lengths at time of capture were calculated to the nearest millimeter for striped bass caught by all gill nets and haul seines during 1976–1979. The annual growth of individual year classes was followed as incremental growth between two consecutive years of sampling.

Mean fresh weights were calculated from random subsamples within body-length strata of 0–400, 401–549, 550–699, 700–899, 900–1,099, and over 1,100 mm TL in 1976 and 200–299, 300–399, 400–499, 500–649, 650–799, 800–1,000, and over 1,000 mm TL in 1977–1979. Fish less than 400 mm TL were weighed to the nearest gram and larger fish were weighed to the nearest 50 g.

Population size.—The population size of striped bass of approximately age V and older within the Hudson River was estimated annually by mark–recapture methods. The estimates were derived for fish equal to or greater than 500 mm TL from gill-net and haul-seine collections thoroughly examined for tagged individuals. A Schumacher–

TABLE 11.—Age and sex composition of striped bass collected in gill nets and haul seines in the Hudson River estuary below km 62 during 19 April–30 May 1976, 10 April–21 May 1977, and 16 April–27 May 1978.[a] Dashes indicate less than 0.05%.

	1976			1977			1978			3-Year mean		
Age	% Male	% Female	Total	% Male	% Female	Total	% Male	% Female	Total	% Male	% Female	Total
II	1.0	0.5	1.5	3.6	2.2	5.8	0.2	–	0.2	1.6	0.9	2.5
III	12.7	15.0	27.7	12.0	8.9	20.9	8.2	4.5	12.7	11.0	9.5	20.3
IV	10.1	11.0	21.1	28.5	23.2	51.7	8.0	8.3	16.3	15.6	14.2	29.8
V	6.4	11.3	17.7	4.9	3.9	8.8	18.0	21.6	39.6	9.8	12.3	22.0
VI	3.3	7.3	10.6	2.5	2.3	4.8	2.7	4.9	7.6	2.9	4.8	7.7
VII	1.3	7.5	8.8	1.4	1.5	2.9	3.1	4.5	7.6	1.9	4.5	6.4
VIII	0.5	1.8	2.3	0.4	1.4	1.8	1.9	4.2	6.1	0.9	2.5	3.4
IX	0.4	0.2	0.7	0.3	0.2	0.5	1.2	2.1	3.3	0.6	0.8	1.5
X	–	2.9	1.9	–	0.4	0.4	0.4	0.6	1.0	0.1	1.3	1.5
XI	0.5	2.5	3.0	0.1	0.5	0.6	0.2	1.2	1.4	0.3	1.4	1.7
XII	–	2.4	2.4	–	0.5	0.5	–	1.4	1.4	–	1.4	1.4
XIII	–	1.1	1.1	0.1	0.4	0.5	0.2	0.6	0.8	0.1	0.7	0.8
XIV	–	0.2	0.2	0.3	0.2	0.5	0.7	0.6	1.3	0.3	0.3	0.7
XV	–	–	–	–	–	–	0.2	0.4	0.6	0.1	0.1	0.2
XVIII	–	0.2	0.2	–	–	–	–	–	–	–	0.1	0.1
Total	36.2	63.9	100.1	54.1	45.6	99.7	45.0	54.9	99.9	45.2	54.8	100.0

[a] Fish in combined gill-net and haul-seine catches were aged and sexed, then calibrated to the size distribution of fish in the seine catch only. Total and seine samples were, respectively, 591 and 268 in 1976, 1,169 and 538 in 1977, and 2,271 and 436 in 1978. The seine catch was too small in 1979 (111 fish) to establish a size distribution for that year.

Eschmeyer multiple census estimate (Ricker 1975) was calculated as

$$\hat{N} = \frac{\Sigma(C_b M_b^2)}{\Sigma(R_b M_b)};$$

$\hat{N}$ = estimated population size;
C_b = total catch during biweekly interval b;
M_b = total number of marked fish available for recapture at midpoint of biweekly interval b;
R_b = number of recaptured fish in C_b.

A 90% confidence interval (CI) for $\hat{N}$ was determined from

$$CI = \frac{\Sigma(C_b M_b^2)}{\Sigma R_b M_b \pm t_{s-2(0.05)} (S^2 \Sigma C_b M_b^2)^{1/2}};$$

$t_{s-2(0.05)}$ = t-value for s sampling intervals ($P_\alpha = 0.10$);

$$S^2 = \frac{\Sigma\left(\dfrac{R_b^2}{C_b}\right) - \dfrac{(\Sigma R_b M_b)^2}{\Sigma(C_b M_b^2)}}{s-2}.$$

Only tagged fish at large at least 2 d prior to recapture were used, which allowed for dispersal of marked fish into the unmarked population. All sampling gears were employed for both marking and recapture, without spatial segregation of marking and recapture effort. Tag loss and tag-

ging-induced mortality were considered to be low because the population estimate encompassed only 3 months, and the cumulative effects of these sources was not expected to be great during this term; therefore, no adjustments to these data were made for these two factors in this analysis. Emigration and immigration were also assumed to be slight during this short time (McLaren et al. 1981).

Results

Age Composition

The population of striped bass larger than 200 mm TL during the 1976–1978 spawning runs contained fish of ages II–XVIII (Table 11). Age-groups III–V were the most abundant. A conspicuous feature of the age distributions was the strong 1973 year class (Klauda et al. 1980), which was age III in 1976.

Maturity

Male striped bass from the Hudson River matured earlier than females. Males began to mature at age II, and three-quarters of them were mature by age IV (Table 12). Females began to mature 2 years later than males. Ninety percent of the females were mature by age VII and all were mature by age XI. The ratios of total-body-weight:gonad-weight that best separated obviously mature and immature fish were 235:1 for males and 70:1 for females. Fish appar-

TABLE 12.—Percentage maturity (N) by age of male[a] and female[b] Hudson River striped bass, March–June 1976–1979.

Age	1976		1977		1978		1979		4-Year mean	
	Male	Female	Male	Female	Male	Female	Male	Female	Male	Female
II	17(6)	0(1)	12(25)	0(24)	0(2)	0(2)	35(20)	0(20)	21(53)	0(47)
III	48(48)	4(25)	35(34)	0(27)	41(37)	0(43)	26(19)	0(14)	40(138)	1(109)
IV	67(33)	7(28)	62(81)	5(76)	88(82)	2(59)	76(71)	0(51)	74(267)	3(214)
V	87(53)	21(56)	70(23)	21(19)	88(114)	16(115)	83(66)	24(50)	85(256)	19(240)
VI	78(45)	47(45)	89(35)	62(48)	84(19)	60(30)	97(63)	69(91)	88(162)	62(214)
VII	100(12)	87(55)	100(18)	90(42)	93(30)	95(58)	78(9)	83(24)	94(69)	90(179)
VIII	100(13)	90(20)	90(10)	92(13)	87(15)	97(36)	100(7)	100(19)	96(45)	95(88)
IX	100(7)	100(4)	100(1)	100(5)	100(3)	100(13)	100(5)	89(9)	100(16)	97(31)
X	100(7)	100(23)	100(2)	100(5)	100(3)	100(6)	100(1)	80(5)	100(13)	97(39)
XI	100(11)	100(18)	100(2)	100(6)		100(3)			100(13)	100(27)
XII	100(3)	100(10)	100(4)	100(8)	100(1)	100(5)	100(3)		100(11)	100(23)
XIII		100(5)	100(1)	100(4)		100(3)		0(1)	100(1)	100(13)
XIV		100(1)	100(1)	100(3)	100(4)	100(2)			100(5)	100(6)
XV		100(1)		100(1)	100(1)	100(1)			100(1)	100(3)
XVI						100(1)			100(1)	100(2)
XVII		100(1)				100(1)				100(2)

[a] Males with a total body-weight:gonad-weight ratio less than 235 were considered mature.
[b] Females with a total body-weight:gonad-weight ratio less than 70 were considered mature.

ently ready to spawn (classified as "ripe and running") usually first appeared in Hudson River collections during the second week of May. Spent fish were collected during the third or fourth week of May, signifying that spawning had begun. Eggs and larvae were first collected at this time (Dey 1981; Boreman and Klauda 1988, this volume).

Fast-growing individuals within a year class tended to mature earlier. For example, the mean lengths and weights of mature striped bass were consistently greater than those of immature fish within the same age-group (Table 13). Fish size

alone did not govern maturation; the largest immature fish collected—a 693-mm male and a 791-mm female—were much larger than the median sizes at maturity, which were 450-mm for males and 600-mm for females.

Changes in the number of mature individuals in each age-group (II–VIII), across years (1976–1979), and for each sex were tested by a multidimensional contingency analysis (Fienberg 1970). For female striped bass, age was the only factor significantly affecting maturity ($\chi^2 = 719.70$; $P < 0.01$). The percentage of females reaching maturity increased the most between the ages of IV

TABLE 13.—Comparisons of length and weight at age between immature and mature Hudson River striped bass. Data are means + SEs (N). In all paired comparisons, mature fish were significantly larger than immature fish (t-test; $P < 0.05$).

Age	Maturity	Males		Females	
		Total length (mm)	Weight (g)	Total length (mm)	Weight (g)
III	Immature	351 ± 8 (50)	472 ± 32 (41)	417 ± 6 (97)	774 ± 37 (78)
	Mature	400 ± 9 (40)	690 ± 66 (26)	483 (1)	(0)
IV	Immature	404 ± 11 (24)	684 ± 55 (19)	466 ± 5 (174)	1,065 ± 52 (109)
	Mature	466 ± 6 (138)	1,082 ± 57 (90)	550 ± 47 (3)	2,500 (1)
V	Immature	458 ± 13 (32)	1,109 ± 127 (27)	528 ± 4 (355)	1,569 ± 54 (189)
	Mature	558 ± 4 (373)	1,854 ± 75 (147)	589 ± 15 (20)	2,594 ± 256 (17)
VI	Immature	515 ± 20 (5)	1,460 ± 182 (5)	546 ± 10 (48)	1,908 ± 168 (27)
	Mature	558 ± 10 (62)	2,307 ± 260 (23)	644 ± 14 (20)	3,150 ± 254 (17)
VII	Immature	528 ± 27 (3)	1,575 ± 375 (2)	561 ± 14 (11)	1,850 ± 161 (3)
	Mature	638 ± 10 (80)	3,393 ± 238 (34)	708 ± 12 (58)	4,492 ± 272 (46)
VIII	Immature	648 ± 23 (3)	3,225 ± 375 (2)	653 ± 118 (2)	4,950 (1)
	Mature	696 ± 18 (31)	4,567 ± 503 (12)	752 ± 14 (41)	5,370 ± 338 (33)

TABLE 14.—Mean age-specific fecundity (number of eggs per mature female in thousands) for Hudson River striped bass collected April–June 1973–1979. Open cells indicate no sample.

Year	Statistic	IV	V	VI	VII	VIII	IX	X	XI	XII	XIII	XIV	XV	XVIII
1973	Mean			451	781	1,549	1,564	1,842		2,351		2,190		
	SE			175	139	125	155	263		356				
	N			2	9	14	9	4		2		1		
1974	Mean		779	727	1,171	1,250	1,498	1,801	1,768					
	SE		227	115	288	86	120	159	354					
	N		3	5	4	15	17	15	5					
1975	Mean	409	645	669	901	949	1,552	1,843	2,056	2,126			2,591	
	SE			39	238	227	200	140	250	263				
	N	1	1	4	2	3	9	11	9	4			1	
1976	Mean		354	765	1,005	1,056	1,798	1,644	2,000	1,918	2,126			
	SE		68	279	101	299	1,119	141	158	188	146			
	N		3	10	24	6	2	14	13	4	4			
1977	Mean			670	578	871	1,552	1,739	2,385	2,440		2,214		
	SE			138	43	118	283	237	43	561		88		
	N			8	15	6	2	2	3	4		2		
1978	Mean	337	557	609	779	958	1,474	1,968	2,182	3,089	3,859	2,753		3,019
	SE		93	145	81	101	241	1,480		477				
	N	1	8	8	30	21	5	2	1	2	1	1		1
1979	Mean		638	649	832	1,094	1,150	1,010			1,346			
	SE		111	41	179	111	168	80						
	N		5	32	7	17	5	3			1			
Years combined	Mean	373	585	664	830	1,141	1,496	1,728	2,022	2,301	2,285	2,342	2,591	3,019
	SE	36	60	50	47	54	77	85	112	180	352	142		
	N	2	20	69	91	82	49	51	31	16	6	4	1	1

and VII. Male striped bass had a significant maturity–age–year interaction ($\chi^2 = 36.54$; $P < 0.01$) generated primarily by differences between the younger age groups.

Fecundity

Mean fecundity ranged from 373,000 eggs per female at age IV to 2.3 million eggs per female at age XIV (Table 14). There were sufficient samples for ages VI–X taken during 1973–1979 for least-squares analyses of log-transformed fecundity values (Sokal and Rohlf 1969). Fecundity varied significantly ($F = 23.8$; $P < 0.01$) across ages but not across years ($F = 1.54$; $P = 0.165$). The greatest increase in fecundity occurred between ages VIII and IX. Fecundity (Fe) was significantly ($P = 0.05$) correlated with body length (TL, mm) and weight (W, g):

$$\log_{10} Fe = 3.82 \log_{10} TL - 5.04; \ r = 0.915;$$
$$\log_{10} Fe = 1.21 \log_{10} W + 1.43; \ r = 0.927.$$

Female stripped bass produced approximately 176,000 eggs/kg of body weight.

Length and Weight

The length–weight relations for Hudson River striped bass age II and older were

males:
$$\log_{10} W = -4.914 + 2.99 \log_{10} TL; \ r = 0.989;$$

females:
$$\log_{10} W = -5.019 + 3.028 \log_{10} TL: \ r = 0.991.$$

The regressions included fish collected during March–June 1976, regardless of spawning condition, and were most heavily representative of fish caught in gill nets. Females were consistently larger than males after age III (Table 15).

During 1976–1979, striped bass incremental growth (ages III–XI) demonstrated a significant year–age interaction ($P < 0.01$), which appeared to be a result of generally slower growth in 1978 than in 1977 and 1976 for ages III–VI and faster growth for age VII. Changes in annual incremental growth of striped bass across ages, years, and sexes were tested in a three-way analysis of

TABLE 15.—Mean total length (mm) and sample size (in parentheses) of male and female striped bass collected in the Hudson River estuary by gill nets (10.2, 12.7, 15.2 and 17.8 cm stretched mesh) during March–June 1976–1979.

Age	1976		1977		1978		1979	
	Male	Female	Male	Female	Male	Female	Male	Female
III	404(64)	408(61)	416(68)	427(45)	399(89)	434(60)	377(9)	397(11)
IV	428(76)	448(71)	455(411)	476(352)	459(244)	481(110)	460(284)	478(164)
V	520(121)	539(104)	479(141)	512(96)	531(632)	559(301)	498(329)	534(184)
VI	570(70)	595(87)	570(118)	642(61)	569(111)	594(93)	586(329)	611(278)
VII	662(23)	710(64)	627(46)	688(51)	620(105)	677(98)	598(51)	655(74)
VIII	741(6)	740(14)	675(9)	763(19)	683(37)	723(49)	716(38)	747(35)
IX	757(11)	874(5)	824(4)	851(9)	718(19)	776(14)	733(20)	764(23)
X	872(8)	941(14)	829(1)	914(7)	790(6)	784(4)	832(5)	811(9)
XI	875(12)	954(16)	899(7)	960(11)	883(3)	940(1)	840(3)	857(3)

variance (Sokal and Rohlf 1969). Mean lengths from gill-net collections during 1976–1979 (Table 15) were used in this analysis because they were based on similar fishing effort each year.

Fork length (FL) of Hudson River striped bass can be derived from total length by

$$FL = -13.313 + 0.969 \ TL.$$

Population Size

The Schumacher–Eschmeyer population estimate of Hudson River striped bass larger than 500 mm TL increased numerically from 1976 to 1979 (90% confidence intervals in parentheses):

> 1976: 102,000 (56,000–548,000);
> 1977: 174,000 (93,000–1,394,000);
> 1978: 188,000 (130,000–336,000);
> 1979: 254,000 (146,000–976,000).

Because the confidence intervals were large, the annual estimates were not significantly different, but a trend of increasing population size would be consistent with recruitment of the 1973 year class. Some fish hatched in 1973 reached 500 mm at age IV in 1977, and recruitment to this size-group was complete by age VI in 1979.

Discussion

Although annual variations in age at maturity and growth have been detected for Hudson River striped bass, there is no evidence of a consistent change in either variable that might be associated with increasing power plant operations and a reduction in striped bass abundance. Other factors in the environment of the striped bass are likely responsible for the observed annual variations, such as variations in food availability or water temperature in the river or in coastal overwintering areas.

Males matured at an earlier age than females in the Hudson River population. Earlier maturation of males than of females is common among fishes.

Bell (1980) proposed that early maturity in female fishes comes at a higher cost than early maturity in males because ovarian maturation diverts more energy from somatic growth than testicular development, and fecundity is related to fish size. Variations in the age of maturation by individual fish may well be inversely related to the amount of somatic growth accumulated.

Age at maturity and age structure are the two life history aspects of the Hudson River population that differ the most from other striped bass populations. The age at which all Hudson River females are mature is 2–4 years greater than that of other populations (Table 16). These differences in the rate of maturity are not clearly a function of latitude, as postulated for the American shad (Carscadden and Leggett 1975). Along the Pacific coast, female striped bass mature 2 years earlier in Oregon than in California. Along the Atlantic coast, females mature 1 year later in South Carolina than in Maryland and 2 years later than in North Carolina. The Oregon and Hudson populations are located at about the same latitude, yet the Hudson River female population is fully mature 3–4 years later. There is some indication of delayed maturation of females in the St. John River, New Brunswick, which has the northernmost population of striped bass, but 100% maturation there is complete 2 years prior to that of the Hudson River. Males also mature later in the Hudson River than elsewhere, the difference in ages of 100% maturity being approximately 3 years (Table 16).

Most estimates of age composition (Table 17) are derived from commercial catches, which may be size selective, or from pound nets and fyke nets, which are considered to be relatively non-size-selective gear (Grant 1974). The available data suggest that recruitment to commercial gear occurs later in the Hudson River than in the Chesapeake Bay and Sacramento–San Joaquin

TABLE 16.—Percentage maturity at age for female and male striped bass in several estuarine systems. Empty cells indicate data were not reported.

System	II	III	IV	V	VI	VII	VIII	IX+	Reference
					Females				
Sacramento–San Joaquin, California			35	87	98	100	100	100	Scofield (1931)
Coos Bay, Oregon		18	68	100	100	100	100	100	Morgan and Gerlach (1950)
Albemarle Sound–Roanoke River, North Carolina		3	78	100	100	100	100	100	Lewis (1962)
Albemarle Sound–Roanoke River, North Carolina		4	94	100	100	100	100	100	Lewis (1962)
Potomac River, Maryland, spawning area		44	79	99	100	100	100	100	Jones et al. (1977)
Potomac River, Maryland, over-wintering area		17	43	86	100	100	100	100	Jones et al. (1977)
Santee–Cooper Reservoir, South Carolina			23	65	85	100	100	100	Scruggs (1955)
St. John River, New Brunswick		0	20	21	82	100			Williamson (1974)
Hudson River, New York		1	3	19	62	90	95	99	TI data (1976–1979)[a]
					Males				
Potomac River, Maryland, spawning area	93	99	100	99	100	100	100	100	Jones et al. (1977)
Potomac River, Maryland, over-wintering area		92	96	100	100	100			Jones et al. (1977)
St. John River, New Brunswick		25		84					Williamson (1974)
Hudson River, New York	21	40	74	85	88	94	96	100	TI data (1976–1979)[a]

[a] Data collected by Texas Instruments during the Hudson River power plant studies.

systems but at approximately the same age as in the St. John and Annapolis rivers of Canada. The age structure of an exploited fish population is a reflection of both natural and fishing mortality.

The fecundity of Hudson River striped bass is similar to that for other striped bass populations (Morgan and Gerlach 1950; Jackson and Tiller 1952; Lewis and Bonner 1966). Based on reported regressions of mean fecundity on body weight, 6-kg females produce approximately 1.0 million eggs in the Roanoke River, 0.9 million eggs in Chesapeake Bay, 1.2 million eggs in Coos Bay (Oregon), and 1.1 million eggs in the Hudson River. Mean fecundities for 14-kg females in the four systems are 2.3 million, 3.2 million, 3.1 million, and 2.4 million eggs, respectively.

It appears unlikely that another large-scale environmental study of the Hudson River, such as this one associated with the power plant case, will ever occur again, but it is important that some of these data again be collected. The four years of this study occurred after a tremendous increase in electrical generating capacity along the river (over 3,700 MW of capacity came on-line during 1974–1976). Population responses to this increased generating capacity may take decades to actually occur, let alone be detected. Therefore, large-scale studies similar to this one should be performed perhaps every 5 years. Additionally, the commercial striped bass fishery in the Hudson River was closed in 1976. The closure certainly decreased the fishing mortality of the Hudson stock, but whether this decrease has forestalled a population collapse such as the Chesapeake stock experienced in the late 1970s and early 1980s (MMES 1987) is unknown.

Acknowledgments

Financial support for this study came from Consolidated Edison Company of New York,

TABLE 17.—Percentage age composition of age-II and older striped bass, sexes combined, in several estuarine systems.

System	Age						Reference
	II	III	IV	V	VI	VIII+	
Chesapeake Bay, Maryland[a]	18	43	13	22	[b]		Tiller (1950)[b]
James River, Virginia	53	18	9	3	4	12	Grant (1974)[c]
York River, Virginia	66	19	6	3	2	5	Grant (1974)
Rappahannock River, Virginia	64	19	6	2	1	8	Grant (1974)
Sacramento–San Joaquin, California[d]		47	23	12	6	12	Collins (1978)
St. John River, New Brunswick		5	14	29	12	40	Williamson (1974)
Annapolis River, Nova Scotia		7	27	31	14	22	Williamson (1974)
Hudson River, New York[e]	2	20	30	22	8	18	This report

[a] From commercial pound-net catches of 1944 and 1945.
[b] Age-VI and older fish.
[c] From commercial pound-net and fyke-net catches of 1967–1971.
[d] From stratified mark–recapture population estimates of 1969–1976; age-II fish are not included.
[e] Mean age composition, 1976–1978: Table 11.

Incorporated, Orange and Rockland Utilities, Incorporated, Central Hudson Gas and Electric Corporation, and the Power Authority of the State of New York. We thank Ronald J. Klauda, John R. Young, Lawrence W. Barnthouse, Douglas S. Vaughan, Robert L. Kendall and two anonymous reviewers for critical reviews and suggestions regarding the content and preparation of this manuscript. We are indebted to Leanna C. Pristash for data-processing assistance and to Deborah A. Hill, who typed and retyped several versions of this paper. We also thank the former field and laboratory personnel at Texas Instruments for their work in data collection.

References

Bell, G. 1980. The costs of reproduction and their consequences. American Naturalist 116:45–76.

Berggren, T. J., and J. T. Lieberman. 1978. Relative contribution of Hudson, Chesapeake, and Roanoke striped bass, *Morone saxatilis*, stocks to the Atlantic coast fishery. U.S. National Marine Fisheries Service Fishery Bulletin 76:335–345.

Boreman, J., and C. P. Goodyear. 1988. Estimates of entrainment mortality for striped bass and other fish species inhabiting the Hudson River estuary. American Fisheries Society Monograph 4:152–162.

Boreman, J., and R. J. Klauda. 1988. Distributions of entrainable life stages of striped bass in the Hudson River estuary, 1974–1979. American Fisheries Society Monograph 4:53–58.

Carscadden, J. E., and W. C. Leggett. 1975. Meristic differences in spawning populations of American shad, *Alosa sapidissima*: evidence for homing to

tributaries in the St. John River, New Brunswick. Journal of the Fisheries Research Board of Canada 32:653–660.

Collins, B. W. 1978. Age composition and population size of striped bass in California's Sacramento–San Joaquin estuary. Period covered: July 1, 1977 through June 30, 1978. California Department of Fish and Game, Project DJF9R-24, Sacramento.

Dew, C. B. 1981. Impact perspective based on reproductive value. Pages 251–255 *in* L. D. Jensen, editor. Issues associated with impact assessment. Ecological Analysts, Towson, Maryland.

Dey, W. P. 1981. Mortality and growth of young-of-the-year striped bass in the Hudson River estuary. Transactions of the American Fisheries Society 110:151–157.

Fienberg, S. E. 1970. The analysis of multidimensional contingency tables. Ecology 51:419–433.

Gardinier, M. N., and T. B. Hoff. 1982. Striped bass diet in the Hudson River estuary. New York Fish and Game Journal 29:152–165.

Grant, G. C. 1974. The age composition of striped bass catches in Virginia rivers, 1967–1971, and a description of the fishery. U.S. National Marine Fisheries Service Fishery Bulletin 72:193–199.

Hamley, J. M. 1975. Review of gill net selectivity. Journal of the Fisheries Research Board of Canada 32:1943–1969.

Jackson, H. W., and R. E. Tiller. 1952. Preliminary observations on spawning potential in the striped bass (*Roccus saxatilis* (Walbaum)). Chesapeake Biological Laboratory, Solomons, Maryland.

Jones, P. W., J. S. Wilson, R. P. Morgan III, H. R. Lunsford, Jr., and J. Lawson. 1977. Potomac River fisheries study striped bass spawning stock assessment. Interpretive report 1974–1976. Chesapeake

Biological Laboratory and Center for Environmental and Estuarine Studies, Solomons, Maryland.

Kahnle, A. W., and R. E. Brandt. 1985. Biology and management of striped bass in New York waters. New York Department of Environmental Conservation, Project AFC-11, Albany.

Klauda, R. J., W. P. Dey, T. B. Hoff, J. B. McLaren, and Q. T. Ross. 1980. Biology of Hudson River juvenile striped bass. Marine Recreational Fisheries 5:101–124.

Lewis, R. M. 1962. Sexual maturity as determined from ovum diameters in striped bass from North Carolina. Transactions of the American Fisheries Society 91:279–282.

Lewis, R. M., and R. R. Bonner. 1966. Fecundity of the striped bass, *Roccus saxatilis* (Walbaum). Transactions of the American Fisheries Society 95:328–331.

MAFMC (Mid-Atlantic Fishery Management Council). 1984. Striped bass fishery management plan. MAFMC, Dover, Delaware.

MAFMC (Mid-Atlantic Fishery Management Council). 1988. Summer flounder fishery management plan. MAFMC, Dover, Delaware.

McLaren, J. B., J. C. Cooper, T. B. Hoff, and V. E. Lander. 1981. Movements of Hudson River striped bass. Transactions of the American Fisheries Society 110:158–167.

McLaren, J. B., R. J. Klauda, T. B. Hoff, and M. Gardinier. 1988. Commercial fishery for striped bass in the Hudson River, 1931–80. Pages 89–123 *in* C. L. Smith, editor. Fisheries research in the Hudson River. State University of New York Press, Albany.

Merriman, D. 1941. Studies on the striped bass (*Roccus saxatilis*) of the Atlantic coast. U.S. Fish and Wildlife Service Fishery Bulletin 50:1–17.

Morgan, A. R., and A. R. Gerlach. 1950. Striped bass studies on Coos Bay, Oregon, in 1949 and 1950. Oregon Fish Commission Contributions 14.

MMES (Martin Marietta Environmental Systems). 1987. Draft ASMFC striped bass management plan. Report to Atlantic States Marine Fisheries Commission, Washington, D.C.

Muessig, P. H., J. B. Hutchinson, L. R. King, R. J. Ligotino, and M. Daley. 1988. Survival of fishes after impingement on traveling screens at Hudson River power plants. American Fisheries Society Monograph 4:170–181.

NMFS (National Marine Fisheries Service). 1984. Marine recreational fishery statistics survey, Atlantic and Gulf coasts, 1979. U.S. Department of Commerce, Washington, D.C.

Ricker, W. E. 1975. Computation and interpretation of biological statistics of fish populations. Fisheries Research Board of Canada Bulletin 191.

Scofield, E. C. 1931. The striped bass of California. California Fishery Bulletin 29:1–84. (Sacramento.)

Scruggs, G. D., Jr. 1955. Reproduction of resident striped bass in Santee–Cooper Reservoir, South Carolina. Transactions of the American Fisheries Society 85:144–159.

Sokal, R. R., and F. T. Rohlf. 1969. Biometry. Freeman, San Francisco.

Tiller, R. E. 1950. A five year study of the striped bass fishery of Maryland, based on analyses of the scales. Chesapeake Biological Laboratory, Solomons, Maryland.

Van Winkle, W., D. Kumar, and D. S. Vaughan. 1988. Relative contributions of Hudson River and Chesapeake Bay striped bass stocks to the Atlantic coastal population. American Fisheries Society Monograph 4:255–266.

Williamson, F. A. 1974. Population studies of striped bass (*Morone saxatilis*) in the Saint John and Annapolis rivers. Master's thesis. Acadia University, Wolfville, Canada.

Young, B. H. 1986. A study of the striped bass in the marine district of New York State. New York Department of Environmental Conservation, Stony Brook.

Young, J. R., and T. B. Hoff. 1988. Age-specific variation in reproductive effort in female Hudson River striped bass. Pages 124–133 *in* C. L. Smith, editor. Fisheries research in the Hudson River. State University of New York Press, Albany.

American Fisheries Society Monograph 4:69–88, 1988

Life History of White Perch in the Hudson River Estuary

RONALD J. KLAUDA,[1] JAMES B. MCLAREN,[2] ROBERT E. SCHMIDT,[3] AND
WILLIAM P. DEY[4]

Texas Instruments Incorporated, Ecological Services Group, Buchanan, New York 10511, USA

Abstract.—The white perch is a year-round resident throughout the 243-km tidal portion of the Hudson River estuary. Life history data were collected from 1972 through 1979 in a 224-km portion of the estuary between the George Washington Bridge at New York City and the Troy Dam near Albany. Adults were widely distributed in brackish and freshwater habitats. A portion of the adult population moved upriver and onshore during April and May, presumably to spawn. Spawning occurred throughout much of the estuary from late April to early June, but peak egg deposition usually occurred from mid-May to early June in the upstream portions of the tidal river when water temperatures were 16 to 20°C. Larvae began to disperse downriver in July during the post–yolk-sac stage. Juveniles (age-0 individuals) dispersed downriver and shoreward in early August, were concentrated in the shore zone until mid to late October, and then moved to the deeper offshore areas of the middle and lower estuary. Growth rates for larvae and juveniles averaged 0.67 mm·d^{-1} and were positively correlated with water temperature. Juvenile weights at the end of the first growing season (November) were negatively correlated with freshwater flows during July and August. The oldest white perch collected were age XI. Individuals older than age VII were uncommon. No single age-group dominated the age structure during the 6-year study period. Males and females began to mature at age II; all males were mature at age IV and all females at age V. All individuals greater than 190 mm total length (TL) were sexually mature. Mean age-specific fecundities ranged from 31,900 eggs for age-II females to 176,000 eggs for age-VI females. Overall mean fecundity of ages-II–VI fish was 88,900 eggs per female. The relationship between TL and fecundity (Fe) was: $\log_{10}$ Fe $= -4.392 + 4.140 \log_{10}$ TL; $r^2 = 0.71$. Females of ages III and IV contributed 65% of the total potential spawn. The 6-year mean abundance of juveniles on 1 August was 32×10^6 (range, 15×10^6 to 50×10^6). September populations of age-I and older white perch during 3 years ranged from 6.7×10^6 to 12.8×10^6. Rates of population decline for post–yolk-sac larvae and early juveniles during 4 years ranged from 10.0%·d^{-1} to 12.8%·d^{-1}. Rates of population decline for older juveniles ranged from 0.5%·d^{-1} to 0.7%·d^{-1}. Annual total mortality rates (A) for age-I and older fish ranged from 0.57 to 0.68. Sport and commercial fishing contributed less than 1% to annual total mortality during the study period. Closure of the fishery in early 1976 due to PCB contamination likely contributed to the low exploitation rates.

The white perch is endemic to coastal waters of eastern North America from Nova Scotia to South Carolina (Woolcott 1962). This euryhaline species inhabits marine, estuarine, and tidal fresh water, although its largest concentrations are in brackish areas (Mansueti 1964). White perch are also successful in landlocked ponds and reservoirs (Greeley 1937; Smith 1939; AuClair 1956, 1960; Taub 1966; Thoits and Mullan 1973). Since the late 1800s, stocking, canal building (Smith

1985), and other human activities extended the original range of white perch north and west into Lake Ontario (Sheri and Power 1968), the waters of Quebec (Scott and Christie 1963), Lake Erie (Larsen 1954; Trautman 1957; Busch et al. 1977), and beyond. Hergenrader (1980) reported that in 1978, white perch populations were present in all coastal states and maritime provinces from South Carolina to Prince Edward Island, inland from Vermont along the Great Lakes to Michigan and Ontario, and in Nebraska; the Nebraska population, introduced in 1964 (Hergenrader and Bliss 1971), may be spreading through the Missouri River drainage.

The white perch is a year-round resident in the 243-km tidal portion of the Hudson River between New York City and the Troy Dam near Albany, New York (Greeley 1937; Perlmutter et al. 1967; McFadden et al. 1978). White perch also make up a large portion of the fish impinged on intake

[1]Present address: The Johns Hopkins University, Applied Physics Laboratory, Environmental Sciences Group, Shady Side, Maryland 20764, USA.

[2]Present address: Beak Consultants Incorporated, 12072 Main Road, Akron, New York 14001, USA.

[3]Present address: Simons Rock of Bard College, Great Barrington, Massachusetts 01230, USA.

[4]Present address: EA Engineering, Science, and Technology, Incorporated, Rural Delivery 2, Goshen Turnpike, Middletown, New York 10940, USA.

screens at power plants that withdraw cooling water from the river (Barnthouse et al. 1983; Barnthouse and Van Winkle 1988, this volume; Mattson et al. 1988, this volume; Muessig et al. 1988, this volume). In recent years, the Hudson River population has contributed little to either the sport or commercial catch (TI 1975; Sheppard 1976); however, the fishery for white perch has been closed in the river since February 1976 due to PCB contamination. The species supports intensive fisheries elsewhere (Mansueti 1964; Busch and Heinrich 1982).

Because of its wide-spread distribution and seasonal dominance in impingement collections, white perch was a focal fish species in near-field (plant vicinity) and far-field (estuary-long) studies sponsored by the Hudson River utilities during the late 1960s and 1970s. The objective of this paper is to discuss the life history of white perch in the Hudson River estuary based on data collected by Texas Instruments Incorporated (TI) between 1972 and 1979 over 224 km of the estuary and in a 12-km segment around the Indian Point generating station (Hutchison 1988, this volume). The data cover adult distributions and movements, time and location of spawning, distributions of larvae and juveniles, growth, age structure of the population, age and size at maturity, fecundity, relative abundance, and mortality.

Methods

The methods described here are summarized from several technical reports prepared for the utilities by TI. These reports, listed in Klauda et al. (1988), Boreman and Klauda (1988, this volume), and Young et al. (1988, this volume), provide detailed descriptions of field, laboratory, and analytical procedures.

Far-field sampling design.—White perch eggs, larvae, juveniles (defined as age-0 individuals prior to 1 January of the calendar year after they were spawned), subadults, and adults were collected during all but the coldest winter months (January, February, March) from 1974 to 1979 between the George Washington Bridge and the Troy Dam. This study area was subdivided into 12 geographical sampling regions, Yonkers through Albany (see Figure 17), which encompass the physiographically distinct lower, middle, and upper zones (see Figure 1; Cooper et al. 1988, this volume). The lower zone (42 km long) is relatively wide and shallow and includes the regions of Yonkers through Croton–Haverstraw. The middle

zone (60 km) is narrow, contains the deepest areas of the estuary, and includes the regions of Indian Point through Poughkeepsie. The upper zone (120 km) includes the regions of Hyde Park through Albany and is moderately wide but shallower than the middle zone. To evaluate depth distributions of early life stages, each sampling region was also divided into three depth or volume strata or compartments (shoal, bottom, channel), as described in Boreman and Klauda (1988). Eggs, larvae, and early juveniles were collected in annual ichthyoplankton surveys that extended from mid-April through mid-August (Boreman and Klauda 1988). Older juveniles, subadults, and adults were collected in a variety of sampling gear (Young et al. 1988).

Near-field sampling design.—Juvenile and older white perch were intensively sampled near the Indian Point generating station from 1972 through 1979 with beach seines, otter trawls, and surface trawls. Samples were collected during the day at seven seine sites and seven trawl transects (Figure 20). A 30.5-m-long by 2.4-m-deep seine

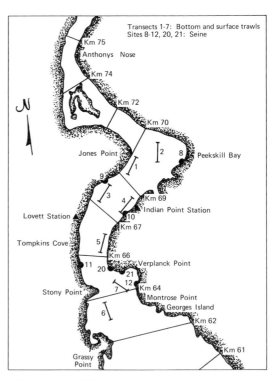

FIGURE 20.—Seining sites and bottom surface trawl transects near the Indian Point generating station, Hudson River. River kilometers are measured from the Battery at the southern tip of Manhattan Island, New York City.

(0.48-cm-mesh net in the bunt) was pulled weekly at each site from April through December approximately 2 h prior to low tide. A 7.8-m (head-rope length) otter trawl with a 3.8-cm stretch-mesh body, 3.3-cm stretch-mesh cod end, and 1.27-cm stretch-mesh cod end liner was towed biweekly along each transect from April through December. From July through December, each trawl transect was also sampled biweekly with a 5.3-m (head-rope length) surface trawl equipped with 3.0-m spreader bars and a 0.6-cm stretch-mesh cod end. The surface trawl was towed by two boats. Both trawls were towed at 1.0 m·s^{-1} for 10 min.

Laboratory procedures.—Samples of eggs, larvae, and early juveniles were preserved in 5% buffered formalin, returned to the laboratory, sorted, identified, and counted. Randomly selected samples of juvenile and older-age fish were placed in 10% buffered formalin in the field and returned to the laboratory for analysis. Age was determined by scale analysis (Jearld 1983). At least 15 scales per fish were removed from the left side of the body above the lateral line between the spinous and soft dorsal fins. A subsample of 3 to 5 nonregenerated scales was selected for age analysis, cleaned in deionized water, and mounted sculpted side up in water between two glass slides (wet mount). The scales were examined with a microprojector at 45× magnification. Annuli were counted to assign an age to each fish. Length-frequency distributions of white perch in year-round collections were analyzed to validate age determinations from scales. —Total length (maximum body length when the lobes of the caudal fin were compressed dorsoventrally, to nearest mm) and wet weight (to nearest 0.1 g) were measured after at least 5 d of preservation to standardize effects of formalin preservation on length and weight measurements (TI 1981). The sex of age-I and older individuals was visually determined during laboratory dissections. Age at maturity was calculated from white perch collected by beach seines, and otter trawls during May and June between river kilometers 48 and 80. Maturity of individuals greater than 100 mm TL was determined by visually examining formalin-preserved gonads. Ovaries were considered immature if they were small and translucent and mature if swollen and opaque.

Fecundity, defined as the number of viable eggs per female, was estimated for ripe-phase females based on a gravimetric method (Snyder 1985). Ovaries were preserved in 10% buffered formalin for at least 1 month before processing. All eggs 0.2 mm in diameter and larger in each female were counted in a weighed subsample (aliquot) of one ovary. The ratio of total weight of the ovaries (to nearest 0.01 g) to the weight of the aliquot (to nearest 0.01 g) was multiplied by the number of eggs 0.2 mm in diameter and larger in the aliquot to estimate the total number of eggs per female. Egg diameters of 200 randomly selected eggs from an ovarian aliquot were measured with an ocular micrometer. The ovarian aliquot was a triangular section (5–8 mm wide by 2–4 mm thick) removed from the right ovary midway along the longitudinal axis.

Fecundity (Fe) estimates were screened with a discriminate function (TI 1980b) to remove females suspected to be in a phase of gonad maturation defined as partially spent. The linear discriminant function (Sokal and Rohlf 1969) was based on the relationship between $\log_{10}$Fe and $\log_{10}$ TL derived from females collected during May. For example, of 84 adult white perch females examined in 1975 to 1977, 7 had low fecundity values and appeared to be outliers. The discriminant function (*D*) was expressed as $D = 35.01 \log_{10}$ Fe $- 155.81 \log_{10}$ TL $+ 185.12$. A value for *D* greater than 0 indicated normal fecundity (i.e., probably not partially spent); a *D* value less than or equal to 0 indicated abnormally low fecundity (i.e., probably partially spent). Application of this discriminant function to females collected in 1975 to 1977 yielded only one fish with a gonad maturation phase classification different from the original visual classification. This single misclassification was near the decision level and probably represented random error.

Calculations.—The distribution and abundance of white perch eggs, larvae, and early juveniles in space and time were described from the average densities of each life stage within each depth stratum (volume compartment) summed across all depth strata to estimate an average density within each of the 12 geographic sampling regions. Densities were calculated as the number of each life stage collected divided by the volume of water sampled and adjusted to 1,000 m^3. Average densities in each depth stratum and region during each week or 2-week period were also multiplied by the volume of water within each depth stratum and region to estimate average population size. Population estimates were then summed across all depth strata and regions during each sampling period to yield a population estimate for the entire study area between river kilometers 19 and 243.

Several approaches were explored to estimate the annual abundance of juvenile white perch. The annual ranks of four abundance indices—seine catch per effort of juveniles in summer, combined juvenile population estimates in summer calculated from beach seine and epibenthic sled surveys, otter trawl catch per effort of yearlings during spring, and otter trawl catch per effort of yearlings during fall—were significantly correlated (range of r^2, 0.78–0.93; $P < 0.05$). The abundance index derived from combined juvenile population estimates in summer drew upon samples from more white perch habitats than the others and was selected to represent annual variations in juvenile abundance (Young et al. 1988).

The size of the fall population of age-I and older white perch in 1977 through 1979 was estimated from spring recaptures of tagged fish released during the previous fall. White perch were also tagged during the spring but not in July and August when handling mortality was unacceptably high. White perch captured throughout the study area were marked with one of two types of individually numbered tags (Floy fingerling or Floy internal anchor). Young et al. (1988) provide more details on this tagging program. Numbers of tagged fish released during the fall of each year were adjusted for short-term (14-d) handling mortality (14.5%) and tag loss (11.0%) based on aquarium and in situ cage studies. The Petersen method (Ricker 1975) was used to estimate the abundance of the age-I and older population (Young et al. 1988).

Growth rates for early juveniles were calculated from far-field seine-survey data collected during July and August in 1973 through 1979 when the juveniles were between 25 mm TL (approximate length when they became accessible to the seine) and 60 mm TL (approximate length reached in August at the end of the rapid growth phase). Mean TL estimates were calculated for juveniles during sampling weeks when at least five individuals were measured. Instantaneous growth rates for juveniles (G_j) were calculated from mean total lengths (TL) with the linear regression

$$\log_e (L_t) = \log_e (L_0) + G_j(X_t);$$

L_t is the mean TL at time t; L_0 is the predicted mean TL on 1 July; X_t is the number of days since 1 July at time t. Estimates of instantaneous growth rates for juveniles were converted to estimates of daily growth rates (G_d, mm·d^{-1}) between 25 mm and 60 mm TL by

$$G_d = \frac{60 - 25}{X_{60} - X_{25}};$$

X_{60} = number of days from 1 July to reach 60 mm TL, or $[\log_e 60 - \log_e (L_0)]/G_j$;

X_{25} = number of days from 1 July to reach 25 mm TL, or $[\log_e 25 - \log_e (L_0)]/G_j$.

Daily growth rates for larval white perch (G_l, mm·d^{-1}) from the time of peak abundance for yolk-sac larvae to the time when the juveniles reached 25 mm TL were calculated by

$$G_l = \frac{25 - L_m}{X_{25} - X_0};$$

L_m is the approximate mean length of yolk-sac larvae (about 3 mm; Mansueti 1964); X_{25} is the number of days from 1 May for juveniles to reach 25 mm TL; X_0 is the mid-date of peak(s) in abundance of yolk-sac larvae. The peak abundance period(s) for yolk-sac larvae was used rather than the peak period(s) of egg deposition because demersal white perch eggs (Mansueti 1964) are more difficult to collect than the yolk-sac larvae. Immediately after hatching, white perch yolk-sac larvae become active, swim to the surface, are attracted to light (Mansueti 1964), and should, therefore, be more available to conventional, towed sampling gear than are the eggs.

Mortality rates for all early life stages combined were estimated from rates of decline in weekly population estimates. We did not attempt to adjust the observed rates of population decline for changes in gear efficiencies associated with larval and juvenile growth and the increased mobility of these life stages. Age-specific or life-stage-specific mortality rates could not be calculated because the necessary length-frequency data were not collected. The spawning period for white perch lasted several weeks in many years, and these samples contained many age groups of newly spawned fish. Thus, the calculated mortality rates should not be viewed as the equivalent of age- or life-stage-specific mortalities.

Instantaneous rates of population decline (U_t) for the early life stages were calculated for the period mid-June through July when post–yolk-sac larvae and early juveniles were most abundant, and for the period August through December for older juveniles, by the linear regression.

$$\log_e (N_t) = \log_e (N_0) - U_t;$$

N_t is the estimated population size at time t; N_0 is the estimated population size at time 0; and t is the number of days between 1 May and the midpoint of the sampling week. Instantaneous rates of population decline (U) were converted to daily rates ($\%\cdot d^{-1}$) by $(1 - e^{-U})$.

Annual total mortality rates (A) for age-I and older white perch were estimated from age-frequency data and catch-curve analysis (Ricker 1975). Age-frequency data were obtained from bottom-trawl catches in the vicinity of the Indian Point generating station from October through December, 1974–1979. The annual catch data from the 1973, 1974, and 1975 year classes were scaled to catches based on a standard number of bottom trawl tows for the time period ($28\cdot month^{-1}$) to standardize effort across years. These three year classes were selected because each had passed through age I and at least four older ages by the end of the study.

The scaled catch data were then fitted to a log-linear model by a least-squares regression technique:

$$\log_e (H_{tc}) = \log_e (H_0) - Z_{tc};$$

H_{tc} is the estimated number of age-tc individuals; H_0 is the estimated number of age-0 individuals; Z is the instantaneous mortality rate (Ricker 1975); tc is age in years. Catch-curve analysis presumes a stable age distribution, no variation in year-class strength, and equal catchability among years. These assumptions were generally satisfied by the 1973, 1974, and 1975 year-class data used in these mortality rate calculations (TI 1980b). Annual total mortality (A) was then determined by $A = 1 - e^{-Z}$ (Ricker 1975).

The estimate of sport-fishing mortality exerted on the white perch population in 1976–1978 was based on the numbers of tags returned by fishermen within 180 d after the fish were released. The mortality estimates were not adjusted for tag loss, handling-related mortality, or fishermen nonresponse.

Results

Adult Distribution and Movements

Far-field sampling began in early April in the lower and middle estuary zones during each study year between 1974 and 1979. Based on a representative year, 1979, white perch adults were most abundant in offshore areas of the middle zone (km 62–122) that were greater than 6 m deep (Figure 21). Adults were also numerous offshore during early April in the lower zone (river kilometers 19 to 61). The abundance of adults in offshore areas of the upper zone (km 123–243) during April or any other month could not be measured because the channel

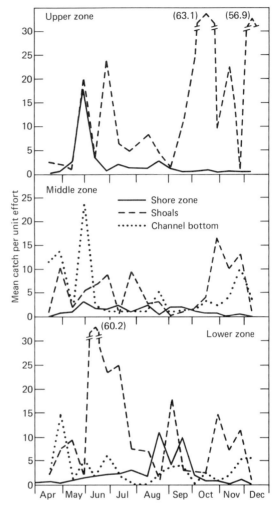

FIGURE 21.—Depth distributions of adult white perch in the upper, middle, and lower estuary zones of the Hudson River, April through mid-December 1979. Shore-zone samples were taken with seines; shoal and channel-bottom samples were taken with bottom trawls.

bottom was not sampled. In early May, adults remained abundant in the deeper offshore areas of the middle zone; from mid to late May, however, catches increased in the shoals and shore areas of the upper zone, at the same time catches decreased in the middle and lower zones. These trends in regional and strata catches during April and May suggest upriver and onshore movements by adult white perch, presumably associated with spawning. Similar catch trends were detected in the Hudson River during spring 1984 between km 42 and km 174 (Normandeau Associates 1985).

Shallow flats, embayments, and tidal creeks suitable for white perch spawning in the Hudson River are most extensive in the Kingston, Saugerties, and Catskill sampling regions (km 138–198) of the upper zone. Normandeau Associates (1985) reported that shallow embayments plus large and small tidal creeks in this upper zone were used as spawning sites in spring 1984, particularly in Stockport Creek (km 194), Esopus Creek (km 162), and Rondout Creek (km 146). Shallow embayments and tidal creeks are relatively sparse in the lower and middle estuary zones. The Croton River (km 54) is the only major tributary stream flowing into the lower zone. The major tributary streams entering the middle zone (Moodna Creek, Fishkill Creek, and Wappingers Creek at km 91, 94, and 107, respectively) offer much less potential spawning habitat than tributary streams entering the upper zone.

The pattern of catches in the three estuary zones from late May through July (Figure 21) suggested that a portion of the adult white perch population moved offshore and downriver after spawning. Offshore movements of adults occurred in all estuary zones from September through November, but an estuary-long gradient in relative abundance that could be interpreted as extensive downriver movements in the fall was not observed in 1979 (Figure 21) or in 1984 (Normandeau Associates 1985). But during fall 1983, a clear shift in adult white perch abundance was observed from the upper to lower zone between September and December that could be clearly

interpreted as downriver movements (Normandeau Associates 1985).

Data collected by TI's white perch tagging program did not support the consistent spring upriver movements and occasional fall downriver movements by adults that were portrayed by the seine and trawl surveys. Over 20,000 adults were tagged and released throughout the study area during spring (April–June) and fall (September–December) of 1976–1980 (Table 18); of these, 451 were recaptured during the same year and season of release (2.2%). These recaptures were used to assess seasonal movements. Tagged fish that were recaptured in different seasons and years were not included in the movements analysis because their lengthy times at large could obscure any seasonal patterns.

Most (383) of the 451 recaptured fish were taken in the same geographic sampling region where they were tagged and released. These fish were defined as nonmovers. Only 68 recaptures were taken in a different region, were assumed to have crossed at least one regional boundary, and were defined as movers. Intake screens at the Indian Point generating station (km 69) collected 62% of the recovered white perch that moved. The TI tagging effort was concentrated in the Croton–Haverstraw and Indian Point regions (km 54–74), near the boundary between the lower and middle estuary zones. The Indian Point intake screens also collected more white perch in this stretch of river than all other gear com-

TABLE 18.—Movements of white perch larger than 150 mm total length within and from the sampling regions in which they were tagged and released during spring (April–June) or fall (September–November) and recaptured during the same season and year of release, Hudson River, 1976–1980.

| | | | Fish that changed regions before being recaptured by | | | |
| | | | Indian Point intake screens | | Fishing gear | |
Year	Number of fish tagged and released	Number of fish recaptured in same region	Moved upriver: number (mean km)	Moved downriver: number (mean km)	Moved upriver: number (mean km)	Moved downriver: number (mean km)
			Fish tagged and recaptured in spring			
1977	4,964	68	13 (13.6)	3 (80.0)	5 (58.6)	7 (37.8)
1978	3,010	10	0	3 (45.9)	0	0
1979	1,735[a]	2	1 (9.7)	2 (56.4)	11 (64.2)	0
1980	2,677	20	0	2 (29.0)	2 (35.4)	0
Total	12,306[a]	100	14 (11.7)	10 (52.8)	8 (86.1)	7 (37.8)
			Fish tagged and recaptured in fall			
1976	11,795	222	9 (14.0)	1 (67.6)	5 (6.2)	5 (6.4)
1977	3,405	28	3 (25.8)	0	0	1 (35.4)
1978	2,529	23	3 (17.7)	0	0	0
1979	2,326	10	1 (14.5)	1 (6.4)	0	0
Total	20,052	283	16 (18.0)	2 (18.5)	5 (6.2)	6 (20.9)

[a] Tagged white perch were not sorted by size in spring 1979 and some were shorter than 150 mm total length.

bined (e.g., about 124,000 and 72,700, respectively, in 1979).

No clear pattern of upriver movements in spring or downriver movements in fall could be detected from the few adults that moved to adjacent sampling regions after being tagged. Downriver movements were longer than upriver movements during both seasons (Table 18). The small number of recaptured fish that moved (68) and the skewed distribution of tag releases and recaptures toward the lower–middle zone boundary limited the usefulness of the tagging program for assessing adult white perch movements.

Time and Location of Spawning

The time and location of white perch spawning in the Hudson River was inferred from the collection of pelagic eggs in the main-stem estuary. White perch eggs are about 1 mm in diameter after they water-harden (Mansueti 1964; Marcy 1976b). They are demersal and adhesive but can be easily dislodged and collected with conventual ichthyoplankton gear (Lippson et al. 1980). If the ratios of these free-floating eggs to attached eggs did not vary among sampling regions, sampling periods, and years, then TI's ichthyoplankton surveys probably provided a reasonable picture of white perch egg distribution within the main-stem of the Hudson River estuary. However, white perch are known to spawn in tidal creeks and shallow embayments of the upper zone (Normandeau Associates 1985), habitats that TI's ichthyoplankton surveys did not sample. The contribution of eggs spawned in tidal creeks and shallow embayments to TI's egg collections in the main-stem estuary could not be measured. Most of the shallow embayments and tidal creeks that were potential spawning sites for white perch are located in the upper zone. Texas Instruments' main-stem sampling program likely underestimated the amount of spawning activity in the upper zone, and the size, but not necessarily the distribution, of weekly egg populations throughout the entire study area.

The spawning season began in late April when water temperatures reached 10 to 12°C and usually extended into early July. Peak spawning activity during some years was contracted into a 2–3-week interval from mid-May through early June (when water temperatures were 16 to 20°C), but peak egg deposition during other years was prolonged over 4–6 weeks when water temperatures increased slowly.

White perch eggs were collected throughout most of the estuarine study area, but spawning activity was apparently greater in the freshwater upper zone, especially in the Saugerties–Albany regions, km 150–243 (Figure 22). The distribution of spawning in the Hudson River agrees with published observations that white perch generally spawn in fresh water, but their eggs are occasionally collected in salinities as high as 2‰ (Mansueti 1961; Miller 1963; Lippson et al. 1980). During peak spawning (mid-May to early June), eggs were most abundant in the bottom stratum, which reflected their demersal characteristics (Figure 23). After peak deposition, a large portion of the eggs was collected in the shoals, perhaps related to the influx of free-floating eggs dislodged from spawning sites in shallow embayments and tidal creeks.

In 1977, we observed an exception to the general upper-zone concentration of white perch spawning: a downriver shift in spawning activity occurred (Figure 24). Egg densities were highest in the lower zone, especially in the Tappan Zee and Croton–Haverstraw sampling regions that are adjacent to and immediately downstream from the confluence with a major tributary, the Croton River at km 54. A large spawn in the Croton River and subsequent transport of dislodged eggs into the main-stem sampling area is one explanation for the large lower-zone peak in egg densities observed in 1977. The importance of tributaries and shallow embayments as white perch spawning sites was not systematically investigated in the Hudson River until spring 1984, when spawning activity in these habitats, especially in the upper-estuary zone, was confirmed (Normandeau Associates 1985).

Distribution of Larvae and Juveniles

White perch yolk-sac larvae were distributed like eggs in most years (Figures 22 and 24), presumably because the yolk-sac larvae possessed limited horizontal mobility (Mansueti 1964; Hardy 1978) and downriver transport was minimal due to their short life-stage duration (3 to 5 d). This similarity supports the inference that TI's egg collections accurately reflected the distribution of spawning activity in the main stem of the Hudson River. Compared to eggs and yolk-sac larvae, post–yolk-sac larvae were more dispersed across the sampling regions in the upper and middle estuary zones (Figures 22 and 24). This pattern suggested either that the mobility of larvae increased and they started to disperse downriver when they were 3.5–4.0 mm long and 3–5 d old, after the yolk was absorbed and active feeding

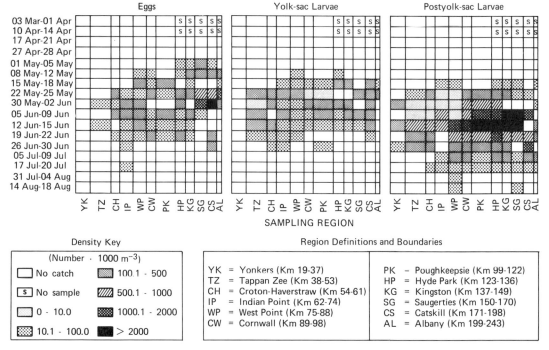

FIGURE 22.—Regional distributions of white perch eggs, yolk-sac larvae, and post–yolk-sac larvae in the Hudson River, 1978. Column widths are proportional to regional water volumes.

began, or that larvae moved from tributaries and embayments to the main-stem estuary. The abundance of post–yolk-sac larvae declined sharply in July due to a combination of mortality, transfor-

mation to the juvenile stage at about 20–25 mm TL, and probable increased gear avoidance.

Juveniles first appeared in late June when water temperatures in the Hudson River reached 22°C,

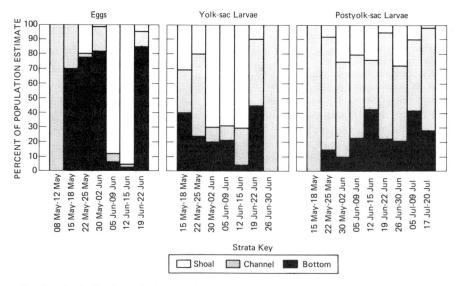

FIGURE 23.—Depth distributions of white perch eggs, yolk-sac larvae, and post–yolk-sac larvae in the Hudson River, 1978.

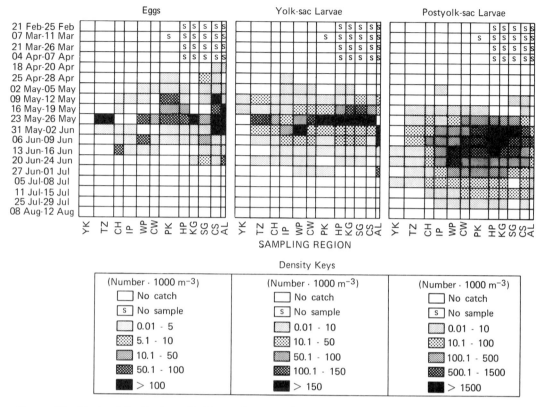

FIGURE 24.—Regional distributions of white perch eggs, yolk-sac larvae, and post–yolk-sac larvae in the Hudson River, 1977. Column widths are proportional to regional water volumes. Regions are defined in Figure 22.

about 6 weeks after post–yolk-sac larvae first appeared. Like post–yolk-sac larvae, most juveniles occurred in the upper and middle estuary zones in July (Figure 25). By early August, juveniles dispersed downriver and shoreward. The population appeared to concentrate in the shore zone during August and September (Figure 26). In mid to late October, the juveniles began to move offshore through the shoals; by early December, they were in water deeper than 6 m. Concurrent with an offshore movement during October and November was an intensified downriver movement into the middle and lower zones of the estuary (Figure 25). Few juvenile white perch were collected in the extreme upper end of the study area after November.

Growth

The general pattern of growth for juvenile white perch during 1975 and 1979 is illustrated with the 1979 data (Figure 27). Growth increased rapidly from about 20 mm TL in late June to almost 60 mm TL by late August, slowed during September

and October, and ceased in November when the population's mean length was about 70 mm TL. The breadths of the length-frequency distributions during all sampling periods in 1979 (Figure 27) reflected the normally protracted spawning period of white perch in the Hudson River.

Growth rates during the larval and early juvenile stages (3–25 mm TL) averaged 0.67 mm·d^{-1} over the 7-year period between 1973 and 1979, and ranged from 0.45 mm·d^{-1} in 1974 to 0.85 mm·d^{-1} in 1976 (Table 19). Annual variations in growth were positively correlated with mean river temperatures during the growth period ($r^2 = 0.94$; $P < 0.01$).

Growth rates during the older juvenile stages (25–60 mm TL) averaged 0.67 mm·d^{-1} from 1973 through 1979, and ranged from 0.57 mm·d^{-1} in 1979 to 0.77 mm·d^{-1} in 1973 (Table 19). Annual variations in growth rates of older juveniles were not significantly correlated with water temperature, freshwater flow, or juvenile white perch abundance. Length–weight relationships were calculated for juvenile white perch collected in

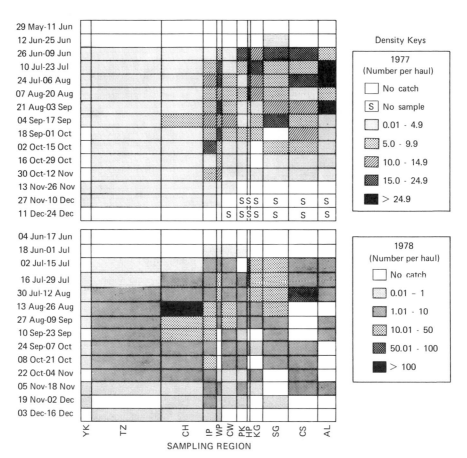

FIGURE 25.—Regional distributions of juvenile white perch in shore-zone areas of the Hudson River, 1977 and 1978. Column widths are proportional to regional shore-zone surface areas. Regions are defined in Figure 22.

seines near the Indian Point generating station during September and October 1975–1979 (Table 20). Based on these regressions, predicted weights (mg) for 70-mm juveniles at the end of the growing season showed a significant negative correlation with mean freshwater flow during July and August ($r^2 = 0.90$; $P = 0.01$).

Total-length data for age-I and older white perch were obtained from seine and bottom-trawl collections near the Indian Point generating station during May and June, at about the time of annulus formation, in 1975–1977 (Table 21). Age-I and -II individuals were significantly ($P < 0.05$) larger in 1976 than in 1975 or 1977. This difference presumably reflected their increased growth as juveniles or yearlings during 1975, a year when water temperatures were higher than average (TI 1980a). Age-III, -IV, and -V females were significantly ($P < 0.05$) larger than males of the same age (Table 22).

Age Structure

Within the age-I and older population of white perch near the Indian Point generating station, age-I and -II fish were most numerous in October–December bottom-trawl catches during 1974–1979 (Table 23). Few white perch older than age V were collected, and individuals older than age VII were uncommon. The oldest individuals, age XI, were caught in 1976. The proportion of total catch by age varied across years (Table 23). The proportion of age-I fish, for example, ranged from 0.33 in 1978 to 0.87 in 1974. Two plausible explanations for this variation were explored.

First, annual differences in these observed proportions of age-I fish may reflect differences in year-class strength. If these proportions were positively correlated with year-class strength, we could more confidently conclude that the proportions presented in Table 23 accurately represented

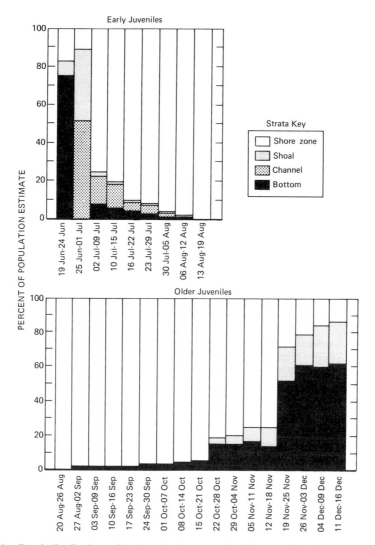

FIGURE 26.—Depth distributions of early and older juvenile white perch in the Hudson River, 1978.

the true age structure of the white perch populations in 1974 through 1979. Data limitations precluded a rigorous test of this hypothesis (TI 1981). But, if annual seine catches of juvenile white perch in July–October accurately represented year-class strength (Young et al. 1988), the proportion of age-I fish collected near Indian Point was positively, but not significantly, correlated with year-class strength ($r^2 = 0.30$; $P > 0.05$). Age-structure profiles for white perch near Indian Point may accurately reflect the population of the entire study area, but a longer time series of age-structure and year-class strength data is needed to establish this point. Other factors also apparently contributed to annual variations in age structure.

Secondly, these variations could have been influenced by annual differences in late-fall depth distributions of the fish. The rate at which age-I white perch moved from the shoals to deeper offshore areas (less accessible to bottom trawls), where they appeared to overwinter (TI 1981), could be negatively correlated with fall water temperatures. If so, the proportional representation of age-I fish in bottom trawl catches should have been lower during warmer-than-average falls (1975, 1978, 1979) than during cooler-than-average falls (1974, 1976, 1977). The proportions of age-I white perch were negatively but nonsignificantly correlated with mean fall water temperatures (October–December) in 1974–1979 ($r^2 = 0.19$; $P > 0.05$). Based on this 6-year data set,

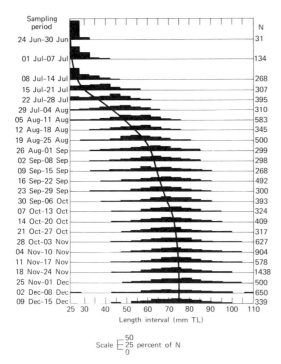

Scale $\left[\begin{matrix}50\\25\\0\end{matrix}\right.$ percent of N

FIGURE 27.—Growth of juvenile white perch in the Hudson River, 1979, expressed as the mean (solid line) and frequency distributions (bars) of total lengths for data collected with all sampling gear.

annual variations in late-fall depth distributions of age-I white perch did not appear to influence TI's estimates of population age structure near the Indian Point generating station.

Age and Size at Maturity

White perch collected during May and June 1972–1977 began to mature at age II when males were 100–110 mm TL and females were 111–120 mm TL. Over 80% of the males and females were mature at age III, but 100% maturity was not attained until age IV for males and age V for females (Table 24). All white perch longer than 190 mm TL were mature.

Fecundity

Fecundity was measured for 106 female white perch, ages II–IV, collected near the Indian Point generating station during May 1975–1977 and during April and May in 1978. Fecundity is difficult to measure for white perch because of the large variation in egg diameters within a single ovary. This variation may be caused by ova retention and resorption, as documented for the related white bass (Ruelle 1977), or because several batches of eggs mature within a single spawning season (Mansueti 1961). Because we could not determine

TABLE 19.—Growth rates for white perch larvae and early juveniles (<25 mm total length, TL) and older juveniles (25–60 mm TL) in the Hudson River, 1973–1979.

	Larvae and early juveniles			Older juveniles	
Year	Period	Mean water temperature (°C)	Growth rate (mm·d^{-1})	Period	Growth rate (mm·d^{-1})
1973	13 Jun–10 Jul	21.9	0.82	10 Jul–24 Aug	0.77
1974	22 May–10 Jul	18.9	0.45	10 Jul–27 Aug	0.74
1975	24 May–26 Jul	20.6	0.67	26 Jul–17 Aug	0.67
1976	12 Jun–08 Jul	22.7	0.85	08 Jul–02 Sep	0.63
1977	24 May–01 Jul	19.9	0.60	01 Jul–23 Aug	0.65
1978	31 May–28 Jun	21.1	0.79	28 Jun–22 Aug	0.64
1979	16 May–28 Jun	19.2	0.50	28 Jun–30 Aug	0.57

TABLE 20.—Relationships of total length (TL, mm) to weight (W, g) for juvenile white perch collected in seines near the Indian Point generating station, Hudson River, during September and October 1975 through 1979.

Year	Sample size	Length–weight relationship	r^2	Predicted weight (g) at 70 mm TL	Mean freshwater flow, July and August[a] (m^3·s^{-1})
1975	340	$\log_{10}W = -5.0255 + 3.0566 \log_{10}TL$	0.96	4.11	235
1976	465	$\log_{10}W = -5.0417 + 3.0497 \log_{10}TL$	0.95	3.85	420
1977	506	$\log_{10}W = -5.0934 + 3.0935 \log_{10}TL$	0.98	4.15	160
1978	531	$\log_{10}W = -4.8514 + 2.9631 \log_{10}TL$	0.97	4.13	150
1979	429	$\log_{10}W = -4.9513 + 3.0249 \log_{10}TL$	0.98	4.27	135

[a] Measured at the Troy Dam near Albany (at river kilometer 243).

TABLE 21.—Mean total length (±SD, mm; N in parentheses) for age-I and -II white perch (male and females combined) collected in the vicinity of the Indian Point generating station, Hudson River, May and June 1975–1977.

Year	Age I	Age II
1975	70.2 ± 8.4 (145)	119.9 ± 11.5 (98)
1976	74.2 ± 8.2 (134)	126.6 ± 9.9 (24)
1977	70.1 ± 9.7 (70)	118.0 ± 9.2 (47)

with certainty which eggs would be spawned during a given season, fecundity estimates were based only on counts of eggs 0.2 mm or larger in diameter, and were assumed to represent only viable eggs.

Mean age-specific fecundities ranged from 31,900 eggs per age-II female to 176,900 eggs per age-VI female (Table 25). Mean fecundity in the age-II to -VI population was 88,900 eggs per female. The relationship between total length (TL) in mm and fecundity (Fe) for females in 1975 through 1977 was

$$\log_{10} \text{Fe} = -4.392 + 4.140 \log_{10} \text{TL}; r^2 = 0.71.$$

The age-specific contribution to spawning was calculated from age-specific fecundity, age structure, and age at maturity. White perch exhibited a typical iteroparous reproductive strategy (Murphy 1968; Giesel 1976). Total reproductive effort in 1978, for example, was spread across age-II through age-VI females, but age-III and -IV females were the primary contributors and accounted for about 65% of the total potential spawn (Table 26).

Relative Abundance

The size of the fall population of age-I and older white perch was estimated in 1977 through 1979 from a mark–recapture program (Young et al. 1988). Petersen estimates of the September populations were 10.7×10^6 in 1977, 12.8×10^6 in 1978, and 6.7×10^6 in 1979.

The annual abundances of juveniles on 1 August in 1974–1979 ranged from 12×10^6 in 1974 to 50×10^6 in 1975. This relatively narrow range may reflect low variability in year-class strength for Hudson River white perch (Young et al. 1988). Average abundance across these 6 years was almost 32×10^6 juveniles.

Mortality

The abundance of early life stages in 1979 increased rapidly from early May through mid-June, declined rapidly through the end of July, and then continued to decline but at a slower rate through mid-December (Figure 28). Similar abundance patterns were observed in 1976 through 1978 (TI 1979, 1980a, 1980b). The rapid increase in abundance from early May to mid-June presumably reflected the recruitment of recently hatched larvae to the sampling gear.

Rates of population decline for post–yolk-sac larvae and early juveniles ranged from $10.0\% \cdot d^{-1}$ in 1979 to $12.8\% \cdot d^{-1}$ in 1978 (Table 27). Rates of decline for older juveniles were predictably lower and ranged from $0.5\% \cdot d^{-1}$ in 1976 to $0.7\% \cdot d^{-1}$ in 1977.

Annual total mortality rates (A) for age-I and older white perch derived from three cohort catch curves were 0.61, 0.57, and 0.68 for the 1973, 1974, and 1975 cohorts, respectively. Based on the recapture of 60 tagged fish by sport fishermen in 1976–1978, sport fishing for white perch in the Hudson River contributed less than 1% to total annual mortality (Table 28). The fishing mortality estimates were not adjusted for tag loss, handling-related mortality, or fisherman nonresponse. But even if these mortality estimates are biased low by a factor of 2 or more, sport-fishing exploitation of white perch in the Hudson River still has been relatively low. Commercial fishing pressure on the white perch population appeared to be even less. Only 12 tags were returned by commercial fishermen between 1976 and 1978. The closure of the

TABLE 22.—Mean total length (±SD, mm; N in parentheses) for male and female white perch, ages III–VII, collected in the vicinity of the Indian Point generating station, Hudson River, May and June 1975–1977 combined.

Sex	Age				
	III	IV	V	VI	VII
Male	153.8 ± 11.9 (93)	171.6 ± 10.3 (83)	180.0 ± 11.1 (81)	190.3 ± 13.6 (23)	233 (1)
Female	158.8 ± 11.1 (96)	173.4 ± 11.7 (144)	187.0 ± 14.7 (96)	199.4 ± 13.9 (19)	216.7 ± 25.5 (3)

TABLE 23.—Proportions of total catch represented by age-I and older white perch collected in bottom trawls near the Indian Point generating station, Hudson River, October–December 1974 to 1979.

	Year					
Age	1974 (N=755)	1975 (N=386)	1976 (N=1157)	1977 (N=284)	1978 (N=515)	1979 (N=179)
I	0.87	0.55	0.67	0.44	0.33	0.67
II	0.05	0.32	0.22	0.40	0.20	0.15
III	0.05	0.06	0.08	0.14	0.27	0.05
IV	0.02	0.05	0.03	0.02	0.15	0.09
V	0.01	0.02	0.01	<0.01	0.06	0.03
VI	<0.01	<0.01	0	0	<0.01	0

Hudson River white perch fisheries in February 1976 contributed to these low exploitation rates.

Discussion

The upriver and onshore movements of adult white perch in the Hudson River during April and May are similar to, but less extensive than, spring spawning movements reported in the Patuxent River, Maryland (Mansuetti 1961) and the Delaware River (Miller 1963). White perch prefer to overwinter in waters deeper than 10 m (AuClair 1956; TI 1981); hence, the magnitude and direction of spring movements could be strongly influenced by the proximity of overwintering areas to spawning sites. In the Hudson River, deep areas appropriate for overwintering are available throughout the estuary except for about 65 km of the upper zone (Cooper et al. 1988). In the Patuxent and Delaware rivers, important overwintering areas are located a substantial distance from preferred spawning sites, which may explain the extensive spring migrations in these rivers. White perch do not exhibit a definite spring spawning run from brackish to fresh water in Connecticut rivers (Whitworth et al. 1968; Marcy 1976a).

With allowances for geographical differences in seasonal water temperature regimes, the time of spawning for Hudson River white perch is similar to those of other white perch populations. Marcy (1976b) reported that white perch spawned during May and June in the Connecticut River when water temperatures ranged from 8.9 to 27°C; eggs were most abundant in shoal areas when water temperatures were between 11 and 21°C. White perch in the Patuxent River spawned in tidal fresh and slightly brackish water in late March to May when water temperatures were between 10 and 15°C (Mansueti 1964). In Nebraska, white perch reached spawning condition during May; most egg deposition occurred from late May to early June when water temperatures ranged from 14.4 to 23.3°C (Zuerlein 1981).

First-year growth of Hudson River white perch was influenced by water temperatures during the larval and early juvenile stages and by freshwater flows during July and August. Growth was enhanced during years when spring water temperatures were above average and summer flows were relatively low. Mansueti (1964) suggested a negative relationship between February–May rainfall and first-year growth of white perch from the Patuxent River. He concluded that heavy rainfall could depress growth by influencing turbidity, solar radiation, temperature, flow, and plankton production in the spawning and nursery areas. Mansueti (1964) reported a significant positive correlation between first-year growth and the number of days in early spring between the times that water temperature first reached 10°C and then

TABLE 24.—Percentages of white perch (N) collected near the Indian Point generating station, Hudson River, that were sexually mature in May and June 1972–1977.

	Year						
Age	1972	1973	1974	1975	1976	1977	Mean
Males							
II	25 (16)	50 (6)	77 (13)	32 (50)	46 (28)	66 (61)	49
III	100 (2)	67 (3)	92 (13)	80 (20)	74 (38)	97 (35)	85
IV	100 (8)	100 (5)	100 (5)	100 (38)	100 (27)	87 (15)	98
V	100 (6)	100 (4)	100 (1)	100 (27)	100 (35)	75 (16)	96
VI+	100 (2)	100 (2)	100 (2)	100 (5)	100 (14)	100 (4)	100
Females							
II	a	24 (17)	19 (26)	18 (60)	18 (33)	39 (36)	24
III	75 (8)	96 (28)	95 (18)	78 (23)	88 (25)	94 (32)	88
IV	100 (18)	96 (28)	95 (18)	95 (43)	95 (21)	96 (23)	96
V	100 (8)	100 (15)	100 (6)	100 (34)	100 (12)	100 (23)	100
VI+	100 (3)	100 (1)	a	100 (9)	100 (3)	100 (5)	100

[a] No samples were examined for these ages.

TABLE 25.—Mean fecundity (±SE, 10^3 eggs per female; N in parentheses) for white perch collected near the Indian Point generating station, Hudson River, April and May 1975–1978.

	Year				All years combined
Age	1975	1976	1977	1978	
II	[a]	32.4 (1)	31.5 (1)	[a]	31.9 ± 0.4 (2)
III	52.9 ± 4.2 (8)	58.4 ± 6.2 (8)	45.1 ± 5.6 (6)	60.7 ± 6.1 (6)	54.5 ± 5.5 (28)
IV	94.3 ± 10.3 (13)	81.3 ± 11.4 (11)	47.2 ± 4.8 (9)	71.6 ± 4.1 (14)	75.5 ± 8.3 (47)
V	104.5 ± 12.5 (8)	124.3 ± 16.9 (9)	103.8 ± 22.8 (3)	75.3 ± 7.4 (5)	105.7 ± 15.0 (25)
VI	153.9 ± 52.7 (2)	289.7 (1)	[a]	110.1 (1)	176.9 ± 44.5 (4)

[a] No fish of this age were examined.

15°C. He also suggested a negative relationship between first-year growth and an index of year-class strength that we did not observe in the Hudson River white perch population.

Compared to other estuarine populations, Hudson River white perch in 1975 to 1977 were about the same length at age as white perch in the James River (Virginia), York River (Virginia), and Albemarle Sound–Roanoke River, North Carolina (through age V), but smaller than white perch in the Thames River (Connecticut), Connecticut River, Delaware River, Patuxent River, and the Hudson River population in 1970 (Table 29). Length at age was substantially greater in several landlocked freshwater populations of white perch than in the Hudson River population (Table 29). The substantial size differences among the adjacent Hudson, Thames, and Connecticut river populations suggest that Hudson River white perch may be overcrowded in some years. Stunted white perch populations have been observed in the Potomac and Susquehanna rivers of Maryland (Richkus et al. 1980).

TABLE 26.—Age-specific contributions to potential egg deposition by female white perch in the Hudson River during 1978.

Age	(A) Relative abundance (age VI = 1)	(B) Percent sexually mature	(C) Relative mean fecundity (age VI = 1)	Index of age-specific contribution to spawn (A × B × C)	Percent of total egg deposition
II	84.3	20	0.18	303	18
III	27.8	82	0.31	707	42
IV	9.2	97	0.43	384	23
V	3.0	100	0.60	180	11
VI	1.0	100	1.00	100	6

Mansueti (1961) observed that longevity of white perch decreases from north to south along the Atlantic coast of North America. In Maine waters, age-XII and -XIII white perch are common; the state record was an age-XVIII individual (AuClair 1960). Age data for the Hudson River population are consistent with Mansueti's observation of a latitudinal gradient in longevity. The oldest white perch collected in the Hudson River were age XI. In the Patuxent River, Maryland, the oldest individuals collected were age X (Mansueti 1961). Conover (1958) reported that the oldest white perch collected in Albermarle Sound, North Carolina, were age VIII. The fast-growing Connecticut River population does not appear to conform to this north–south gradient; Marcy and Richards (1974) reported a maximum observed age of only VII.

The age structure of the Hudson River white perch population is similar to the age structure of populations in Quabbin Reservoir, Massachusetts (Taub 1966), Patuxent River (Mansueti 1961), and Lake Ontario (Sheri and Power 1968). No single age group was dominant in any of these four populations, suggesting that extreme variations in year-class strength are not characteristic of reasonably healthy white perch populations. The relatively low variability in year-class strength for Hudson River white perch contrasts sharply with the wide fluctuations in year-class strength of its congener, the striped bass, in the Hudson River and elsewhere (Klauda et al. 1980).

Based on our data and on Holsapple and Foster (1975), Hudson River white perch begin to mature at the same age and at about the same size as Patuxent River and Lake Ontario populations, but

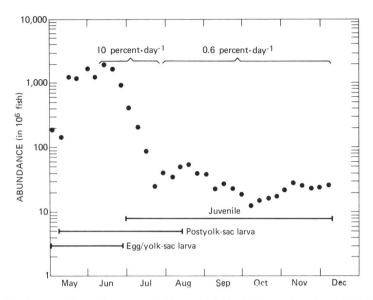

FIGURE 28.—Abundances of the early stages of white perch in the Hudson River, 1979. Periods when life-stages were present are indicated by horizontal lines near the bottom, and estimated daily mortality over two time intervals is indicated near the top.

maturity is not completed in the Hudson River population until a later age. All males and females in the Patuxent River were mature at ages II and IV respectively (Mansueti 1961). Busch and Heinrich (1982) reported that 32% of males and 8% of females collected in Lake Ontario were mature by age II, and that all age-IV fish were mature. Because of their rapid growth, white perch stocked in Nebraska matured at age I and at 102–127 mm TL, much younger but at about the same length as in the Hudson River and many other white perch populations (Zuerlein 1981). Bath and O'Connor (1982) collected Hudson River white perch in the vicinity of the Indian Point generating station and reported that "most fish were sexually mature by their second year," a different conclusion from ours and one not supported by published data.

Because of variability in egg diameters within a single white perch ovary and differences among workers in their criteria for identifying mature eggs, comparisons of fecundities among white perch populations are difficult to interpret. Even comparisons within the Hudson River population across years can be puzzling. Holsapple and Foster (1975) reported that the average number of eggs spawned by age-III females in 1972 was 20,676; the average for this age group during our 1975–1978 study was 54,500 eggs. The fecundity of age-III white perch in the Hudson River did not

likely increase by a factor of 2.6 between 1972 and 1975–1978. Methodology differences are a more plausible explanation.

Barnthouse et al. (1983) emphasized that a reliable index of white perch year-class strength is needed for the Hudson River population. They stressed that such an index should be based on data collected with different sampling gear. The index derived from the study data discussed in Young et al. (1988), and presented in this paper as the combined juvenile population estimate, was based on catches of juvenile white perch in two gears that sampled shore-zone and shoal habitats.

TABLE 27.—Rates of population decline for white perch larvae and juveniles in the Hudson River, 1976–1979.

Year	Time period	Rate of decline ($\%\cdot d^{-1}$)
1976	20 Jun–31 Jul	12.6
	1 Aug–11 Dec	0.5
1977	6 Jun–13 Aug	10.7
	14 Aug–29 Oct	0.7
1978	12 Jun–30 Jun	12.8
	27 Aug–23 Sep	0.2
	24 Sep–23 Oct	0.7
	24 Oct–18 Nov	0.9
1979	11 Jun–28 Jul	10.0
	29 Jul–15 Dec	0.6

TABLE 28.—Number of white perch tagged and released (M) in the Hudson River during spring (April–June) and fall (July–December), 1976–1978, and recaptured (R) by sport fishermen within 180 d after release.

Total length at release	Variable	Year							
		1976		1977		1978		All years combined	
		Spring	Fall	Spring	Fall	Spring	Fall	Spring	Fall
≤150 mm	M	4,501	10,737	5,710	1,939	2,279	4,498	12,490	17,174
	R	6	4	1	0	1	0	8	4
	(R/M) × 100	0.13	0.04	0.02		0.04		0.03	0.02
>150 mm	M	8,022	11,795	4,964	3,405	3,010	2,529	15,996	17,729
	R	26	7	12	0	1	2	39	9
	(R/M) × 100	0.32	0.06	0.24		0.03	0.08	0.24	0.05

This was the preferred index for this paper. Future studies in the Hudson River should strive to verify the reliability of this index for monitoring trends in white perch year-class strength.

Daily mortalities of white perch post–yolk-sac larvae and juveniles in the Hudson River are relatively low compared to most freshwater species and many marine species, but are generally similar to mortality rates for anadromous fishes (Dahlberg 1979). The calculated mortality rates for Hudson River white perch ($10.0–12.8\%\cdot d^{-1}$ for post–yolk-sac larvae and early juveniles, $0.5–0.7\%\cdot d^{-1}$ for older juveniles) are similar to daily mortality rates for these life stages of Hudson River striped bass (Dey 1981; TI 1981). We are not aware of any published reports of daily mortality rates for the early life stages in other white perch populations.

Annual total mortality rates for age-I and older white perch in the Hudson River (57 to 68%) are slightly higher than mortality rates reported for populations in the York River (St. Pierre and Davis 1972), the Patuxent River (Mansueti 1961), and the Delaware River near Artificial Island (Wallace 1971). Mortality rates in the Hudson River population, which occupies an estuary laden with domestic and industrial pollution (Mueller et al. 1982), are slightly lower than those for white perch in the James River, a heavily polluted system (St. Pierre and Davis 1972).

In summary, white perch use the entire tidal portion of the Hudson River estuary for spawning, nursery, and overwintering activities. The wide spatial and temporal distributions of the early life stages (eggs, larvae, juveniles) offer a plausible explanation for the relatively small annual variations in reproductive success. A widely dispersed distribution pattern would tend to minimize the effects of localized or temporary fluctuations in key environmental factors, abiotic or biotic, on early life stage growth and survival.

TABLE 29.—Mean back-calculated total lengths (mm) at annulus formation for several white perch populations (male, female, and unsexed fish combined). Published standard lengths (SL) were converted to total lengths (TL) by the equation TL = 1.57 + 1.2 SL.

Population	Age						
	I	II	III	IV	V	VI	VII
Estuarine							
Thames River[a]	91	159	204	238	256	261	
Connecticut River[b]	87	197	225	255	278	308	340
Hudson River[c]	72	121	156	172	184	194	225
Hudson River[d]	90	150	183	200	213	223	234
Delaware River[e]	83	134	158	174	186	196	206
Patuxent River[f]	89	137	164	183	198	219	239
James River[g]	75	120	150	174	192	208	225
York River[g]	78	118	146	172	192	213	233
Albemarle Sound[h]	74	112	150	183	211	234	
Fresh water							
Quabbin Reservoir[i]	91	155	208	234	253	269	282
Oneida Lake[j]	89	190	226	244	257	269	307
Bay of Quinte[k]	77	128	165	190	211	227	245
Lake Ontario[l]	87	148	193	217	235	252	269
Cross Lake[m]	91	142	168	170	216	208	241
Wagon Train Reservoir[n]	109	178	229	274			

[a] Whitworth et al. (1975).
[b] Marcy and Richards (1974).
[c] Present study.
[d] Bath and O'Connor (1982).
[e] Wallace (1971) from Marcy and Richards (1974).
[f] Mansueti (1961).
[g] St. Pierre and Davis (1972).
[h] Conover (1958).
[i] Taub (1966).
[j] Alsop and Forney (1962).
[k] Sheri and Power (1969) from Busch and Heinrich (1982).
[l] Busch and Heinrich (1982).
[m] Richards (1960) from Zuerlein (1981).
[n] Zuerlein (1981).

Acknowledgments

These studies were supported by Consolidated Edison Company of New York, Incorporated; Power Authority of the State of New York; Orange and Rockland Utilities, Incorporated; and Central Hudson Gas and Electric Corporation.

We are grateful to the men and women who labored countless hours during day and night to collect, process, and compile the data. Lawrence W. Barnthouse, Douglas S. Vaughan, and two anonymous referees provided constructive reviews of the manuscript. Jackie Klein typed the manuscript.

References[5]

Alsop, R. G., and J. L. Forney. 1962. Growth and food of white perch in Oneida Lake. New York Fish and Game Journal 9:133–136.

AuClair, R. P. 1956. The white perch, *Morone americana* (Gmelin), in Sebasticook Lake, Maine. Master's thesis. University of Maine, Orono.

AuClair, R. P. 1960. White perch in Maine. Maine Department of Inland Fisheries and Game, Augusta.

Barnthouse, L. W., R. J. Klauda, and D. S. Vaughan. 1988. Introduction to the monograph. American Fisheries Society Monograph 4:1–8.

Barnthouse, L. W., and W. Van Winkle. 1988. Analysis of impingement impacts on Hudson River fish populations. American Fisheries Society Monograph 4:182–190.

Barnthouse, L. W., W. Van Winkle, and D. S. Vaughan. 1983. Impingement losses of white perch at Hudson River power plants: magnitude and biological significance. Environmental Management 7:355–364.

Bath, D. W., and J. M. O'Connor. 1982. The biology of white perch, *Morone americana*, in the Hudson River estuary. U.S. National Marine Fisheries Service Fishery Bulletin 80:599–610.

Boreman, J., and R. J. Klauda. 1988. Distributions of early life stages of striped bass in the Hudson River estuary, 1974–1979. American Fisheries Society Monograph 4:53–58.

Busch, W. N., D. H. Davies, and S. J. Nepszy. 1977. Establishment of white perch, *Morone americana*, in Lake Erie. Journal of the Fisheries Research Board of Canada 34:1039–1041.

Busch, W. N., and J. W. Heinrich. 1982. Growth and maturity of white perch in Lake Ontario. New York Fish and Game Journal 29:206–208.

Conover, N. R. 1958. Investigations of white perch, *Morone americana* (Gmelin), in Albermarle Sound and the lower Roanoke River, North Carolina. Master's thesis. North Carolina State College, Raleigh.

Cooper, J. C., F. R. Cantelmo, and C. E. Newton. 1988. Overview of the Hudson River estuary. American Fisheries Society Monograph 4:11–24.

Dahlberg, M. D. 1979. A review of survival rates for fish eggs and larvae in relation to impact assessment. U. S. National Marine Fisheries Service Marine Fisheries Review 41(3):1–12.

Dey, W. P. 1981. Mortality and growth of young-of-the year striped bass in the Hudson River estuary. Transactions of the American Fisheries Society 110:151–157.

Giesel, J. T. 1976. Reproductive strategies as adaptations to life in temporally heterogeneous environments. Annual Review of Ecology and Systematics 7:57–79.

Greeley, J. R. 1937. Fishes of the area with annotated list. Pages 45–103 *in* A biological survey of the lower Hudson watershed. New York State Conservation Department, Supplement to 26th Annual Report, Albany.

Hardy, J. D., Jr. 1978. *Morone americana* (Gmelin), white perch. U.S. Fish and Wildlife Service Biological Services Program FWS/OBS-78/12 (volume 3):71–85.

Hergenrader, G. L. 1980. Current distribution and potential for dispersal of white perch (*Morone americana*) in Nebraska and adjacent waters. American Midland Naturalist 103:404–406.

Hergenrader, G. L., and Q. P. Bliss. 1971. The white perch in Nebraska. Transactions of the American Fisheries Society 100:734–738.

Holsapple, J. G., and L. E. Foster. 1975. Reproduction of white perch in the lower Hudson River. New York Fish and Game Journal 22:122–127.

Hutchison, J. B., Jr. 1988. Technical descriptions of Hudson River electricity generating stations. American Fisheries Society Monograph 4:113–120.

Jearld, A., Jr. 1983. Age determination. Pages 301–324 *in* L. A. Nielsen and D. L. Johnson, editors. Fisheries techniques. American Fisheries Society, Bethesda, Maryland.

Klauda, R. J., W. P. Dey, T. B. Hoff, J. B. McLaren, and Q. E. Ross. 1980. Biology of juvenile Hudson River striped bass. Marine Recreational Fisheries 5:101–123.

Klauda, R. J., P. H. Muessig, and J. A. Matousek. 1988. Fisheries data sets compiled by utility-sponsored research in the Hudson River estuary. Pages 7–88 *in* C. L. Smith, editor. Fisheries research in the Hudson River. State University of New York Press, Albany.

Larsen, A. 1954. First record of the white perch (*Morone americana*) in Lake Erie. Copeia 1954:154.

Lippson, A. J., and seven coauthors. 1980. Environmental atlas of the Potomac estuary. Maryland Department of Natural Resources, Power Plant Siting Program, Annapolis.

Mansueti, R. J. 1961. Movements, reproduction and mortality of the white perch, *Roccus americanus*, in the Patuxent estuary, Maryland. Chesapeake Science 2:142–205.

Mansueti, R. J. 1964. Eggs, larvae, and young-of-the-year white perch, *Roccus americanus*, with comments on its ecology in the estuary. Chesapeake Science 5:3–45.

Marcy, B. C., Jr. 1976a. Fishes of the lower Connecticut River and the effects of the Connecticut Yankee plant. American Fisheries Society Monograph 1:61–113.

Marcy, B. C., Jr. 1976b. Planktonic fish eggs and larvae

[5]See Table 1 for sources of legal documents and unpublished reports pertaining to the Hudson River.

of the lower Connecticut River and the effects of the Connecticut Yankee plant including entrainment. American Fisheries Society Monograph 1:115–139.

Marcy, B. C., Jr., and F. P. Richards. 1974. Age and growth of the white perch *Morone americana* in the lower Connecticut River. Transactions of the American Fisheries Society 103:117–120.

Mattson, M. T., J. B. Waxman, and D. A. Watson. 1988. Reliability of impingement sampling designs: an example from the Indian Point station. American Fisheries Society Monograph 4:161–169.

McFadden, J. T., Texas Instruments, and Lawler, Matusky & Skelly Engineers. 1978. Influence of the proposed Cornwall pumped storage project and steam electric generating plants on the Hudson River estuary, revised. Report to Consolidated Edison Company of New York.

Miller, L. W. 1963. Growth, reproduction and food habits of the white perch, *Roccus americanus* (Gmelin) in the Delaware River estuary. Master's thesis. University of Delaware, Newark.

Mueller, J. A., T. A. Gerrish, and M. C. Casey. 1982. Contaminant inputs to the Hudson–Raritan estuary. NOAA (National Oceanic and Atmospheric Administration), OMPA (Office of Marine Pollution Assessment) Technical Memorandum 21.

Muessig, P. H., J. B. Hutchison, Jr., L. R. King, R. J. Ligotino, and M. Daley. 1988. Survival of fishes after impingement on traveling screens at Hudson River power plants. American Fisheries Society Monograph 4:170–181.

Murphy, G. I. 1968. Patterns in life history and the environment. American Naturalist 102:391–403.

Normandeau Associates. 1985. Final report for the 1983–1984 Hudson River white perch stock assessment study. Report to Orange and Rockland Utilities.

Perlmutter, A., E. E. Schmidt, and E. Leff. 1967. Distribution and abundance of fish along the shores of the lower Hudson River during the summer of 1965. New York Fish and Game Journal 14:47–75.

Richards, W. J. 1960. The life history, habits and ecology of the white perch, *Roccus americanus* (Gmelin) in Cross Lake, New York. Master's thesis. Syracuse University, Syracuse, New York.

Richkus, W. A., J. K. Summers, T. T. Polgar, and A. F. Holland. 1980. A review and evaluation of fisheries stock management models, part I-text. Martin Marietta Corporation, Baltimore.

Ricker, W. E. 1975. Computation and interpretation of biological statistics of fish populations. Fisheries Research Board of Canada Bulletin 191.

Ruelle, R. 1977. Reproductive cycle and fecundity of white bass in Lewis and Clark Lake. Transactions of the American Fisheries Society 106:67–76.

Scott, W. B., and W. J. Christie. 1963. The invasion of the lower Great Lakes by the white perch, *Roccus americanus* (Gmelin). Journal of the Fisheries Research Board of Canada 20:1189–1195.

Sheppard, J. D. 1976. Valuation of the Hudson River fishery resources: past, present and future. New York Department of Environmental Conservation, Albany.

Sheri, A. N., and G. Power. 1968. Reproduction of white perch, *Roccus americanus* in the Bay of Quinte, Lake Ontario. Journal of the Fisheries Research Board of Canada 25:2225–2231.

Sheri, A. N., and G. Power. 1969. Validity of the scale method and back calculations for estimating age and growth of white perch, *Morone americana*. Pakistan Journal of Zoology 1:97–111.

Smith, C. L. 1985. The inland fishes of New York State. New York State Department of Environmental Conservation, Albany.

Smith, M. W. 1939. Fish populations of Lake Jesse, Nova Scotia. Nova Scotian Institute of Science 19:389–427.

Snyder, D. E. 1985. Fish eggs and larvae. Pages 165–197 *in* L. A. Nielsen and D. L. Johnson, editors. Fisheries techniques. American Fisheries Society, Bethesda, Maryland.

Sokol, R. R., and F. J. Rohlf. 1969. Biometry. W. H. Freeman, San Francisco.

St. Pierre, R. A., and J. Davis. 1972. Age, growth and mortality of the white perch, *Morone americana,* in the James and York Rivers. Chesapeake Science 13:272–281.

Taub, S. H. 1966. Some aspects of the life history of the white perch, *Roccus americanus* (Gmelin), in Quabbin Reservoir, Massachusetts. Master's thesis. University of Massachusetts, Amherst.

TI (Texas Instruments). 1975. First annual report for the multiplant impact study of the Hudson River estuary, volumes 1 and 2, text and appendices. Report to Consolidated Edison Company of New York.

TI (Texas Instruments). 1979. 1976 year class report for the multiplant impact study of the Hudson River estuary. Report to Consolidated Edison Company of New York.

TI (Texas Instruments). 1980a. 1977 year class report for the multiplant impact study of the Hudson River estuary. Report to Consolidated Edison Company of New York.

TI (Texas Instruments). 1980b. 1978 year class report for the multiplant impact study of the Hudson River estuary. Report to Consolidated Edison Company of New York.

TI (Texas Instruments). 1981. 1979 year class report for the multiplant impact study of the Hudson River estuary. Report to Consolidated Edison Company of New York.

Thoits, C. F., and J. W. Mullan. 1973. A compendium of the life history and ecology of the white perch *Morone americana* (Gmelin). Massachusetts Division of Fish and Game, Fishery Bulletin 24, Boston.

Trautman, M. 1957. The fishes of Ohio. Waverly Press, Baltimore.

Wallace, D. C. 1971. Age, growth, year class strength, and survival rates of white perch, *Morone americana* (Gmelin), in the Delaware River in the vicinity of Artificial Island. Chesapeake Science 12:205–218.

Whitworth, W. R., P. L. Berrier, and W. T. Keller.

1968. Freshwater fishes of Connecticut. State Geological and National History Survey of Connecticut, Bulletin 101:3–134.

Whitworth, W. R., D. R. Gibbons, J. H. Heuer, W. E. Johns, and R. E. Schmidt. 1975. A general survey of the fisheries resources of the Thames River watershed, Connecticut. Storrs Agricultural Experiment Station Bulletin 435:1–37.

Woolcott, W. S. 1962. Infraspecific variation in the white perch, *Roccus americana* (Gmelin). Chesapeake Science 3:94–113.

Young, J. R., R. J. Klauda, and W. P. Dey. 1988. Population estimates for juvenile striped bass and white perch in the Hudson River estuary. American Fisheries Society Monograph 4:89–101.

Zuerlein, G. 1981. The white perch in Nebraska. Nebraska Game and Parks Commission, Nebraska Technical Series 8, Lincoln.

American Fisheries Society Monograph 4:89–101, 1988

Population Estimates for Juvenile Striped Bass and White Perch in the Hudson River Estuary

JOHN R. YOUNG,[1] RONALD J. KLAUDA,[2] AND WILLIAM P. DEY[3]

Texas Instruments Incorporated
Ecological Services Group, Buchanan, New York 10511, USA

Abstract.—Population sizes for juvenile striped bass and white perch in the tidal portion of the Hudson River were estimated with Petersen mark–recapture and density-extrapolation methods during 1974–1979. The principle difficulty for mark–recapture estimates was that marked fish did not mix randomly with the rest of the population for several months; by the time valid recapture samples could be taken, some juvenile striped bass (but not white perch) had emigrated from the estuary. Potential biases of population estimates due to marking mortality, loss of marks, emigration, and natural mortality were evaluated. Annual Petersen estimates for juvenile fish ranged from 1.3×10^6 to 12×10^6 for striped bass and from 13×10^6 to 205×10^6 for white perch. Half-widths of 95% confidence intervals averaged 57% of the estimate for striped bass and 25% for white perch. Without large recapture samples from impingement collections, estimates would have been much less precise. In the density-extrapolation method, catches by beach seines and epibenthic sleds, which sampled known areas and volumes, were expanded to the total riverine habitat occupied by juveniles. Extrapolation estimates were adjusted for gear efficiencies and diel changes in fish distribution based on separate studies. Extrapolation estimates generally agreed with mark–recapture estimates for striped bass but were well below the mark–recapture estimates for white perch. Mark–recapture estimates were generally more precise than density extrapolation and appear to be more accurate; however, valid estimates could not always be calculated. Large recapture samples provided by impingement collections were key to success of the mark–recapture program. Density extrapolation can also produce useful estimates but is extremely sensitive to adjustments for sampling distribution and efficiency. Without a check against the mark–recapture estimates, the accuracy of density extrapolation is completely unknown.

Assessments of the impact of power generating facilities on fish populations often require estimates of absolute fish abundance. Large estuarine systems such as the tidal Hudson River are difficult to sample effectively for this purpose because, among other problems, of their large geographic extent, their diversity of habitats, their dynamic physics and chemistry, and the anadromous nature of some of their fish species. In this paper, we describe the approaches used to obtain population estimates for juveniles of key Hudson River species in the face of these problems. We emphasize the practical aspects of this research; the underlying theory has been adequately addressed by Seber (1973), Ricker (1975), and others. Several sampling and analytical problems

were not overcome during the Hudson River study, and we hope that examination of these shortcomings will lead to better impact assessments in the future.

The principal fish species of concern during the Hudson River study were striped bass, the most valuable sport and commercial species, and white perch, one of the most abundant species to which most of the population research was directed. These two species, though closely related taxonomically, have markedly different life histories in the Hudson River (Hoff et al. 1988; Klauda et al. 1988, both this volume). Perhaps the key difference for estimating juvenile population size is that striped bass emigrate from the study area at the end of the first growing season, but white perch are lifelong residents. Population estimates were also generated for Atlantic tomcod, but their unique life history (McLaren et al. 1988, this volume) necessitated a separate sampling program. For both striped bass and white perch, the goal was to estimate absolute abundance—numbers of individuals in the estuary—for juvenile fish, the life stage most vulnerable to impingement at power plants. Indexes of relative abundance

[1]Present address: Consolidated Edison Company of New York, Incorporated, Environmental Affairs, 4 Irving Place, New York, New York 10003, USA.

[2]Present address: The Johns Hopkins University, Applied Physics Laboratory, Environmental Sciences Group, Shady Side, Maryland 20764, USA.

[3]Present address: EA Engineering, Science and Technology, Rural Delivery 2, Goshen Turnpike, Middletown, New York 10940, USA.

TABLE 30.—Sampling gear used to estimate juvenile population sizes of Hudson River striped bass and white perch in mark–recapture and density-extrapolation programs.

Sampling Gear	Size	Time	Mark–recapture		Density-extrapolation stratum
			Marking	Recapture	
Beach Seine	30 m	Day	×	×	Shore zone
	61 m	Day–night	×	×	a
	152 m	Day–night	×	×	a
Box trap	1×1×2 m	Day–night	×	×	a
Bottom trawl	8 m	Day		×	a
	4 m	Day		×	a
Tucker trawl	1 m²	Night		×	Channel
Epibenthic sled	1 m²	Night		×	Bottom, shoal, channel
Impingement[b]		Day–night		×	a

a Not used for density-extrapolation estimates.
b At power plant intake screens.

would have required less sampling effort and would have provided useful information about population trends, but only estimates of absolute abundance allowed conditional mortality rates to be calculated for the "exploitation" caused by power plants (Barnthouse and Van Winkle 1988, this volume). Two types of estimates of absolute abundance were made annually for juveniles of each species during 1974–1979: one was derived from the recapture of marked fish, and the other was based on expansions of sampling catches per unit area or volume to the total estuarine area occupied by juveniles, here called density extrapolation.

Sampling Programs

Sampling programs used to estimate population sizes of juvenile fishes in the Hudson River estuary from 1974 through 1979 included beach-seine, bottom-trawl, epibenthic-sled, Tucker-trawl, box-trap, and impingement sampling (Table 30). These sampling programs were designed to determine distributions, abundances, movements, and life history characteristics of the fish fauna of the estuary; hence, some of the sampling designs were not necessarily optimal in terms of location and timing for obtaining population estimates. However, allocation of sampling effort was generally higher in areas where striped bass juveniles were more abundant because of the interest in this species. The beach-seine sampling program was the major source of fish that were marked; additional seine sampling in areas of high fish density was used within the mark–recapture study program to supplement the marking and recapture effort. All other sampling programs, including trawling and impingement sampling, provided additional recapture effort.

Beach Seine

The beach-seine survey provided data on abundance, distribution, and population characteristics of fishes in the shore zone. Samples were collected with a 30.5-m × 2.4-m seine (0.48-cm-mesh net in bunt). During this survey, 100 samples were collected biweekly during the day from April through June and September through December, and 100 samples were collected weekly during the day in July and August. In 1974, nighttime sampling was also conducted in some regions. The samples were allocated on a stratified random basis to each of the 12 regions (Figure 29). From 1974 through 1978, sample allocations were keyed to the distributions of juvenile white perch and striped bass in previous years. In 1979, samples were allocated to each region based on the amount of shore-zone area (0–3 m deep) in each region; the minimum regional allocation was five samples. Additional sampling with 30.5-m, 61-m, and 152-m seines was conducted solely for the mark–recapture effort in regions where striped bass and white perch were abundant, or where few fish had been marked during the beach-seine survey.

During sampling, one end of the seine was fixed at the shoreline while the boat was backed slowly away, perpendicular to the shore. When the entire seine had played out over the bow, the free end of the net was towed to shore so that the distance between the ends of the net at the shoreline was approximately 18 m for the 30.5-m net. Based on this methodology, each set was assumed to sweep 450 m².

Juvenile striped bass and white perch were fin-clipped (two fins per fish) and released at least 100 m from shore during the fall marking period (generally late August or September through November).

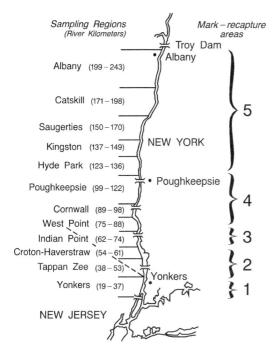

FIGURE 29.—Locations of 12 geographic regions and 5 mark–recapture areas (with river boundaries) used during the field sampling program, Hudson River estuary.

Summer months were excluded because previous studies had shown low survival rates for fish marked during these months (TI 1977). Results of 14-d survival tests of marked fish were used to adjust the monthly number of marked fish released for mortality caused by handling and marking.

The combinations of fins to be clipped were determined so that fish marked in each month (September, October, or November) and in each of the five mark–recapture areas (Figure 29) could be identified upon recapture. The combination of fins always consisted of one median fin and one lateral fin. The first dorsal fin was used as the median fin in September and the anal fin in October, because the hard-rayed fins regenerate more slowly than the soft-rayed second dorsal. Single fin clips and paired lateral fin clips were found to be unsuitable, particularly for white perch, which seem to have a relatively high incidence of naturally anomalous fins in the Hudson River (TI 1978a). Suspected fin-clipped recaptures were preserved for verification.

Bottom Trawls

Recapture effort for the marked fish in offshore areas was provided by a bottom trawl survey conducted annually in a 75-km segment between the Tappan Zee Bridge and Poughkeepsie during 1974–1979. The otter-type trawl had a 7.8-m head-rope length and 3.8-cm stretched-mesh netting with a 1.3-cm stretched-mesh cod end cover. Trawl doors were 0.8 m wide $\times$ 1.2 m long. Tows were conducted at a speed of 1.3 $\mathrm{m\cdot s}^{-1}$ for 5 min against the prevailing current. During the trawl survey, which was designed so that the distribution and abundance of fish in deeper areas of the middle estuary could be defined, 38 fixed stations in areas more than 6 m deep were sampled biweekly during the day from April through November. All juvenile striped bass and white perch collected were checked for fin clips.

Additional recapture effort was provided in 1979 by a second bottom trawl survey conducted with 3.6-m (head-rope length) trawl constructed of a 3.8-cm stretched-mesh body and a 3.3-cm stretched-mesh cod end with a 0.6-cm stretched-mesh liner. Trawl doors were 0.3 m wide and 0.6 m long. Tow speed was 1.5 $\mathrm{m\cdot s}^{-1}$ against the prevailing current and tows lasted for 10 min. Approximately 100 randomly selected daytime samples were taken from 12 regions in 1.5–6-m-deep offshore areas biweekly from April through mid-December. Samples were allocated to each region in proportion to shoal volume (areas 6 m or less in depth).

Epibenthic Sleds

Epibenthic sleds were used to sample the shoal (3–6 m deep) and bottom strata (deepest 3 m in water greater than 6 m deep) at night from early July through mid-December in 1974–1979. Sleds had a fixed 1-m² mouth opening and were equipped with a 3-mm-mesh net and a conical fyke inside the cod end. During the standard 5-min tow, an electronic flowmeter measured tow speed and a digital flowmeter measured the volume of water sampled. Tow speed was 1.5 $\mathrm{m\cdot s}^{-1}$. The sled was deployed in 7 of the 12 regions (Yonkers through Poughkeepsie) in 1974–1978 and in all 12 regions in 1979. Samples (100 per biweekly survey) were randomly allocated to the shoal and bottom strata in each region based on the distribution of juvenile striped bass in previous years.

Box Traps

Box traps were used to supplement marking and recapture effort when juvenile fish were abundant in the shore zone. Traps were 1 $\times$ 1 $\times$ 2 m with two 6-m wings (1-cm mesh) and one 15-m lead (1-cm mesh). Trapping effort was generally conducted in the middle areas of the estuary between river kilometers 54 and 98.

Tucker Trawls

During early August, 1-m² Tucker trawls (0.5-mm mesh) were used to sample in the channel stratum, from the water surface to within 3 m of the bottom in water greater than 6 m deep. This gear actually sampled fish eggs and larvae as part of the April–August ichthyoplankton program and was used to estimate juvenile abundance simply because it provided the only data available for the channel stratum. After use of Tucker trawls was discontinued in August, channel densities were estimated from bottom or shoal sampling with epibenthic sleds. Due to their small mouth openings, deployment locations, and 1-m·s⁻¹ tow speed, Tucker trawls did not sample juvenile striped bass and white perch very effectively.

Impingement Sampling

Impingement sampling was conducted daily at the Indian Point station (Mattson et al. 1988, this volume) and on a scheduled basis at Bowline Point (Hutchison and Matousek 1988, this volume) and other plants. All striped bass and white perch collected were counted and checked for fin clips.

Mark–Recapture Estimates

The Petersen method was chosen to estimate sizes of the juvenile populations because it is flexible with regard to the timing of the recapture sample. Unlike multiple census estimates, it allows time for the error caused by nonrandom mixing of the marked fish into the population to decline. Chapman's modified Petersen estimate (Ricker 1975) is statistically unbiased in most situations when numbers of recaptures are small, i.e., less than 10:

$$N = \frac{(M + 1)(C + 1)}{R + 1};$$

N = adjusted Petersen population estimate;
M = number of fish marked;
C = number of fish examined in the recapture sample;
R = number of marked fish recaptured.

Confidence intervals around the estimates were calculated by considering the number of recaptures as a Poisson variable (Ricker 1975).

Mark–recapture methods of population estimation require that several basic assumptions are met to yield unbiased estimates. Very few mark–recapture studies can meet all of the necessary assumptions; thus, adjustments must be made to the data so that the assumptions are at least approximately true. Many of these adjustments have been reported in the fisheries literature (see Seber 1973 and Ricker 1975 for thorough reviews of mark–recapture methods), and are now well-accepted techniques. The basic assumptions required for the Peterson estimate, and methods used to meet them for the Hudson River study, are outlined in the following paragraphs.

Assumption (1). Marked and unmarked fish suffer the same mortality. Short-term mortality of marked fish was assessed by holding samples of marked and unmarked fish (generally at least 20 of each) in laboratory tanks for 14 d. The selection of specific test results to use for adjustment factors was somewhat subjective. Often catches sufficiently large to use for mortality tests could not be obtained near the laboratory facility, and test fish had to be transported back to the laboratory from distant regions. The stress of the long trip was often greater than that of the marking process, resulting in high mortality of marked and control fish. Substitution of adjustment factors from previous months or years was sometimes necessary.

Monthly adjustment factors for each species generally were 0.75 or greater. No adjustment was attempted for errors caused by sustained higher mortality of marked fish. Reduced survival of fish marked by fin clips has been documented by Coble (1971) and Mears and Hatch (1976), but Shetter (1952) found no significant differences in the growth or survival of marked and unmarked fish.

Assumption (2). Marked and unmarked fish are equally vulnerable to fishing or sampling. Unlike many tags, fin clips are generally not thought to affect the catchability of marked fish, particularly when sufficient time is allowed between marking and recapture sampling.

Assumption (3). Marked fish do not lose their marks. Errors due to continuous mark loss (in this case, fin regeneration) were reduced by marking the fish near the end of the growing season, when regeneration is slow, and by completing the recapture effort before extensive growth and fin regeneration would be likely. In addition, use of the spiny-rayed first dorsal and anal fins in the early marking months helped prolong the recognizability of recaptured fish.

The need both to mark in the fall and to clip spiny fins in warmer months was demonstrated by the rapid regeneration of clipped second dorsal fins (EA 1985). For hatchery-reared striped bass, recognizability of second-dorsal clips after 6 weeks was only 5% for fish marked in August but approximately 60% for fish marked in September.

Assumption (4). Marked fish either become randomly mixed with unmarked fish or the recapture samples are selected randomly from the entire population. The assumption that marked and unmarked fish mix randomly, or that the recapture sample is taken randomly from the entire population, was the most difficult assumption to satisfy. Other studies have encountered problems with randomization, even in systems much smaller than the Hudson River (Schumacher and Eschmeyer 1943; Cooper 1952; Swingle and Smitherman 1966; Van Den Avyle 1976). Nonrandom distribution of marked fish or of recapture effort distort the fraction of marked fish in the recapture sample (*R/C*). Population estimates derived from nonrandom samples will be biased.

To satisfy the randomization assumption, it is not necessary that both the marking sample and the recapture sample be random, one of the two is sufficient. However, in order to mark sufficient numbers of fish, it was necessary to concentrate the marking efforts in areas where fish were abundant. In the Hudson River, it was possible to seine effectively only on beaches without obstructions; thus, only a small, nonrandomly selected fraction of the total available shore-zone habitat could be sampled at all. Although fish were released at least 100 m off shore after they were marked, this procedure was not sufficient to promote adequate mixing of marked and unmarked fish. Recaptures of marked fish during the fall generally indicated that the fish had moved very little, not enough to assure that they had dispersed randomly throughout the population. In 1977, for example, very little movement of marked striped bass between areas was detected during September–November. Of 1,010 recaptured fish, only 38 (4%) were found outside the area in which they were originally marked, even though emigration was suspected to have begun in late summer. This overall lack of movement was consistent in all years.

When the mixing of marked and unmarked fish is slow, it may be possible to delay the recapture effort until, or to select an unbiased subset of the recapture data when, *R/C* values have stabilized.

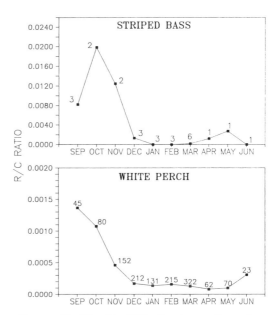

FIGURE 30.—Monthly *R/C* ratios (proportions of recaptured marked fish, *R*, in the catch, *C*) for juvenile striped bass and white perch marked in the Hudson River estuary during September–November 1979 and recaptured during September 1979–June 1980. Monthly catches (*C*, in thousands) are indicated above the data points.

However, a delay to allow more thorough mixing may increase the effect of other types of error.

The proportions of marked fish in samples (*R/C*) during the fall and early winter indicated that mixing of marked and unmarked fish was not sufficient for an unbiased population estimate (Figure 30). The observed *R/C* values climbed rapidly during marking, then declined rapidly over the months following marking. Ideally, the proportion of marked fish in recapture samples should decline as the fish become mixed, and should stabilize at the true proportion of marked fish in the population (*M/N*). Because *R/C* values continued to decline during the fall, no mark–recapture estimate would be appropriate in this time period.

Two lines of supplementary evidence, in addition to the lack of demonstrated interregional fish movement, supported the view that marked striped bass and white perch did not mix rapidly with the unmarked ones. First, there were hints that the fish may home. Both species were caught primarily in beach seines, and any tendency of the fish to return to the place of initial capture could greatly inflate *R/C* values from field sampling. Such a tendency to return to the place of initial

TABLE 31.—Recapture rates (R/M)[a] during December 1978–June 1979 for juvenile white perch fin-clipped in the Hudson River estuary during September–November 1978.

Marking area (river kilometers) or statistic	Marking month								
	September			October			November		
	M	R	R/M	M	R	R/M	M	R	R/M
1 (19–37)	3	1	0.333	3	1	0.333	9	3	0.333
2 (38–61)	5,458	66	0.012	1,073	37	0.034	270	30	0.111
3 (62–74)	1,553	54	0.035	1,144	102	0.089	1,554	111	0.071
4 (75–122)	2,412	44	0.018	1,244	75	0.060	511	38	0.074
5 (123–243)	468	11	0.024	58	3	0.052	32	1	0.031
Total	9,894	176	0.018	3,522	218	0.062	2,376	183	0.077
χ^2			53.8			32.5			14.4
P			0.001			0.001			0.001

[a] M = number of fish marked, adjusted for marking and handling mortality of 0% in September, 30% in October, and 2.5% in November, and for fish recaptured during September through November 1978; R = number recaptured during December 1978–June 1979.

capture is well documented for centrarchids (e.g., Larimore 1952; Hasler and Wisby 1958) and some evidence exists for homing in other *Morone* species beside the two considered here (Hasler et al. 1958; Horrall 1961). In the Hudson, five juvenile striped bass were recaptured at the site of original capture after they had been transported approximately 8 km upriver, from Hastings to the Tappan Zee Bridge (Byron Young, New York State Department of Environmental Conservation, personal communication). Second, difference in R/C ratios between field (primarily beach-seine) samples and impingement collections indicated that mixing was not complete even after several months.

When the R/C ratio changes rapidly, it is not possible to obtain a reliable estimate of population size. We found, however, that the R/C ratio generally stabilized after several months. For white perch, impingement collections provided a large recapture sample that could be used to evaluate mixing. By January, R/C values generally had declined to a low level and remained relatively stable over several months (Figure 30). Movement data for this period also indicated that mixing had occurred, because recapture rates for fish marked in different regions (R/M) were also relatively stable (Table 31).

The assumption of randomization is supported if either R/C ratios are constant through time or R/M ratios are similar for fish marked in all areas of the river. In practice, small sample sizes for the recapture sample (C) in particular regions or sampling gear, or small numbers of fish marked (M) in particular regions (Table 31), resulted in small expected values for χ^2-tests of the equality of R/C or R/M ratios. When sample sizes were large,

however, the test nearly always indicated significant differences among ratios $(P \leq 0.05)$. Thus, we were forced either to make no estimate at all due to the significant χ^2 result, or try to find the recapture period that produced the best compromise among mixing, adequate sample sizes, and satisfaction of other assumptions.

Assumption (5). All marked fish in the recapture sample are recognized. Incomplete reporting of recaptured fish was minimized by using only recaptures from sources (environment contractors) that were known to report all recaptures. Field procedures were designed so that any fish with abnormal-appearing fins was returned to the laboratory for verification by a single individual who applied specific criteria to decide whether or not each fish actually had been marked during the program. Although these steps caused many such fish to be rejected, this additional effort was preferable to the errors that could arise by allowing each field person to judge the validity of suspected recaptures.

Occasionally even microscopic examination was not sufficient to confidently decide whether a fish had been clipped or simply had a fin anomaly. X-ray examination, with mammography equipment, was then used to examine both the external and internal fin structures, which sometimes helped to arrive at a decision. Quality-control procedures (methodical reexamination of samples) for recognition of marked fish from Indian Point impingement collections, which contributed a large portion of the recapture sample, were also applied in laboratory processing.

Assumption (6). Recruitment to the population is negligible during the mark–recapture study.

Population estimates were made for the youngest age-group in the population, which could be easily distinguished from older fish on the basis of length. Data collection was completed prior to the appearance of the next year class. Fish from other populations of striped bass and white perch would not be found in the Hudson during the juvenile phase.

Assumption (7). No emigration occurs from the study area (i.e., the system is closed). The assumption of a closed system is a likely source of error in estimating population size of juvenile fish in estuaries. Many anadromous fish, like striped bass, use the estuary principally during the first growing season, then move out of the estuary to continue their life cycle. In a large estuary like the Hudson River, it is extremely difficult to mark and recapture a sufficient number of fish between the time the fish are large enough to mark and the time emigration begins. If the sampling effort is intensive enough to capture a sufficient number of fish, adequate randomization (mixing of marked and unmarked fish) may not have occurred.

In the Hudson River, striped bass reached a size at which they could be safely marked, about 50 mm, near the end of August, when water temperatures were also low enough that excessive marking-related mortality did not occur. This is also approximately the time at which emigration from the river begins (Dey 1981), so it was not possible to obtain a mark–recapture estimate of year-class size prior to emigration. The month-specific fin clips, however, allowed an evaluation of emigration and permitted selection of marking data that could satisfy the closed-system assumption. Separate estimates could be calculated for each month if sufficient numbers of fish were recaptured.

We evaluated the potential bias to population estimates due to emigration of juvenile striped bass by examining the ratio of recaptured fish to total number marked (R/M) for each marking month. During the fall, recapture rates generally were highest for fish marked and released in September, then declined progressively through October and November. However, in the following 6 months, January–June, recapture rates were generally highest for fish marked and released in November and lowest for those marked in September. For example, in the fall of 1978, recapture rates were 9.8%, 7.6%, and 3.2% for the three marking months. This pattern was expected because fish marked in September were exposed to much more recapture effort after they were marked than were fish marked in November. In the following 6-month period, however, recapture rates were 0.5%, 1.4%, and 1.6% for the fish marked in September, October, and November. We believe that these differences in the recapture rates reflect, to a great degree, the differing proportions of the fish present in each of the fall months that remained in the estuary over the winter. This pattern is also consistent with greater mortality of fish marked earlier in the season, and was also seen for white perch, which do not emigrate from the river. For white perch, however, the differences in recapture rates among the marking months were generally less distinct.

To overcome the differences in monthly recapture rates, whether due to emigration or to mortality, separate estimates were made for each month when the number of recaptures was sufficient. In some years, estimates could be made for every marking month individually; in other years certain months were omitted or pooled. If R/C ratios never stabilized sufficiently or too few recaptures were available, no estimates were made. Selection of months to use for marking and recapture effort was made each year based on the stability of R/M and R/C ratios.

Striped bass population estimates ranged from 1.3 million for the October–November period of 1974 to 12.0 million in September 1978 (Table 32). For 1978, the only year when separate monthly estimates could be calculated, the declining number of fish in the river due to emigration and mortality was clearly indicated. White perch population estimates were larger than those for striped bass, ranging from 12.9 million juveniles in November 1977 to 205.1 million in September 1976 (Table 33). September population estimates were more often calculated for white perch because recaptures from September months were higher than for striped bass even though the number of fish marked was not substantially greater. Precision of the mark–recapture estimates was generally good; confidence interval half-widths averaged 57% of the estimate for striped bass and 25% for white perch. The precision of these estimates, however, was due primarily to the large recapture sample provided by impingement collections, particularly for white perch. With R/M and R/C ratios as low as they were in this study, the amount of field sampling to obtain sufficient recaptures for usable population estimates without impingement data would have been prohibitive.

TABLE 32.—Mark–recapture data for population estimates (N) for juvenile Hudson River striped bass; CI is 95% confidence interval.

Marking				Recapture in next year[a]				Estimate (millions of fish): N (CI)
Year and month	Number marked	14-D survival	Adjusted number marked	Recapture period[b]	Marks available	Recapture sample	Recaptures	
1974								
Aug–Sep		0.75–1.0	4,731					
Oct–Nov	4,451	0.77	3,427	Mar–Jun	3,281	3,119	7	1.3 (0.7–2.6)
1975								
Aug–Sep		0.84–1.0	8,520					
Oct	4,612	0.77	3,551 ⎱	Mar–Jun	5,757	8,186	20	2.2 (1.5–3.6)
Nov	3,114	0.77	2,398 ⎰					
1976								
Sep	4,449	0.95	4,226					
Oct	2,930	0.77	2,256					
Nov	2,124	0.77	1,635					
1977								
Sep	6,624	0.95	6,293					
Oct	6,788	0.75	5,091 ⎱	Dec–Apr	7,718	9,109	9	7.0 (3.9–14.1)
Nov	3,777	0.75	2,833 ⎰					
1978								
Sep	10,076	0.95	9,522	Jan–Jun	8,588	63,000	44	12.0 (9.0–16.4)
Oct	2,241	1.00	2,241	Jan–Jun	2,071	63,000	28	4.5 (3.0–6.7)
Nov	1,143	1.00	1,143	Jan–Jun	1,106	63,000	18	3.7 (2.4–6.0)
1979								
Sep	728	0.40	291					
Oct	537	1.00	537 ⎱	Dec–Jun	1,054	16,600	7	2.2 (1.1–4.6)
Nov	545	1.00	545 ⎰					

[a] Marking months without corresponding recapture values yielded insufficient or inappropriate data for a population estimate. Braces group marking months for which recapture data were pooled to produce a population estimate.

[b] Some recapture periods included December of the marking year.

Density Extrapolation

Density-extrapolation estimates were calculated by multiplying the mean catch per area swept for beach-seine samples or catch per volume for epibenthic-sled samples by the estimated strata surface area or volume to obtain a standing crop for each stratum (Figure 31) in each region sampled. For regions in which the shoal or bottom stratum was not sampled, density estimates for sampled strata were substituted. The standing crops were adjusted to account for differences in diel distribution of sampling effort (beach-seine sampling was done during daytime and sled sampling at night) and catch efficiency of the gear. Combined standing crops were obtained by summing across the different regions and sampling strata, and standard errors of the estimates were calculated by the methods of TI (1981).

Catch efficiency of the 30.5-m beach seine was estimated from experimental seining conducted inside an enclosure (TI 1978b, 1979). Density estimates from seine samples inside a 50-m × 50-m enclosure were compared to mark–recapture estimates of the same enclosed population. Estimated night catch efficiency was 25.5% for juvenile striped bass and 18.2% for white perch (TI 1979). Catch

efficiency of the epibenthic sled was examined by comparing sample densities at tow speeds of 1.5 and 3.0 m·s^{-1}. However, catch data were too variable to distinguish a change in efficiency at higher speed (TI 1980a). The apparently conservative (high) value of 50% was selected based on the range of efficiencies reported by Kjelson and Colby (1977). Seine catches were also multiplied by the average ratio of night and day catches from a comparison study conducted in 1974. Night:day ratios were 2.136 for striped bass and 1.685 for white perch. The combination of efficiency factors used here represents the culmination of several studies and was applied to 1975–1979 data.

Density-extrapolation estimates generally produced a declining trend in population abundance over the fall months. For striped bass, peak abundance usually occurred in mid or late August and ranged from about 18 to 50 million juveniles (Figure 32). Patterns of population decline, reflecting both mortality and emigration, were variable among years, but population estimates had usually stabilized by November at less than 5 million. When mark–recapture estimates were also available, the trends in abundance were similar; however, the density-extrapolation estimates

TABLE 33.—Mark–recapture data for population estimates (N) for juvenile Hudson River white perch; CI is 95% confidence interval.

Year and month	Number marked	14-D survival	Adjusted number marked	Recapture period[b]	Marks available	Recapture sample	Recaptures	Estimate (millions of fish): N (CI)
	Marking			Recapture in next year[a]				
1974								
Aug–Sep		0.67–0.83	2,034					
Oct–Nov	3,032	1.00	3,032	Jan–Mar	2,994	89,679	13	19.2 (11.6–34.0)
1975								
Aug–Sep		0.70–0.83	14,407					
Oct	6,461	0.95	6,138	Jan–Jun	6,060	210,338	30	41.1 (29.1–60.1)
Nov	3,583	0.95	3,404					
1976								
Sep	15,959	1.00	15,959	Jan–Jun	14,968	671,387	48	205.1 (155.5–276.9)
Oct	7,095	0.95	6,740					
Nov	2,175	0.95	2,066					
1977								
Sep	5,287	1.00	5,287	Dec–May	5,014	495,500	61	40.1 (31.3–51.2)
Oct	7,085	0.75	5,314	Dec–May	5,212	495,500	95	26.9 (22.1–32.8)
Nov	4,920	0.75	3,690	Dec–May	3,618	495,500	138	12.9 (10.9–15.2)
1978								
Sep	10,903	1.00	10,903	Dec–Jun	10,394	1,300,000	176	76.3 (65.9–88.4)
Oct	5,293	0.70	3,705	Dec–Jun	3,523	1,300,000	218	20.9 (18.3–23.9)
Nov	2,501	0.98	2,439	Dec–Jun	2,375	1,300,000	183	16.8 (14.5–19.4)
1979								
Sep	3,240	0.20	648					
Oct	4,763	0.58	2,739	Dec–Jun	2,659	1,300,000	52	65.2 (50.0–85.1)
Nov	3,651	0.89	3,235	Dec–Jun	3,208	1,300,000	78	52.8 (42.4–65.7)

[a] Marking months without corresponding values yielded insufficient or inappropriate data for a population estimate.
[b] Some recapture periods included December of the marking year.

were often less precise than the mark–recapture estimates, as indicated by the wider approximate 95% confidence limits.

For white perch juveniles, the agreement between the extrapolated standing crops and mark–recapture estimates was poor (Figure 33). Mark–recapture estimates were generally far higher than density-extrapolation estimates, which were more erratic than the density-extrapolation estimates for striped bass. White perch often exhibited a late-fall increase in the density-extrapolation estimate due to a rise in epibenthic-sled catches. The extreme difference between the mark–recapture estimates and density extrapolation in September suggests that seine efficiency may have been substantially overestimated. Previous comparisons based on a 6% seine efficiency (TI 1978b) produced much closer agreement between estimates (TI 1980b, 1980c).

Conclusions

Mark–recapture estimates for juvenile fishes in the Hudson River estuary were difficult to make. Although it is generally better to restrict mark–recapture studies to a short time period, the large geographic extent of the estuary, wide distribution of the fish, and large population sizes forced

marking and recapture efforts to extend over several months. Use of time and area-specific fin clips allowed the needed flexibility in choosing appropriate time periods for valid population estimates. This flexibility was a key factor that contributed to the success of the program, because data were sufficiently different each year to force adjustments during data analysis.

The mark–recapture program was able to generate acceptable estimates both for white perch,

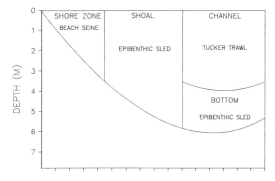

FIGURE 31.—Diagrammatic representation of a Hudson River cross-section depicting strata designations and sampling gear for density-extrapolation estimates of fish population sizes.

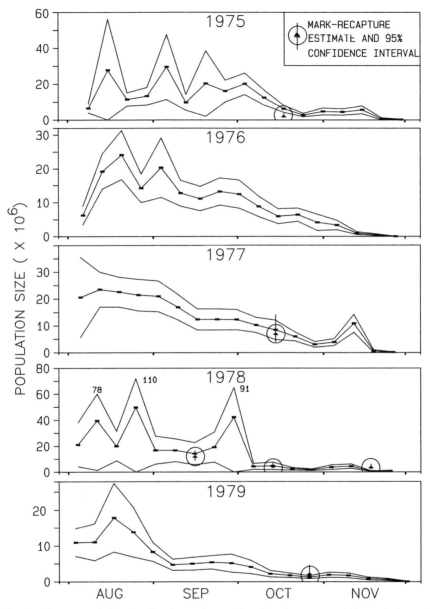

FIGURE 32.—Density-extrapolation (small points and curves) and mark–recapture (▲) population estimates with approximate 95% confidence limits for juvenile striped bass in the Hudson River estuary, 1975–1979. In the 1978 panel, numbers represent the (off-scale) confidence bounds.

which are lifelong residents of the estuary, and for the anadromous striped bass. Emigration of juvenile striped bass during the marking period generally prevented population estimates for the entire cohort and restricted estimates to those fish remaining in the river. Again, the time-specific nature of the fin clips allowed valid estimates for fish that remained in the river.

Reliance solely upon field sampling would not have provided sufficient number of recaptures, subsequent to mixing of marked and unmarked fish, for valid estimates. The availability of impingement collections, which provided relatively large numbers of fish for the recapture sample, was the principal reason for the program's success. The large recapture sample was necessary

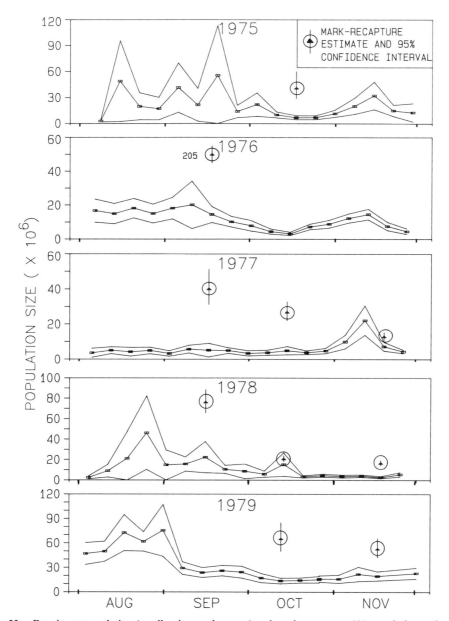

FIGURE 33.—Density-extrapolation (small points and curves) and mark–recapture (▲) population estimates with approximate 95% confidence limits for juvenile white perch in the Hudson River estuary, 1975–1979. In the 1976 panel, the number represents the (off-scale) confidence bounds.

due to the small fraction of the population that was marked.

Slow mixing of marked and unmarked fish caused the recapture sampling effort to extend as long as 10 months beyond marking. This is the practical limit for fin clips because fin regeneration makes marked fish increasingly hard to recognize as the fish begin rapid growth the following summer. The tendency for striped bass and white perch to move downriver during the winter was the key to obtaining a reasonably random recapture sample, even though a vast majority of the sample was taken from impingement collections. Movements during the warmer months were not sufficient to randomize the marked and unmarked fish.

Density-extrapolation estimates, which included adjustments for diel distribution differences and sampling gear efficiency, generally agreed well with the mark–recapture estimates for striped bass and helped fill in periods for which mark–recapture estimates could not be made. When the two methods did not agree, as often occurred for white perch, no obvious reason was apparent for the discrepancy, although overly optimistic estimates of catch efficiency are suspected. In situations as this, the most reasonable estimate must be selected based on other information and experience.

Recommendations

For future impact assessments, researchers who need to estimate absolute population size may be faced with a choice between mark–recapture and density-extrapolation estimates. If the required assumptions are satisfied, mark–recapture techniques can provide an accurate and precise estimate of the population, though our experience on the Hudson River has shown that their success is difficult to predict. Density-extrapolation techniques require standardized quantitative sampling in randomized designs, as well as estimates of catch efficiency. Without an independent estimate of population size, accuracy of density extrapolation is difficult to assess.

Researchers considering mark–recapture analyses to estimate estuarine juvenile populations should note the following points.

(1) Prior knowledge of seasonal movement and emigration patterns of target species is necessary to effectively plan the temporal and geographic extent of the program. Natural movements by the fish should be substantial if marked fish are to become adequately distributed within the population. For anadromous species, estimates of the population prior to emigration may not be possible.

(2) The target species must be hardy enough to withstand the marking process. Fin clips are easily applied and effective marks that may not greatly affect survival. Tests should be done to assess short-term marking-related mortality.

(3) Regeneration of fin clips, or loss of other types of tag, must also be assessed if the study will extend over periods of rapid fish growth. Use of spiny-rayed fins for marking will reduce regeneration rates.

(4) Combinations of fins used for marking should be varied among regions and time periods to assist in evaluating assumptions.

(5) Large numbers of marked fish must be recaptured, preferably from a variety of locations and with different sampling gear than was used for the marking effort. Single-location capture, such as by impingement at industrial water intakes, can meet mark–recapture needs for the recapture sample if prior movements have dispersed marked fish within the population.

(6) A large effort will have to be committed to a program that may supply little useful information on species other than the target species. During dedicated mark–recapture sampling, priority must be given to marking, or examining, as many of the target species as possible. Accurate counts and life history data for other species may not be compatible with the mark–recapture goals. Nonrandom distribution of sampling effort may also limit the utility of data.

Key factors to consider in evaluating the potential success of a density-extrapolation program are the following:

(1) Density extrapolation requires prior knowledge of all habitats occupied by the target population and an effective sampling gear that can be used in a standardized quantitative manner in each habitat.

(2) Sampling must be done according to a random or stratified random design and must be intensive enough within each stratum to assess precision.

(3) Special studies should be designed to assess catch efficiency of the gear. Precise estimates of efficiency may not be practical but the order of magnitude of efficiency can at least be determined.

(4) The sampling effort should be repeated several times to determine whether or not a population trend follows natural patterns of mortality and emigration.

(5) A mark–recapture analysis should be considered for at least a one-time calibration of the density-extrapolation method.

Acknowledgments

These studies were funded by Central Hudson Gas and Electric Corporation, Consolidated Edison Company of New York, Incorporated, New York Power Authority, and Orange and Rockland Utilities, Incorporated. We also acknowledge the contributions of many who participated in the collection, analysis, and presentation of the data contained in this manuscript.

References[4]

Barnthouse, L. W., and W. Van Winkle. 1988. Analysis of impingement impacts on Hudson River fish pop-

---

[4]See Table 1 for sources of legal documents and unpublished reports pertaining to the Hudson River.

ulations. American Fisheries Society Monograph 4:182–190.

Coble, D. W. 1971. Effects of fin clipping and other factors on survival and growth of smallmouth bass. Transactions of the American Fisheries Society 100:460–477.

Cooper, G. P. 1952. Estimation of fish populations in Michigan lakes. Transactions of the American Fisheries Society 81:4–16.

Dey, W. P. 1981. Mortality and growth of young-of-the-year striped bass in the Hudson River estuary. Transactions of the American Fisheries Society 110:151–157.

EA. 1985. Hudson River striped bass hatchery 1984 overview. EA Engineering, Science and Technology, Middletown, New York.

Hasler, A. D., R. M. Horrall, W. J. Wisby, and W. Braemer. 1958. Evidence for a sun-orientation mechanism in fishes. Limnology and Oceanography 3:353–361.

Hasler, A. D., and W. Wisby. 1958. The return of displaced largemouth bass and green sunfish to a "home" area. Ecology 39:289–293.

Hoff, T. B., J. B. McLaren, and J. C. Cooper. 1988. Stock characteristics of Hudson River striped bass. American Fisheries Society Monograph 4:59–68.

Horrall, R. M. 1961. A comparative study of two spawning populations of the white bass, *Roccus chrysops* (Rafinesque), in Lake Mendota, Wisconsin, with special reference to homing behavior. Doctoral dissertation. University of Wisconsin, Madison.

Hutchison, J. B., and J. A. Matousek. 1988. Evaluation of a barrier net used to mitigate fish impingement at a Hudson River power plant intake. American Fisheries Society Monograph 4:280–285.

Kjelson, M. A., and D. R. Colby. 1977. The evaluation and use of gear efficiencies in the estimation of estuarine fish abundance. Pages 416–424 *in* M. Wiley, editor. Estuarine processes, volume 2. Academic Press, New York.

Klauda, R. J., J. B. McLaren, R. E. Schmidt, and W. P. Dey. 1988. Life history of white perch in the Hudson River estuary. American Fisheries Society Monograph 4:69–88.

Larimore, R. W. 1952. Home pools and homing behavior of smallmouth black bass in Jordan Creek, Illinois. Illinois Natural History Survey Biological Notes 28:1–12.

Mattson, M. T., J. B. Waxman, and D. A. Watson. 1988. Reliability of impingement sampling designs: an example from the Indian Point station. American Fisheries Society Monograph 4:161–169.

McLaren, J. B., T. H. Peck, W. P. Dey, and M. Gardinier. 1988. Biology of Atlantic tomcod in the Hudson River estuary. American Fisheries Society Monograph 4:102–112.

Mears, H. C., and R. W. Hatch. 1976. Overwinter survival of fingerling brook trout with single and multiple fin clips. Transactions of the American Fisheries Society 105:669–674.

Ricker, W. E. 1975. Computation and interpretation of biological statistics of fish populations. Fisheries Research Board of Canada Bulletin 191.

Schumacher, F. X., and R. W. Eschmeyer. 1943. The estimate of fish populations in lakes or ponds. Journal of the Tennessee Academy of Science 18:228–249.

Seber, G. A. F. 1973. The estimation of animal abundance and related parameters. Hafner Press, New York.

Shetter, D. S. 1952. The mortality and growth of marked and unmarked lake trout fingerlings in the presence of predators. Transactions of the American Fisheries Society. 81:17–34.

Swingle, W. E., and R. O. Smitherman. 1966. Estimation of bass numbers in a farm pond prior to draining with electro-shocking and angling. Proceedings of the Annual Conference Southeastern Association of Game and Fish Commissioners 19:246–253.

TI (Texas Instruments). 1977. 1974 year class report for the multiplant impact study of the Hudson River estuary, volumes 1 and 2, text and appendices. Report to Consolidated Edison Company of New York.

TI (Texas Instruments). 1978a. 1975 year class report for the multiplant impact study of the Hudson River estuary. Report to Consolidated Edison Company of New York.

TI (Texas Instruments). 1978b. Catch efficiency of 100-ft (30-m) beach seines for estimating of young-of-the-year striped bass and white perch in the shore zone of the Hudson River estuary. Report to Consolidated Edison Company of New York.

TI (Texas Instruments). 1979. Efficiency of a 100-ft beach seine for estimating shore zone densities at night of juvenile striped bass, juvenile white perch, and yearling and older (≥150 mm) white perch. July 1979. Report to Consolidated Edison Company of New York.

TI (Texas Instruments). 1980a. Report on 1978–1979 studies to evaluate catch efficiency of the 1.0-m^2 epibenthic sled. Report to Consolidated Edison Company of New York.

TI (Texas Instruments). 1980b. 1977 year class report for the multiplant impact study of the Hudson River estuary. Report to Consolidated Edison Company of New York.

TI (Texas Instruments). 1980c. 1978 year class report for the multiplant study of the Hudson River estuary. Report to Consolidated Edison Company of New York.

TI (Texas Instruments). 1981. 1979 year class report for the multiplant impact study of the Hudson River estuary. Report to Consolidated Edison Company of New York.

Van Den Avyle, M. J. 1976. Analysis of seasonal distribution patterns of young largemouth bass (*Micropterus salmoides*) by use of frequency-of-capture data. Journal of the Fisheries Research Board of Canada 33:2427–2432.

American Fisheries Society Monograph 4:102–112, 1988
© Copyright by the American Fisheries Society 1988

Biology of Atlantic Tomcod
in the Hudson River Estuary

James B. McLaren,[1] Thomas H. Peck,[2] William P. Dey,[3]
and Marcia Gardinier[4]

Texas Instruments Incorporated
Ecological Services Group, Buchanan, New York 10511, USA

Abstract.—The adult Atlantic tomcod population of the Hudson River estuary consisted of three age-groups. Age-I fish (11–13 months of age) composed 92% to 99% of the population during the five spawning seasons of 1975–1976 through 1979–1980. Most of the remaining spawners were age II; age-III Atlantic tomcod were extremely rare (<0.1% of the total). The sex ratio approached 1:1; annual differences in observed sex ratios probably were due to sex-related distributional differences. Growth rates of larvae prior to 15 May were positively correlated with water temperature. Growth of juveniles was fastest from mid-May through June, after which it was suppressed by water temperatures approaching the upper lethal limit of 26.5°C. By September, both gonadal and somatic growth had resumed, and maturation was achieved by 11 months of age. Spawning occurred during December and January. Mean population fecundity ranged annually from 12,900 to 19,800 eggs per female. Mean fecundity and egg size (diameter and weight) were positively correlated with spawner size, and were postulated as possible mechanisms for modulating annual variations in reproductive potential. Atlantic tomcod primarily eat invertebrates, but juveniles and adults occasionally will consume fish eggs (including their own species'), larvae, and juveniles.

The Atlantic tomcod was one of the three species of major concern during the Hudson River power plant case. The species is one of the most common in the Hudson River estuary and, due to its concentration in the lower portion of the estuary, it is particularly subject to entrainment and impingement at the several power plants sited there (Barnthouse and VanWinkle 1988; Boreman and Goodyear 1988, both this volume). Despite its prevalence in the Hudson River and other northwestern Atlantic estuaries (Bigelow and Schroeder 1953), the Atlantic tomcod has rarely been the subject of published scientific studies. This paper summarizes studies of Atlantic tomcod biology along the Hudson River lasting from December 1973 through February 1980.[5]

The Atlantic tomcod exhibits a life history very different from that of the striped bass (Hoff et al. 1988, this volume) and white perch (Klauda et al.

1988a, this volume). Atlantic tomcod spawn under the ice during winter, months before the spawning of striped bass, white perch, and the majority of fish species in the Hudson River. Young Atlantic tomcod, therefore, are initially subject to a constantly cold thermal environment and have little competition for food from larvae of other species. The lower half of the Hudson River and waters of the New York City metropolitan area are nursery areas for the young. In late fall, adults move upstream from lowest portions of the estuary to spawn near shore, generally from Tappan Zee to Newburgh Bay (river kilometers 38–88; see Figure 17). Building upon this very general description of Atlantic tomcod distribution and movements, this paper will describe the population dynamics in greater detail.

Methods

Field and laboratory procedures.—Atlantic tomcod were caught in 0.9-m × 0.9-m × 1.8-m box traps during the spawning seasons (December–February) of 1975–1976 through 1979–1980. The traps were fished without leads in 1 to 5 m of water from piers or other structures where access was possible during winter ice conditions. Approximately 20 traps were set during each spawning season between km 29 and km 155. Most trap sites were the same each year.

Traps were tended every 24 h (weekdays only).

---

[1]Present address: Beak Consultants Incorporated, 12072 Main Road, Akron, New York 14001, USA.

[2]Present address: 2318 Cotswold Court, Fort Collins, Colorado 80526, USA.

[3]Present address: EA Engineering, Science and Technology, Rural Delivery 2, Goshen Turnpike, Middletown, New York 10940, USA.

[4]Present address: IBM Corporation, Sterling Forest, New York 10979, USA.

[5]Studies have continued, and reports have been issued by Battelle (1983) and NAI (1984, 1986).

Most Atlantic tomcod caught were anesthetized with tricaine (MS-222), measured to the nearest millimeter total length (TL), either fin-clipped (one or two fins) or tagged with a Carlin-type tag beneath the first dorsal fin, and released in the area of capture but away from the traps (or the replacement of traps was delayed for several hours) to minimize immediate recapture. When conditions prevented marking the entire catch, the unmarked portion was examined for marks, counted, and released.

Once a week, the entire catch from each of several standard trap sites (common sites across the years) was brought to the laboratory in fresh condition. The catch from each trap was sorted into eight 25-mm length strata ($\leq$125, 126–150, . . ., 251–275, $\geq$276 mm TL). A random subsample of 20 fish per stratum was measured for length and weighed; sexes were determined and otoliths were removed from fish larger than 150 mm TL. All fish 150 mm or smaller were assumed to belong to the youngest age group, based upon preliminary analysis. The total number of fish and the sex ratio in each length stratum were recorded. Age determinations were made for the subsampled fish from the dark bands on whole otoliths viewed with reflected light, either when the otoliths were removed or later after they had soaked in water.

Fifteen ovaries per length stratum were removed each spawning season from females showing no evidence of egg loss through handling or partial spawning. The ovaries were placed in Simpson's modified version of Gilson's fluid and ruptured to allow preservation. Later, the total weight of each ovary was recorded and eggs in a weighed (about 2-g) subsample of each ovary were counted; total fecundity was extrapolated by weight proportion to the combined ovaries. Mean egg diameter in these fish was determined by aligning a random sample of eggs continuously in a single row along a measured distance (approximately 30 mm) and counting the number of eggs in that distance. Mean egg weight was the weight of an ovary subsample divided by the number of eggs.

During May through early December, Atlantic tomcod were caught with a 7.8-m otter bottom trawl (1.27-cm-mesh cod end). Total catches from one or more hauls per week were brought to the laboratory, and 80 juveniles per haul were randomly sampled for length and weight measurements. Generally, sex ratios were determined for one sample per week, June–September. Ten go-

nads per sex from fish caught during August–November were used to determine time and rate of maturation. Gonads were removed, stored in 10% formalin and later drained until damp and weighed. Otoliths were removed from subsampled fish caught between May and November for age determinations and to monitor the development of otolith bands as a partial validation of the aging technique.

Larval and older Atlantic tomcod were collected approximately weekly from early spring through mid-August 1975–1977 between km 22 and km 224 with an epibenthic sled and a Tucker trawl, each with a 1-m^2 opening (Boreman and Klauda 1988, this volume).

The contents of 46 stomachs of adults and 486 stomachs of juveniles were examined from collections made in December 1973 and in May–October 1974, respectively. All samples came from the area between the George Washington Bridge (km 20) and Cornwall-on-Hudson (km 88). Adult tomcod stomachs were injected with 10% formalin, excised, and subsequently stored in 10% formalin. Juvenile fish were preserved whole in 10% formalin upon capture.

Analytical procedures.—Age and length compositions of the spawning population were determined from the stratified random subsamples of box-trap catches. An age–length distribution was generated from the subsampled fish. This age–length relationship then was weighted by the overall length composition to provide an estimate of the overall age composition of the spawning population.

Mean lengths and weights of juvenile Atlantic tomcod caught in bottom trawls were plotted visually by date of capture from early May through November. Specific growth rates (Ricker 1975) were calculated for four observed growth phases: prior to 15 May, 16 May–30 June, 1 July–15 September, and 16 September–31 December. Mean lengths and weights for 15 May, 30 June, and 15 September were read from linear regressions of $\log_{10}$length or $\log_{10}$weight on time (days) each year, 1974–1979, and specific growth rates were based on differences in length or weight on these successive dates. For the period prior to 15 May, the initial size was assumed to be zero on 1 January, the approximate date of peak spawning. For the final period, length measurements of adults caught by box traps were used, and growth rates were calculated separately for males and females. Growth rates for weight were not estimated for the fall population because of potential distortions caused by gonad maturation and spawning.

Mean annual growth in length of Atlantic tomcod during their second and third year of life was estimated from the length-frequency data used for age composition. Annual instantaneous growth rates of these fish were calculated according to Ricker (1975).

Population size was estimated for juveniles by a catch-extrapolation method similar to that described by Young et al. (1988, this volume), and for spawning adults by mark–recapture methods. The extrapolation method used density estimates derived from Tucker-trawl and epibenthic-sled sampling and volume estimates for individual river strata to produce standing crop estimates. Mark–recapture estimates for the 1975–1976, 1977–1978, 1978–1979, and 1979–1980 adults were made by the modified Petersen method (Ricker 1975). The tendency for Atlantic tomcod to move downriver after spawning enabled us to have simultaneous marking and recovery efforts in separate locations. The marking region was designated as km 75–144, the area of highest box-trap catches. The recovery region was km 19–74. Recovery of marked fish was from box traps, intake screens at the Indian Point nuclear generating station (km 66), and bottom trawls. No adjustments were made for tagging mortality or tag loss over the short (2-month) recovery period, because marked fish held in the laboratory for 14 d had nearly 100% survival and tag retention. For purposes of mark–recapture estimation of population size, the Hudson River system was assumed to be closed. This is a tenable assumption because no other large population of the species has been documented in the vicinity of the Hudson River.

If a Petersen estimate from spatial rather than temporal stratification of data is to be valid, all fish must have either the same probability of being captured in the marking area or the same probability of being captured in the recovery area (Ricker 1975). These conditions were not met for the 1976–1977 adult population, as shown by chi-square analysis of ratios of marked to unmarked fish through time (TI 1979, 1980a, 1980b). A Schaefer (1951) estimate for stratified populations, therefore, was applied to the 1976–1977 data, based on 2-week release and recovery periods from December through mid-February. The Schaefer estimate was adjusted for the estimated fraction of the population residing in the release area for longer than 2 weeks, and a method for calculating the standard error of the adjusted estimate was derived (TI 1979).

Age-specific fecundity was estimated as

$$Fe_j = \frac{\sum_{i=1}^{8} N_{ij}\, Fe_i}{\sum_{i=1}^{8} N_{ij}} \; ;$$

Fe_j = mean fecundity for age j (j = 1 or 2);
N_{ij} = number of females in 25-mm length group i of age j collected during the spawning season;
Fe_i = mean fecundity for length group i.

Annual egg deposition (E) was estimated as

$$E = N\, P_t\, Fe;$$

$$Fe = \text{population mean fecundity} = \sum_{j=1}^{2} Fe_j\, P_j \, ;$$

P_j = estimated proportion of all females that are age j;
N = estimated adult population size;
P_t = proportion of females in the total spawning population.

Total annual mortality rates (A) were calculated directly from estimates of egg deposition and adult population estimates for each of the five year classes 1975–1979:

$$A = 1 - \frac{N_A}{E_1};$$

E_1 = estimated egg deposition by age-I females;
N_A = estimated adult spawning stock produced by that egg deposition.

To investigate the pattern of mortality during the first year of life, daily rates of population decline were calculated from estimates of juvenile abundance within the sampling area. The total annual mortality rates for Atlantic tomcod during their second and third years were estimated for individual year classes and by sex from the age-composition and adult abundance estimates of successive years.

Frequency of occurrence (percentage of stomachs containing a particular food item) and percent frequency (average percentage of all food items in a stomach) were calculated on a monthly basis for food habits analysis.

Results and Discussion

Age Composition and Sex Ratios

It was apparent from otolith analysis that the adult populations of Atlantic tomcod in the Hud-

TABLE 34.—Estimated numbers of Atlantic tomcod, by age and sex, collected in the Hudson River estuary during the 1975–1976 through 1979–1980 spawning seasons.

Spawning season	Age, males			Age, females		
	I	II	III	I	II	III
1975–1976	4,961	88	0	1,283	40	0
1976–1977	8,881	603	1	3,029	241	1
1977–1978	2,169	32	0	869	79	0
1978–1979	5,135	35	0	1,912	40	1
1979–1980	4,292	116	0	2,214	86	0

son River estuary consisted of three age-groups. The most abundant group was 11–13 months of age (hereafter referred to as age I for simplicity); it made up 92–99% of the population during the five spawning seasons sampled (Table 34). Age-II fish (23–25 months old) composed almost the entire remainder of the spawning population. Age-III fish (35–37 months old) were extremely rare (<0.1% of the total population).

After larvae transformed to juveniles in March–May, growth was rapid (see below), and otoliths examined in the spring had a uniform opaque band. The hyaline band formed during the summer, reflecting slowed growth, and this band was distinct by approximately mid-August. By mid-September, otoliths began to develop a second opaque band, which continued forming through winter and the following spring. The hyaline annuli, therefore, were set down during the summer, unlike those of many other temperate fish species, which form in winter. A corresponding annulus was also observed on scales. Inasmuch as scale analysis was complicated by spawning checks deposited during the winter, scales were not used for age determination.

The sex ratio of adults was difficult to estimate accurately from box-trap catches. Males apparently preceded females to spawning areas along shore, where box traps were set, and remained longer after peak spawning activity (an example is in Figure 34). As a result, the ratio of males to females varied dramatically over short intervals.

In summer trawl samples of juveniles, sex ratios were close to 1:1, though males tended to predominate (Table 35). Significant differences in the sex ratio of juveniles among years ($\chi^2 = 27.52$; df = 4; $P < 0.05$) may have been caused by year-to-year geographic shifts of the population along the river. There are no data on longitudinal trends in sex ratio, but smaller Atlantic tomcod tended to be further upriver than larger ones (TI 1980b) and males tended to be smaller at age than

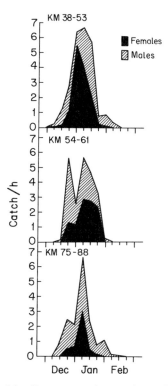

FIGURE 34.—Box-trap catches per hour of male and female Atlantic tomcod from river km 38 through km 88, Hudson River estuary, December 1978–March 1979. Total catch rates are represented by the upper curves in each panel.

females, at least among mature fish (Table 36). In 1977 and 1979, the population was further downriver than in other sampling years (Klauda et al. 1988b), which placed its upriver end closer to our sampling area. It was in 1977 and 1979 that samples contained the highest proportion (approximately 60%) of juvenile males (Table 35).

Growth

Three phases of first-year growth were evident for Hudson River Atlantic tomcod (Figure 35): a summer phase of little or no growth separated the rapid growth phases of spring and fall. By December, the mean total length of age-I fish ranged from 141 to 158 mm for males and 156 to 179 mm for females over the 1975–1979 year classes. A period of reduced growth during summer had also been observed for Atlantic tomcod in an earlier Hudson River study (Dew and Hecht 1976). Older fish could only be sampled when they aggregated on the spawning ground in winter; there, age-II males reached 214–222 mm and age-II females

TABLE 35.—Numbers of males and females and sex ratios for juvenile Atlantic tomcod collected in the Hudson River estuary, June–September 1975–1979. Sex ratios differed significantly among years ($\chi^2 = 27.52$; df = 4; $P < 0.05$).

Measure	1975	1976	1977	1978	1979	Combined
Male/female	289/240	592/558	650/451	515/498	480/323	2,526/2,070
Sex ratio	1.2:1	1.1:1	1.4:1	1.0:1	1.5:1	1.2:1

were 251–258 mm long (Table 36). Only three age-III tomcod were measured: two females of 298 and 265 mm, and a male of 298 mm.

Water temperatures apparently played a major role in controlling first-year growth. After an egg incubation period of approximately 36 to 42 d (Hardy and Hudson 1975), 5-mm yolk-sac larvae began a period of growth highly correlated with prevailing water temperatures. At peak densities in March, yolk-sac larvae experienced temperatures of 2–5°C. As temperatures rose during the period from hatching until 15 May (to 12–15°C), growth of larvae ranged annually from 0.243 mm·d^{-1} to 0.360 mm·d^{-1} or 0.0006 g·d^{-1} to 0.0073 g·d^{-1} (Table 37). Larval growth showed a significant ($r = 0.94$; $P = 0.006$) positive correlation annually with mean water temperature, measured at the City of Poughkeepsie Water Works, during

this period for the years 1974 through 1979. Growth in weight also showed a positive but nonsignificant correlation with water temperature ($r = 0.77$; $P = 0.076$).

Juvenile growth rates during the late spring were more rapid and less correlated ($P > 0.05$) annually with temperature than those of larvae. From 16 May through 30 June, they ranged from 0.692 to 1.008 mm·d^{-1} or 0.0469 to 0.0754 g·d^{-1} (Table 37).

Summer water temperatures exceeded optimal values for growth of juvenile Atlantic tomcod. They ranged from 25 to 27°C in river regions occupied by the fish; the reported upper lethal temperature is 26.5°C (EA 1978). Growth rates annually were very low during this period: −0.001 to 0.255 mm·d^{-1} and −0.0081 to 0.0514 g·d^{-1} (Table 37). In this temperature range, growth tended to be slower at higher

TABLE 36.—Mean total lengths (mm) of male (M) and female (F) Atlantic tomcod collected in the Hudson River estuary. Ages and sample sizes are in parentheses.

Spawning season	Sex	Year class					
		1974	1975	1976	1977	1978	1979
1975–1976	M	217.3 (II:78)	156.1 (I:4,983)				
	F	253.2 (II:42)	171.2 (I:1,283)				
1976–1977	M		214.4 (II:607)	141.1 (I:8,871)			
	F		251.3 (II:239)	156.1 (I:3,027)			
1977–1978	M			221.4 (II:34)	157.7 (I:2,197)		
	F			257.5 (II:80)	176.8 (I:876)		
1978–1979	M				222.3 (II:36)	154.9 (I:5,154)	
	F				257.0 (II:40)	179.0 (I:1,915)	
1979–1980	M					216.8 (II:116)	152.2 (I:4,252)
	F					256.8 (II:86)	175.0 (I:2,214)

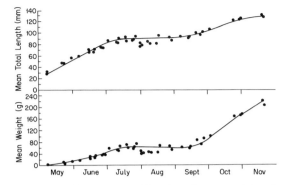

FIGURE 35.—Mean total length and mean weight of Atlantic tomcod juveniles collected by bottom trawls and epibenthic sled during 1979, Hudson River. Curves were fitted by eye.

water temperature, but correlations were not significant ($P < 0.05$). Significant correlations ($r = 0.83$, $P = 0.04$ for length; $r = 0.94$, $P < 0.01$ for weight) were observed between summer growth and mean freshwater flow measured at Green Island Dam. These correlations may have been related to increased nutrient input with increasing freshwater flow.

As temperatures declined in the fall, juveniles resumed rapid growth. Growth rates ranged from 0.412 to 0.731 mm·d^{-1} for males and 0.552 to 0.910 mm·d^{-1} for females (Table 37). This time period encompassed both somatic and gonadal growth.

Maturation and Fecundity

Hudson River Atlantic tomcod of both sexes began to mature at 9 months of age and were capable of spawning at 11 months. Gonad weight, as a percentage of total body weight, began to increase in September and reached an asymptote by late November or early December. No immature Atlan-

tic tomcod were collected during the December–February spawning period.

The mean fecundity for the population (ages I and II combined) ranged from 12,900 to 19,800 eggs per female (Table 38). This population mean closely resembled the mean fecundity of age-I fish each year because of the relative abundance of this age group. Based on the estimated age structure for individual sampling years, age-I Atlantic tomcod contributed 85% to 97% of annual egg production. The contribution of age-III Atlantic tomcod was considered negligible.

Fecundity was significantly correlated with total length of females after logarithmic transformation (Table 38). This correlation emphasized the influence of first-year growth on the reproductive potential for the population. Fecundity was lowest during the 1976–1977 spawning season, when the mean length of age-I fish was considerably less than in other years (Table 38).

In addition to annual differences in mean fecundity, yearly variation in egg size (diameter and weight) was also evident. Significant differences in egg diameters were found among years ($F = 16.14$; $P < 0.001$) as well as among length strata ($F = 8.90$; $P = 0.001$). Egg diameters were smallest during the 1976–1977 spawning season and largest during 1979–1980; there were no detectable differences among the three remaining spawning seasons. Overall egg diameters varied nearly twofold (range, 0.70–1.36 mm), differences that are magnified with respect to volume, weight, and yolk content of eggs (e.g., a 25% increase in diameter from 0.80 to 1.00 mm results in a 95% volume increase). Mean egg size was related to fish size; in 1978–1979, for example, egg diameter was correlated ($P < 0.01$) with both length and weight of females ($r = 0.52$ and 0.50, respectively), as was egg weight ($r = 0.53$ in both cases).

TABLE 37.—Growth rate estimates for larval and juvenile Atlantic tomcod in the Hudson River estuary, 1974–1979.

	Larvae		Juveniles					
	Prior to 15 May		Spring 16 May–30 Jun		Summer 1 Jul–15 Sep		Fall 16 Sep–31 Dec	
Year	Length (mm·d^{-1})	Weight (g·d^{-1})	Length (mm·d^{-1})	Weight (g·d^{-1})	Length (mm·d^{-1})	Weight (g·d^{-1})	Male length (mm·d^{-1})	Female length (mm·d^{-1})
1974	0.310	0.0073	0.840	0.0739	0.158	0.0242		
1975	0.290	0.0038	0.760	0.0469	0.240	0.0337	0.598	0.740
1976	0.360	0.0044	0.692	0.0754	0.255	0.0514	0.412	0.552
1977	0.295	0.0027	0.712	0.0545	0.094	0.0044	0.731	0.910
1978	0.243	0.0006	1.008	0.0738	−0.001	−0.0081	0.665	0.890
1979	0.291	0.0036	0.873	0.0712	0.080	−0.0003	0.565	0.725

TABLE 38.—Age-specific mean fecundity (eggs per fish) of, proportion of total eggs spawned by, and correlation of fecundity with total length of age-I and -II female Atlantic tomcod in the Hudson River estuary. All correlations (r) were significant ($P < 0.01$).

Spawning season	Age (years)	Mean fecundity (Fe)	N	Mean total length (mm TL)	Proportion of total spawn	Log_{10}Fe versus $\text{log}_{10}\text{TL}: r$
1975–1976	I	12,500	56	171	0.96	
	II	41,500	24	253	0.04	
	I + II	13,400	80			0.919
1976–1977	I	10,400	50	156	0.85	
	II	43,800	42	251	0.15	
	I + II	12,900	92			0.949
1977–1978	I	16,900	64	177	0.85	
	II	52,000	40	257	0.15	
	I + II	19,800	104			0.954
1978–1979	I	16,600	99	179	0.97	
	II	43,600	21	257	0.03	
	I + II	17,200	120			0.896
1979–1980	I	14,400	64	175	0.90	
	II	46,400	41	257	0.10	
	I + II	15,600	105			

Population Size and Mortality Rates

Mark–recapture estimates of the spawning stock size during five consecutive winters ranged from a high of 10.4×10^6 in 1976–1977 to a low of 2.5×10^6 in the following season, 1977–1978 (Table 39). When combined with age structure, sex ratio, and age-specific mean fecundity data for corresponding years, the estimated total egg deposition ranged from 22×10^9 to 65×10^9 eggs, if all eggs from all females were spawned.

Based on estimated fish abundance north of the George Washington Bridge, there were two distinct periods of population decline, approximately equal in duration from the time of egg deposition to the maturation of the young Atlantic tomcod (Figure 36). The separation between them occurred in late June to early July. Rates of decline

TABLE 39.—Mark–recapture population estimates[a] and confidence intervals (CI) for Atlantic tomcod spawning stocks in the Hudson River estuary and estimated egg deposition.

Spawning season	Population (10^6 fish)	95% CI	Egg deposition (10^9 eggs)
1975–1976	3.68	3.00–4.51	22
1976–1977	16.19 (10.41)[b]	11.08–21.30	65
1977–1978	2.53	1.73–3.87	21
1978–1979	6.04	4.42–8.51	51
1979–1980	9.13	6.36–13.60	57

[a] Modified Petersen method for all years except 1976–1977, when the Schaefer method was used.
[b] Adjusted for residency in release area.

in the first period were very high, but decreased after early summer (Table 40). The observed decline was probably a combined function of mortality, increasing avoidance of sampling gear by fish, and emigration by juveniles downriver to unsampled areas of the lower estuary. The true population size of juveniles from mid-spring through summer likely was much higher than the part estimated between km 22 and km 224. The true mortality rate from spawning through the end of spring then would be lower than our estimate.

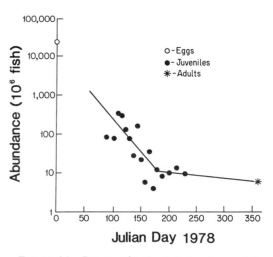

FIGURE 36.—Pattern of estimated abundance of the 1978 year class of Atlantic tomcod in the Hudson River estuary. Lines were fitted by eye.

TABLE 40.—Rates of population decline of juvenile Atlantic tomcod in Hudson River estuary during specified time periods, 1975–1979.

Year	Period	Rate of population decline ($\%\cdot d^{-1}$)
1975	1 Jan–1 Jul	3.3
	2 Jul–1 Dec	1.1
1976	1 Jan–15 Jun	3.8
	16 Jun–31 Dec	0.6
1977	1 Jan–24 Jun	4.6
	25 Jun–31 Dec	1.5
1978	1 Jan–2 Jul	3.7
	3 Jul–31 Dec	0.3
1979	1 Jan–23 Jun	5.2
	24 Jun–31 Dec	0.2

Total annual mortality (A) for age I, estimated between egg deposition and spawning by age-I fish 1 year later, ranged from 0.99954 to 0.99996 (Table 41). The year of highest mortality (1977) was the year of greatest calculated egg deposition. A positive relationship between egg deposition and subsequent mortality was evident from the remaining years. Summer water temperatures (mid-June to mid-September) may affect mortality, in addition to their effect on growth reported earlier. Mean summer water temperatures were positively and significantly correlated with total age-I mortality ($r = 0.99$; $P = 0.002$) for the 5 years.

High annual mortality rates continued through the second year of life, ranging from 0.810 to 0.980 (Table 41). Fishing mortality was an inconsequential part of this total mortality. Rates of tag returns by anglers ranged only from 0.05 to 0.51% during this 5-year period. There is no commercial fishery specifically aimed at Atlantic tomcod, and the sport fishery in the river is primarily limited to winter and spring.

TABLE 41.—Estimates of total mortality for Atlantic tomcod, and mean summer water temperatures (mid-June–mid-September), in the Hudson River estuary, 1975–1979.

Year	Mortality Age I	Mortality Age II	Mean water temperature
1975	0.99989	0.980	24.2°C
1976	0.99954	0.810	23.2°C
1977	0.99996	0.992	24.2°C
1978	0.99972	0.974	23.7°C
1979	0.99983	0.950	24.0°C

TABLE 42.—Frequencies of occurrence[a] and percent frequencies[b] of principal food items found in 41 adult Atlantic tomcod captured in trap nets during December 1973 in the Hudson River.

Food Item	% Occurrence	% Frequency
Polychaeta	2.4	0.1
Cladocera	0.0	0.0
Calanoida	0.0	0.0
Cyclopoida	0.0	0.0
Harpacticoida	0.0	0.0
Chirodotea sp.	7.3	2.0
Cyathura polita	4.9	0.3
Neomysis sp.	36.6	11.5
Gammarus sp.	63.4	76.9
Monoculodes sp.	4.9	4.0
Leptocheirus sp.	0.0	0.0
Crangon sp.	17.1	2.9
Chaoborus sp.	0.0	0.0
Chironomidae (larvae)	9.8	0.5
Chironomidae (pupae)	0.0	0.0
Striped bass	2.4	0.4[c]
Morone sp.	4.9	0.3
Fish eggs	2.4	0.4
Fish (unidentified)	2.4	0.1
Other	7.3	0.4

[a] Percentage of all stomachs that contained the item.
[b] Percentage of all prey represented by the item.
[c] One Atlantic tomcod (208 mm total length) contained three striped bass.

Food Habits

Juvenile Atlantic tomcod fed primarily upon invertebrates. Calanoid copepods were their prevalent prey, occurring in 86% of stomachs. By July, the diet was more diverse, primarily consisting of harpacticoid copepods, *Cyathura polita*, *Neomysis* sp., *Gammurus* sp., *Monoculodes* sp., and *Chaoborus* sp.. In June, one white perch larva was found in a juvenile Atlantic tomcod stomach and, in July, one unidentified fish larva was found. A detailed analysis was presented by Nittel (1976). A dietary shift by July was also noted by Grabe (1978) for juvenile Atlantic tomcod collected in the Hudson River from Haverstraw Bay during 1973–1975. Grabe reported amphipods, mysids, and isopods to be the major diet items for juveniles from July through December.

Adult Atlantic tomcod demonstrated greater piscivory than did juveniles, but invertebrates were still predominant in their diet. *Gammarus*, *Neomysis*, *Crangon*, and chironomid larvae had the highest frequencies of occurrence (Table 42), and these plus *Monoculodes* and *Crangon* were most numerous in stomachs. One adult Atlantic tomcod stomach contained three juvenile striped bass. Unidentified fish and fish identifiable to genus *Morone* together composed 0.4% of all food items (Table 42). Fish eggs, most likely Atlantic tomcod eggs, were found in one of the stomachs.

Only five (11%) of the 46 adults examined had empty stomachs. Grabe (1980) also found adult Atlantic tomcod to be piscivorous occasionally in the Hudson River; fish species consumed included Atlantic tomcod (eggs, larvae, and juveniles), bay anchovy, alewives, blueback herring, and American eels (presumably elvers).

Life History Strategy

Hudson River Atlantic tomcod are short-lived and have relatively low fecundity compared to striped bass (Hoff et al. 1988) and white perch (Klauda et al. 1988a). They are essentially semelparous, the majority (>90%) spawning only once. This life history pattern suggests that Hudson River Atlantic tomcod are subjected to a predictable environment in their early life stages or to an unpredictable environment as adults (Murphy 1968; Schaffer 1974; Pianka and Parker 1975). The result of semelparity is often a widely varying population size reflecting year-to-year conditions. We postulate that Atlantic tomcod in the Hudson River have evolved density-dependent mechanisms to modulate annual variations in population size. These mechanisms would involve intraspecific competition and its effect on energy investment for growth and reproduction.

Fecundity and egg size were correlated with spawner size, implying that each spawning female would have a higher reproductive potential during years of better-than-average growth. A larger egg could increase the survivability of larval fish (Bagenal 1969; Ware 1975; Pitman 1979). A causal relationship between population density and fish growth, establishing density dependence and possibly intraspecific competition, could not be conclusively demonstrated. There was a suggestion of a negative relationship between fall juvenile growth and a density index (age-I spawning population size). The correlation was not significant (male growth: $r = -0.84$, $P = 0.07$; female growth: $r = -0.76$, $P = 0.14$) for the limited 5-year data set. Mean fecundity was also lowest and egg diameters were smallest during the 1976–1977 spawning season, when population size was greatest and the mean length of females was the least of the 5 years.

Summer water temperatures appear to be highly influential on both growth and survival. Hudson River Atlantic tomcod undergo a period of stress, characterized by reduced growth and possibly increased mortality, when optimal temperatures are exceeded. The Hudson River population is at the southern edge of the species' range for major breeding populations (Dew and

TABLE 43.—Range of lengths of young-of-the-year Atlantic tomcod at several locations in northeastern North America.

Location	Time	Total-length range (mm)
St. Mary Bay, Nova Scotia[a]	Mid-Aug	83–120
Peck's Cove, New Brunswick[b]	Late Sep	40–80
Woods Hole, Massachusetts[c]	Aug	80–120
Weweantic River, Massachusetts[d]	mid-Aug	85–90
Pine Orchard, Connecticut[e]	Aug	71–110
Hudson River, New York[f]	mid-Aug	55–106

[a] Cox (1921).
[b] M. J. Dadswell, Canada Department of Fisheries and Oceans, personal communication.
[c] Lux and Nichy (1971).
[d] Howe (1971).
[e] Merriman (1947).
[f] 1977 data.

Hecht 1976; Grabe 1978), and may be prevented from successfully dispersing to the south by excessive summer temperatures. Populations to the north appear to be less affected by midsummer temperatures, but annual growth rates might be slower overall and longevity increased. Cooler water temperatures in more northern latitudes may delay hatching. Cox (1921) observed larval Atlantic tomcod approximately 6–13 mm long on 10 May 1918 in the Miramichi estuary, New Brunswick. However, Peterson et al. (1980) observed Atlantic tomcod eggs in a New Brunswick estuary hatching during March and April. They hypothesized the existence of a later spawning that may coincide more closely with the timing of Cox's observations. In the Hudson River, Atlantic tomcod larvae were present in February; by mid-May they had reached the juvenile stage and ranged in length from 25 to 45 mm. Although larger at this stage than more northerly juveniles, their slowed growth, caused by warm temperatures, left them about the same size as northern Atlantic tomcod by late summer (Table 43).

Data on growth of more northern populations after the first summer of life are sparse and possibly complicated by aging techniques. The Atlantic tomcod literature generally lacks validation of age determination, the importance of which was emphasized by Beamish and McFarlane (1983). The result is an apparent contradiction of reported age structures for geographically similar populations and questionable growth curves. Roy et al. (1975) reported Atlantic tomcod up to age VII in the St. Lawrence estuary and Chaleur Bay; M. J. Dadswell (personal communication) has observed a maximum age of III for fish in Cumberland Basin, New Brunswick; and Cox

(1921) reported age-IV and older Atlantic tomcod in spawning populations in New Brunswick. Contrasting age distributions also may be the result of variations in sampling techniques and up to 60 years of elapsed time between studies. If Atlantic tomcod were accurately aged during the earlier studies, growth beyond the first year appears to be slower at higher latitudes. The maximum length reported for the St. Lawrence estuary (Roy et al. 1975) was 353 mm TL (compared to 338 mm in the Hudson River), but Atlantic tomcod larger than 300 mm TL were rare.

Slower growth at more northern latitudes may cause later maturity. Cox (1921) observed maturity to first occur at age III in New Brunswick. Other researchers did not report the minimum age of maturity, but the smallest females used for fecundity analysis were 170 mm for Massachusetts (Schaner and Sherman 1960) and 180 mm for the St. Lawrence River (Vladykov 1955). The fecundity of females at these two locations was approximately the same as that of females of the same size in the Hudson River.

The life history of the Hudson River population may, therefore, represent the extreme end of a temperature-related latitudinal gradient of growth, age, and maturity characteristics. Other Atlantic tomcod populations also rely most heavily on a single age group for spawning, but rapid growth in the Hudson River might result in earlier maturity. Age analysis that is substantiated by rigorous validation procedures is required for other Atlantic coast populations.

Acknowledgments

The studies were jointly financed by Central Hudson Gas and Electric Corporation, Consolidated Edison Company of New York, Incorporated, Orange and Rockland Utilities, Incorporated, and the New York Power Authority. R. J. Klauda, D. S. Vaughn, and L. W. Barnthouse graciously reviewed the original manuscript. We especially thank all the former employees of the Ecological Services Group of Texas Instruments Incorporated, who contributed to field, laboratory, and data processing efforts during the 8-year study.

References[6]

Bagenal, T. B. 1969. Relationship between egg size and fry survival in brown trout *Salmo trutta L.* Journal of Fish Biology 1:349–353.

----

[6]See Table 1 for sources of legal documents and unpublished reports pertaining to the Hudson River.

Batelle. 1983. 1980 and 1981 year class report for the Hudson River estuary monitoring program. Report to Consolidated Edison Company of New York.

Bigelow, H. B., and W. C. Schroeder. 1953. Fishes of the Gulf of Maine. U.S. Fish and Wildlife Service Fishery Bulletin 53.

Barnthouse, L. W., and W. Van Winkle. 1988. Analysis of impingement impacts on Hudson River fish populations. American Fisheries Society Monograph 4:182–190.

Beamish, R. J., and G. A. McFarlane. 1983. The forgotten requirement for age validation in fisheries biology. Transactions of the American Fisheries Society 112:735–743.

Boreman, J., and C. P. Goodyear. 1988. Estimates of entrainment mortality for striped bass and other fish species inhabiting the Hudson River estuary. American Fisheries Society Monograph 4:152–160.

Boreman, J., and R. J. Klauda. 1988. Distributions of early life stages of striped bass in the Hudson River estuary, 1974–1979. American Fisheries Society Monograph 4:53–58.

Cox, P. 1921. Histories of new food fishes. V. The tomcod. Bulletin of the Biological Board of Canada 5.

Dew, C. B., and H. H. Hecht. 1976. Ecology and population dynamics of Atlantic tomcod (*Microgadus tomcod*) in the Hudson River estuary. *In* Hudson River Ecology, 4th symposium. Hudson River Environmental Society, New Paltz, New York.

EA (Ecological Analysts). 1978. Hudson River thermal effects studies for representative species: final report. Report to Central Hudson Gas and Electric Corporation, Consolidated Edison Company of New York, and Orange and Rockland Utilities.

Grabe, S. A. 1978. Food and feeding habits of juvenile Atlantic tomcod, *Microgadus tomcod,* from Haverstraw Bay, Hudson River. U.S. National Marine Fisheries Service Fishery Bulletin 76:89–94.

Grabe, S. A. 1980. Food of age 1 and 2 Atlantic tomcod, *Microgadus tomcod,* from Haverstraw Bay, Hudson River, New York. U.S. National Marine Fisheries Service Fishery Bulletin 77:1003–1006.

Hardy, J. D., Jr., and L. L. Hudson. 1975. Descriptions of the eggs and juveniles of the Atlantic tomcod, *Microgadus tomcod.* Chesapeake Biological Laboratory, Reference 75–11, Solomons, Maryland.

Hoff, T. B., J. B. McLaren, and J. C. Cooper. 1988. Stock characteristics of Hudson River striped bass. American Fisheries Society Monograph 4:59–68.

Howe, A. B. 1971. Biological investigations of Atlantic tomcod (*Microgadus tomcod*) (Walbaum), in the Weweantic River estuary, Massachusetts, 1967. Master's thesis. University of Massachusetts, Amherst.

Klauda, R. J., J. B. McLaren, R. E. Schmidt, and W. P. Dey. 1988a. Life history of white perch in the Hudson River estuary. American Fisheries Society Monograph 4:69–88.

Klauda, R. J., R. E. Moos, and R. E. Schmidt. 1988b. Life history of Atlantic tomcod, *Microgadus tomcod,* in the Hudson River estuary, with emphasis on spatio-temporal distribution and movements. Pages 216–251 *in* C. L. Smith, editor. Fisheries research in the Hudson River. State University of New York Press, Albany.

Lux, F. E., and F. E. Nichy. 1971. Number and lengths, by season, of fishes caught with an otter trawl near Woods Hole, Massachusetts, September 1961 to December 1962. U.S. Department of Commerce National Marine Fisheries Service Special Scientific Report Fisheries 822.

Merriman, D. 1947. Notes on the midsummer ichthyofauna of a Connecticut beach at different tide levels. Copeia 1947:281–286.

Murphy, G. I. 1968. Patterns in life history and the environment. American Naturalist 102:391–403.

NAI (Normandeau Associates). 1984. Abundance and stock characteristics of the Atlantic tomcod (*Microgadus tomcod*) spawning populations in the Hudson River, winter 1982–83. Report to Consolidated Edison Company of New York.

NAI (Normandeau Associates). 1986. Abundance and stocking characteristics of the Atlantic tomcod (*Microgadus tomcod*) spawning population in the Hudson River, winter 1983–1984. Report to Consolidated Edison Company of New York.

Nittel, M. 1976. Food habits of Atlantic tomcod (*Microgadus tomcod*) in the Hudson River. *In* Hudson River ecology, 4th symposium. Hudson River Environmental Society, New Paltz, New York.

Peterson, R. H., P. H. Johansen, and J. L. Metcalfe. 1980. Observations on early life stages of Atlantic tomcod, *Microgadus tomcod*. U.S. National Marine Fisheries Service Fishery Bulletin 78:147–158.

Pianka, E. R., and W. S. Parker. 1975. Age-specific reproductive tactics. American Naturalist 109:453–464.

Pitman, R. W. 1979. Effects of female age and egg size on growth mortality in rainbow trout. Progressive Fish-Culturist 41:202–204.

Roy, J. M., G. Beaulieu, and G. Labrecque. 1975. Observations sur le poulamon *Microgadus tomcod* (Walbaum) de l'estuaire du Saint-Laurent et de la Baie des Chaleurs. Ministere de L'Industrie et du Commerce, Cahiers d'information, 70. Quebec City.

Ricker, W. E. 1975. Computation and interpretation of biological statistics of fish populations. Fisheries Research Board of Canada Bulletin 191.

Schaefer, M. B. 1951. Estimation of the size of animal populations by marking experiments. U.S. Fish and Wildlife Service Fishery Bulletin 52:189–203.

Schaffer, W. M. 1974. Optimal reproductive effort in fluctuating environments. American Naturalist 108:783–790.

Schaner, E., and K. Sherman. 1960. Observations on the fecundity of the tomcod, *Microgadus tomcod* (Walbaum). Copeia 1960:347–348.

TI (Texas Instruments). 1979. 1976 year class report for the multiplant impact study of the Hudson River estuary. Report to Consolidated Edison Company of New York.

TI (Texas Instruments). 1980a. 1977 year class report for the multiplant impact study of the Hudson River estuary. Report to Consolidated Edison Company of New York.

TI (Texas Instruments). 1980b. 1978 year class report for the multiplant impact study of the Hudson River estuary. Report to Consolidated Edison Company of New York.

Vladykov, V. D. 1955. Fishes of Quebec. Cods. Quebec Department of Fisheries, album 4. Quebec City.

Ware, D. M. 1975. Relation between egg size, growth, and natural mortality of larval fish. Journal of the Fisheries Research Board of Canada 32:2503–2512.

Young, J. R., R. J. Klauda, and W. P. Dey. 1988. Population estimates for juvenile striped bass and white perch in the Hudson River estuary. American Fisheries Society Monograph 4:89–101.

American Fisheries Society Monograph 4:113–120, 1988
© Copyright by the American Fisheries Society 1988

Technical Descriptions of Hudson River Electricity Generating Stations

Jay B. Hutchison, Jr.

Orange and Rockland Utilities, Incorporated
One Blue Hill Plaza, Pearl River, New York 10965

Abstract.—Six fossil-fueled and one nuclear electricity generating plants are sited along the Hudson River estuary between kilometers 8 and 228, measured from the river mouth. Their aggregate rated capacity is 5,798 MW of electricity; operating at that capacity they would withdraw cooling water from the river at the rate of 1.5×10^9 m³/d and reject heat at the rate of 155×10^9 kcal/d. Three of these plants, the fossil-fueled Roseton and Bowline and the nuclear Indian Point facilities, account for 75% of total rated capacity, 62% of maximum water withdrawal, and 79% of potential heat rejection. These three plants and a proposed (but never built) pumped-storage facility at Cornwall, all sited between km 60 and 106, were the focus of environmental litigation. Operational characteristics of the seven functioning plants are summarized in this paper.

Seven electricity generating plants with once-through cooling and a total net rated capacity of 5,798 MW of electricity are located on the Hudson River estuary between river kilometers 8 and 228 (Figure 37). In addition to these existing plants, a pumped-storage hydroelectric plant was proposed for Cornwall (km 93), although it was never built. This paper presents a brief technical description of each of these plants.

In the course of environmental litigation that occurred in the late 1960s and throughout the 1970s, most attention was focused on the three large, new plants between km 60 and km 106—Indian Point, Bowline Point, and Roseton—and the proposed Cornwall project. These facilities are described in the accounts that follow. The technical descriptions emphasize the power plant characteristics that were important for assessing impacts on the Hudson River biota: location on the river; generation capacity; and intake design, through-plant transit, and discharge of heated water. These and other data are tabulated in Appendixes 1–4. Additional details are given in CHG&E (1977), McFadden (1978), and ORU (1977).

Existing Power Plants

Albany steam electric generating station, operated by Niagara Mohawk Power Corporation, is on the west bank of the river at river km 228 in the tidal freshwater portion where salinity is less than 0.5‰ (Cooper et al. 1988, this volume). The plant has four fossil-fueled (oil) generating units totaling 400 MW net capacity. Its shoreline water intake is just below the river surface and its shoreline discharge is at the surface. The maximum cooling-water withdrawal rate is 1,332 m³/min.

Danskammer Point generating station, operated by Central Hudson Gas and Electric Corporation (CHG&E), is on the west bank at km 107 in the normally tidal freshwater portion of the river (less than 0.5‰ salinity). Its four fossil-fueled (oil or gas) generating units total 474 MW net capacity. The plant's cooling system has an intake canal from the north and a surface discharge into a cove immediately to the south (Figure 38). The maximum cooling water withdrawal rate is 1,117 m³/min.

Roseton generating station, operated by CHG&E, is on the west bank at km 106 in the normally tidal freshwater river reach (less than 0.5‰ salinity). Each of its two fossil-fueled units (oil or gas) has 600 MW net capacity. The cooling-water system is common to both units except there are two condenser units beneath each turbine unit. An intake structure containing four circulating water pumps supplies a single intake pipe that transports river water to the plant (Figure 39). The maximum withdrawal rate is 2,421 m³/min; from approximately October through May, the plant operates at 65% of maximum flow. At the plant, a tunnel system distributes the water to the individual condensers (two for each unit) and then collects the heated water for return to the river through a single discharge pipe. At the shoreline, a seal well joins the discharge pipe to a submerged diffuser for final discharge to the river approximately 125 m from shore.

Indian Point nuclear generating station is on the east bank at km 69 in the oligohaline stretch of river (0.5–5‰ salinity). Units 1 and 2 are owned

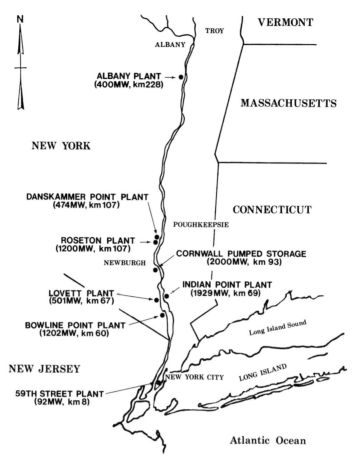

FIGURE 37.—Locations of Hudson River electricity generating plants. Capacities are in megawatts of electricity (MW). Distances are measured from the Battery in New York City. The proposed facility at Cornwall was not built; the other facilities are operational.

by Consolidated Edison Company of New York (ConEd), but unit 1 has been closed since 1 November 1974; unit 3 is operated by the New York Power Authority (NYPA). Units 2 and 3 have net rated capacities of 864 and 965 MW, respectively.

The shoreline intake structures of both units 2 and 3 have seven separate intake bays, one bay for each of the six circulating water pumps and a partitioned bay for six service water pumps (Figure 40). The two units withdraw water at 3,294 m³/min; from approximately November through April the plants operate at 60% of maximum flow. After passing through the plant, the cooling water enters a discharge canal and is released to the river south of unit 3 via an outfall structure consisting of 12 submerged ports.

Lovett generating station, operated by Orange and Rockland Utilities, Incorporated (ORU), is on the west bank of the river at km 67 in the oligohaline portion of the river (0.5‰ salinity). Its five fossil-fueled (oil or gas) generating units total 501 MW net capacity. The five units have a combined maximum withdrawal of 1,200 m³/min. Units 1 and 2 have a common intake canal, whereas the other three units have shoreline intakes (Figure 41). The effluents from units 1, 2, and 3 are combined to form a common surface discharge to the south. Unit 4 has a single-port submerged discharge with an outlet to the south. Unit 5 has a skimmer wall and discharges cooling water to the north.

Bowline Point generating station, operated by ORU, is on the west bank at km 60 in the mesohaline river reach (5–18‰ salinity). Each of its two fossil-fueled (oil or gas) generating units has 600 MW net capacity. Each unit has a separate cooling water system; however, all pumps are

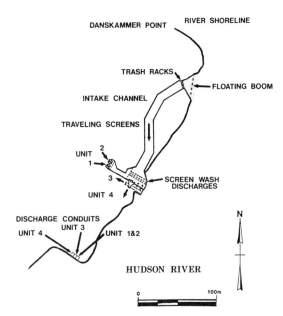

FIGURE 38.—Danskammer Point generating station cooling-water system.

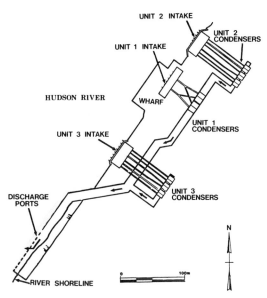

FIGURE 40.—Indian Point generating station cooling-water system.

contained in a common intake structure located in a 49-hectare embayment called Bowline Pond (Figure 42). The intake structure contains six circulating water pumps, three supplying water to the intake pipe of each unit. The six pumps have a combined maximum flow of 2,907 m³/min when both units are operating. Flow can be varied at the plant; from approximately October

through May the facility operates at 67% of maximum flow. At the plant, the cooling water for each unit is distributed to the individual condensers (four in each unit) and the heated water returns to the river through a discharge

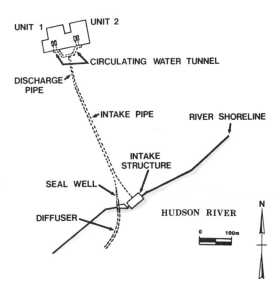

FIGURE 39.—Roseton generating station cooling-water system.

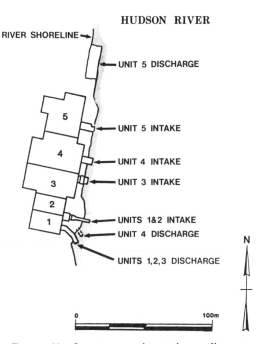

FIGURE 41.—Lovett generating station cooling-water system.

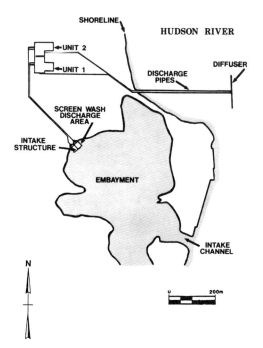

FIGURE 42.—Bowline Point generating station cooling-water system.

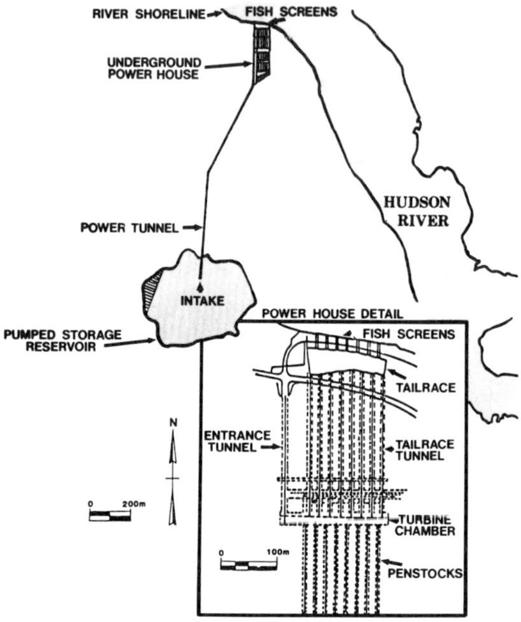

FIGURE 43.—Proposed Cornwall pumped-storage water system.

pipe and submerged diffusers located approximately 425 m from shore.

59th Street generating station, operated by ConEd, is on the east bank at km 8 in the polyhaline portion of river (18–20‰ salinity). Its five fossil-fueled (oil or gas) generating units total 92 MW net capacity. The plant has a shoreline intake and a shoreline surface level discharge. The five units have a combined maximum cooling-water flow of 636 m³/min.

Cornwall pumped-storage plant was proposed by ConEd to be located on the west bank at km 93 in the normally tidal freshwater stretch of river (less than 0.5‰ salinity). Its generating units were to total 2,000 MW net capacity. The plant was planned to provide electricity to the ConEd system during peak energy demand periods. Plans to construct this plant were abandoned in 1980 as one condition of the Hudson River settlement agreement. The main features of the plant were an upper reservoir, an intake structure in the upper reservoir, a pressure conduit to convey water between the upper reservoir and the Hudson River, an underground power plant housing eight reversible pump-turbine units and an intake structure on the Hudson River (Figure 43). The volume of water stored in the 97-hectare upper reservoir would provide the 2,000 MW generating capacity

for approximately 12 h. During pumping, water would be withdrawn from the river at a maximum rate of approximately 35,300 m³/min.

Discussion

Roseton and Bowline Point are jointly owned plants, and co-owners receive a percentage share of plant generation capacity. Roseton ownership is CHG&E (30%), ConEd (40%), and Niagara Mohawk (30%). Bowline Point ownership is ConEd (67%) and ORU (33%).

The Indian Point plant normally operates at 100% generation capacity; the other plants may experience daily operating load changes that vary from approximately 50% to 100% of total generation capacity, depending upon system electrical demand or economic considerations. The fossil-fuel plants undergo scheduled major maintenance outages in the spring or fall that last for several weeks. Indian Point, which operates on an approximately 18-month fuel cycle, undergoes major maintenance during refueling outages. All plants experience periodic unscheduled outages for repairs. Maintenance outages at Bowline Point, Indian Point, and Roseton now are scheduled to coincide with the terms of the Hudson River settlement agreement (Barnthouse et al. 1988, this volume).

Acknowledgments

Information for this section was assembled during the 1970s by personnel from Consolidated Edison Company of New York, Central Hudson Gas and Electric Corporation, New York Power Authority, Orange and Rockland Utilities, and various consulting groups associated with the utilities' Hudson River studies. I appreciate critical reviews of the manuscript by Ronald J. Klauda, Lawrence W. Barnthouse, and Douglas S. Vaughan.

References[1]

Barnthouse, L. W., J. Boreman, T. L. Englert, W. L. Kirk, and E. G. Horn. 1988. Hudson River settlement agreement: technical rationale and cost considerations. American Fisheries Society Monograph 4:267–273.

CHG&E (Central Hudson Gas & Electric Corporation). 1977. Roseton generating station: near-field effects of once-through cooling system operation on Hudson River biota.

Cooper, J. C., F. R. Cantelmo, and C. E. Newton. 1988. Overview of the Hudson River Estuary. American Fisheries Society Monograph 4:11–24.

McFadden, J. T., editor. 1978. Influence of the proposed Cornwall pumped storage project and steam electric generating plants on the Hudson River estuary with emphasis on striped bass and other fish populations. Report to Consolidated Edison Company of New York.

NYPP (New York Power Pool). 1980. Report of member electrical systems of the New York Power Pool and the Empire State Electric Energy Research Corporation pursuant to section 5-112 of the energy law of New York State.

ORU (Orange and Rockland Utilities). 1977. Bowline Point generating station: near-field effects of once-through cooling system operation on Hudson River biota.

---

[1]See Table 1 for sources of legal documents and unpublished reports pertaining to the Hudson River.

Appendix 1

General site descriptions for the existing electricity generating plants on the Hudson River estuary and for the proposed Cornwall facility.

Plant (operating utility)[a]	Fuel type[b]	Location: river km[c] (bank)	Generating unit	Net rated capacity[d] (MW)	Date of commercial operation	Maximum cooling water flow (m³/min)
Albany steam station (NiMo)	Fossil	228.5 (west)	1	100	1952	333.1
			2	100	1952	333.1
			3	100	1953	333.1
			4	100	1954	333.1
			All	400		1,332.4
Danskammer Point (CHG&E)	Fossil	107.0 (west)	1	61	1951	79.5
			2	65	1955	159.0
			3	122	1959	310.4
			4	226	1967	567.8
			All	474		1,116.7
Roseton (CHG&E)	Fossil	106.2 (west)	1	600	1974	1,213.1
			2	600	1974	1,213.1
			All	1,200		2,426.2
Indian Point (ConEd: units 1 and 2; NYPA: unit 3)	Nuclear	69.2 (east)	1	0	1962[e]	0
			2	864	1973[f]	3,294.0
			3	965	1976	3,294.0
			All	1,829		6,588.0
Lovett (ORU)	Fossil	67.6 (west)	1	19	1949	95.4
			2	20	1951	95.4
			3	63	1955	159.7
			4	197	1966	394.8
			5	202	1969	454.2
			All	501		1,199.5
Bowline Point (ORU)	Fossil	60.3 (west)	1	602	1972	1,453.4
			2	600	1974	1,453.4
			All	1,202		2,906.9
59th Street (ConEd)	Fossil	8.0 (east)	All (5)	92	1918	635.9
Cornwall pumped-storage (ConEd)		93 (west)	All (8)	2,000	Proposed	35,300

[a] NiMo = Niagara Mohawk Power Corporation; CHG&E = Central Hudson Gas and Electric Corporation; ConEd = Consolidated Edison Company of New York, Incorporated; NYPA = New York Power Authority; ORU = Orange and Rockland Utilities, Incorporated.

[b] Fossil fuel burned is oil or natural gas.

[c] Kilometers upriver from the Battery at the southern tip of Manhattan.

[d] NYPP (1980).

[e] Commercial operation was suspended in 1974; the unit was rated at 285 MW and it withdrew water at the rate of 1,203.6 m³/min.

[f] Unit 2 began full-load commercial operation in 1974.

Appendix 2

Cooling-water intake systems for five of the seven existing electrical plants located between km 60 and 107 of the Hudson River estuary and for the proposed Cornwall facility. NA: not applicable.

Plant	Generating unit	Number of circulating water pumps	Travelling screen number, type, and mesh size	Normal screen operation[a]	Screen wash water pressure (kg/cm^2)	Cooling-water intake characteristics
Danskammer Point	All (4)	9	12, conventional, 0.95 cm	Intermittent	7	Shoreline canal (137 m long, 10 m wide, 3 m deep)
Roseton	All (2)	4	8, conventional, 0.95 cm	Intermittent	7	Shoreline
Indian Point	2 and 3	12	12, conventional, 0.95 cm; 0.95-cm fixed screens at unit 2	Intermittent	7	Shoreline
Lovett	All (5)	10	11, conventional, 0.95 cm	Intermittent	7	Shoreline
Bowline Point	All (2)	6	6, conventional, 0.95 cm	Intermittent	2–4	Embayment (49 hectares)
Cornwall pumped-storage	All (8)	8	12, fixed, 0.95 cm	Intermittent	NA	Shoreline

[a] Screens may be operated with continuous rotation when debris-loading or icing potential is high; intermittent screen operation varies at each facility and ranges from rotation and washing once every 4 h to once every 24 h.

Appendix 3

Characteristics of the cooling-water systems at the three largest Hudson River electricity generating plants and proposed for the Cornwall facility. NA: not available.

Characteristic	Roseton	Indian Point	Bowline Point	Cornwall pumped-storage
Intake systems				
Generating unit	All (2)	2 and 3	All (2)	All (8)
Maximum intake approach velocity (m/s)	0.18	0.2–0.4	0.20	0.21
System pressure change (kg/cm^2)[a]	0.02–1.4	0.6–1.1	0.37–0.89	1.8–36.9[b]
System velocity ranges (m/s)	0.9–3.9	1.1–2.5	2.5–3.0	1.5–5.5
Transit time (s)[a]	164–252	30–50	150–220	1,000[b]
Discharge systems				
System pressure range (kg/cm^2)[c]	−0.27–0.67	−0.18–1.2	0.02–0.29	7.0–36.9[d]
Discharge pipe (canal) water velocity (m/s)[c]	1.2–3.9	1.0–1.9	1.4–3.0	1.5–5.5
Transit time (s)[c]	210–336	280–1,130	315–470	700[d]
Maximum discharge water velocity at diffusers (m/s)	4.4	3.1	4.6	NA

[a] Values portray range for various cooling water flow rates and pipe sizes and configurations from intake pumps to condenser inlet.

[b] Value during pumping to upper reservoir.

[c] Values portray range for various cooling water flow rates and pipe sizes and configurations from condenser inlet to point of discharge to river.

[d] Values during generation: release of water from upper reservoir through turbines.

Appendix 4

Cooling-water discharge systems at existing electricity generating plants of the Hudson River estuary and at the proposed Cornwall Facility. NA: not applicable or available.

Plant	Generating unit	Maximum temperature rise (°C)	Heat rejection (10^9 kcal/d)	Size and length of pipe, canal or tunnel	Discharge type
Albany steam station	1–4	5.7[a]	2.75[a]	NA	Surface discharge common to all units
Danskammer Point	1	10.0	2.29	NA	Surface, varies by unit
	2	10.0	2.29		
	3	10.0	4.46		
	4	10.0	8.16		
Roseton	1 and 2	9.9[a]	17.74[a]	Single pipe, 3.6 m diameter, 457 m long	Submerged diffuser common to both units
Indian Point	2	8.1	38.56	Common canal, 475 m long, 5.5 m to 12.1 m wide, 3 m deep	Submerged ports common to both units
	3	8.9	42.08		
Lovett	1–3	8.3[b]	4.41[b]	NA	Units 1–3, common surface discharge; units 4 and 5, submerged
	4	12.2	6.96		
	5	10.0	6.53		
Bowline Point	1 and 2	8.3[a]	17.06[a]	Separate pipe, 3.2 m diameter[a]	Submerged diffuser[a]
59th Street	All (5)	3.7	2.09	NA	Surface
Cornwall	All (8)	NA	NA	2,765 m	Surface

[a] Each unit.
[b] Units 1–3 combined.

American Fisheries Society Monograph 4:121–123, 1988

SECTION 3: ENTRAINMENT AND IMPINGEMENT IMPACTS

Introduction

DOUGLAS S. VAUGHAN[1]

Environmental Sciences Division, Oak Ridge National Laboratory
Post Office Box 2008, Oak Ridge, Tennessee 37831, USA

Two major areas of concern are discussed in section 3: (1) the extent of the direct impact on fish populations resulting from entrainment and impingement at Hudson River power plants; and (2) the effect of biological compensation on the ability to predict the long-term consequences of entrainment and impingement on fish populations. The final paper in this section considers the contribution of striped bass spawned in the Hudson River to the Atlantic stock, and it serves as a framework for extending entrainment and impingement impacts on Hudson River striped bass to the Atlantic stock. Power plant impact (or just impact) refers to the effect of power plant operations on year classes or populations, as distinct from the effect on individual organisms, which is referred to as mortality.

Both physical and biological data are needed to obtain direct entrainment and impingement impact estimates for year classes (Figure 44). Estimates of direct impact may then be combined with a life-cycle model to obtain long-term impact estimates. In addition to age-specific life history data, the life-cycle model requires the incorporation of biological compensation to more realistically assess long-term impacts on fish populations. Compensation in the present context refers to changes in life history parameters (e.g., reproduction and survival) that tend to dampen population effects of additional mortality on any life stage of fish. The following definitions of the early life stages of fish were used consistently during the Hudson River power plant case: (1) egg, or the embryonic life stage commencing with fertilization and lasting until hatching; (2) yolk-sac larva, or the transitional life stage from hatching through development of a complete and functional digestive system; (3) post–yolk-sac larva, or the life

stage from initial development of a complete and functional digestive system (regardless of degree of yolk or oil retention) to transformation to the juvenile life stage; and (4) juvenile (also referred to as young of year), or the life stage beginning when the individual acquires the full complement of adult fin rays and extending to age I (i.e., through 31 December of the year in which a fish was spawned). Yearling (age I) and older (age II+) fish are generally considered to be adults.

Entrainment is the process by which small fish (e.g., fish eggs and larvae, usually less than 50 mm long) are drawn into the cooling water intakes of power plants and pass through the meshes of the debris screens. Estimates of the numbers of fish eggs and larvae entrained require such physical data as plant flows and river morphometry and such biological data as far-field distributions and life stage durations (see papers in Section 2 of this monograph). Two other factors are important in determining how many fish eggs and larvae are killed as a result of power plant entrainment. The W-factor, or ratio of egg or larva density in the power plant intake to that in the river nearby, is included in entrainment models primarily to account for the effects of nonuniform distribution of organisms in the river on the number entrained (Boreman et al. 1982). Organisms may concentrate in shallow nearshore water from where the cooling water may be primarily withdrawn, which produces a W-factor greater than 1; W-factors less than 1 may also occur. The entrainment mortality factor, denoted as either the f-factor or simply as f, is the fraction of entrained live organisms killed as a result of passage through the plant. Physical, thermal, and chemical stresses encountered by entrained organisms as they pass through the power plant cooling system can cause mortality. Additional mortality may occur in the thermal plume created by the heated discharge as it enters the river. Muessig et al. discuss the f-factor in this section.

---

[1]Present address: National Marine Fisheries Service, Southeast Fisheries Center, Beaufort Laboratory, Beaufort, North Carolina 28516, USA.

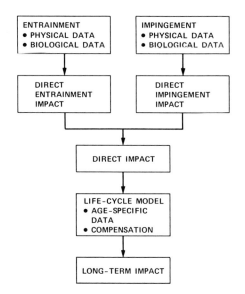

FIGURE 44.—Conceptual framework for the analysis of population-level impacts due to entrainment and impingement of fish at power plants (modified from Figure 2 in Christensen et al. 1981).

Due to competing sources of mortality (e.g., natural mortality) during the entrainment process, an approach suggested from classical fisheries science (Ricker 1975) is used to obtain estimates of direct entrainment impact. This approach is based on conditional mortality rates, or the fraction of an initial population that would be killed by some agent during the year if no other sources of mortality operated. Conditional entrainment mortality rates are used as estimates of the direct impact of power plants on individual year classes. In this section, Christensen and Englert discuss the historical development of entrainment models for Hudson River striped bass. Englert and Boreman compare two approaches for estimating conditional entrainment mortality rates and investigate the evolution of input parameters with changes in estimated conditional entrainment mortality rates. Boreman and Goodyear use a data-intensive methodology to estimate conditional entrainment mortality rates for fish populations entrained at Hudson River power plants.

Impingement refers to the entrapment of larger organisms (e.g., juveniles and adult fish longer than 50 mm) on the surface of power plant debris screens. An assessment of impingement impact on fish populations requires plant flow information (Hutchison 1988, this volume), in addition to estimates of impingement rates (Mattson et al.,

this section) and of impingement survival (Muessig et al.). Barnthouse and Van Winkle employ an empirical model for obtaining estimates of the conditional impingement mortality rates.

Conditional power plant mortality rates from entrainment or impingement are independent of conditional natural and fishing mortality rates, which represent the fraction of each year class lost to natural causes or to fishing. Impacts on fish populations, expressed as conditional mortality rates, are particularly useful for three major reasons. (1) They can be compared and combined with other conditional mortality rates such as those estimated for other plants, other life stages, or other types of environmental stress. (2) Conditional mortality rates can be entered directly into life-cycle models for assessing potential long-term impacts on fish populations. (3) Provided that density-dependent mortality is low relative to density-independent mortality, conditional entrainment and impingement mortality rates are approximately equal to the fractional reductions in year-class abundance due to entrainment and impingement. Barnthouse et al. (1984) noted that the direct impact assessment approach became extremely useful during the settlement negotiations, when the primary concern was with the relative effectiveness of alternative schemes for reducing impacts due to entrainment and impingement. Englert et al. (1988, this volume) describe how conditional entrainment mortality rates were used to schedule reductions in withdrawal of cooling water and maintenance shutdowns of power plant units for minimizing entrainment impacts.

Conditional entrainment and impingement mortality rates also may be extrapolated to obtain estimates of long-term impacts. In addition to life-table data, an appropriate estimate of biological compensation is necessary for making meaningful long-term projections. Biological compensation represents the ability of a fish population to offset, either in whole or in part, reductions in numbers caused by an increase in mortality (e.g., as a result of entrainment and impingement). Compensation results from density-dependent processes; that is, processes, such as reproduction or survival, that control the size of a population as a function of density of that population. Goodyear (1980) described several density-dependent processes (both compensatory and depensatory) found in fish populations. In general, compensatory processes have a stabilizing effect on a population because they tend to increase mortality or decrease reproduction as population size increases. Depensatory processes are destabilizing for a population because they tend to increase mortality or decrease reproduction as population size decreases.

A search for evidence of compensation in Hudson River fish populations formed part of the rationale for collecting life history data for selected fish populations, as described in Section 2. Major emphasis of the search for compensatory processes dealt with striped bass, white perch, and Atlantic tomcod through the study of age structure, age of maturity, fecundity, growth, condition, and mortality. The search emphasized early life history stages (i.e., eggs, yolk-sac larvae, and post–yolk-sac larvae) during which compensatory processes were thought most likely to occur. Sissenwine et al. (1984) identified two mechanisms (cannibalism and competition for limited resources) that form the biological rationale for expecting compensation. However, the limited number of years for which data were collected and the high variability in parameter estimates obtained resulted in an inability to quantify the compensatory process under investigation.

Stock–recruitment models, which describe the numerical relationship between stock size and subsequent recruits into the stock, were developed by fisheries scientists (Ricker 1954, 1975; Beverton and Holt 1957) and are used to represent compensatory processes in fish populations. In this section, Savidge et al., Lawler, Christensen and Goodyear, and Fletcher and Deriso discuss the usefulness of stock-recruitment models as applied to fish populations in the Hudson River estuary. Because of the controversial nature of these papers on compensation, their authors were offered a chance to rebut one another, and the comments submitted are included in this section. Goodyear discusses the management implications of using stock–recruitment models in particular, and of assuming compensatory processes in general.

Finally, Van Winkle et al. treat the contribution of Hudson River striped bass to the Atlantic stock, based on an intensive 1-year sampling effort in 1975 by Texas Instruments Incorporated. This paper provides a framework for extending the effect of entrainment and impingement impacts by Hudson River power plants to the Atlantic coastal stock.

Christensen et al. (1981) discussed the convergence and divergence of opinion through time regarding the various aspects of power plant impacts on Hudson River fish populations. The papers that follow reflect both the convergence and divergence of methods, results, and conclusions over the years prior to the settlement.

Acknowledgments

Preparation of this manuscript was supported by the Ecological Research Division, Office of Health and Environmental Research, U.S. Department of Energy, under contract DE-AC05-840R21400 with Martin Marietta Energy Systems, Incorporated. This is publication 2996, Environmental Sciences Division, Oak Ridge National Laboratory.

References

Barnthouse, L. W., and five coauthors. 1984. Population biology in the courtroom: the lesson of the Hudson River controversy. BioScience 34:14–19.

Beverton, R. J. H., and S. J. Holt. 1957. On the dynamics of exploited fish populations. Fishery Investigations, Series II, Marine Fisheries, Great Britain Ministry of Agriculture, Fisheries and Food, 19.

Boreman, J., and seven coauthors. 1982. Entrainment impact estimates for six fish populations inhabiting the Hudson River estuary, volume 1. The impact of entrainment and impingement on fish populations in the Hudson River estuary. Oak Ridge National Laboratory, ORNL/NUREG/TM-385/V1, Oak Ridge, Tennessee.

Christensen, S. W., W. Van Winkle, L. W. Barnthouse, and D. S. Vaughan. 1981. Science and the Law: confluence and conflict on the Hudson River. Environmental Impact Assessment Review 2:63–68.

Englert, T. L., J. Boreman, and H. Y. Chen. 1988. Plant flow reductions and outages as mitigative measures. American Fisheries Society Monograph 4:274–279.

Goodyear, C. P. 1980. Compensation in fish populations. Pages 253–280 in C. H. Hocutt and J. R. Stauffer, editors. Biological monitoring of fish. D. C. Heath, Lexington, Massachusetts.

Hutchison, J. B., Jr. 1988. Technical descriptions of Hudson River electrical generating stations. American Fisheries Society Monograph 4:113–120.

Ricker, W. E. 1954. Stock and recruitment. Journal of the Fisheries Research Board of Canada 11:559–623.

Ricker, W. E. 1975. Computation and interpretation of biological statistics of fish populations. Fisheries Research Board of Canada Bulletin 191.

Sissenwine, M. P., W. J. Overholtz, and S. H. Clark. 1984. In search of density dependence. Pages 119–137 in B. R. Melteff and D. H. Rosenberg, editors. Proceedings of the workshop on biological interactions among marine mammals and commercial fisheries in the southeastern Bering Sea. University of Alaska, Alaska Sea Grant Report 84-1, Fairbanks.

American Fisheries Society Monograph 4:124–132, 1988

Advances in Field and Analytical Methods for Estimating Entrainment Mortality Factors

PAUL H. MUESSIG AND JOHN R. YOUNG[1]

EA Science, and Technology
Rural Delivery 2, Goshen Turnpike, Middletown, New York 10940, USA

DOUGLAS S. VAUGHAN[2]

Environmental Sciences Division, Oak Ridge National Laboratory
Post Office Box 2008, Oak Ridge, Tennessee 37831, USA

BARRY A. SMITH

EA Engineering, Science, and Technology, Incorporated
221 Oakcreek Drive, Westgate Park, Lincoln, Nebraska 63528, USA

Abstract.—The hearings and settlement negotiations in the Hudson River case focused on various modeling input factors for calculating reduction in year-class strength of important fish species in the Hudson River due to power plant operations. One of the inputs that received considerable attention, and about which the various parties to the case came to general agreement, was the entrainment mortality factor, or *f*-factor, the probability that an entrained live organism will be killed as a result of its passage through a power plant in the condenser cooling water. Entrainment mortality includes death due to temperature increases and death due to mechanical processes. The convergence of views about the *f*-factor arose from increased understanding of the sources of estimation error, improvements in gear for collection of fish larvae, and development of a model for the thermal component of entrainment mortality. Improvements in sampling-gear design reduced potential biases in mortality estimates, reduced estimates of mechanical mortality from 0.923 to 0.412, and improved the statistical power to detect entrainment mortality and the precision of mortality estimates.

Entrainment mortality factors (*f*-factors), probabilities that entrained live organisms will be killed by passage through a power plant in the condenser cooling water, were the subject of considerable scrutiny during the hearings and settlement negotiations in the Hudson River case. Early studies indicated a high potential for mechanical damage of organisms during entrainment (Marcy 1975) and the assumption of 100% mortality was widely accepted. These early empirical studies were conducted with standard conical plankton nets that potentially introduced several inherent biases. Throughout the 1970s, consultants for the Hudson River utilities focused their attention on refinement of field gear and sampling methodology (EA 1979a, 1979c, 1981a, 1982). Sampling gear evolved from standard plankton nets to larva tables to rear-draw plankton flumes. During this time, consultants to the U.S. Environmental Protection Agency concentrated on the refinement of theoretical and analytical definitions of the potential biases, assumptions, and precision of entrainment mortality estimates (Boreman and Goodyear, 1981; Boreman et al. 1982; Vaughan and Kumar 1982). The research led to increased understanding of the sources of error, improvements in sampling gear, development of a model for the thermal component of entrainment mortality, and convergence of positions with respect to *f*-factors among the various parties to the case (Christensen et al. 1981). The objective of this paper is to trace the progression in sampling technology and its effect on the precision of mortality estimates. The extensive entrainment data base at the Indian Point power plant is used as a case study.

Entrainment Mortality Estimates

There are three general types of stress on entrained organisms: chemical, thermal, and mechanical. Chemical stress is an unimportant factor in the present context because biocides generally

---

[1]Present address: Consolidated Edison Company of New York, Incorporated, 4 Irving Place, New York, New York 10003, USA.

[2]Present address: National Marine Fisheries Service, Southeast Fisheries Center, Beaufort Laboratory, Beaufort, North Carolina 28516, USA.

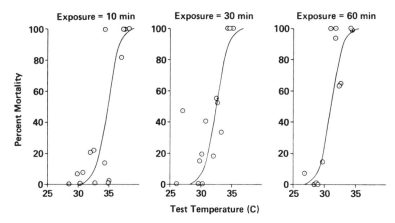

FIGURE 45.—Example of predictive thermal mortality curves for striped bass larvae. Curves are for exposure durations of 10, 30, and 60 min and an acclimation temperature of 18°C. Also shown are empirical data (circles) from laboratory thermal tolerance tests (from Jinks et al. 1978).

are not used to control fouling organisms in the cooling systems of Hudson River power plants during the period when entrainable fish are abundant. Thus, the probability of entrainment mortality, f, can be calculated from the probabilities of mortality from thermal stress, f_t, and mechanical stress, f_m, under the assumption that the two probabilities are independent:

$$f = 1 - (1 - f_t)(1 - f_m).$$

The mortality terms f_t and f_m were estimated from independent study programs, then combined for the overall mortality estimates.

Thermal mortality factors were estimated, for commonly entrained taxa, from thermal tolerance studies (Kellogg and Jinks 1985). Acclimation temperature, exposure temperature, exposure duration, and fish length were all important in predicting thermal tolerance (Jinks et al. 1978; Kellogg et al. 1984). Results of these studies were used to develop thermal mortality curves to predict f_t (Figure 45).

Estimates of mechanical mortality were derived from in-plant sampling programs conducted during periods when thermal stress should have been negligible. If few organisms dead from natural causes are present in intake waters (universally assumed), the proportion of organisms alive in intake samples (P_I) estimates the probability that a fish will survive sampling, and the proportion alive in discharge samples (P_D) estimates the probability that the fish will survive the combined effects of sampling and mechanical stress of entrainment; thus,

$$f_m = 1 - (P_D/P_I).$$

The validity of this formula depends on the assumptions that organisms have the same probability of surviving being sampled from intake and discharge waters, and that their probabilities of surviving sampling and of surviving the mechanical stress of entrainment are independent. Violation of these assumptions, which could not be tested empirically, could result in a biased estimate of mortality associated with entrainment.

Direct entrainment mortality may be immediate or latent. Latent mortality is delayed mortality from physical or physiological damage, estimated by the differential mortality of entrained organisms versus control organisms exposed to similar sampling and thermal stresses. Mortality of entrained organisms from extrinsic factors (e.g., predation) is more difficult to assess; it is not within the scope of this paper, but it has been addressed in some limited work by EA (1979b), Van Winkle et al. (1979), and Cada et al. (1981).

Latent mortality is assessed by comparison of the proportion of initial survivors alive at some fixed time interval (e.g., 24 h) after collection at the intake and discharge points. If mortality differences are found between control and entrained samples, both immediate and latent mortality are incorporated in P_I or P_D:

$$P = \frac{l_t + s_t}{l_i + s_i + d_i};$$

l = number of fish alive initially, i, or at time t;

s = number stunned initially, i, or at time t;
d = number dead initially.

Estimates of f_m can range from -1 to $+1$. Negative values result when P_I is less than P_D because of sampling variation or a greater sampling stress at the intake than at the discharge station. This estimator should only be used under relatively constant conditions of power plant operation, for which a constant value of f_m may be expected to exist for a particular fish of a given age. Field data were subdivided, when sample sizes were adequate, to obtain separate estimates of f_m for different discharge water temperatures and plant operational modes.

Sources of Variability and Bias

The estimates of mechanical mortality are subject to several sources of variability and bias. These were considered in the design of sampling equipment and of analytical methods to predict total mortality factors.

The number of organisms collected at each sampling station and the proportion of organisms found alive in the intake samples (P_I) strongly control the variability of the mechanical entrainment mortality estimate (Vaughan and Kumar 1982). In particular, increasing the sample size or increasing the proportion of organisms found alive in the intake sample, through a reduction of stress associated with sampling, will (1) decrease the minimum entrainment mortality that can be detected as significantly greater than zero, (2) narrow the width of the confidence interval about the entrainment mortality rate, and (3) increase the statistical probability of detecting entrainment mortality when it exists (power). In general, a change in P_I has a greater effect on statistical power than an equivalent proportional change in sample size.

Boreman and Goodyear (1981) discussed several potential sources of bias associated with sampling. Most of these biases are related to differential susceptibility among species or development stages to capture by the gear, differential extrusion from the gear, and differential susceptibility to sampling mortality between intake and discharge stations. It is important that sampling conditions at the two stations be as similar as possible, particularly with respect to water velocity through the sampling gear, sampling duration, and handling procedures.

Bias may also be associated with the location of the intake and discharge sampling stations with respect to the beginning and end of thermal and mechanical stresses induced by the power plant. For example, during the early years of the study, discharge samples were collected before the actual point of cooling-water discharge to the river; thus sampled organisms had not been subjected to the entire entrainment process.

Developments in Field Sampling

Through attempts to reduce potential for bias and to increase precision of entrainment mortality estimates, studies on the Hudson River produced many new developments in sampling gear. Equipment development addressed three major problems: extrusion of organisms through the net meshes of the collection apparatus, differential sampling mortality, and capture efficiency.

Early entrainment mortality estimates were derived from collections with standard conical plankton nets suspended in the intake and discharge waters of the plant (Marcy 1971; Lauer et al. 1974). One major disadvantage of standard plankton nets is that survival of captured organisms is greatly influenced by water velocity through the nets (O'Connor and Schaffer 1977). Unless velocities are equivalent for intake and discharge samples, which rarely occurs, entrainment mortality estimates cannot be correctly adjusted for sampling mortality. Velocity differences also affect extrusion and the ability of organisms to avoid the net; generally, avoidance will be greater and extrusion lower at lower velocities. One means of limiting sampling mortality is to maximize the ratio of filtration area to mouth-opening area; however, space limitations at sampling stations frequently require short nets (small R-values).

The larva table (Figure 46), developed in 1974, is basically a flume modified specifically for collection of planktonic organisms (McGroddy and Wyman 1977). Ecological Analysts first used the larva table to study mortality of entrained ichthyoplankton at Hudson River power plants in 1977 and made several improvements in the design and operation of the device during the next 4 years. Water was pumped from the cooling-water system to the table, which provided equivalent velocities and sampling stresses for both intake and discharge samples. With early versions of the table, however, the pump itself was a major part of the sampling stress. Further, the volume of water sampled and, consequently, the number of organisms collected, were much smaller than they were with plankton nets. Thus, although the va-

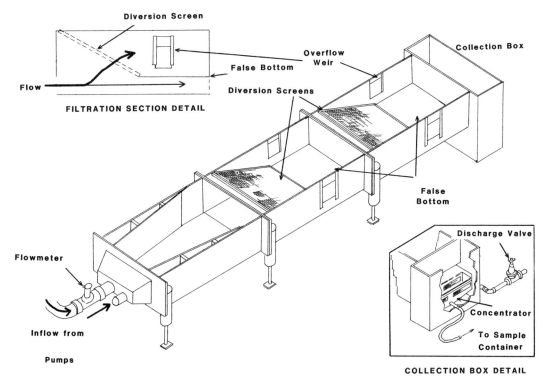

FIGURE 46.—Larva table collection system used in the survival study at Indian Point station, 1978.

lidity of the survival estimates increased, statistical power and the precision of estimates remained low. Studies with various pumps and flow rates led to reduced sampling mortality and improved survival estimates (EA 1979c). The final evolution—elimination of the pump and the mortality associated with it—occurred as another problem was being solved. Because the larva table was large and bulky, sampling could not be conducted at the end of the cooling water system at plants with offshore discharges. In 1978, EA tested a modified table at the offshore diffuser discharge from the Bowline Point plant. A scaled-down flume was floated at the discharge so that the velocity of

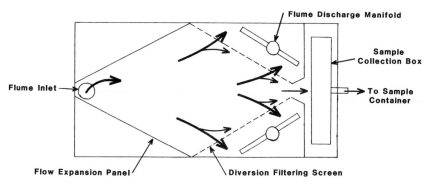

FIGURE 47.—Floating pumpless larva flume (top view) used at the discharge station at Indian Point generating station, 1980.

water exiting the diffuser carried entrained water to the sampler without use of a pump.

In order to have a comparable intake sample, EA developed a floating rear-draw larva-sampling flume in 1979 (Figure 47). The pump was moved downstream of the filtration screen, where it pulled water across the system instead of forcing water and entrained organisms into it. Thus, the full cooling-system effect could be evaluated with comparable floating systems that minimized sampling mortality.

Case History: Entrainment at Indian Point

All developmental stages of the sampling gear described above were used at the Indian Point generating station, which we use as a proxy for experience elsewhere. For illustrative purposes, we concentrate on post–yolk-sac larvae of striped bass, a life stage and species of particular interest in the power plant case. We also limit discussion to situations in which discharge temperatures were less than 33°C, when mortality of striped bass larvae due to thermal stress should be minimal (Kellogg et al. 1984). There rarely were any significant differences in latent mortality between intake and discharge samples (EA 1982). All these conditions allow the development of sampling gear to be evaluated in terms of immediate, mechanically induced, entrainment mortality.

TABLE 44.—Entrainment data from various collection gears used at the Indian Point generating station to estimate mechanical mortality factors (f_m) for post–yolk-sac larvae of striped bass.

Gear	Year	Variable[a]	Intake	Discharge	f_m[b]
Net	1978	P	0.548	0.042	0.923
		N	217	1,106	
		V	0.289	0.897	
	1979	P	0.194	0.185	0.046
		N	165	541	
		V	0.261	0.658	
Larva table	1977	P	0.610	0.483	0.208
		N	806	518	
	1978	P	0.447	0.263	0.412
		N	423	551	
Rear-draw and pumpless flumes	1979	P	0.500	0.684	−0.368
		N	64	114	
	1980	P	0.951	0.559	0.412
		N	142	207	

[a] P = proportion surviving; N = number of organism in sample; V = mean through net mouth velocity (m/s).
[b] $f_m = 1 - (P_{\text{discharge}}/P_{\text{intake}})$.

During the late 1970s, the sample methods described earlier were tested at Indian Point. Entrainment survival studies conducted with 0.5-m-diameter stationary plankton nets during 1978 and 1979 were difficult to evaluate. Average water velocities at the discharge station were 2.5–3 times greater than at the intake (Table 44), introducing considerable bias due to differential mortality and extrusion of, and avoidance by, striped bass larvae. The estimates of mortality from mechanical stress, f_m, for these 2 years were 0.923

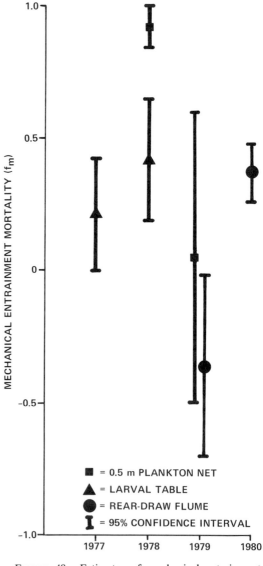

FIGURE 48.—Estimates of mechanical entrainment mortality (f_m) with 95% confidence intervals for post–yolk-sac larvae of striped bass at the Indian Point generating station, 1977–1980.

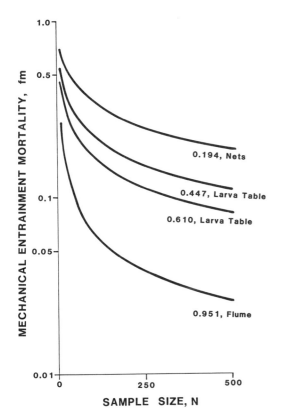

FIGURE 49.—Minimum detectable mechanical entrainment mortality (f_m) as functions of sample size and observed survival proportion at the cooling-water intake (number along curve) for various intake sampling gears.

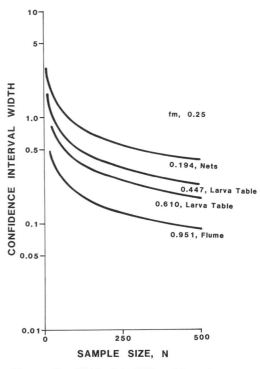

FIGURE 50.—Width of the 95% confidence interval for a mechanical entrainment mortality (f_m) of 0.25 as functions of sample size and observed survival proportion at the cooling-water intake (number along curve), for various intake sampling gears.

and 0.046; however, the high sampling mortality (intake) observed in 1979 resulted in an extremely wide confidence interval (Figure 48) and low precision of the survival estimate.

During 1977 and 1978, pumps were used to deliver samples from intake and discharge locations to a larva table. Although this gear was effective in reducing differences in velocity-induced sampling stress between intake and discharge stations, which was apparent for stationary plankton nets (O'Connor and Shaffer 1977), the sampling pumps were thought to cause significant sampling mortality of more fragile species and life stages. Survival proportions for striped bass larvae in intake samples, P_I, were better on the average than they were with plankton nets, but still only 0.610 and 0.447 (Table 44).

The pumpless and rear-draw plankton-sampling flumes used in 1979 indicated considerably higher survival of larvae in the plant discharge than had been observed with the larva tables. However,

unanticipated differences in sampling stress between the rear-draw (intake) and pumpless (discharge) flumes caused a greater sampling effect at the intake than at the discharge station and resulted in a negative estimate of f_m (Table 44). This differential sampling stress, moreover, was related to the length of larvae, a finding confirmed by experiments conducted with hatchery-reared striped bass (EA 1981b). The rear-draw (intake) flume caused higher mortality for 4–10-mm-long striped bass larvae than did the pumpless (discharge) flume; differences in sampling mortality declined as larvae approached 11 mm. These findings, therefore, precluded the use of intake samples to adjust for sampling effects in the 1979 estimates of entrainment mortality.

It was suspected that subtle differences in configuration of the two sampling flumes induced spacially variable flows and related mortality. Modifications were made to the screen and pump placement in the rear-draw flume to enhance flow uniformity and improve comparability. These modifications appeared to provide the anticipated reduction in sampling effects because survival at the intake station

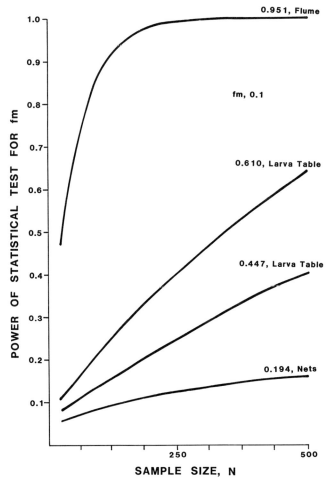

FIGURE 51.—Power of the statistical test of the null hypothesis H_0: $f = 0$, when mechanical entrainment mortality (f_m) is 0.10 and P_α is 0.05, as functions of sample size and observed survival proportion at the cooling-water intake (number along curve), for various intake sampling gears.

improved dramatically in 1980 ($P_I = 0.951$; Table 44) thus reducing the bias in estimates of f_m.

Precision of Entrainment Mortality Estimates

Improvements in sampling techniques at Hudson River power plants enhanced the precision of f_m estimates, as Vaughan and Kumar (1982) have shown. Based on intake survival rates (P_I) at Indian Point, the minimum statistically detectable level of entrainment mortality was reduced nearly an order of magnitude (Figure 49), from 0.27 to 0.04 (for a sample of 250 larvae), when intake survival improved from 0.194 (plankton nets) to 0.951 (rear-draw flume). A concurrent reduction in width of a 95% confidence interval around f_m was also realized (Figure 50).

The power of the statistical test for f_m, that is, the probability of detecting real differences between intake and discharge mortality (entrainment mortality), also increased substantially with the improved sampling techniques. Power is positively affected by increases in both sample size and intake survival (Figure 51); however, the modifications in sampling gear had opposite effects on N and P_I. Although the introduction of the larva table sharply reduced sampling related mortality effects, the sample volume collected in a given period was much lower than for nets and the associated sample sizes were smaller than for the plankton nets. Generally, this trade-off of higher survival for smaller sample size increased the power of the estimate of entrainment mortality, because changes in P_I have a greater effect on power than changes in N (Figure 51).

Technical Lessons

An earlier appreciation of the need to minimize sampling-induced mortality might have lead to an earlier development of the larva table. However, the earliest investigators of entrainment survival before the Hudson River studies matured had concluded that entrainment mortality was generally high or close to 100%. Thus, the need to reduce sampling effects at Hudson River power plants was clouded by an early assumption of 100% entrainment mortality. A critical review of existing data and preconceived ideas was needed to open the way for the advances made in entrainment mortality assessment during the late 1970s on the Hudson River. Once it was recognized that some species and life stages have a considerable survival probability, analytical considerations of precision highlighted the need for lower sampling-induced mortality and larger sample sizes. Analyses of bias demonstrated the need to make sampling procedures and differential efficiencies more similar at intake and discharge stations, and to develop other techniques for reducing potential bias in estimates of entrainment mortality.

Acknowledgments

Work conducted by EA authors was supported in part by Central Hudson Gas and Electric Corporation, Consolidated Edison Company of New York, Incorporated, Orange and Rockland Utilities, Incorporated, and the New York Power Authority. Efforts by D. Vaughan were supported by the Office of Nuclear Research, U.S. Nuclear Regulatory Commission, under interagency agreement DOE 40-550-75 with the U.S. Department of Energy, under contract W-7405-eng-26 with Union Carbide Corporation, and by the U.S. Environmental Protection Agency, Region II, under interagency agreement DOE 40-1054-79 with the U.S. Department of Energy.

References[3]

Boreman, J., and C. P. Goodyear. 1981. Biases in the estimation of entrainment mortality. Pages 79–89 in L. D. Jensen, editor. Issues associated with impact assessment: proceedings of the fifth national workshop on entrainment and impingement. Ecological Analysts, Sparks, Maryland.

---

[3]See Table 1 for sources of legal documents and unpublished reports pertaining to the Hudson River.

Boreman, J., and seven coauthors. 1982. Entrainment impact estimates for six fish species inhabiting the Hudson River estuary, volume 1. The impact of entrainment and impingement on fish population in the Hudson River estuary. Oak Ridge National Laboratory, ORNL/NUREG/TM-385/V1, Oak Ridge, Tennessee.

Cada, G. F., J. S. Suffern, K. D. Kumar, and J. A. Solomon. 1981. Investigations of entrainment mortality among larval and juvenile fishes using power plant simulator. Pages 111–122 in L. D. Jensen, editor. Issues associated with impact assessment: proceedings of the fifth national workshop on entrainment and impingement. Ecological Analysts, Sparks, Maryland.

Christensen, S. W., W. Van Winkle, L. W. Barnthouse, and D. S. Vaughan. 1981. Science and law: confluence and conflict on the Hudson River. Environmental Impact Assessment Review 2:63–88.

EA (Ecological Analysts). 1979a. A review of entrainment study methodologies: abundance and survival. Report to Empire State Electric Energy Research Corporation.

EA (Ecological Analysts). 1979b. Effects of heat shock on predation of striped bass larvae by yearling white perch. Report to Central Hudson Gas and Electric Corporation, Consolidated Edison Company of New York, Orange and Rockland Utilities, and Power Authority of the State of New York.

EA (Ecological Analysts). 1979c. Ichthyoplankton entrainment survival and abundance sampling study: research and development of gear and methods. Report to Empire State Electric Energy Research Corporation.

EA (Ecological Analysts). 1981a. Entrainment survival studies. Report to Empire State Electric Energy Research Corporation.

EA (Ecological Analysts). 1981b. Indian Point generating station entrainment survival and related studies. Annual report (1979) to Consolidated Edison Company of New York and Power Authority of the State of New York.

EA (Ecological Analysts). 1982. Indian Point generating station entrainment survival and related studies. Annual report (1980) to Consolidated Edison Company of New York and Power Authority of the State of New York.

Jinks, S. M., T. Cannon, D. Latimer, L. Claflin, and G. Lauer. 1978. An approach for the analysis of striped bass entrainment survival at the Hudson River power plants. Pages 343–350 in L. D. Jensen, editor. Fourth national workshop on entrainment and impingement. EA Communications, Melville, New York.

Kellogg, R. L., and S. M. Jinks. 1985. Short-term thermal tolerance of 10 species of Hudson River ichthyoplankton. New York Fish and Game Journal 32:41–52.

Kellogg, R. L., R. J. Ligotino, and S. M. Jinks. 1984. Thermal mortality prediction equations for entrainable striped bass. Transactions of the American Fisheries Society 113:794–802.

Lauer, G. J., and eight coauthors. 1974. Entrainment

studies on Hudson River organisms. Pages 37–82 *in* L. D. Jensen, editor. Proceedings of the second entrainment and intake screening workshop. Electric Power Research Institute, Report 15, Palo Alto, California.

Marcy, B. C., Jr. 1971. Survival of young fishes in the discharge canal of a nuclear power plant. Journal of the Fisheries Research Board of Canada. 28:1057–1060.

Marcy, B. C., Jr. 1975. Entrainment of organisms at power plants, with emphasis on fishes—an overview. Pages 89–106 *in* S. B. Saila, editor. Fisheries and energy production: a symposium. Heath, Lexington, Massachusetts.

McGroddy, P. M., and R. L. Wyman. 1977. Efficiency of nets and a new device for sampling living fish larvae. Journal of the Fisheries Research Board of Canada 34:571–574.

O'Connor, J. M., and S. A. Schaffer. 1977. The effects of sampling gear on the survival of striped bass ichthyoplankton. Chesapeake Science 18:312–315.

Van Winkle, W., S. W. Christensen, and J. S. Suffern. 1979. Incorporation of sublethal effects and indirect mortality in modeling population-level impacts of a stress, with an example involving power-plant entrainment and striped bass. Oak Ridge National Laboratory, ORNL/NUREG/TM-288, Oak Ridge, Tennessee.

Vaughan, D. S., and K. D. Kumar. 1982. Entrainment mortality of ichthyoplankton: detectability and precision of estimates. Environmental Management 6:155–162.

American Fisheries Society Monograph 4:133–142, 1988

Historical Development of Entrainment Models for Hudson River Striped Bass

SIGURD W. CHRISTENSEN

Environmental Sciences Division, Oak Ridge National Laboratory
Post Office Box 2008, Oak Ridge, Tennessee 37831-6036, USA

THOMAS L. ENGLERT

Lawler, Matusky, & Skelly Engineers
One Blue Hill Plaza, Pearl River, New York 10965, USA

Abstract.—In the mid-1960s, concerns surfaced regarding entrainment and impingement of young-of-the-year (age-0) striped bass by electric power generating facilities on the Hudson River. These concerns stimulated the development of increasingly complex models to evaluate the impacts of these facilities. The earliest simplistic formulas, based on empirical data, proved inadequate because of conceptual shortcomings, incomplete development, and lack of data. By 1972, complex transport models based on biological and hydrodynamic principles had been developed and applied by scientists representing both the utilities and the government. Disagreements about the acceptability of these models spurred the development of even more complex models. The entrainment models stimulated the collection of substantial amounts of field data to define the spatial distributions and entrainment survival of early life stages. As the difficulties of accounting for the movement of early life stages from hydrodynamic principles became more evident and as more field data became available, simpler empirical modeling approaches became both practical and defensible. Both empirical and hydrodynamic modeling approaches were applied during the U.S. Environmental Protection Agency's hearings on the Hudson River power case (1977–1980). The main lessons learned from the experience with entrainment–impingement modeling are that complex mechanistic models are not necessarily better than simpler empirical models for young fish, and that care must be taken to construct even the simple models correctly.

In the early 1960s, when the Hudson River power plant case began with analyses of the impact of Consolidated Edison's proposed Cornwall pumped-storage facility, it became evident that estimation of power plant impact would be a complex problem. Distribution, movement, and transport of the early life stages of striped bass were largely unknown. Field studies performed during 1966–1967 provided some information on these (HRPC, undated); however, important questions regarding power plant impact remained unanswered.

• How would the power plant withdraw its water from the river, i.e., from what portion of the water body would organisms be drawn into the plant?

• Would organisms be transported into the plant or onto the intake screens in direct proportion to the power plant flow?

• How are the organisms distributed in the near-field area around the plant?

• Would organisms survive passage through the plant?

• What proportion of the total population or standing crop of young-of-the-year fish would be exposed to entrainment?

• With increased swimming ability, would older life stages be able to avoid entrainment?

• How do the river hydrodynamics and behavior of the organisms affect the distribution of organisms in the Hudson and their exposure to power plant withdrawal?

• Would organisms survive impingement on the intake screens?

Various models were developed during the ensuing years to predict the percent reduction (PR) in numbers of yearling fish due to power plant operation, defined as

$$\text{PR} = 100 \left(\frac{Y_0 - Y_p}{Y_0} \right);$$

Y_0 is the number of age-0 fish surviving to become yearling fish without the impact; Y_p is the number of age-0 fish surviving to become yearling fish with the impact. The objective of the power plant studies prior to the 1977–1980 settlement hearings conducted by the U.S. Environmental Protection Agency (EPA) was usually to estimate the PR (of yearlings) from observations of the distribution and abundance of earlier life stages. By informal consensus, the primary objective in the EPA hearings (under Section 316[b] of the Federal Water Pollution Control Act Amendments of 1972: U.S. Congress 1972) was to establish the conditional mortality rate due to entrainment and impingement. This is similar to the previous percentage reduction concept, but it does not consider nonlinear natural mortality rates due to compensatory (density-dependent) mortality. (See Vaughan 1988, this volume, for definitions of terms relating to mortality rates and compensation.)

The history of development of models of the Hudson River striped bass can be divided into three distinct eras based on the power plants and regulatory procedures that were of primary concern:

1965–1968 studies related to the Cornwall pumped-storage facility;

1972–1976 studies related to licensing hearings for Indian Point units 2 and 3;

1977–1980 multiplant studies for the EPA 316(b) hearings.

As discussed in more detail below, at least one model formulation was developed and applied during each of these eras. The formulations range from the very simple calculations that were the basis for the Hudson River fisheries investigations (HRFI) model (HRPC, undated), developed as part of the Cornwall studies, to the extremely complex formulation of the "real-time life-cycle" model (LMS 1975), applied by the utility consultants during the 316(b) studies. The models developed during each of the three eras are discussed below.

Cornwall Pumped-Storage Facility

The Cornwall pumped-storage facility, proposed by Consolidated Edison Company of New York (ConEd) in the early 1960s, generated public opposition. One concern was the project's potential impact on striped bass eggs, larvae, and young juveniles resulting from withdrawal of water from the Hudson River at an average daily rate of 170 m³/s. Although striped bass were known to spawn in the vicinity of Cornwall, comprehensive data on the distributions of early life stages were lacking. As part of the Hudson River fisheries investigations (HRFI), a 3-year sampling program was conducted to characterize these distributions. A relatively simple formula, presented without

derivation, was applied to seasonally averaged data to estimate separately the fraction of eggs and larvae subject to removal from the Hudson River estuary, which extends from the Battery upstream some 243 km to Troy Dam (HRPC, undated):

$$\frac{N_1 Q_w f R}{N_2 Q_t};$$

N_1 is the density of organisms near the point where water would be withdrawn; N_2 is the average density of organisms in the Cornwall area (defined as the reach extending from 10.1 km above to 8.8 km below the proposed facility site); Q_w is the projected instantaneous rate of water withdrawal by the plant when actually pumping (510 m³/s); f (= 8/24) is the fraction of each 24-h day the plant would be withdrawing water; Q_t is the mean tidal flow rate (regardless of direction; i.e., the average of the flow rate during ebb and during flood tide) in the Hudson at Cornwall (2,832 m³/s); R is the fraction ". . . of the total number of [organisms in the given life stage] in the estuary [that] were in the Cornwall area." (HRPC, undated). The results were presented as estimates of annual percentage removal from the river's population. Less than 1% of egg production and 3% of larva production were predicted to be subject to loss. With a method based on area rather than mean tidal flow, it was calculated that up to 6.2% of young juveniles could also be removed.

The HRFI formula was the first entrainment model of which we are aware in which a quantitative method was applied to data to estimate the fractional reduction of fish in early life stages due to an electric power generating facility. It was conceptually flawed because it used mean tidal flow without allowing for multiple exposures of organisms to a power plant and because the definition of R was an oversimplification. The part of the equation concerned with flow uses average plant withdrawal flow in the numerator and mean tidal flow in the denominator. This provides an estimate of entrainment risk, for organisms passing in front of the plant, per half-tidal cycle (about 6.2 h). Such organisms could, however, be susceptible to entrainment several times as the tides moved them back and forth past the plant. If the freshwater flow (net downstream flow) were used in place of the net tidal flow or if allowance were made for repeated exposure of the organisms due to tidal oscillation, the formula would be improved. It could then be decomposed into (1) the fraction of net river flow diverted into the plant,

(2) a *W*-factor (expressing the ratio of concentration of organisms in withdrawn water to the mean concentration in the river cross section), and (3) a residual term, *R*, to represent the fraction of total annual production of the life stage that would pass in front of the plant and thus become vulnerable. The term *R* clearly needs to be related to the duration of the life stage. Eggs hatch a day or two after spawning occurs (Boreman and Klauda 1988, this volume), so only those spawned near the plant could be entrained. The post–yolk-sac larva stage, however, lasts for weeks, and larvae originating a considerable distance from the plant could become vulnerable. The term *R* in the HRFI formula was simply the ratio of the estimated mean standing crop in an arbitrary "Cornwall region" to the estimated mean riverwide standing crop. Only by happenstance would such a quantity represent vulnerability. By 1972, after work with entrainment models had been done in connection with Indian Point unit 2, flaws in the HRFI approach were recognized (Clark 1972; USNRC 1975). This perception was a key factor in the reopening of hearings on Cornwall, which, in turn, may have affected the ultimate decision not to build the plant.

At least two lessons can be learned from the experience with the HRFI entrainment model.

First, an entrainment model that is not supported by a mathematical derivation should be immediately suspect. A set of simple derived equations that can be used to achieve part of what the HRFI model attempted first became available in published form nearly a decade later (Goodyear 1977). The second lesson is that the entrainment problem often seems simpler than it really is. The HRFI formula's oversimplification in the handling of organism movement and in specifying the vulnerable subpopulation led to what came to be considered as substantial underestimates of the potential impact (Carter 1974).

Indian Point Units 2 and 3

The entrainment–impingement issue figured prominently in the 1972–1973 licensing hearings and in the final environmental impact statement for Indian Point unit 2 (USAEC 1972). Six models were developed in connection with this case. The Clark model, the Quirk, Lawler and Matusky (QLM) "completely mixed model" (Lawler 1972b), the Oak Ridge National Laboratory (ORNL) "probability" and FOCAL models (USAEC 1972), the GDYEAR model (a revised version of FOCAL: Goodyear 1973), and the QLM one-dimensional "transport" model (Lawler 1972a) exemplify a

TABLE 45.—A comparison of entrainment models with respect to their treatment of some key physical and biological phenomena in the Hudson River.

	Method of treating key phenomena					
	Space		Transport and movement			
Entrainment model[a]	Longitudinal segments	Vertical layers	Vertical migration	Longitudinal movement	Time[b]	Aging
HRFI model (HRPC, undated)	1	1	No	Mean tidal flow	Development	Life stage
Clark model (Clark 1972)	1	1	No	None	Development	Life stage
QLM "completely mixed" model (CMM; Lawler 1972b)	1	1	No	Uniform	Explicit	Cohorts
ORNL "probability" model (USAEC 1972)	1	1	No	Mean tidal flow and freshwater flow	Development	None
ORNL FOCAL model (USAEC 1972)	19	2	Yes	Upper and lower mean tidal flows	Explicit	Cohorts
ORNL GDYEAR model (Goodyear 1973)	18	2	Yes	Upper and lower mean tidal flows	Explicit	Cohorts
QLM "transport" model (TM; Lawler 1972a)	8	1	No	Freshwater flow	Explicit	Cohorts
ORNL STRIPE model (Eraslan et al. 1976)	76	1	No	Freshwater flow	Explicit	Life stage
LMS "real-time life-cycle" model (RTLC; LMS 1975)	29	2	Yes	Tidal flows	Explicit	Cohorts
TI "empirical entrainment" methodology (EEM; TI 1975)	12	1	No	Empirical	Explicit	Life stage and total young of year
NPPT–ORNL "empirical transport" model (ETM; Boreman et al. 1981)	12	1	No	Empirical	Explicit	Cohorts

[a] Key to initialisms and abbreviations: HRFI: Hudson River fisheries investigations; LMS: Lawler, Matusky & Skelly Engineers; NPPT: National Power Plant Team (U.S. Fish and Wildlife Service); ORNL: Oak Ridge National Laboratory; QLM: Quirk, Lawler & Matusky Engineers; TI: Texas Instruments Incorporated.

[b] Developmental time is related to stage of biological development. Explicit time has a calendar basis.

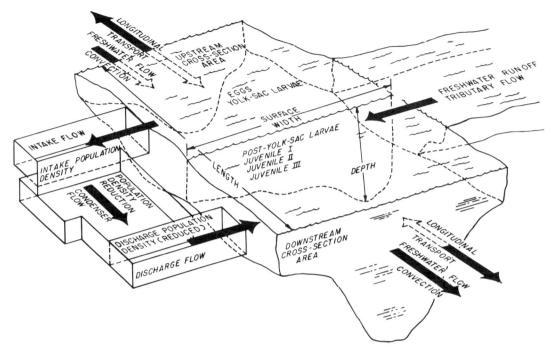

FIGURE 52.—Schematic diagram of a segment in a hydrodynamic transport model for striped bass entrainment, illustrating some of the processes simulated by the model such as longitudinal transport and power plant water diversions (modified from Eraslan et al. 1976).

variety of modeling approaches to the entrainment–impingement problem (Table 45). We have selected four main phenomena to characterize these and other Hudson River entrainment models. With respect to space, the models differ in the number of segments (divisions along the length of the river) and in the number of vertical layers considered. In models with only a single compartment, losses of organisms to power plants are compared with estuary-averaged densities or standing crops (usually calculated externally to the model). Transport and movement relate to vertical migration (a feature restricted to two-layered models) and to the means used to locate and move organisms longitudinally in the model. Longitudinal movement of organisms may be based on "real-time" tidal flows, on mean tidal flow, on freshwater flow, or on empirical changes in distributions of organisms from one sample to the next. Time is either based on the developmental sequence of young of year or, much more commonly, made explicit (i.e., on a calendar basis). The main advantage in making time explicit is that variation in power-plant withdrawals and entrainment survival probabilities can more easily be modeled. Finally, treatment of aging

may be limited to differentiating life stages, or a more complex treatment of cohorts (individuals of approximately the same age) may be attempted. Although field data typically distinguish only among life stages, establishing cohorts within an entrainment model simplifies reconciliation of the aging process with an explicit treatment of time. A simpler classification is to distinguish empirical models from hydrodynamic ones. Empirical models rely entirely on field data to establish the distributions and relative abundances of the life stages of young fish. In contrast, hydrodynamic models involve a combination of field data and physical (e.g., hydrodynamic transport) and biological (e.g., vertical migration) processes to define distribution and abundance (Figure 52).

The simplest of these Indian Point models, the Clark (empirical) model (Clark 1972), estimated the loss of fish of each life stage based on samples taken in the vicinity of Indian Point and compared those losses with estimates of production based on whole-river samples. Estimating production from standing crop data is not a straightforward procedure. To do so formally requires information on the ratio of live to dead organisms among those sampled, the magnitude and pattern of mortality within the life stage, and the duration of the life stage. Further development of the Clark model might have given rigorous consideration to these factors, but attention became focused instead on other, more complex models. The Clark model is interesting nonetheless, because it was relatively simple and well founded conceptually in many ways. The Clark and the ORNL "probability" models were the last in this case to be implemented without a computer (a footnote in Clark's testimony presenting his model explains that slide rule accuracy prevails throughout), although the empirical models could all have been employed with a simple calculator if the data sets had not been so much more extensive.

Egg production and life-stage-specific durations and mortality rates were input to the QLM "completely mixed" model (Lawler 1972b). Mortality rates for juveniles could be modified in accordance with a "compensatory" function such that mortality was reduced at low population densities and increased at high densities. A simple first-order-decay cohort model kept track of the riverwide population densities of the life stages at risk from entrainment (eggs, yolk-sac larvae, post–yolk-sac larvae, and young juveniles) as well as older juveniles and adults. Estimates of organisms killed were based on the riverwide population densities, which were assumed to be the same as those in the cooling-water flow to the plant. An adjustment to allow for survival of some entrained organisms (the f-factor discussed in Muessig et al. 1988, this volume) was included in the model equations but was not implemented. The entrainment–impingement model was linked to an adult life-cycle model, which propagated the effects of entrainment and impingement through the adult population (Figure 53).

The unsatisfied assumption underlying the QLM "completely mixed" model—uniform spatial distributions—was soon remedied with presentation in written testimony (Lawler 1972a) of the utilities' primary entrainment–impingement model in the Indian Point unit 2 hearings. In this QLM "transport" model, a spatial dimension had been added to the "completely mixed" model; the Hudson was divided into eight equal longitudinal segments. Hydrodynamic terms moved early life stages of striped bass with the freshwater flow in the Hudson. The biological treatment was also more sophisticated, involving age-structured cohorts within life stages. Biological compensation was also a major feature of the model. In this QLM one-dimensional "transport" model, the

estimates of the number of organisms killed at each plant were now based on the model population in the segment where the plant was located, modified by factors (currently termed W-factors) accounting for local distribution phenomena as well as f-factors allowing for some entrainment survival.

An interesting example of the contrasting model formulations developed during this period, and their influence on the predictions of exposure to entrainment impact, can be found by comparing the QLM "transport" with the ORNL FOCAL (USAEC 1972) and GDYEAR (Goodyear 1973) models (Table 45). These models viewed the influence of hydrodynamic transport on the organisms in two different ways. The QLM "transport" model assumed that organisms were transported downstream with the freshwater flow in the Hudson. However, in the ORNL FOCAL and GDYEAR models, the vertical differences in the Hudson flows caused by saltwater intrusion from the ocean were simulated explicitly (Figure 54). In these models, organisms were transported with a net downstream flow in the upper layer and a net upstream flow in the lower layer. In addition, vertical diurnal migration of the larval stages of striped bass was included in these models' formulations. The coupling of the upper and lower layer flows with the diurnal migration of the larvae had the net effect of circulating the organisms in front of the power plant, resulting in multiple exposures to entrainment. Because of the loop-like trajectories the organisms followed, these ORNL models came to be known as the "endless belt" models.

During this phase of model development, only a small amount of field data was available. Consequently, calibration and verification of the models were limited, and the precise phenomena affecting transport and movement of the organisms were not well documented.

In extensive and heated hearings in late 1972 and early 1973, the relative merits and weaknesses of the ORNL "endless belt" models and the QLM "transport" model were argued. In addition, the W-factors (which adjusted vulnerability to plant intakes for uneven cross-sectional distributions of organisms in the river), the f-factors for entrainment mortality, and the implementation of compensatory mortality were contested. In the limited time available, the origins of the order-of-magnitude difference between the (typically) 30 to 50% annual losses predicted by the ORNL model and the 2.5 to 4% losses pre-

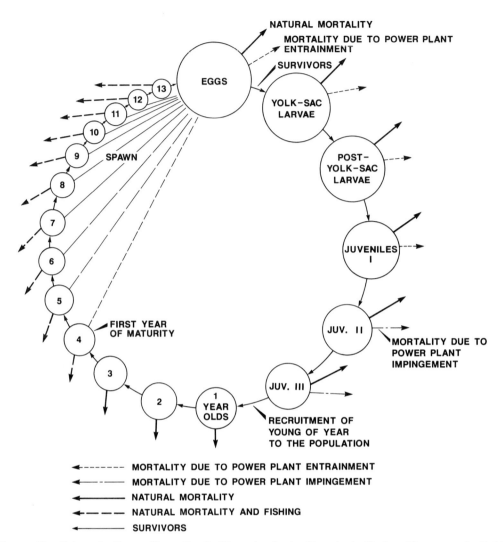

FIGURE 53.—Schematic diagram illustrating the life cycle of striped bass in the Hudson River, as embodied in a linked entrainment—impingement—life cycle model. Decreasing sizes of circles indicate decreasing population size, although not to scale. The right half of the diagram represents the entrainment—impingement portion of the model; the left half represents the adult life-cycle portion. Dashed lines connecting 4- to 6-year old adults to eggs denote the increasing fraction of females that are mature at these ages (modified from Lawler 1972a).

dicted by the QLM "transport" model were explored in part through sensitivity analysis (Lawler 1973). Differences in W-factors, f-factors, and compensation accounted for a large part of the differences in the results; with W- and f-factors set to unity and no compensation, the QLM "transport" model predicted a 15% reduction in yearlings due almost entirely to entrainment (Lawler 1973). Differences in model structure and in the values of the other parameters (particularly those determining longitudinal distributions) presumably accounted for the remaining differences. Although

there was considerable debate about the models themselves, the W- and f-factors were also highly controversial topics. Their importance to the magnitude of predicted impact from any model was clear, and a substantial effort was made in the next several years to develop better gear (especially to estimate f) and to collect more and better data to estimate these factors (Muessig et al. 1988). Sampling to determine riverwide longitudinal distributions also continued.

The period 1974–1976 saw a second phase of model development, spurred by licensing actions

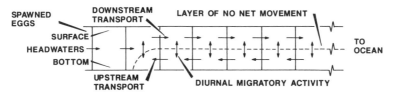

FIGURE 54.—Schematic diagram of the Oak Ridge National Laboratory FOCAL model (modified from USAEC 1972). Arrows indicate movement of entrainable stages of striped bass longitudinally and vertically; dashed line represents the boundary between fresh water (above) and saline water (below).

related to Indian Point unit 3. The two mechanistic models developed during this period, the "real-time life-cycle" model (LMS 1975) and the STRIPE model (Eraslan et al. 1976), were clearly outgrowths of the earlier phase of modeling development. Interestingly, utility consultants developed the RTLC model, which included some of the same mechanisms originally incorporated into the earlier ORNL FOCAL and GDYEAR models, while government consultants at ORNL developed the STRIPE model, which was conceptually similar to the earlier "transport" model developed by utility consultants.

The Atomic Safety and Licensing Appeal Board's (1974) decision on Indian Point unit 2 in April 1974 stated that the utilities' approach in the QLM "transport" model had been more realistic and recommended that a fresh look be taken by the government. As part of this, ORNL increased the effort already under way to develop STRIPE, its own hydrodynamic transport model (Eraslan et al. 1976). Like the QLM "transport" model, it used one-dimensional tidally averaged hydrodynamic transport to move the organisms, coupled with a biological model to keep track of organisms by life stage up to the yearling stage, but it differed in numerous details. Most runs of this age-0 model were done without biological compensation; instead, compensation was incorporated into the overall impact analysis by means of a separate life-cycle model for adult striped bass (Van Winkle et al. 1974; DeAngelis et al. 1978). By this time, data from Texas Instruments' Hudson River sampling program were available for model calibration. Because these data indicated considerably slower net downstream movement of eggs and larvae than predicted from hydrodynamic principles, convective transport defect flux (CTDF) terms were incorporated in the model to retard organism movement in relation to water. The ORNL transport model was applied in the final environmental statement for Indian Point unit 3 (USNRC 1975). Transport avoidance factors, analogous to the CTDF terms, had been added by QLM to its "transport" model by late 1974 (Lawler 1974).

A stipulation agreement in the Indian Point unit 3 proceedings forestalled (indefinitely, as it turned out) the expected battle of the ORNL and QLM hydrodynamic models. Lawler, Matusky & Skelly Engineers (LMS, formerly QLM) developed a more complex, two-dimensional transport model (Figure 55), the "real-time life-cycle" (RTLC) model (LMS 1975). The RTLC model took the concepts introduced in the earlier ORNL FOCAL and GDYEAR models one additional step by including the real-time simulation of flows in the upper and lower layers of the Hudson and coupling this with a simulation of the vertical migration phenomenon. The organism distributions ob-

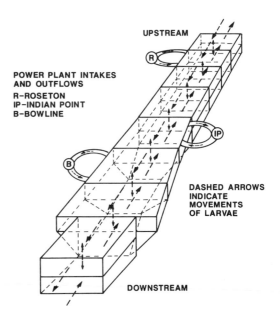

FIGURE 55.—Schematic diagram of the two-dimensional model developed by Lawler, Matusky & Skelly Engineers (LMS 1975). This model is conceptually similar to the FOCAL and GDYEAR models (Figure 54; Table 45), although the transport and movement aspects differ substantially, and the complexity of the two-dimensional model is much greater.

served in the field data indicated that larvae were transported downstream at a rate even slower than that predicted by the RTLC model. However, it was not clear whether this was the result of a deficiency in the simulation, difficulties in field sampling, differential mortality of the organisms at different locations in the Hudson, or combinations of these.

Following the development of STRIPE, government consultants at ORNL took a different tack: they began working with the QLM "transport" model (now the LMS "transport" model). This was done in large part to try to defuse the inevitable controversy about which model was better, and to focus instead on the more important biological input values. After performing a sensitivity analysis of the model (Van Winkle et al. 1976), ORNL used it (including a modification of the compensation function) in a study on the impact of multiple power plants conducted for the U.S. Army Corps of Engineers (Barnthouse et al. 1977).

Meanwhile, as field data became more abundant and of higher quality, modelers began to turn to empirical models that used the data directly and did not rely on the mechanistic formulations used in the other models. Empirical entrainment and impingement methodologies were developed and used by Texas Instruments (TI) in its multiplant impact study (TI 1975). For these, TI used the data it had been collecting on age-0 striped bass distributions throughout the estuary since 1973. The losses of each life stage were calculated at weekly intervals based on f- and W-factors, plant water withdrawals, and estimated densities in 21-km reaches centered on the plants. These losses were compared with total standing crops of all life stages. In essence, the data, which were used by hydrodynamic modelers to calibrate terms in the hydrodynamic transport models so these models could predict distributions, were used in the empirical models directly to define those distributions. The U.S. Fish and Wildlife Service's National Power Plant Team (NPPT) and ORNL worked together to develop an alternative empirical model. This model (Boreman et al. 1978, 1981) was similar in concept to models previously used in adjudicatory hearings on the Summit power station in 1974 and 1975 (Christensen et al. 1975). Cohorts of organisms were followed in space and time, enabling data on spatial and temporal distributions by life stage to be incorporated into the estimation of entrainment mortality. Both the TI and the NPPT–ORNL empirical models were used in the final Hudson

River power case hearings (Englert et al. 1988, this volume), representing an alternative approach to the complex "real-time life-cycle" transport model. While these two empirical models were based on considerably different formulations, they both used the field data directly and did not include any hydrodynamics or mechanistic transport. The coexistence of empirical and transport models here paralleled the situation in 1972 with Indian Point unit 2, although much had been learned since then and the empirical approach was no longer in danger of being overwhelmed.

Common Ground for Entrainment Models

All the entrainment models shared a set of common needs. Foremost was the need for data to define the W- and f-factors. Whether such factors were explicitly or implicitly incorporated or ignored in a particular model, they had a necessary role in providing reliable estimates of entrainment. Recognition of this need stimulated the collection of a great deal of useful data on W- and f-factors prior to and during the EPA hearings.

Second, all the models shared a need for data on the spatial distribution of life stages. The empirical models literally operate on such data, and the transport models require them for calibration and validation. Continuation and refinement of the extensive riverwide data collection effort made possible a much more robust application of entrainment models, both for striped bass and for other species, than would have been possible if only a single year's data had existed.

Entrainment models must be soundly based conceptually. As Table 45 shows, there are many ways of constructing entrainment models. What is more important than the particular method chosen, however, is that the model be demonstrably sound mathematically. Models that cannot be derived mathematically (e.g., the HRFI model) should not be trusted. Of course, a model's derivability does not ensure the reliability of its estimates. These depend on the degree of consistency of the mathematical assumptions with natural processes, and on the implementer's ability to determine "correct" parameter values.

Finally, there is the question of the relative utility of empirical versus hydrodynamic models (Swartzman et al. 1978; Barnthouse et al. 1984). The real impetus for developing a complex hydrodynamic model lies in the capability of such a model to predict the consequences of conditions that cannot be observed, such as different spawning patterns or variations in the magnitude and timing of freshwater runoff. In this case, the hydrodynamic models have been applied only to those years for which sampling data exist. To do this, it has been necessary to use factors to adjust for the failure of early life stages to be transported with water as would dissolved material. The need for these factors complicated the calibration of the models and also made them harder to justify. How well such models simulated actual fish transport and movement could always have been argued, especially given inevitable anomalies in the ichthyoplankton field data. Because the field data were available anyway, empirical models required a minimum number of assumptions, were easy to explain and defend, and could be inexpensively run for different species and scenarios. They were, therefore, useful in the Hudson River power case, particularly in the settlement negotiation process during 1979–1980. At present, however, complex hydrodynamic models have more than a full decade of development behind them. These models can be thought of as experimental tools with which one can try to understand the factors involved in the observed distributions of organisms. It is possible to vary not only parameter values, but also the physical mechanisms being modeled. It will be interesting to see whether the extensive effort that has gone into the development and application of mechanistic models will lead eventually to the ability to both account for and predict the movement of young-of-the-year fish.

Acknowledgments

L. W. Barnthouse, R. J. Klauda, P. H. Muessig, W. Van Winkle, D. S. Vaughan, and an anonymous reviewer critically reviewed the manuscript and provided helpful suggestions. S. W. Christensen's contribution was supported by the Office of Nuclear Reactor Regulation, U.S. Nuclear Regulatory Commission, under interagency agreement DOE 40-544-75 with the U.S. Department of Energy (DOE); by the Division of Biomedical and Environmental Research, DOE, under contract DE-AC05-840R21400 with Martin Marietta Energy Systems, Incorporated; and by the Office of Research and Development, U.S. Environmental Protection Agency, under interagency agreement DOE 40-740-78 (EPA 79-D-X0533). This is publication 2916 of the Environmental Sciences Division, Oak Ridge National Laboratory.

References[1]

Atomic Safety and Licensing Appeal Board (U.S. Atomic Energy Commission). 1974. Decision (ALAB-188) in the matter of Consolidated Edison Company of New York, Inc. (Indian Point station, unit no. 2). Docket 50-247.

Barnthouse, L. W., J. Boreman, S. W. Christensen, C. P. Goodyear, W. Van Winkle, and D. S. Vaughan. 1984. Population biology in the courtroom: the lesson of the Hudson River controversy. BioScience 34:14–19.

Barnthouse, L. W., and 12 coauthors. 1977. A selective analysis of power plant operation on the Hudson River with emphasis on the Bowline Point generating station. Oak Ridge National Laboratory, ORNL/TM-5877 (volumes 1 and 2), Oak Ridge, Tennessee.

Boreman, J., C. P. Goodyear, and S. W. Christensen. 1978. An empirical transport model for evaluating entrainment of aquatic organisms by power plants. U.S. Fish and Wildlife Service Biological Services Program FWS/OBS-78/90.

Boreman, J., C. P. Goodyear, and S. W. Christensen. 1981. An empirical methodology for estimating entrainment losses at power plants sited on estuaries. Transactions of the American Fisheries Society 110:253–260.

Boreman, J., and R. J. Klauda. 1988. Distributions of early life stages of striped bass in the Hudson River estuary, 1974–1979. American Fisheries Society Monograph 4:53–58

Carter, L. J. 1974. Con Edison: endless storm king dispute adds to its troubles. Science (Washington, D.C.) 184:1353–1358.

Christensen, S. W., W. Van Winkle, and P. C. Cota. 1975. Effect of Summit power station on striped bass populations. Testimony (March 1975) before the U.S. Atomic Energy Commission in the matter of Delmarva Power and Light Co. (Summit power station, units 1 and 2). Dockets 50-450 and 50-451.

Clark, J. R. 1972. Effects of Indian Point units 1 and 2 on Hudson River aquatic life. Testimony (October 30, 1972) before the U.S. Atomic Energy Commission in the matter of Consolidated Edison Company of New York, Inc. (Indian Point station, unit no. 2). Docket 50-247.

DeAngelis, D. L., and six coauthors. 1978. A generalized fish life-cycle population model and computer program. Oak Ridge National Laboratory, ORNL/TM-6125, Oak Ridge, Tennessee.

Englert, T. L., and J. Boreman. 1988. Historical review of entrainment impact estimates and the factors influencing them. American Fisheries Society Monograph 4:143–151.

Eraslan, A. H., and six coauthors. 1976. A computer simulation model for the striped bass young-of-the-year population in the Hudson River. Oak Ridge National Laboratory, ORNL/NUREG-8, Oak Ridge, Tennessee.

---

[1]See Table 1 for sources of legal documents and unpublished reports pertaining to the Hudson River.

Goodyear, C. P. 1973. Probable reduction in survival of young-of-the-year striped bass in the Hudson River as a consequence of the operation of Danskammer, Roseton, Indian Point units 1 and 2, Lovett, and Bowline steam electrical generating stations. Testimony (February 8, 1973) before the U.S. Atomic Energy Commission in the matter of Consolidated Edison Company of New York, Inc. (Indian Point station, unit no. 2). Docket 50-247.

Goodyear, C. P. 1977. Mathematical methods to evaluate entrainment of aquatic organisms by power plants. U.S. Fish and Wildlife Service Biological Services Program FWS/OBS-76/20.3.

HRPC (Hudson River Policy Committee). Undated [1968]. Hudson River fisheries investigations (1965–1968). Report to Consolidated Edison Company of New York.

Lawler, J. P. 1972a. Effect of entrainment and impingement at Indian Point on the population of the Hudson River striped bass. Modifications and additions to testimony of April 5, 1972. Testimony (October 30, 1972), before the U.S. Atomic Energy Commission in the matter of Consolidated Edison Company of New York, Inc. (Indian Point station, unit no. 2). Docket 50-247.

Lawler, J. P. 1972b. The effect of entrainment at Indian Point on the population of the Hudson River striped bass. Testimony (April 5, 1972) before the U.S. Atomic Energy Commission in the matter of Consolidated Edison Company of New York, Inc. (Indian Point station, unit no. 2). Docket 50-247.

Lawler, J. P. 1973. Responses to questions . . . on the sensitivity of the model presented in the testimony of October 30, 1972, on the effect of entrainment and impingement at Indian Point on the population of Hudson River striped bass. Testimony (February 5, 1973) before the U.S. Atomic Energy Commission in the matter of Consolidated Edison Company of New York, Inc. (Indian Point station, unit no. 2). Docket 50-247.

Lawler, J. P. 1974. Effect of entrainment and impingement at Cornwall on the Hudson River striped bass population. Testimony (October 1974) before the Federal Power Commission in the matter of Consolidated Edison Company of New York, Inc. Project 2338.

LMS (Lawler, Matusky & Skelly Engineers). 1975. Report on development of a real-time, two-dimensional model of the Hudson River striped bass population. Report to Consolidated Edison Company of New York.

Muessig, P. H., J. R. Young, D. S. Vaughan, and B. A. Smith. 1988. Advances in field and analytical methods for estimating entrainment mortality factors. American Fisheries Society Monograph 4:124–132.

Swartzman, G. L., R. B. Deriso, and C. Cowan. 1978. Comparison of simulation models used in assessing the effects of power-plant induced mortality on fish populations. University of Washington, College of Fisheries, Center for Quantitative Science, UW-NRC-10, Seattle.

TI (Texas Instruments). 1975. First annual report for the multiplant impact study of the Hudson River estuary, volumes 1 and 2. Report to Consolidated Edison Company of New York.

U.S. CONGRESS. 1972. Federal Water Pollution Control Act Amendments of 1972 (Public Law 92-500). Pages 816 et seq. in U.S. statutes at large 86. U.S. Government Printing Office, Washington, D.C.

USAEC (U.S. Atomic Energy Commission). 1972. Final environmental statement related to operation of Indian Point nuclear generating plant, unit no. 2, volumes 1 and 2. Docket 50-247.

USNRC (U.S. Nuclear Regulatory Commission). 1975. Final environmental statement related to operation of Indian Point nuclear generating plant, unit no. 3, volumes 1 (NUREG-75/002) and 2 (NUREG-75/003). Docket 50-286.

Van Winkle, W., S. W. Christensen, and G. Kauffman. 1976. Critique and sensitivity analysis of the compensation function used in the LMS Hudson River striped bass models. Oak Ridge National Laboratory, ORNL/TM-5437, Oak Ridge, Tennessee.

Van Winkle, W., B. W. Rust, C. P. Goodyear, S. R. Blum, and P. Thall. 1974. A striped-bass population model and computer programs. Oak Ridge National Laboratory, ORNL/TM-4578, Oak Ridge, Tennessee.

Vaughan, D. S. 1988. Introduction [to entrainment and impingement impacts]. American Fisheries Society Monograph 4:121–123.

American Fisheries Society Monograph 4:143–151, 1988

Historical Review of Entrainment Impact Estimates and the Factors Influencing Them

Thomas L. Englert

Lawler, Matusky & Skelly Engineers, One Blue Hill Plaza
Pearl River, New York 10965, USA

John Boreman

National Marine Fisheries Service, Northeast Fisheries Center
Woods Hole Laboratory, Woods Hole, Massachusetts 02543, USA

Abstract.—Estimates of entrainment mortality have played a key role in the Hudson River power plant case since the late 1960s. The estimates of entrainment mortality of striped bass due to the operation of Indian Point nuclear plant were first presented by utility and government consultants in 1972. These estimates were derived with the first of many generations of mathematical models. Assumed values had to be used for some of the critical parameters because of the limited amount of data available. The government estimates were much higher as a result of the intentional use of conservative assumptions. As more data became available and model formulations evolved, fewer assumptions were needed, the government estimates declined, and the gap between estimates from the two groups of consultants narrowed markedly. Several factors contributed to the convergence of model estimates, but a key factor was field data showing that more than half of the entrained striped bass survive passage through the plant. The convergence of estimates of entrainment mortality was a key ingredient in the eventual formulation of the settlement agreement and showed that progress had been made in resolving an important portion of the heavily contested issues.

Estimates of entrainment mortality have played a key role in the licensing and adjudicatory hearings regarding the Hudson River power plants since the late 1960s. The controversy, which came to be known as the Hudson River power plant case, was initially centered almost exclusively on the question of power plant impact on the Hudson River striped bass population. Over the period of a decade, wide-ranging estimates of entrainment and impingement mortality of striped bass were generated. At least at first, the estimates generated by the consultants to the regulatory agencies were considerably higher than those provided by the consultants to the utilities. In this paper, we discuss in detail the reasons for the initially large discrepancy in the impact estimates from the two groups of consultants and chronicle the eventual near convergence of these estimates. Elsewhere in this monograph, Christensen and Englert (1988) discuss the evolution of the various modeling techniques used in the Hudson River power plant case and Boreman and Goodyear (1988) summarize the estimates of entrainment mortality generated in the final written testimony presented by consultants to the U.S. Environmental Protection Agency (EPA) in May 1979.

After the Indian Point unit 2 hearing in 1973, it became evident that differences in impact esti-

mates by utility and government consultants were largely the result of differences in estimates for the input parameters to the models rather than in the model formulations themselves. As the case evolved, the modeling techniques were refined by both groups of consultants; it usually happened that when the same input parameters were specified for alternative models, from either group, the resulting impact estimates were very close. However, the precise values for the input parameters to the model remained in dispute. Attempts to estimate the long-term impact on the fishery were further confounded by the controversy regarding the potential for offset of the power plant impact by changes in density-dependent growth, survival, and fecundity rates, referred to collectively as compensation or compensatory response. This topic is addressed later in this monograph by Lawler (1988) and Fletcher and Deriso (1988). Our principal concern here will be with estimates of conditional entrainment mortality (i.e., density-independent mortality due to power plant operations). The discussion will also be limited to striped bass because this species has been modeled since the beginning of the case and was the one that received most attention in the formulation of the settlement agreement.

Input Parameters

A simplified formulation of an entrainment model is:

$$\frac{dc}{dt} = -K_n c - \frac{Q_p}{V} W \cdot f \cdot c; \tag{1}$$

c = organism concentration (number/m^3);
t = time (days);
K_n = natural mortality rate;
Q_p = power plant flow (m^3/d);
V = river segment volume (m^3);
W = withdrawal factor;
f = entrainment mortality factor.

From this simplified model, it is evident that two parameters derived from biological field data have a direct influence on the predictions of impact—withdrawal ratio (W) and the through-plant mortality factor (f). The W-ratio was developed to account for differences between concentrations of organisms in the water withdrawn by the plant and the average river water concentration in front of the plant. This ratio is important because the estimates of river concentrations are computed by the models in terms of cross-sectional average or segment concentrations. Multiplying these model-calculated concentrations by the W-ratio provides an estimate of the withdrawal concentration (i.e., the concentration of organisms in the cooling water).

This concentration is multiplied by f, the fraction of the entrained organisms that do not survive passage through the plant. Further multiplication of this product by the plant flow provides an estimate of the number of organisms cropped by the plant during a specified time period. Thus, while the modeled spatial and temporal distributions in the river and resultant exposure of organisms to entrainment are influenced principally by the model formulation of organism transport, the estimated cropping of these organisms by the plant is determined by the input parameters W and f. As this became evident during the course of the studies and the Hudson River hearings, there was considerable debate regarding the appropriate values for these parameters. At first, data were lacking and assumptions were necessary. The evolution of input parameters during the 1970s (Table 46) reflected several important developments in the technical aspects of the case, including

(1) the accumulation of several years of data from which the parameters could be evaluated,
(2) the development of improved sampling techniques such as the larva table (McGroddy and Wyman 1976) for determining through-plant mortality, and
(3) the formulation and implementation of a variety of data analysis procedures for calculating W-ratios and through-plant mortality (f-factors).

Before 1972, it was assumed that W and f were equal to unity (i.e., that the intake concentration was the same as the cross-sectional average concentration in front of the plant) and that none of the entrained organisms survived passage through the plant. Field measurements conducted as early as 1973 indicated that these assumed values resulted in overestimates of through-plant mortality, because W and f were both measured to be less than 1.0 under some conditions.

Utility consultants first used field data collected at Indian Point by New York University staff to estimate W and f in 1972. After 1973, government consultants began to use the field data collected by utility consultants to calculate their own estimates of W and f. As discussed in more detail below, use of the field data by government consultants was an essential factor in the eventual convergence of the estimates of entrainment impact from the two bodies of consultants.

Once there were data available to estimate W and f, debate began regarding the reliability of the data and the computational procedures to be used in analyzing them. Oak Ridge National Laboratory (ORNL) reviewed the data collected and analyzed by utility consultants. In two reports—the Indian Point unit 3 final environmental statement (USNRC 1975) and the Bowline Point Corps of Engineers study (Barnthouse et al. 1977)—ORNL concluded that W was between 0.5 and 1.0. In the final environmental statement for Indian Point unit 3, ORNL computed estimates of f from the utility data base that agreed closely with those computed by utility consultants. For Bowline Point, however, ORNL computed its own estimates of f from utility data for some life stages, and these differed considerably from those computed by the utility consultants, for example, for juveniles—0.69 (ORNL) versus 0.16 (utility consultants).

In 1979 testimony during the hearing conducted pursuant to Section 316(b) of the Federal Water Pollution Control Act Amendments, consultants to EPA introduced several innovative approaches for computing the W-ratio and also proposed alternative ways of analyzing the data on through-plant mortality. The resulting values of the W-

ratio varied widely; in some cases, values considerably greater than 1.0 were computed. However, the overall average values for Indian Point ranged from 0.7 to 1.0 (Table 46), and most values for yolk-sac and post–yolk-sac larvae were less than 1.0. Larva-table data collected during 1975–1977 were used to develop estimates of the f-factor. Analyses of these data by EPA consultants resulted in values ranging from 0.18 to 0.55 for the larval stages (Boreman et al. 1979).

Combined Influence of W and f

Further insight into the evolution and eventual convergence of the values for model input parameters can be gained from an examination of the product of the W-ratio and the f-factor developed during the course of the studies. This product is important because all the model formulations use it when the number of organisms cropped by entrainment is computed. For example, integration of the simplified model presented in equation (1) gives

$$c = c_0 e^{-(K_n + \frac{Q_p}{V} W \cdot f \cdot t)}; \qquad (2)$$

c_0 = initial organism concentration (number/m^3).

The percent reduction, R, in the population as a result of power plant operation can then be computed from

$$R = \frac{c_0 e^{-K_n t} - c_0 e^{-(K_n t + \frac{Q_p}{V} W \cdot f \cdot t)}}{c_0 e^{-K_n t}} \times 100; \qquad (3)$$

$$R = (1 - e^{-\frac{Q_p}{V} W \cdot f \cdot t}) \times 100. \qquad (4)$$

Based on this simplified model, the percent reduction from power plant operations (i.e., the impact) increases exponentially with the product of W, f, and t. Because t represents the time duration of exposure of the organisms to entrainment, it is typically estimated based on the duration of a given life stage. Thus, eggs that hatch after 3 d are exposed to entrainment for a much shorter duration than other life stages, so estimates of entrainment impact typically are smaller for eggs than for other life stages. Further, the W and f factors for eggs carry less weight in the overall estimate of entrainment impact. To account for this, a plant impact index (PII) was

computed, as shown in Table 46. This index combines the individual life stage values for W and f into a single parameter that has weighted the influences of life stage durations on the percent reduction as expressed in equation (4).

Estimates of the PII computed from data used by the utility consultants were consistently around 0.3 with the exception of the assumed value of 1.0 used in the simple models (Table 46). Although the PII has remained approximately the same, the individual values of W and f contributing to the PII values have varied considerably. Overall, however, an increase in one of the parameters, W or f, was offset by a decrease in the other. Estimates of the PII computed from W- and f-values used by government consultants ranged from 0.26 to 1.0. However, when values used prior to 1974 were eliminated, the range of PII values computed from data used by both groups of consultants narrowed considerably. The utility consultant values ranged from 0.12 to 0.33, while the government consultant values ranged from 0.26 to 0.70. The value used in the settlement agreement, 0.26, falls toward the upper end of the range of values used by the utility consultants and at the lower end of the range used by the government consultants.

Of the two factors, W and f, the f-factor contributed most significantly to reduction of the PII from 1.0 in 1973 to the 0.26 value used in the settlement agreement. While W estimates declined from assumed values of 1.0 to computed values of 0.8 in the settlement agreement, the f-factor declined from an assumed value of 1.0 to values of 0.4 for larvae and 0.2 for juveniles, the life stages exposed for the longest periods of time to entrainment. As discussed by Muessig et al. (1988, this volume) the lower estimates of through-plant mortality, f, were the result of improved methods of measuring entrainment mortality (i.e., the larva table). By reducing sampling mortality, the larva table permitted measurement of through-plant mortality values, showing that a considerable percentage of the entrained organisms survived plant passage.

Predictions of Entrainment Impact

To provide a basis for comparison of model results spanning more than a decade, all the predictions of impact summarized in Table 46 are expressed in terms of conditional mortality (i.e., independent of any other source of mortality and in the absence of compensation) caused by entrainment of striped bass at Indian Point units 1 and 2. These were the only units considered in

TABLE 46.—Summary of model input parameters and striped bass conditional entrainment mortalities due to operation of Indian Point units 1 and 2.

Model[a]	Year	Funding source	Life stage	Withdrawal ratio (W)	Through-plant mortality (f)	$W \cdot f$	Plant impact index (PII)[b]	Conditional mortality[c] (%)	References and notes
HRFI	1968	Utilities	All	1.0	1.0	1.0	1.0		HRPC (undated): no estimate given for Indian Point
CMM	1972	Utilities	All	1.0	1.0	1.0	1.0	19	Lawler (1972): conditional mortality due to Indian Point based on Table 8 and Figure 7 of that document
Probability	1972	Government	All	1.0	1.0	1.0	1.0		USAEC (1972), page iii, gives the probability of entrainment but does not provide an estimate of conditional mortality
FOCAL	1972	Government	All	1.0	1.0	1.0	1.0	40	USAEC (1972), page iii, item j.2, says entrainment at Indian Point units 1 and 2 will result in 30–50% reduction in striped bass larvae
GDYEAR	1973	Government	All	1.0	1.0	1.0	1.0	32	Goodyear (1973): 32% is the average of the estimates of impact due to Indian Point units 1 and 2 presented in Goodyear's Table 1
Transport	1973	Utilities	Eggs	0.4	1.0	0.40			Lawler (1973): case 31 of Table 1 presents an estimated conditional mortality of 4.3% due to entrainment at Indian Point units 1 and 2
			Larvae	0.4	1.0	0.40	0.27	4	
			Juveniles	0.2	0.5	0.10			
Transport	1974	Utilities	Eggs	0.6	0.8	0.48			Lawler (1974): the 5% conditional mortality is an estimated value based on the ratio of the PII values for the 1974 and 1973 applications of the transport model
			Larvae	0.2	0.6	0.12	0.32	5	
			Juveniles	0.8	0.7	0.56			
STRIPE	1975	Government	Eggs	0.75[d]	0.8	0.60			USNRC (1975): the estimated impact of entrainment at Indian Point units 1, 2, and 3 is 21–31% (page vi and case 11, Table V-20); after removal of the effect of unit 3 and use of the ratio of plant flows, the midpoint of this range of impacts is 16%
			Larvae	0.75[d]	0.6	0.45	0.49	16	
			Juveniles	0.75[d]	0.7	0.53			
Transport (ORNL application)	1977	Government	Eggs	1.0[e]	0.7	0.70			Barnthouse et al. (1977): Table 5.8.9 indicates that the decrease in multiplant conditional mortality with implementation of closed-cycle cooling at Indian Point units 2 and 3 would be 9.1% (29.7%−20.6%), including impingement effects; because impingement effects are typically close to the closed-cycle impact, the estimated conditional entrainment mortality at Indian Point units 2 and 3 can be assumed to be about 9% and the effect of units 1 and 2 to be about 8%, based on the ratio of plant flows
			Larvae	1.0	0.7	0.70	0.70	8	
			Juveniles	1.0	0.7	0.70			

TABLE 46.—Continued.

Model[a]	Year	Funding source	Life stage	With- drawal ratio (W)	Through- plant mortality (f)	W · f	Plant impact index (PII)[b]	Condi- tional mortality[c] (%)	References and notes
TI	1977	Utilities	Eggs	0.5	0.8	0.40			McFadden and Lawler (1977):
			Larvae	0.5	0.6	0.30	0.33	8	Table 2-VI-1 gives an average
			Juveniles	0.5	0.7	0.35			conditional mortality of 5.8% for entrainment at Indian Point unit 2; scaling this value up based on plant flow gives about 8% for the combined effect of Indian Point units 1 and 2
RTLC	1977	Utilities	Eggs	1.5	0.6	0.90			McFadden and Lawler (1977):
			Larvae	0.7	0.2	0.14	0.12	3	results in Table 3-VIII-1 show a
			Juveniles	0.2	0.1	0.02			conditional entrainment mortality of 5–8% due to Indian Point units 2 and 3, Bowline units 1 and 2, and Roseton units 1 and 2 combined; because the major portion of this impact occurs at Indian Point, the corresponding conditional mortality due to entrainment at Indian Point units 1 and 2 is 3%
ETM	1979	Govern- ment	Eggs	0.8	0.7	0.56			Boreman et al. (1979): Appendix G-4
			Larvae	0.8	0.4	0.32	0.26	7	gives an average conditional
			Juveniles	0.8	0.2	0.16			entrainment mortality estimate of 10.6% for Indian Point units 2 and 3 based on the f-factors presented here; correction of this value for the flow difference gives an estimate of 7% conditional mortality for entrainment at Indian Point units 1 and 2
ETM (settlement agreement)	1980	Govern- ment and utili- ties	Eggs	0.8	0.7	0.56			Settlement agreement runs indicated
			Larvae	0.8	0.4	0.32	0.26	7	a conditional entrainment
			Juveniles	0.8	0.2	0.16			mortality of about 10% for Indian Point units 2 and 3; based on relative plant flows, the corresponding value for Indian Point units 1 and 2 is about 7%
ETM	1983	Utilities	Eggs	0.9	0.4	0.36			LMS (1983): Tables 3.0-1 and 3.0-8
			Larvae	0.8	0.3	0.24	0.16	4	indicate the estimated conditional
			Juveniles	0.2	0.2	0.04			entrainment mortality due to Indian Point units 2 and 3 is about 5.5%; the corresponding value for units 1 and 2 is 3.8%, based on the relative magnitude of the plant flows

[a]HFRI = Hudson River fisheries investigations; CMM = completely mixed model; ORNL = Oak Ridge National Laboratory; TI = Texas Instruments; RTLC = real-time life-cycle; ETM = empirical transport model.

[b]PII = average of $W \cdot f$ weighted by durations of the life stages in days to reflect their relative importance to conditional entrainment mortality:

$$PII = [3(W \cdot f)_{eggs} + 36(W \cdot f)_{larvae} + 28(W \cdot f)_{juveniles}]/67.$$

[c]Conditional entrainment mortality is the fractional reduction in the young-of-the-year population due to power plant operations in the absence of other sources of mortality and of compensatory responses.

[d]Middle of the 0.5–1.0 range used in the final environmental statement for Indian Point unit 3.

[e]ORNL used the 0.5–1.0 range, but conditional mortality was not computed for the 0.5 value.

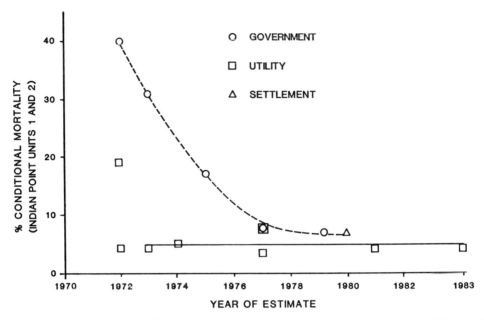

FIGURE 56.—Convergence of conditional entrainment mortality estimates by government and utility consultants for striped bass at Indian Point nuclear power plant.

some of the earlier modeling studies and it was also more practical to scale back results of later multiplant studies to these units alone than to attempt to extrapolate earlier results to the multiplant situation. More recent studies did not provide direct estimates of the effects of Indian Point units 1 and 2. In these cases, the results for Indian Point unit 2 alone or units 2 and 3 together were scaled up or down based on the appropriate ratio of power plant flows. Prediction of conditional entrainment mortality at Indian Point units 1 and 2 ranged from 4 to 40%, an order-of-magnitude difference (Table 46). This wide spectrum resulted from differences in both the model formulations and the parameter values used in various studies.

Over the period 1972–1979, predictions of conditional entrainment mortality by the government consultants declined from 40% to about 7% (Figure 56). Their initial estimates were intentionally conservative; in the absence of field data, worst-case assumptions were made regarding input parameters. In 1974, the Atomic Safety and Licensing Appeal Board (ASLAB) urged the government consultants to take a more realistic approach to the estimates and, with the increasing availability of data collected by the utilities, their subsequent estimates of impact declined substantially. As part of their independent review and analysis of the utility data, the government con-

sultants raised important concerns that were incorporated by the utility consultants in the design and analysis of future studies. This resulted in a data base that was more acceptable to both groups.

The first modeling study performed by the utility consultants in early 1972, prior to the availability of field data, involved a "completely mixed model." This used the same conservative input parameters employed by the government consultants and produced a conditional mortality estimate of 19% for Indian Point units 1 and 2. In late 1972 and early 1973, the utility consultants estimated an entrainment impact of only 4%, based on input parameters calculated from field data and a transport modeling approach. Their subsequent modeling studies through 1983 resulted in consistent predictions of 4–8% mortality (Figure 56). This consistency was largely a result of their early incorporation of empirical values for W and f into the models, coupled with a somewhat fortuitous counterbalancing of changes in these parameters.

The accumulation and averaging of several years of data, which was done in developing model input parameters for the settlement agreement runs, helped to lessen the uncertainty in the parameter values, and also lessened the influence of a single year's data. Nevertheless, the esti-

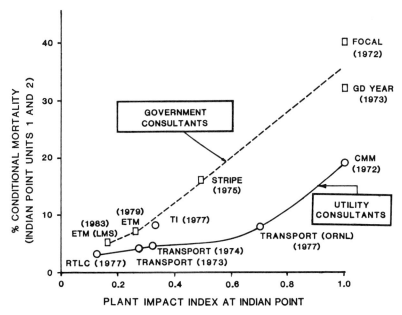

FIGURE 57.—Relationship of estimated conditional entrainment mortality of striped bass to the plant impact index at Indian Point nuclear power plant. CMM = completely mixed mode; ETM = empirical transport model; LMS = Lawler, Matusky & Skelly; ORNL = Oak Ridge National Laboratory; RTLC = real-time life-cycle; TI = Texas Instruments.

mates of impact for Indian Point units 1 and 2, based on the settlement agreement input parameters, agreed closely with earlier estimates derived from data from individual years. Subsequent to the settlement agreement, consultants to the utilities computed estimates of conditional mortality with the empirical transport model and average impact parameters based on data collected during 1979–1980 (LMS 1983). These runs produced estimates nearly identical to those derived in earlier studies by utility consultants, and were reasonably close to those derived in the settlement agreement analysis.

Influence of Model Formulations

From the preceding discussions regarding input parameters used by two groups of consultants and the eventual agreement on values that approached those used in earlier studies by the utility consultants, the convergence of estimates of impact shown in Figure 56 was not unexpected. However, it must be recognized that, in addition to changes in input parameters, a variety of model formulations were used over the course of the studies. Of interest here is the role of the changing model formulations in the eventual convergence of the estimates of entrainment impact.

When estimates of conditional entrainment mortality from the various studies are compared graphically with changes in the plant impact index, some interesting patterns emerge (Figure 57). Estimates of conditional mortality for various values of the PII fall on two trend lines, one through estimates generated principally from models formulated by the consultants to the utilities and the other through results from the models formulated by the government consultants. Based on the differences in the two curves, it is evident that model formulations played an important role in the estimates of conditional mortality generated from the various studies. The shapes of the two lines and differences of the conditional mortality estimates at the same value of the PII are also of interest.

First, regarding the shape of the two curves, the line through the results from the government models is closer to the slightly convex shape expected based on equation (4), that is, it is closer to what would result if all the conditional mortality values were generated by the same model. The line through the results from the utility models is concave instead of convex, suggesting that fundamental changes in the formulation of the utility models occurred over the course of the studies.

Indeed, the change from the completely mixed model formulation, in which all organisms have an equal probability of being entrained, to the transport model, in which the susceptibility to entrainment was determined by the spatial distribution of the organisms, represented a substantial change in the formulation of the utility models. Results of a sensitivity analysis (Lawler 1973) indicated that with PII = 1.0, the conditional entrainment mortality estimated from the transport model was 14%. Using this result in lieu of the completely mixed model estimate results in a curve whose shape is closer to that expected when the results are all from one model. However, given that the estimates of conditional mortality at PII values of 0.3–1.0 are all from the transport model, one would expect the results to follow the theoretical curve more closely. This would be the case if the conditional mortality estimate from the ORNL application of the transport model (PII = 0.7) were 11–12%, instead of 8%.

Comparison of the differences in estimates of conditional mortality from the utility- and agency-sponsored models at various levels of PII provides some additional insights. The apparently large difference in estimates of conditional mortality from the STRIPE (conditional mortality = 16% at PII = 0.49) and transport models (conditional mortality = 14% at PII = 1.0) is of interest, for example, because these models are similar conceptually. Both models are based on a one-dimensional, tidally averaged simulation of the effects of the river flow on distribution of the planktonic life stages. There were subtle differences in the ways in which the models handled transport and aging of the organisms, and the STRIPE model, as originally implemented, allowed organisms to be transported out of the system, thereby possibly increasing the estimate of conditional mortality; however, the precise reason for the difference in estimates from the STRIPE and transport models is not apparent.

Discussion

The convergence in predictions of conditional entrainment mortality (Figure 56), which played a key role in laying the groundwork for the settlement agreement, was the result of a gradual evolution of modeling approaches, sampling methods, and data analysis techniques acceptable to both sides. The summary of input parameters and estimates of impact presented here demonstrates that a principal reason for the decline in the government consultants' estimates of impact

was their acceptance of estimates of through-plant mortality obtained from larva-table data collected at the power plants. By reducing sampling mortality, the larva table demonstrated that a considerable percentage of the entrained organisms survived passage through the plant.

Model formulations also played a role in the convergence of impact estimates. After development of several generations of complex mechanistic models that simulated the aging, growth, and transport of the young-of-the-year striped bass, the parties agreed, during settlement negotiations, that the empirical transport model, developed by government consultants, would be used to predict the effectiveness of various mitigating measures such as flow reductions and outages. This model used the extensive field data on spatial and temporal distributions of the early life stages directly, without complex simulations, and thus helped to focus the issues (Christensen and Englert 1988).

By the time of the settlement negotiations, the modeling and field studies had evolved to include five species and three power plants. However, the striped bass remained the species of primary concern and the same modeling, data collection, and analysis methods were applied to the other species as were applied to striped bass.

Given the complexity of the problem, it is in some ways remarkable that the estimates of impact from the government and utility consultants did eventually converge. It was the combined influences of the availability of many years of field data, the improvements in sampling techniques, the use of empirical models, and the focusing on conditional mortality rather than on long-term estimates of impact that led to final resolution of the technical issues. The modeling studies had served their purpose by providing a measure of impact that reflected the complex interaction of a host of factors and by defining the relative importance of each component of the analysis.

References[1]

Barnthouse, L. W., and twelve coauthors. 1977. A selective analysis of power plant operation on the Hudson River with emphasis on the Bowline Point generating station, volumes 1 and 2. Oak Ridge National Laboratory, ORNL/TM-5877, Oak Ridge, Tennessee.

Boreman, J., L. W. Barnthouse, D. S. Vaughan, C. P. Goodyear, and S. W. Christensen. 1979. Entrainment impact estimates for six fish populations in-

[1]See Table 1 for sources of legal documents and unpublished reports pertaining to the Hudson River.

habiting the Hudson River estuary. Report to U.S. Environmental Protection Agency, Region 2.

Boreman, J., and C. P. Goodyear. 1988. Estimates of entrainment mortality for striped bass and other fish species inhabiting the Hudson River estuary. American Fisheries Society Monograph 4:152–160.

Christensen, S. W., and T. L. Englert. 1988 Historical development of entrainment models for Hudson River striped bass. American Fisheries Society Monograph 4:133–142.

Fletcher, R. I., and R. B. Deriso. 1988. Fishing in dangerous waters: remarks on a controversial appeal to spawner–recruit theory for long-term impact assessment. American Fisheries Society Monograph 4:232–244.

Goodyear, C. P. 1973. Probable reduction in survival of young-of-the-year striped bass in the Hudson River as a consequence of the operation of Danskammer, Roseton, Indian Point units 1 and 2, Lovett, and Bowline steam electrical generating stations. Testimony (February 8, 1973) before the U.S. Atomic Energy Commission in the matter of Consolidated Edison Company of New York, Inc. (Indian Point station, unit no. 2). Docket 50-247.

HRPC (Hudson River Policy Committee). Undated [1968]. Hudson River fisheries investigations (1965–1968). Report to Consolidated Edison Company of New York.

Lawler, J. P. 1972. The effect of entrainment at Indian Point on the population of the Hudson River striped bass. Testimony (April 5, 1972) before the U.S. Atomic Energy Commission in the matter of Consolidated Edison Company of New York, Inc. (Indian Point station, unit no. 2). Docket 50-247.

Lawler, J. P. 1973. Responses to questions . . . on the sensitivity of the model presented in the testimony of October 30, 1972, on the effect of entrainment and impingement at Indian Point on the population of Hudson River striped bass. Testimony (February 5, 1973) before the U.S. Atomic Energy Commission in the matter of Consolidated Edison Company of New York, Inc. (Indian Point station, unit no. 2), Docket 50-247.

Lawler, J. P. 1974. Effect of entrainment and impingement at Cornwall on the Hudson River striped bass population. Testimony (October 1974) before the Federal Power Commission in the matter of Consolidated Edison Company of New York, Inc. Project 2338.

Lawler, J. P. 1988. Some considerations in applying stock–recruitment models to multiple-age spawning populations. American Fisheries Society Monograph 4:204–218.

LMS (Lawler, Matusky & Skelly Engineers). 1983. 1979 and 1980 data analyses and application of empirical models of Hudson River fish populations. Jointly financed by Consolidated Edison Company of New York, Orange and Rockland Utilities, Central Hudson Gas and Electric, and New York Power Authority.

McFadden, J. T., and J. P. Lawler, editors. 1977. Supplement I to influence of Indian Point unit 2 and other steam electric generating plants on the Hudson River estuary, with emphasis on striped bass and other fish populations. Report to Consolidated Edison Company of New York.

McGroddy, P. M., and R. L. Wyman. 1976. Efficiency of nets and a new device for sampling live fish larvae. Journal of the Fisheries Research Board of Canada 34:571–574.

Muessig, P. H., J. R. Young, D. S. Vaughan, and B. A. Smith. 1988. Advances in field and analytical methods for estimating entrainment mortality factors. American Fisheries Society Monograph 4:124–132.

USAEC (U.S. Atomic Energy Commission). 1972. Final environmental statement related to operation of Indian Point nuclear generating plant, unit no. 2, volumes 1 and 2. Docket 50-247.

USAEC (U.S. Atomic Energy Commission). 1974. Tide tables. U.S. Department of Commerce, Washington, D.C.

USNRC (U.S. Nuclear Regulatory Commission). 1975. Final environmental statement regulated to operation of Indian Point nuclear generating plant, unit no. 3, volumes 1 (NUREG-75/002) and 2 (NUREG-75/003). Docket 50-286.

American Fisheries Society Monograph 4:152–160, 1988

Estimates of Entrainment Mortality for Striped Bass and Other Fish Species Inhabiting the Hudson River Estuary

JOHN BOREMAN

National Marine Fisheries Service, Northeast Fisheries Center
Woods Hole Laboratory, Woods Hole, Massachusetts 02543, USA

C. PHILLIP GOODYEAR[1]

U.S. Fish and Wildlife Service, National Fisheries Center—Leetown
Kearneysville, West Virginia 25430, USA

Abstract.—An empirically derived age-, time-, and space-variant equation was used to estimate entrainment mortality at power plants for seven fish species inhabiting the Hudson River estuary. Entrainment mortality is expressed as a conditional rate, which is the fractional reduction in year-class strength due to entrainment if other sources of mortality are density-independent. Estimates of the conditional entrainment mortality, based on historical and projected once-through cooling operation of five power plants, were 11–22% for striped bass, 11–17% for white perch, 5–7% for Atlantic tomcod, 14–21% for American shad, 4–11% for river herring (alewife and blueback herring combined), and 35–79% for bay anchovy. Closed-cycle cooling (natural-draft cooling towers) at three of the power plants (Indian Point, Bowline Point, and Roseton) would reduce entrainment mortality of striped bass by 50–80%, of white perch by 75–80%, of Atlantic tomcod by 65–70%, of American shad by 80%, of river herring by 30–90%, and of bay anchovy by 45–80%. The life stages most vulnerable to entrainment mortality were post–yolk-sac larva and entrainable-size juvenile.

The potential impact of entrainment mortality on striped bass in the Hudson River estuary stimulated the development of a series of mathematical models (Englert and Boreman 1988, this volume). Modeling was necessary because, at the time, there were no simultaneous measurements of the number of entrained organisms and the number of organisms of the same age remaining in the estuary. Attempts to simulate organism movement patterns through modeling, however, resulted in poor approximations to field data. Among other reasons, the biological parameters used in the attempts were too simple to mesh with the hydrodynamic detail necessary to describe the physical system (Swartzman et al. 1978). Recognizing this limitation, we decided to use an empirically derived age-, time-, and space-variant entrainment equation, termed the empirical transport model (ETM), which calculates a conditional entrainment mortality rate (Boreman et al. 1981). This conditional rate is the fractional reduction in year-class strength due to entrainment if other sources of mortality are density-independent. We used the ETM to estimate entrainment

mortality for striped bass, white perch, Atlantic tomcod, American shad, alewife, blueback herring, and bay anchovy. Due to the difficulty in distinguishing larvae of blueback herring and alewife in field samples, these species were jointly categorized as "river herring" for entrainment impact assessment. The life histories of these species in the Hudson River were described by McFadden et al. (1977) and Boreman (1981).

This paper presents estimates of the conditional entrainment mortality of the seven fish species that we submitted as direct testimony in the Hudson River power plant case. The estimates reflect the levels of entrainment mortality experienced by the fish species in 1974 and 1975, and the levels that would be experienced with two projected conditions of cooling-water withdrawal by power plants: once-through water flow at all plants, and closed-cycle flow (natural-draft cooling towers) at Bowline Point, Indian Point, and Roseton. Projections were run with 1974 and 1975 biological values separately, because we did not feel that an average of the values for the 2 years would necessarily represent future biological conditions in the estuary. Englert and Boreman (1988) compare these estimates with those derived by other methods in the Hudson River power plant case.

---

[1]Present address: National Marine Fisheries Service, Southeast Fisheries Center, Miami Laboratory, 75 Virginia Beach Drive, Miami, Florida 33149, USA.

Empirical Transport Model

The mathematical basis of the ETM, its assumptions, and its limitations are found in Boreman et al. (1981). Among the more important assumptions are that (1) the data used to establish the spatial and temporal distributions of organisms are accurate, (2) organisms redistribute instantaneously among regions of the water body between life stages, but do not move among regions within each life stage, (3) power plant effects on organisms do not alter their overall pattern of distribution within the water body, (4) organism distribution parameters are estimated from field measurements of the entire standing crop of each entrainable life stage, and (5) natural mortality of a given life stage is uniform within the modeled system. The ETM does not account for possible density-dependent natural mortality that may act to reduce or magnify the effects of reduction in year-class strength caused by entrainment.

Sufficient data were available (except for Atlantic tomcod) to use the multiple-cohort version of the ETM, which incorporates life-stage distributions that are averaged over the entire entrainment period for all species. In this version, the effect of entrainment mortality on a cohort is independent of the effect of entrainment mortality on other cohorts. The total entrainment mortality (m_T) is obtained by summing over the entrainment survival rates for all cohorts and subtracting from 1:

$$m_T = 1 - \sum_{s=1}^{S} R_s \prod_{j=0}^{J} \prod_{l=1}^{L} \sum_{k=1}^{K} D_{kl} \exp(-E_{s+j,kl} C_{jl} t); \quad (1)$$

m_T = total conditional entrainment mortality rate;

s = week 1, 2, 3, . . . , S of the spawning period;

R_s = proportion of total eggs deposited that are spawned in week s;

j = age 0, 1, 2, . . . , J (in weeks);

l = life stage 1, 2, 3, . . . , L;

k = river region 1, 2, 3, . . . , K;

D_{kl} = average proportion of the total standing crop of life-stage l individuals in region k during the entrainment period;

$E_{s+j,kl}$ = instantaneous entrainment mortality rate constant of life-stage l individuals during week $s + j$ in region k (units: per day);

C_{jl} = proportion of age-j individuals in life-stage l;

t = duration of the model time step (1 week).

Calculation of the instantaneous entrainment mortality rate (E) requires specification of power plant flow rates, region volumes, and susceptibility of individuals within a region to withdrawal by power plants as well as their subsequent mortality due to plant passage:

$$E_{s+j,kl} = \frac{P_{s+j,k} f_{s+j,kl} W_{s+j,kl}}{V_k} ; \quad (2)$$

$P_{s+j,k}$ = rate of water withdrawal by power plants in region k during week $s + j$ (units: per day);

$f_{s+j,kl}$ = fraction of life-stage l individuals entering the intake that are eventually killed by plant passage during week $s + j$ in region k;

$W_{s+j,kl}$ = ratio of the average intake density to average region density of life-stage l individuals during week $s + j$ in region k;

V_k = volume of region k.

The single-cohort version was used for Atlantic tomcod because egg deposition data were unavailable:

$$M_T = 1 - \prod_{j=0}^{J} \prod_{l=1}^{L} \sum_{k=1}^{K} D_{kl} \exp(-E_{kl} C_{jl} t). \quad (3)$$

Physical Input Data

Regions used in the ETM were chosen to coincide with the sampling scheme employed by the staff of Texas Instruments (TI) in their 1974 and 1975 ichthyoplankton surveys of the Hudson River estuary. Region volumes (Table 47), V in equation (2), were calculated by TI from surface areas and depths recorded on U.S. Geological Survey maps of the river (TI 1975).

Due to the tidal characteristic of the Hudson River estuary, power plants may withdraw water from more than one region. Each power plant's water withdrawal is assumed to be directly related to the proportion of the daily tidal cycle when water "belonging" to a particular region is in front of the power plant's intake. The total daily withdrawal flow for each power plant (Table 47) is, therefore, distributed among the regions in the ETM according to the proportion of each region's water volume that would pass in front of the intake during a daily tidal excursion (21 km: TI

TABLE 47.—Locations, estimated water volumes, and proportions of power plant withdrawal flows from each of the geographic regions used in the empirical transport model. Asterisks mark the region in which each power plant is located.

Region	Location (river kilometers from the Battery)	Volume $(10^8 m^3)$	Source proportions of water withdrawn at				
			Bowline Point	Lovett	Indian Point	Roseton	Danskammer Point
Yonkers	19–37	2.29	0	0	0	0	0
Tappan Zee	38–53	3.22	0.271	0	0	0	0
Croton Haverstraw	54–61	1.48	0.358*	0.369	0.298	0	0
Indian Point	62–74	2.08	0.371	0.549*	0.562*	0	0
West Point	75–88	2.07	0	0.082	0.140	0	0
Cornwall	89–98	1.40	0	0	0	0.273	0.196
Poughkeepsie	99–122	2.98	0	0	0	0.727*	0.804*
Hyde Park	123–136	1.65	0	0	0	0	0
Kingston	137–149	1.41	0	0	0	0	0
Saugerties	150–170	1.76	0	0	0	0	0
Catskill	171–198	1.61	0	0	0	0	0
Albany	199–243	0.71	0	0	0	0	0

1975). These proportions were used in the ETM to determine the values of the *P*-parameter.

The actual water withdrawal flow rates of the Hudson River power plants during the entrainment period for striped bass in 1974 and 1975 are expressed as averages of the daily flow rates during each week (Table 48). Projected once-through and closed-cycle flows at each plant (Table 49) do not include three of the five units at Lovett, two of the four units at Danskammer, and Indian Plant unit 1 because they were expected to be used very little, if at all, in the future (Barnthouse et al. 1977).

Biological Input Data

Data pertaining to striped bass are used to illustrate the derivation of the biological parameter values used in the ETM to estimate entrainment mortality for the seven fish species. Parameter values for the other species are in Boreman et al. (1982).

Riverwide distribution and abundance.—The 1974 and 1975 average distributions of the ichthyoplankton life stages of striped bass in the Hudson River (*D* in equation 1) are listed in Table 50. These distributions are based on time-integrated abundance estimates derived from data collected by TI staff in their survey programs. Only 1974 data were available for American shad and only 1975 data on the larva and juvenile life stages were available for Atlantic tomcod. Details of the TI sampling programs are in TI (1977).

The abundance patterns were obtained directly from the longitudinal river survey program; juvenile abundance patterns were also based on combined abundance estimates from the beach-seine survey program. Use of beach-seine data in deriv-

ing juvenile distribution patterns allowed recognition of movement of juveniles into the shore zone, defined by TI (1977) as that part of the river less than 6 m deep. The entrainment period ended when the riverwide average lengths of juveniles in the beach-seine survey exceeded 50 mm, considered the maximum entrainable size.

Egg deposition (*R* in equation 1) was estimated from the longitudinal river survey data; the weekly abundance of eggs was divided by the total riverwide abundance summed over all weeks eggs were present (Table 51). Average durations of the egg and yolk-sac larva stages of the seven species were assumed to be directly related to water temperature during the period of their occurrence in the Hudson River. Durations of these life stages were derived from relationships between water temperature and development published in the open literature (summarized in Boreman 1981). The durations of the post–yolk-sac larva and juvenile stages were assumed to be less influenced by water temperature and more by growth because the fish actively seek food by this time. Therefore, the durations of these life stages were based on the period between the first appearance of each life stage and the first appearance of the subsequent life stage in field samples. In 1974, the average life stage durations for striped bass were approximately 2.5 d for eggs, 7 d for yolk-sac larvae, and 28 d each for post–yolk-sac larvae and juveniles up to 50 mm. In 1975, durations were 2 d for eggs, 5.5 d for yolk-sac larvae, and 28 d for post–yolk-sac larvae and juveniles.

Intake densities and plant passage mortality.— The *W*-factor (equation 2) is the ratio of the density of organisms in the power plant intake water to the average density in an idealized cross

TABLE 48.—Mean daily water withdrawals (averaged weekly) actually made by Hudson River power plants during the striped bass entrainment season, 1974 and 1975.

	Mean water withdrawals (10^3 m³/d) at					
	Bowline		Indian Point			Danskammer
Year and week	Point	Lovett	Unit 1	Unit 2	Roseton	Point
1974						
Apr 29–May 5	1,722	1,223	1,234	3,124	894	1,307
May 6–May 12	1,722	1,223	710	2,750	527	1,131
May13 –May 19	1,613	1,223	947	1,832	359	1,209
May 20–May 26	0	1,271	1,638	3,976	235	1,341
May 27–Jun 2	0	1,174	1,621	4,629	489	1,335
Jun 3–Jun 9	1,898	1,296	1,638	4,432	373	1,134
Jun 10–Jun 16	2,492	1,180	1,640	3,768	334	1,013
Jun 17–Jun 23	3,447	1,333	1,633	3,914	349	1,165
Jun 24–Jun 30	3,155	1,223	1,414	4,190	400	1,145
Jul 1–Jul 7	3,236	1,343	1,511	4,430	345	1,368
Jul 8–Jul 14	2,822	1,285	892	4,433	380	1,409
Jul 15–Jul 21	3,445	1,034	1,646	3,913	1,168	1,477
Jul 22–Jul 28	3,445	1,222	1,646	3,666	1,168	1,654
Jul 29–Aug 4	3,445	1,159	1,646	1,300	1,203	1,686
Aug 5–Aug 11	3,183	1,291	1,532	3,599	2,084	1,629
Aug 12–Aug 18	3,445	1,350	1,646	4,510	1,556	1,527
Aug 19–Aug 25	3,639	1,251	1,646	4,613	2,238	1,627
Aug 26–Sept 1	3,565	1,410	1,646	4,688	2,279	1,573
1975						
May 11–May 17	1,447	760	796	3,913	2,657	882
May 18–May 24	1,401	1,367	312	3,915	3,058	880
May 25–May 31	1,401	1,223	185	4,117	3,004	1,076
Jun 1–Jun 7	2,368	1,223	446	4,492	2,953	1,427
Jun 8–Jun 14	2,802	1,197	713	4,582	2,274	1,481
Jun 15–Jun 21	2,802	1,081	556	4,714	2,474	1,551
Jun 22–Jun 28	2,848	1,278	450	4,703	3,058	1,656
Jun 29–Jul 5	2,802	1,060	187	4,453	2,824	1,619
Jul 6–Jul 12	2,802	1,151	222	4,643	2,824	1,292
Jul 13–Jul 19	2,802	1,167	107	4,688	2,932	1,424
Jul 20–Jul 26	3,353	1,253	565	4,659	3,011	1,487
Jul 27–Aug 2	3,429	1,359	867	1,756	3,058	1,519
Aug 3–Aug 9	3,445	1,250	705	117	2,722	1,634
Aug 10–Aug 16	3,445	1,205	32	4,033	2,125	1,544
Aug 17–Aug 23	3,362	1,221	25	4,390	2,324	1,547
Aug 24–Aug 30	3,356	1,216	270	4,645	2,279	1,590

section of the river in front of the intake. Wherever possible, W-factors were estimated from simultaneous sampling of ichthyoplankton in the plant and in the river transect in front of the plant intake. Because differences between sampling gears and deployment techniques used in the in-plant and transect sampling introduced unknown, but potentially important, biases into the W-factor estimates, two alternative computational techniques were employed (Boreman et al. 1982): (1) a modification of the method used by utility consultants (McFadden and Lawler 1977), which calculates the W-factor as a seasonally averaged ratio of the intake and river densities; and (2) a method that accounts for differential efficiencies of sampling in the intake and river. The W-factors were approximated from data on riverwide depth distributions of ichthyoplankton when sample sizes obtained from in-plant or transect sampling were insufficient for calculating meaningful values (Boreman et al. 1982). Due to possible day–night differences in depth distribution among life stages, species, and location in the river, W-factors were estimated separately for year, day, night, life stage, species, and power plant (Table 52).

Two methods were used to estimate the f-factor, or fraction of organisms entering the intake that are eventually killed by plant passage (f-parameter in equation 2). One method relied on samples of live and dead organisms obtained from the power plant intakes and discharges by means of nets or larva tables. Descriptions of the gear and their deployment are provided by McGroddy and Wyman (1977) and Muessig et al. (1988, this volume). Estimates of the f-factors for striped bass, based on net and larva-table data, were 0.66 for eggs, 0.55 for yolk-sac larvae, 0.40 for post–yolk-sac larvae, and 0.18 for juveniles.

TABLE 49.—Projected daily water withdrawals by Hudson River power plants during once-through (OT) and closed-cycle (CC) cooling operations.

Month	Cooling mode	Projected water withdrawals (10^3 m³/d) at				
		Bowline Point	Lovett	Indian Point	Roseton	Danskammer Point
Apr	OT	1,543	917	8,122	2,061	482
	CC	48		508	78	
May	OT	2,524	1,061	6,470	2,644	720
	CC	78		434	70	
Jun	OT	3,816	1,164	6,442	3,034	945
	CC	87		439	74	
Jul	OT	4,186	1,133	9,130	3,482	1,062
	CC	87		543	87	
Aug	OT	4,196	1,176	9,123	3,482	1,237
	CC	87		545	87	
Sep	OT	3,657	1,030	7,919	2,730	1,179
	CC	83		542	38	
Oct	OT	2,544	903	6,971	1,780	1,084
	CC	77		484	34	

The other method for estimating mortality due to plant passage used laboratory studies of the effects of temperature to derive a mathematical model of thermally induced mortality for striped bass, white perch, and river herring (Boreman et al. 1982). The thermal model uses a regression equation to estimate the thermal component of entrainment mortality as a function of acclimation (or ambient) temperature, exposure duration (or transit time), and exposure temperature. The thermal model results are then combined with estimates of the mechanical component of entrainment mortality to give a combined estimate of the f-factor. The ranges of f-factors used for striped bass, based on the thermal model, were 0.5–0.6 for yolk-sac larvae, 0.2–0.5 for post–yolk-sac larvae, and 0.1–0.3 for juveniles. Due to a lack of samples, the f-factor for eggs was the value derived from the net data (0.66).

The f-factors for larvae and juveniles incorporated 24-h latent direct mortality, determined from laboratory survival studies, and indirect mortality that was 10% of the direct mortality. Latent direct mortality is the inevitable mortality directly caused by the entrainment experience, and indirect mortality is mortality of organisms that survive the entrainment experience but are rendered more vulnerable to other stresses (e.g., disease, predation, or starvation). Incorporation of indirect mortality was recommended by Van Winkle et al. (1979) to account for the next higher and next lower trophic levels. The mathematical derivation of the thermal model, the sources of data and equations used to estimate the f-factors from net and larva-table data, and the incorporation of latent direct and indirect mortality were discussed by Boreman et al. (1982).

Results and Discussion

Among the seven species considered, historical (1974 and 1975) and projected estimates of conditional entrainment mortality were greatest for bay anchovy and least for river herring (Table 53). Differences in life history characteristics affect the relative entrainment vulnerabilities of these species (Boreman 1981). Bay anchovies concentrate in the power plant regions of the river; they are vulnerable to entrainment for the longest period of time (June–October); and they are pelagic. River herring have almost opposite characteristics: their vulnerable life stages are concentrated in the uppermost regions of the estuary; they are present for a relatively short time (May–August); and their earliest stages are demersal. American shad, which also spawn upriver from the power plants, move into the power plant regions as late larvae and juveniles (McFadden et al. 1977); these life stages account for almost all of the entrainment mortality suffered by the population. Post–yolk-sac larvae and juveniles account for most of the entrainment mortality of striped bass, white perch, and Atlantic tomcod, due to the longer presence of these life stages in the power plant regions.

TABLE 50.—Distributions of entrainable life stages of striped bass (percent of total numbers) throughout the Hudson River, 1974 and 1975.

Year	Region	Eggs	Yolk-sac larvae	Post–yolk-sac larvae	Juveniles
1974	Yonkers	0.00	0.14	0.07	0.62
	Tappan Zee	0.09	2.65	3.84	28.41
	Croton Haverstraw	18.20	10.07	6.72	23.10
	Indian Point	23.97	11.43	23.87	5.01
	West Point	36.54	10.39	21.71	2.72
	Cornwall	2.63	22.12	12.28	11.38
	Poughkeepsie	4.12	35.30	18.70	2.74
	Hyde Park	3.99	5.45	4.04	2.52
	Kingston	5.85	1.40	6.47	9.43
	Saugerties	1.88	0.73	1.26	4.60
	Catskill	2.55	0.29	1.03	7.40
	Albany	0.18	0.03	0.01	2.07
1975	Yonkers	0.00	0.05	0.51	2.05
	Tappan Zee	0.32	4.34	2.24	32.27
	Croton Haverstraw	6.43	9.97	9.31	26.84
	Indian Point	35.71	23.02	34.46	9.13
	West Point	38.24	23.62	20.19	2.05
	Cornwall	9.22	11.12	10.99	6.64
	Poughkeepsie	4.99	12.82	13.65	8.59
	Hyde Park	2.36	9.75	3.98	2.53
	Kingston	0.48	2.38	3.39	3.73
	Saugerties	0.98	2.83	0.82	4.24
	Catskill	1.19	0.09	0.46	1.72
	Albany	0.08	0.01	0.00	0.21

The utilities' estimates of total entrainment mortality of striped bass were 8.1% for 1974 and 11.9% for 1975 (McFadden and Lawler 1977). The utilities' estimates reflected operation of Indian Point unit 2, Bowline Point, and (in 1975 only) Roseton. Estimates derived with the ETM for similar operating conditions were lower than the utilities' estimates for 1974 (6.3–7.6%; Boreman et al. 1982) and slightly higher in 1975 (12.8–

TABLE 51.—Temporal distributions of egg deposition by Hudson River striped bass, 1974 and 1975.

Year and week	Proportion of eggs deposited
1974	
Apr 29–May 5	0.25
May 6–May 12	12.26
May13–May 19	41.89
May 20–May 26	38.75
May 27–Jun 2	5.41
Jun 3–Jun 9	0.47
Jun 10–Jun 16	0.46
Jun 17–Jun 23	0.34
Jun 24–Jun 30	0.17
1975	
May 11–May 17	3.09
May 18–May 24	49.52
May 25–May 31	41.15
Jun 1–Jun 7	4.70
Jun 8–Jun 14	0.08
Jun 15–Jun 21	0.52
Jun 22–Jun 28	0.94

13.0%; Table 53). Estimates of total striped bass entrainment mortality projected by utility consultants for future operation of Bowline Point, Indian Point, and Roseton were 5.8% with the 1974 data base and 8.1% with the 1975 data base (McFadden and Lawler 1977), lower than the ETM estimates by approximately eight to nine percentage points (14.1–17.4%, Table 53).

The entrainment mortality estimates for bay anchovy and Atlantic tomcod probably do not reflect rates for the entire river populations of these species. Longitudinal distribution data collected in 1975 and 1976 revealed that the vulnerable life stages of these species were concentrated in the two lowermost sampling regions, suggesting that undefined fractions of the populations were below river km 22 (Yonkers region). Therefore, estimates of the conditional entrainment mortality rates for these species are applicable to only that portion of each population that remained above river km 22 during the entrainment period. We decided to include the ETM estimates for these species in our testimony because they gave an indication of the degree to which the standing crops of the two species might be reduced in the power plant regions by entrainment mortality. Both species are important food items in the diet of other fish, particularly striped bass (McFadden et al. 1977) and bluefish (TI 1976).

TABLE 52.—Ranges of W-factors[a] for Hudson River striped bass used as input to the empirical transport model.

| Power plant | Life stage | W, by parameter derivation method[b] | | |
		Modified utility	Gear bias correction	River data
Bowline Point	Egg	0.00–0.28		
	Yolk-sac larva	0.10–0.36		
	Post–yolk-sac larva	0.08–0.60		
	Juvenile			0.78
Lovett	Egg	0.16–0.98	0.32–0.59	
	Yolk-sac larva	0.14–4.16	1.15–1.16	
	Post–yolk-sac larva	0.04–3.31	1.45–3.37	
	Juvenile			0.80
Indian Point	Egg	0.49–1.61	0.16–1.22	
	Yolk-sac larva	0.28–0.98	0.77–0.86	
	Post–yolk-sac larva	0.39–1.40	0.91–0.98	
	Juvenile			0.84
Roseton	Egg	1.95–3.31	0.24–0.66	
	Yolk-sac larva	0.76–1.66	0.76–0.98	
	Post–yolk-sac larva	0.29–2.20	0.55–0.88	
	Juvenile			0.73
Danskammer Point	Egg	0.39–6.82	0.94–1.18	
	Yolk-sac larva	0.48–2.58	0.52–0.88	
	Post–yolk-sac larva	0.16–2.34	0.68–0.99	
	Juvenile			0.81

[a]Ratio of average intake density to average regional density.
[b]See Boreman et al. (1982) for explanation of methods.

Two natural-draft cooling towers each at the Bowline, Indian Point, and Roseton power plants would reduce riverwide entrainment mortality of striped bass by 50–80%, of white perch by 75–80%, of Atlantic tomcod by 65–70%, of American shad by 80%, of river herring by 30–90%, and of bay anchovy by 45–80%. At the Bowline, Indian Point, and Roseton facilities alone, cooling towers would reduce entrainment mortality of striped bass by 90–95%, white perch by 90%, Atlantic tomcod by 70%, American shad by 95%. These projected reductions in entrainment mortality caused by a change in technology, coupled with the projected reductions in impingement mortality (Barnthouse and Van Winkle 1988, this volume), were the reasons the U.S. Environmental Protection Agency sought construction of cooling towers to minimize adverse impact on fish populations in the Hudson River (Christensen et al. 1981). Terms of the settlement agreement did not include the requirement for cooling towers. The ETM was used during settlement negotiations as a basis for determining the relative reduction in entrainment mortality caused by planned outages and cooling water flow reductions (Englert et al. 1988, this volume).

We originally designed the ETM to coincide with the format of the field data pertaining to the distribution and abundance of ichthyoplankton in the Hudson River. If the ETM is to be used in future estuarine assessments, we suggest that the ichthyoplankton sampling regions be centered around the power plant intake(s) and encompass a longitudinal river distance greater than the daily tidal excursion. We also recommend that the estuary surveys encompass the entire area where vulnerable life stages of the species of interest are expected to be found.

We suggest that particular care be taken in collection of data pertaining to the ratio of intake to river densities of organisms (W-factor) and to the mortality of organisms due to plant passage (f-factor). Our model runs indicated that the conditional entrainment mortality rate is particularly sensitive to these parameters, and sampling bias is a major source of error in the parameter estimates (Boreman and Goodyear 1981; Muessig et al. 1988). Efforts should be concentrated on life stages that are most vulnerable to entrainment mortality. For the seven fish species discussed in this paper, the most important life stages are post–yolk-sac larva and entrainable-size juvenile. Adequate data on entrainment of juveniles are especially difficult to obtain due to the juveniles' ability to avoid sampling gear and to their lower numbers relative to younger life stages. Neverthe-

TABLE 53.—Estimates of conditional entrainment mortality at Hudson River power plants, expressed as percentages, for fish species inhabiting the Hudson River estuary. Estimates were generated by the empirical transport model.

Species	Historical		Projected	
	1974	1975	Once-through cooling[a]	Closed-cycle cooling
All power plants				
Striped bass	11.1–14.5	18.2–18.4	16.0–21.7 (15.7–24.4)	4.4–8.3
White perch	10.9–11.7	13.0–13.6	15.7–17.1 (15.2–18.9)	3.2–4.1
Atlantic tomcod		5.2–8.4	6.6–7.3 (5.3–10.5)	2.3
American shad	13.6		20.5	4.1
River herring[b]	3.5–4.1	6.1–11.2	6.2–11.1	1.3–4.2
Bay anchovy	54.1–77.8	34.6–46.0	44.3–78.6	12.7–25.2
Bowline, Indian Point units 1–3, and Roseton				
Striped bass	6.8–8.0	12.8–13.0	14.1–17.4 (14.0–19.1)	2.2–3.3
White perch	6.9–7.8	8.7–9.6	13.4–15.3 (13.5–18.0)	1.3–1.4
Atlantic tomcod		4.3–7.1	6.0–7.0 (4.0–6.6)	1.8–1.9
American shad	8.6		17.9	0.9
River herring[b]	1.8–2.0	4.0–5.9	4.7–7.9	0.3–0.4
Bay anchovy	36.2–65.1	25.9–36.6	38.1–75.3	2.2–7.9

[a]Estimates in parentheses incorporate thermal model mortality (f) factors.
[b]Alewife and blueback herring combined.

less, entrainment impacts may be seriously under-estimated if most data collecting is directed primarily at the younger life stages.

Acknowledgments

Helpful suggestions and encouragement were provided by Lawrence Barnthouse, Oak Ridge National Laboratory, Douglas Vaughan, National Marine Fisheries Service, and Ronald Klauda, The Johns Hopkins University.

References[2]

Barnthouse, L. W., and twelve coauthors. 1977. A selective analysis of power plant operation on the Hudson River with emphasis on the Bowline Point generating station, volumes 1 and 2. Oak Ridge National Laboratory, ORNL/TM-5877, Oak Ridge, Tennessee.
Barnthouse, L. W., and W. Van Winkle. 1988. Analysis of impingement impacts on Hudson River fish populations. American Fisheries Society Monograph 4: 182–190.
Boreman, J. 1981. Life histories of seven fish species that inhabit the Hudson River estuary. National Marine Fisheries Service, Northeast Fisheries Center, Woods Hole Laboratory Document 81-34, Woods Hole, Massachusetts.
Boreman, J., and C. P. Goodyear. 1981. Biases in the estimation of entrainment mortality. Pages 79–89 in L. D. Jensen, editor. Issues associated with impact assessment. Ecological Analysts, Sparks, Maryland.
Boreman, J., C. P. Goodyear, and S. W. Christensen. 1981. An empirical methodology for estimating entrainment losses at power plants sited on estuaries. Transactions of the American Fisheries Society 110:255–262.
Boreman, J., and seven coauthors. 1982. Entrainment impact estimates for six fish populations inhabiting the Hudson River estuary, volume 1. The impact of entrainment and impingement of fish populations in the Hudson River estuary. Oak Ridge National Laboratory, ORNL/NUREG/TM-385/V1, Oak Ridge, Tennessee.

[2]See Table 1 for sources of legal documents and unpublished reports pertaining to the Hudson River.

Christensen, S. W., W. Van Winkle, L. W. Barnthouse, and D. S. Vaughan. 1981. Science and the law: confluence and conflict on the Hudson River. Environmental Impact Assessment Review 2:63–88.

Englert, T. L., and J. Boreman. 1988. Historical review of entrainment impact estimates and the factors influencing them. American Fisheries Society Monograph 4:143–151.

Englert, T. L., J. Boreman, and H. Y. Chen. 1988. Plant flow reductions and outages as mitigative measures. American Fisheries Society Monograph 4:274–279.

McFadden, J. T., and J. P. Lawler. 1977. Influence of Indian Point unit 2 and other steam electric generating plants on the Hudson River estuary, with emphasis on striped bass and other fish populations, supplement 1. Report to Consolidated Edison Company of New York.

McFadden, J. T., Lawler, Matusky & Skelly Engineers, and Texas Instruments. 1977. Influence of Indian Point unit 2 and other steam electric generating plants on the Hudson River estuary, with emphasis on the striped bass and other fish populations. Report to Consolidated Edison Company of New York.

McGroddy, P. M., and R. L. Wyman. 1977. Efficiency of nets and a new device for sampling live fish larvae. Journal of the Fisheries Research Board of Canada 34:571–574.

Muessig, P. H., J. R. Young, D. S. Vaughan, and B. A. Smith. 1988. Advances in field and analytical methods for estimating entrainment mortality factors. American Fisheries Society Monograph 4:124–132.

Swartzman, G. L., R. B. Deriso, and C. Cowan. 1978. Comparison of simulation models used in assessing the effects of power-plant-induced mortality on fish populations. U.S. Nuclear Regulatory Commission, NUREG/CR-0474, Washington, D.C.

TI (Texas Instruments). 1975. First annual report for the multiplant impact study of the Hudson River estuary. Report to Consolidated Edison Company of New York.

TI (Texas Instruments). 1976. Predation by bluefish in the lower Hudson River. Report to Consolidated Edison Company of New York.

TI (Texas Instruments). 1977. 1974 year-class report for the multiplant impact study of the Hudson River estuary. Report to Consolidated Edison Company of New York.

Van Winkle, W., S. W. Christensen, and J. S. Suffern. 1979. Incorporation of sublethal effects and indirect mortality in modeling population-level impacts of a stress, with an example involving power-plant entrainment and striped bass. Oak Ridge National Laboratory, ORNL/NUREG/TM-288, Oak Ridge, Tennessee.

American Fisheries Society Monograph 4:161–169, 1988
© Copyright by the American Fisheries Society 1988

Reliability of Impingement Sampling Designs:
An Example from the Indian Point Station

MARK T. MATTSON

Normandeau Associates, Incorporated, 25 Nashua Road
Bedford, New Hampshire 03102, USA

JEFFREY B. WAXMAN

Consolidated Edison Company of New York
4 Irving Place, New York, New York 10003, USA

DAN A. WATSON

ARCO Exploration Company
Post Office Box 2819, PRC 1011, Dallas, Texas 75221, USA

Abstract.—A 4-year data base (1976–1979) of daily fish impingement counts at the Indian Point electric power station on the Hudson River was used to compare the precision and reliability of three random-sampling designs: (1) simple random, (2) seasonally stratified (by 3-month periods), and (3) empirically stratified (based on trends in daily impingement variation). The precision of daily impingement estimates improved logarithmically for each design as more days in the year were sampled. Simple random sampling was the least, and empirically stratified sampling was the most precise design, and the difference in precision between the two stratified designs was small. Computer-simulated sampling was used to estimate the reliability of the two stratified-random-sampling designs. The 95% confidence limits about the sample mean daily impingement count enclosed the true mean in 93–95% of the simulations when sampling intensity exceeded 20–30% of the available sampling days a year. A seasonally stratified sampling design was selected as the most appropriate reduced-sampling program for Indian Point station because (1) reasonably precise and reliable impingement estimates were obtained using this design for all species combined and for eight common Hudson River fish by sampling only 30% of the days in a year (110 d), and (2) seasonal strata may be more precise and reliable than empirical strata if future changes in annual impingement patterns occur. The seasonally stratified design applied to the 1976–1983 Indian Point impingement data showed that selection of sampling dates based on daily species-specific impingement variability gave results that were more precise, but not more consistently reliable, than sampling allocations based on the variability of all fish species combined. The latter approach was used subsequently: random sampling effort totaling 110 d/year at each Indian Point unit was allocated among four seasonal strata by Neyman apportionment. For the years 1982–1986, this reduced design achieved 4–29% levels of precision (100 · SE/mean) at unit 2 and 9–17% at unit 3, generally similar to the predicted levels of 8–9% in most years. Low precisions were associated with unexpectedly high impingement variability of weakfish, and with an extended power outage that reduced sampling effort.

Data collected at the Indian Point electric power station have provided one of the most complete records of fish impingement for any power plant (Stupka and Sharma 1977). Since units 2 and 3 went on-line in 1973 and 1976, respectively, regulatory agencies required continuous collection and sampling from intake structures at 24-h intervals to provide daily counts of the number of fish impinged. This continuous sampling regime, however, may be more than adequate to reliably estimate impingement rates. Studies by Johnston (1976), Murarka and Bodeau (1977), Kumar and Griffith (1978), Murarka et al. (1978), and El-Shamy (1979) have each suggested

that precise estimates of fish impingement may be obtained with a less-than-daily sampling schedule.

The allocation of sampling effort (number of days) required to obtain a specified level of precision is related primarily to the timing and magnitude of variation in daily impingement counts. Times of year with higher variation among daily impingement counts may require daily sampling, whereas low-variation periods may require less frequent sampling to obtain the same degree of precision. In this study, we empirically determined the statistical precision and reliability of three selected sampling designs: simple random, stratified random with empirical strata, and stratified random with seasonal strata of 3 months each. The effectiveness of each design was compared among various levels of sampling effort based on the Indian Point impingement data base for 1976–1979.

Precision and reliability of reduced-sampling designs may be affected by the number of sampling days selected at various times of the year, particularly if the impingement rates of individual fish species are of interest and some species are seasonally variable in abundance or exposure to impingement. The effectiveness of selecting sampling days at the "optimum" time of year ("optimum" design) to maximize precision and reliability of annual impingement estimates for each of eight common fish species was determined for 1976–1983. The results of this program were compared to those of a sampling plan in which sampling days were selected throughout the year based on the annual variation in abundance of all fish species combined ("general" design). Finally, a reduced-sampling design was implemented at Indian Point on 1 July 1981; precision of the reduced design was determined for the years 1982–1986 and compared to values predicted from the aforementioned analyses.

Methods

Impingement data used in this study were obtained from the Indian Point intake structure, approximately 68 km upriver from the Battery at the southern tip of Manhattan Island in New York City (Hutchison et al. 1988, this volume). Daily counts of impinged fish during 1976–1979 were selected for evaluation of sampling designs because units 2 and 3 of the generating station were in continuous operation for the greatest number of days; unit 1 has been out of operation since 1974. Each of the two operating units uses six water circulation pumps; prior to 1981, each pump operated at full capacity (530 m^3/min) from 1 April through 31 December and at 60% capacity during the remaining months. Cooling water pumped by each circulator at unit 2 first passes through a fixed intake screen (9.5-mm mesh) at the river's edge, then through a coarse trash bar rack (102-mm spacing), and finally through a traveling screen (9.5-mm mesh) in the intake forebay before it reaches the intake pump. Unit 3 has trash bars at the river's edge, but no fixed screens, and has traveling screens in the intake forebay.

Numbers of fish impinged at each unit per day (total count for all species combined) were obtained from the product of the daily impingement rate (number of fish/10^{6}m^3) and the pumping capacity of all six circulator pumps (i.e., 4.5 × 10^{6}m^3/d at 100% pumping capacity and 2.75 × 10^{6}m^3/d at 60% pumping capacity). Daily counts, therefore, represented the maximum number of fish that could have been impinged if all six pumps operated continuously for each 24-h period. Maximum daily counts were used as the sampling units because they (like impingement rates) were considered relatively independent of operational variation in the daily volume of cooling water circulated, and they were discrete numbers (counts) representative of a typical sampling unit (Cochran 1977) rather than continuous variables (rates).

To assure maximum daily impingement counts were not underestimated due to the loss of fish because of ice floes, unscheduled screen washes, trash loading, equipment failure, plant maintenance schedules, etc., daily impingement data were examined to select days for which accurate impingement rates could be calculated. Selection criteria applied to the 1976–1979 data base were as follows.

(1) The circulator pump associated with each screen washed at a given time must have operated continuously since the last time the screen was washed. Otherwise, back pressure resulting from starting and stopping of pumps could cause the loss of an unknown number of impinged fish.

(2) The duration of a collection period for any screen washed at a given time must not have exceeded 28 h (between 0800 and 1200 hours on one day through 0800 to 1200 hours on the following day). Sampling durations of approximately 1 d were necessary because preliminary time series analysis with auto-regressive moving-average models (Box and Jenkins 1976) suggested that significant short-term periodicities in the data base may bias impingement rates derived from longer periods.

(3) Each screen was washed and the associated circulator pump was operating at the time of washing, or the pump was not operated and the screen was not washed. Unusual operating conditions at any screen at a unit (e.g., an unscheduled wash due to high debris accumulation, an abnormal collection, etc.) resulted in exclusion of that day's rate from this analysis.

(4) Air-curtain operation at any unit excluded that day's impingement rate. Air bubblers, evaluated during 1976 as mitigation devices to exclude fish from the cooling-water intakes (TI 1980), were found to wash impinged fish from the fixed

TABLE 54.—Number of operating days at the Indian Point station, 1976–1979, and the number of days selected for inclusion in the impingement data base.

Year	Unit 2		Unit 3	
	Days of operation	Days selected	Days of operation	Days selected
1976	193	41	329	164
1977	303	214	303	213
1978	274	166	317	248
1979	278	230	262	210
Total	1,048	651	1,211	835

TABLE 55.—Seasonal and empirical strata for stratified random sampling of the 1976–1979 impingement data base at Indian Point units 2 and 3. The number of days in stratum h is N_h.

Stratum h	Unit 2		Unit 3	
	Interval	N_h	Interval	N_h
	Seasonal strata			
Winter	1 Jan–31 Mar	92	1 Jan–31 Mar	92
Spring	1 Apr–30 Jun	90	1 Apr–30 Jun	90
Summer	1 Jul–30 Sep	91	1 Jul–30 Sep	91
Fall	1 Oct–31 Dec	92	1 Oct–31 Dec	92
	Empirical strata			
1	1 Jun–31 Aug	92	1 Jun–31 Aug	92
2	1 Sep–30 Sep	30	1 Sep–30 Sep	30
3	1 Oct–27 Nov	58	1 Oct–10 Dec	71
4	28 Nov–15 Mar	108	11 Dec–31 Jan	52
5	16 Mar–7 Apr	23	1 Feb–25 Mar	53
6	8 Apr–31 May	54	26 Mar–15 Apr	21
7			16 Apr–31 May	46

screens (this affected only unit 2 prior to 1 April 1976).

Selection based on these four criteria removed 38% of the possible sampling days at unit 2 and 31% of the days at unit 3 (Table 54).

Three sampling designs were compared: simple random, stratified random with seasonal strata, and stratified random with empirically determined strata. Simple random sampling was chosen because it is a basic design against which the others could be compared. Formulae for the simple random mean daily impingement count, $\bar{y}_{ran}$, and the standard error, $SE(\bar{y}_{ran})$, followed Cochran (1977):

$$\bar{y}_{ran} = \sum_{i=1}^{n} y_i/n; \qquad (1)$$

$$SE(\bar{y}_{ran}) = \frac{S}{\sqrt{n}} \sqrt{\frac{N-n}{N}}; \qquad (2)$$

y_i = the maximum number of fish impinged in the ith sampling unit (day);
n = number of days randomly selected;
N = total number of days available;
S = standard deviation for the "parent" population.

Formulae for the stratified mean daily impingement count, $\bar{y}_{st}$, and standard error, $SE(\bar{y}_{st})$, also followed Cochran (1977):

$$\bar{y}_{st} = \sum_{h=1}^{L} N_h \bar{y}_h/N; \qquad (3)$$

$$SE(\bar{y}_{st}) = \frac{1}{N} \sqrt{\sum_{h=1}^{L} N_h(N_h - n_h) \frac{S_h^2}{n_h}}; \qquad (4)$$

$\bar{y}_h$ = simple random mean daily impingement count in the hth stratum;
n_h = number of days randomly selected in the hth stratum;
N_h = total number of days available in the hth stratum;

L = number of strata;
S_h^2 = variance for the hth stratum in the "parent" population.

Different strategies were used with the two stratified designs regarding the selection of strata within a year and the allocation of sampling effort among these temporal strata. Seasonal strata were selected because historical data (TI 1980) suggested that the movement pattern and exposure to impingement were seasonal for numerically important species in the Hudson River such as white perch, Atlantic tomcod, and blueback herring. For the stratified design with seasonal strata, each year was divided into four time periods of 3 months each (Table 55). Empirical strata were determined by examining the 1976–1979 data base to identify periods of high and low variation in daily impingement counts. A running-median, curve-smoothing technique (Tukey 1977) was used to separate residual variation from annual impingement patterns. Visual examination of scatter plots of this residual variation permitted identification of six relatively distinct temporal strata at unit 2 and seven empirical strata at unit 3 (Table 55).

Precision of each sampling design was compared by plotting the standard error and coefficient of variation (100 · SE/mean) for each of 12 sampling fractions (fraction of a year sampled: 5, 10, 20, 25, 30, 40, 50, 60, 70, 75, 80, and 90%). For the two stratified designs, Neyman allocation (Cochran 1977) was used to distribute sampling effort among seasonal or empirical strata. This method allocated sampling effort in direct proportion to both the number of days and the variance in each stratum by the formula

$$n_h = n \frac{N_h S_h}{\Sigma N_h S_h};$$ (5)

S_h = standard deviation for the hth stratum in the "parent" population.

The reliability of each sampling design was the percentage of times that the calculated 95% confidence interval (based on Student's t-distribution for $P_\alpha = 0.05$ and effective degrees of freedom: Cochran 1977) enclosed the "true" mean daily impingement count for the "parent" population. Computer-simulated sampling involved repeated application of each of the two stratified designs until 1,000 randomly selected samples of size n were drawn from the parent population for each sampling fraction. The impingement count for each day in sample n was selected without replacement; however, the values were placed back into the parent population before the next sample was drawn. The percentage of times the sample confidence interval failed to include the true mean was considered an estimate of the type-I error probability (P_α). The bias of each design was estimated by the percentage of times that the sample mean daily impingement count, with 95% confidence limits, significantly underestimated or overestimated the true mean.

The 1976–1983 Indian Point data base of actual daily impingement counts was used to evaluate seasonal variability in the impingement of individual fish species. Unlike the analyses conducted for 1976–1979, which were based on maximum daily counts of fish species impinged on selected days, actual daily impingement counts for all operating days were included in the 1976–1983 data base. Stratified random sampling based on four seasonal strata and Neyman sample allocations derived from 1976–1979 stratum-variance estimates for all fish species combined is termed a "general" design. The term general means sample allocations are based on total impingement counts for all species combined and are subject to variance of one or more fish species in each stratum. The general design allocates samples among strata to maximize precision in the estimate of all fish species combined. This general design was contrasted with a seasonal design with sample allocations based on 1976–1979 stratum-variance estimates for individual fish species, termed an "optimum" design. The optimum design involves a Neyman sample allocation that is optimum for

each species without regard for variation in daily impingement counts of other species. The optimum design provides the most precise estimate of mean daily impingement for each species, without regard for other species. Eight common fish species were selected for comparison of optimum and general designs during 1976–1983. Precisions of general and optimum designs were compared by examining the ratio (general:optimum) of standard errors or reliabilities of the two stratified designs at the same sampling fraction. These ratios can be termed the design effect or "efficiency" of the sampling design when expressed as a percentage (Cochran 1977), and are hereafter referred to as design efficiency.

Results

Sampling Precision and Reliability (1976–1979)

Precision improved logarithmically (SE decreased) for each sampling design as sample size (n) increased at both unit 2 and unit 3 (Figure 58). Simple random sampling was the least precise, stratified random sampling with four seasonal strata was intermediate in precision, and stratified random sampling with empirical strata was the most precise design at both units. However, the relationship between sampling design, sampling effort, and precision differed between units. At unit 2, both stratified designs were considerably more precise than the simple random design, particularly for larger sampling fractions. The two stratified designs responded similarly to increasing sampling effort (curves were nearly parallel) except at intermediate fractions, when the precision of empirical stratification was disproportionately higher than for the seasonal design. In contrast, the curves for unit 3 were less widely separated among the three designs, and the change in precision for the seasonal design more closely resembled that of the simple random design than that of the empirical design.

The reliability of both stratified designs converged on the value $1 - P_\alpha = 0.95$ (95%) as sampling fractions exceeded 20% for unit 2 and 30% for unit 3 (Table 56). Differences in reliability between empirical and seasonal stratification were not apparent. A positive skewness in the "parent" population data base at each unit resulted in a slight negative bias in the estimated sample means. The sample mean tended to underestimate the true mean for all levels of sampling effort because the underlying parent population

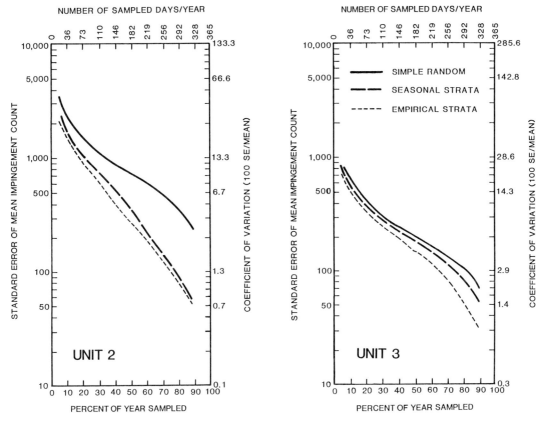

FIGURE 58.—Relationship between precision (standard error SE or coefficient of variation) of mean estimated maximum daily impingement at Indian Point units 2 and 3, 1976–1979, and percent of the year sampled by three sampling designs.

was asymmetric and had proportionally more days with low impingement counts than with high counts. The degree of bias was greater at sampling fractions below the 95% convergence point than at higher fractions.

Estimated reliability decreased slightly as the level of sampling effort exceeded 40–60% (Table 56). This result was an artifact of the interaction between the Neyman allocation method for stratified sampling and the computer-simulated sampling approach. At sampling fractions between 40 and 60%, one or more high-variance strata received 100% allocations (all days were selected). Selection of all possible days in a stratum removed the contribution of the stratum variance to the computed standard error of the mean daily impingement count ($N_h = n_h$; therefore, $N_h - n_h = 0$ in equation 4). Confidence intervals were calculated, therefore, with variance values derived from a decreasing number of strata representing a subset of each "parent" population.

Increased skewness in the parent population subset resulted in the observed slight decrease in reliability.

Species Design Efficiencies (1976–1983)

Precision increased logarithmically with increasing sampling fraction for each of eight fish species at Indian Point unit 2 and unit 3, but the relationship differed among species and sampling designs. General designs, with Neyman sample allocations derived from seasonal variation in impingement counts of all species combined, were usually less precise for each sampling fraction than designs with optimum sample allocations for each species. Three patterns were observed in the relationship between precision, number of days sampled, and sampling design (optimum or general). For striped bass and white perch, precision for optimum and general designs exhibited parallel improvement; at each level of sampling effort (percentage of the year sampled) the optimum

TABLE 56.—Relative bias for two stratified-random-sampling designs, expressed as the percentage of times that the calculated mean daily impingement count, with 95% confidence limits, was higher or lower than the true mean.[a]

Sampling fraction (percent of year)	Seasonal design			Empirical design		
	Percent missing low	Percent missing high	Total percent missing	Percent missing low	Percent missing high	Total percent missing
	Unit 2					
5	11.4	0.0	11.4	12.9	0.4	13.3
10	11.2	0.2	11.4	9.6	0.2	9.8
20	10.0	0.5	10.5	9.1	0.4	9.5
25	8.8	0.9	9.7	6.5	0.5	7.0
30	5.2	0.5	5.7	5.5	1.0	6.5
40	5.2	0.7	5.9	5.3	1.3	6.6
50	4.3	1.1	5.4	4.2	0.8	5.0
60	4.1	0.7	4.8	3.7	1.6	5.3
70	3.4	1.2	4.6	4.1	1.7	5.8
75	5.7	1.6	7.3	5.1	0.9	6.0
80	4.6	1.3	5.9	4.4	1.4	5.8
90	4.3	1.1	5.4	5.2	1.2	6.4
	Unit 3					
5	13.1	0.3	13.4	8.7	0.1	8.8
10	8.5	0.3	8.8	6.8	1 0	7.8
20	5.2	0 7	5.9	5.9	1.1	7.0
25	6.2	0.7	6.9	5.1	1.5	6.6
30	6.7	0.3	7.0	4.8	0.8	5.6
40	4.7	1.3	6.0	4.9	1.2	6.1
50	5.6	1.7	7.3	3.6	1.1	4.7
60	4.7	0.8	5.5	4.9	1.3	6.2
70	5.7	1.9	7.6	5.0	1.2	6.2
75	5.3	1.5	6.8	5.0	1.2	6.2
80	5.9	0.7	6.6	5.4	1.5	6.9
90	5.8	0.3	6.1	6.8	1.2	8.0

[a]Results were based on 1,000 sampling simulations for each design at each fractional sampling of the 1976–1979 Indian Point impingement data base.

design was between 30 and 100% more precise than the general design. At the 30% sampling fraction (110 d/year), the optimum design was 10% (unit 3) and 60% (unit 2) more precise than the general design for striped bass, and it was 40% more precise for white perch at both units (Table 57). For Atlantic tomcod, American shad (unit 2 only), blueback herring (unit 2 only), alewife (unit 2 only), bay anchovy, and weakfish, the optimum design exhibited greater improvement with increasing sample size than the general design. For American shad, blueback herring, and alewife at unit 3, precision of the optimum design initially diverged from the general design, as seen by the relatively high efficiency at the 30% sampling fraction (Table 57), and subsequently converged at large sampling fractions (more than 30% per year). This divergence–convergence pattern was the result of seasonal impingement of these clupeid species during the fall stratum; both designs required complete sampling of all the days available due to the high variability.

TABLE 57.—Design efficiencies[a] of stratified random sampling at the Indian Point station. Sample allocations of 110 d/year were based either on seasonal variation in impingement of selected fish species (optimum design) or on seasonal variation in impingement of all fish species combined (general design), 1976–1983.

	Unit 2 efficiency		Unit 3 efficiency	
	Precision	Reliability	Precision	Reliability
All species combined	1.0	1.0	1.0	1.0
Striped bass	1.6	1.3	1.1	1.0
White perch	1.4	1.2	1.4	1.3
Atlantic tomcod	3.6	1.1	1.9	0.8
American shad	2.7	2.3	2.1	1.3
Blueback herring	1.7	3.0	7.4	0.4
Alewife	1.5	1.6	2.2	0.9
Bay anchovy	4.3	1.1	2.3	1.0
Weakfish	10.8	0.8	14.7	1.1

[a]Efficiency is, for reliability, (general design reliability/optimum design reliability) or, for precision, (general SE/optimum SE).

Weakfish exhibited the greatest improvement in precision (Table 57) because this species was impinged in high numbers during August in the summer stratum of 1983, and that stratum was saturated with sampling effort (NAI 1984). During the 1976–1979 period, weakfish was not among the fish species contributing more than 1% to the annual total impingement at the Indian Point station (TI 1980). In 1983, however, weakfish was the second most abundant fish species in impingement collections, and contributed approximately 20% of the annual total impingement (NAI 1984). The recent (1983) increase in weakfish abundance during the summer stratum was not forecast by Neyman sample allocations derived from the 1976–1979 data base. The summer stratum was inadequately sampled according to the general design, but the optimum design for weakfish allocated the most sampling effort to the summer stratum and, therefore, this species exhibited the greatest design efficiency. With the exception of 1983, which may have been an unusually dry year with saltwater intrusion well above Indian Point during the summer, weakfish have contributed less than 2% annually to Indian Point impingement at both unit 2 and unit 3 between 1976 and 1986 (NAI 1986, 1987).

No consistent differences in reliability were found between general and optimum designs for each of the eight common fish species impinged at units 2 and 3 between 1976 and 1983 (Table 57). In the few instances when the 95% confidence interval failed to enclose the true mean, the estimated mean was lower than the true mean, for most fish species, due to a positive skewness in the data base frequency distribution (e.g., weakfish at unit 2 and blueback herring at unit 3). For white perch at unit 2, however, the sample mean impingement count with 95% confidence limits tended to overestimate the true mean.

Discussion

Although the cooling-water intakes at both units 2 and 3 were located in the same bulkhead, daily impingement counts at unit 2 were more variable and demonstrated greater improvement in precision with increasing sample size than those at unit 3. Unit 2 has both fixed and traveling screens in the intake and relatively low and variable collection efficiency (as defined by the percentage of marked fish released near the intake screens that were recovered during the screen wash after 24-h: TI 1980). In contrast, unit 3 has only traveling screens and relatively high and constant collection efficiency. The double screen system at unit 2 is apparently washed less efficiently and with more variability than the one-screen system at unit 3 and this high variability was reflected in the data.

Stratified sampling with Neyman sample allocation was more efficient than simple random sampling at both units 2 and 3 (Figure 58). Less apparent were the advantages of empirical over seasonal stratification. Intuitively, the most precise and reliable stratified design would dictate selection of empirical strata based on variance patterns in the data base. However, an empirical design may be relatively sensitive to future changes in annual impingement patterns or operation modes (e.g., scheduled outages) because empirical strata were defined and evaluated from the same data set. Short-duration empirical strata with high variability would receive a relatively large sampling effort that may not always coincide with annual periods of high variability.

A stratified-random-sampling design provided precise and reliable estimates of daily impingement at Indian Point station when sampling effort was reduced below continuous daily monitoring. Murarka and Bodeau (1977) generally observed greater improvements in precision with stratified systematic random sampling than with stratified random sampling. Stratified systematic random sampling is often considered practical and relatively easy to implement by field crews because sampling can be regularly scheduled within each stratum (e.g., every third day). However, a systematic component to the sampling design was not appropriate for Indian Point because short-term (1–4-d and 9–14-d) periodicities were apparent in the data, which could have biased impingement counts when the systematic sampling interval matched periodic peaks in abundance. The cause for the apparent periodicity is unknown, but may be related to tidal currents or diel activities of fishes near the station.

Stratified sampling with seasonal strata is probably the most appropriate reduced-sampling design for the Indian Point station. The four seasonal strata generally coincided with annual impingement patterns but were also somewhat arbitrary or independent with respect to empirical variance patterns in the historical data base. A seasonal design, therefore, may be relatively robust with respect to future changes in impingement patterns (e.g., a temporal change in abundance or impingement variability of a fish species compared to the pattern observed between 1976 and 1979). An appropriate level of sampling effort for a seasonally stratified design would be 30% (110 d/year). These analyses suggest that little gain in reliability will be achieved at greater effort (Table 56) and the coefficient of variation will have declined to approximately 10% at a 30% sampling effort.

Although this seasonal design is based on daily variation in the total count of all species combined

TABLE 58.—Seasonal numbers of daily impingement collections for stratified random sampling of 30% of the days in a year at Indian Point units 2 and 3, based on daily impingement variation during 1976–1979 or 1976–1983.

| | Unit 2 | | Unit 3 | |
Stratum	1976–1979	1976–1983	1976–1979	1976–1983
Winter (Jan–Mar)	30	23	27	35
Spring (Apr–Jun)	10	8	18	20
Summer (Jul–Sep)	11	11	31	31
Fall (Oct–Dec)	59	68	34	24
Total	110	110	110	110

The adequacy of a seasonally stratified sampling design for estimates of impingement by individual species depends on the contribution of single species to the seasonal pattern of variation. Seasonal strata consistently dominated by single species are most effectively sampled by the Neyman sample allocation scheme. Periodic examination for trends in stratum variances and their relationship with Neyman sample allocations would permit adjustment of the seasonally stratified sampling design to future changes in impingement patterns at the Indian Point station.

A reduced-sampling program at the Indian Point station should use a seasonally stratified design and a general allocation scheme, based on patterns of variation for all fish species combined. Coefficients of variation for this design are predicted to average approximately 8–9% at both units, and 95% confidence intervals about the annual mean daily impingement estimate should enclose the true yearly mean between 92 and 93% of the time. Revised sample allocations (based on 1976–1983 data) for units 2 and 3 (Table 58) closely resemble the design implemented 1 July 1981 through 31 December 1984 (based on 1976–1979 data), suggesting that the historical pattern of impingement variation has not changed since 1976–1979. Therefore, the seasonally stratified design with revised sample allocations should produce effective estimates of fish impingement at the Indian Point station.

(general design), it provided relatively precise and reliable impingement estimates when tested on several common Hudson River fish (Table 57). Clupeids, particularly blueback herring at unit 2, were impinged at highly variable rates and were estimated with the least precision and reliability. High fall variation in blueback herring impingement due to exposure of downriver-migrating schools to the impingement structure appeared primarily responsible for the relatively large number of sampling days allocated to the fall season. In contrast, white perch represented 46–50% of the fish impinged annually and dominated collections during the winter season, yet impingement of this species had relatively low variation and was estimated precisely and reliably.

TABLE 59.—Precision achieved by stratified random sampling with seasonal strata at Indian Point station, 1982–1986.[a]

| Year | Operational days (N) | Sampling days (n) | Daily number of impinged fish | | Design efficiency[c] (%) |
			Mean	CV[b]	
		Unit 2			
1982	304	63	2,983	10.3	91
1983	340	111	2,414	29.2	32
1984	238	94	1,428	13.0	60
1985	364	277	1,969	4.2	186
1986	285	108	2,585	12.2	64
		Unit 3			
1982	135	36	5,157	14.1	59
1983	48	22	614	17.2	48
1984	306	110	1,663	9.4	82
1985	266	77	1,359	11.7	66
1986	357	111	995	8.4	92

[a]Neyman sample allocations based on variance estimates from the 1976–1979 impingement data base were used for 1982–1984. Revised Neyman sample allocations based on the pooled 1976–1983 data base were used for 1985–1986.

[b]CV = coefficient of variation = $100 \cdot$ SE/mean.

[c]Efficiency = $100 \cdot$ predicted CV/actual CV. The predicted CV for 1982–1984 was 9.4 at unit 2 and 8.3 at unit 3, based on stratified random sampling of 110 d/year with seasonal strata by Neyman allocation and seasonal variances that were derived from daily impingement counts for all fish species combined during 1976–1979. The predicted CV for 1985–1986 was 7.8 at unit 2 and 7.7 at unit 3, based on a sampling of 110 d/year and variances that were derived from daily impingement counts for all fish species combined during 1976–1983.

A good test of any sampling design is its actual precision during an independent period compared to the predicted precision derived from a previous period. At the Indian Point station, the stratified-sampling design based on a Neyman allocation of 110 impingement-sampling days among four seasonal strata in each year was implemented on 1 July 1981. This sampling effort relied on variance estimates derived from the 1976–1979 data base. For the three complete years stratified-impingement sampling was conducted by this design, the actual design efficiency was 32–91% at unit 2 and 48–82% at unit 3 (Table 59). Unpredicted high impingement rates for weakfish during the summer explained the low efficiency at unit 2 during 1983, and a relatively low sampling effort due to extended outage explained the low efficiency at unit 3 during 1983. For 1985–1986, stratified sampling was conducted according to revised sample allocations based on 1976–1983 data, and efficiency ranged between 64 and 186%. The high efficiency attained at unit 2 during 1985 was due to sampling 76% of the days that year instead of 30%. If 1983 is excluded, and years are examined in which the sampling design allocation of nearly 30% (110 d/year) was achieved, it appears that the actual sampling design has been between 60 and 92% as precise as was predicted.

References[1]

Box, G. E. P., and G. M. Jenkins. 1976. Time series analysis forecasting and control, revised edition. Holden-Day, Oakland, California.

Cochran, W. G. 1977. Sampling techniques, 3rd edition. Wiley, New York.

El-Shamy, F. M. 1979. Impingement sampling frequency: a multiple population approach. Environmental Science and Technology 13:315–320.

Hutchison, J. B., Jr. 1988. Technical descriptions of Hudson River electricity generating stations. American Fisheries Society Monograph 4:113–120.

Johnston, E. M. 1976. On the choice of a sampling frequency for fish impingement at the Cook plant. Preliminary report to the American Electric Power Service Corporation, Palo Alto, California.

Kumar, K. D., and J. S. Griffith. 1978. Temporally stratified sampling programs for estimation of fish impingement. Pages 281–289 in L. D. Jensen, editor. Fourth national workshop on entrainment and impingement. EA Communications, Melville, New York.

Murarka, I. P., and D. J. Bodeau. 1977. Sampling designs and methods for estimating fish-impingement losses at cooling-water intakes. Argonne National Laboratory Report, ANL/ES-60, Argonne, Illinois.

Murarka, I. P., S. A. Spigarelli, and D. J. Bodeau. 1978. Statistical comparison and choices of sampling designs for estimating fish impingement at cooling water intakes. Pages 267–280 in L. D. Jensen, editor. Fourth national workshop on entrainment and impingement. EA Communications, Melville, New York.

NAI (Normandeau Associates, Incorporated). 1984. Hudson River ecological study in the area of Indian Point. 1983 annual report to Consolidated Edison Company of New York and New York Power Authority.

NAI (Normandeau Associates, Incorporated). 1986. Hudson River ecological study in the area of Indian Point. 1985 Annual Report to Consolidated Edison Company of New York and New York Power Authority.

NAI (Normandeau Associates, Incorporated). 1987. Hudson River ecological study in the area of Indian Point. 1986 Annual Report to Consolidated Edison Company of New York and New York Power Authority.

Stupka, R. C., and R. K. Sharma. 1977. Survey of fish impingement at power plants in the United States, volume 3. Estuaries and coastal waters. Argonne National Laboratory, Argonne, Illinois.

TI (Texas Instruments). 1980. Hudson River ecological study in the area of Indian Point. Annual report (1979) to Consolidated Edison Company of New York, Inc.

Tukey, J. W. 1977. Exploratory data analysis. Addison-Wesley, Reading, Massachusetts.

[1]See Table 1 for sources of legal documents and unpublished reports pertaining to the Hudson River.

American Fisheries Society Monograph 4:170–181, 1988

Survival of Fishes after Impingement on Traveling Screens at Hudson River Power Plants

PAUL H. MUESSIG

EA Science and Technology
Rural Delivery 2, Box 91, Goshen Turnpike
Middletown, New York 10940, USA

JAY B. HUTCHISON, JR.

Orange and Rockland Utilities, Incorporated
One Blue Hill Plaza, Pearl River, New York 10965, USA

LAURENCE R. KING

Environmental Affairs Consulting Services
18 Sharon Street, Sidney, New York 13838, USA

REBECCA J. LIGOTINO

EA Science and Technology
Rural Delivery 2, Box 91, Goshen Turnpike

MARTIN DALEY

Central Hudson Gas and Electric Corporation
284 South Avenue, Poughkeepsie, New York 12602, USA

Abstract.—We examined the survival of Hudson River fishes, juveniles and adults, after they had been impinged on continuously rotated traveling screens at the Bowline Point and Danskammer Point power plants. Survival of principal species was similar at the two plants, and estimates of survival improved as monitoring stress was reduced. Adjusted for survival of control fish, survival over 84–108 h after fish were recovered from the screens was highest for Atlantic tomcod, striped bass, and white perch (50–90%) and lowest for bay anchovy, alewife, and blueback herring; other species showed intermediate survival. Survival of striped bass and white perch was positively correlated with water temperature in winter and with conductivity in spring and fall. Continual rotation of the screens, which shortens the average time that fish are impinged, increased survival over that associated with intermittent rotation.

Impingement of fishes on power plant intake screens is an important source of mortality associated with plant operations (Hansen et al. 1977; Jensen 1977, 1978; Christensen et al. 1981; Barnthouse and Van Winkle 1988, this volume). Debris and macroorganisms at a power plant intake usually are filtered from the water withdrawn for cooling first with course (8.9-cm spacing) bar racks and then with vertically rotating traveling screens (0.95-cm mesh). Organisms that collect on, and are removed from, the screens constitute the impinged population. Traveling screens are rotated continuously or intermittently (every 2 h or 4 h), and the impinged material is washed off and returned by a sluiceway system to the source waterbody.

Survival rates of the impinged fish population are important considerations when power plant effects on the water body are to be mitigated.

King et al. (1978) showed that fish losses at several Hudson River plants could be mitigated if the traveling screens are rotated and washed continuously, and they demonstrated that mortality usually increased as time between screen rotation and wash increased. King et al. (1978) presented estimates of postimpingement survival for young-of-the-year (age-0) white perch and striped bass and adult (age-1+) Atlantic tomcod at the Bowline Point, Danskammer Point, and Roseton plants during 1976 and 1977. We refined those estimates and extended the study through 1981 at the Bowline Point and Danskammer Point plants, incorporating improved methodology. We also related postimpingement survival to water temperature and conductivity.

The data in our study reflect the continuous rotational mode of traveling screen operation.

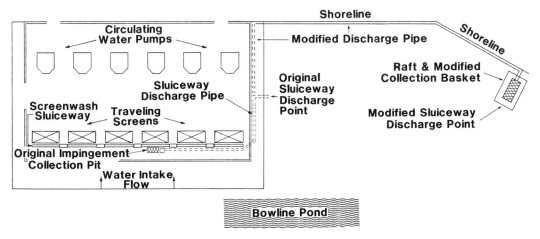

FIGURE 59.—Bowline Point intake structure.

Peak fish impingement at Hudson River plants occurs during winter, when many of the common migratory and resident species overwinter in the lower estuary. High impingement may also occur during major migrations of alewife, herrings, shad, white perch, and striped bass downriver in fall and upriver in spring. Our data characterize the range of postimpingement survival values for the dominant species found in the lower Hudson River estuary, including striped bass, white perch, American shad, alewife, blueback herring, gizzard shad, and Atlantic tomcod.

Field Methods

Impingement survival sampling was undertaken at the plant intakes (Hutchison 1988, this volume) near the traveling screens. Sampling surveys (consisting of data composited from multiple samples) were conducted over a 24-h period once per week or on alternate weeks, depending on fish

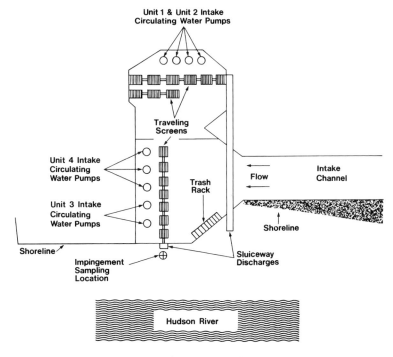

FIGURE 60.—Danskammer Point intake structure.

TABLE 60.—Chronological development of methods for studies of postimpingement fish survival at the Bowline Point plant intake.

Intake design	Fish	Time	Fish collection and handling
Original	Impinged	1975–Jun 1980	Transport from screens: fish washed off the screens were collected immediately in the collection basket in front of the screen; without the collection basket in place, fish were carried 10 m through a 31-cm-diameter pipe and discharged 1 m from the intake structure Collection period: 15–30 min Collection facility: 12.7-mm-mesh basket, 193 cm long × 97 mm wide × 61 cm deep, immersed (to reduce stress on fish) in a collection pit within the intake building Monitoring: usually 12 h, 36 h, 60 h, and 96–108 h after collection
	Control		Seined or trawled from the Hudson River, held at least 3 d before tests, and fed daily during the holding period
		1975–Jun 1978	Introduced as a group to the collection basket and exposed to the screenwash flow for 0, 1, 10, 15, 20, and 30 min; estimated survival was the average unweighted survival for the five exposure periods.
		Jul 1978–present	Introduced to the collection basket in equal numbers at 1-min intervals for a 5-min collection period (to reduce stress on fish); estimated survival represented the average exposure of 2.5 min
Modified	Impinged	Jul 1980–present	Transport from screens: new sluiceway added to bypass original collection pit and pipe; fish washed off screens were collected at the end of a 51-cm-diameter pipe 43 m from the intake Collection period: 5 min (to reduce stress on fish) Collection facility: collection basket of the same mesh as in the 1975–Jun 1980 studies but dimensions were 110 cm long × 110 cm wide × 150 cm deep; basket with removable trays was supported by a floating raft (390 cm long × 670 cm wide) and the basket was immersed (to reduce stress on fish). Monitoring: same as for original design, 1975–Jun 1980
	Control	Jul 1980–present	Same as for the original design, Jul 1978–Jun 1980

abundance. Impinged fish were collected at a point either in the sluiceway system or where the sluiceway discharged into the waterbody (Figures 59, 60). Field methods at the Bowline Point and Danskammer Point plants varied considerably during the study because physical configurations at the plant intakes were modified by the utility to reduce impingement mortality and because changes were made in experimental procedures to minimize added stress on the fish (Tables 60, 61). The original intake sluiceway system at Bowline Point (Figures 59, 61). was modified to return impinged material farther away from the intake, so it would be less likely to recirculate through the intake, and the impingement collection basket was redesigned (Figure 62). The collection location at the Danskammer Point plant was the same throughout the study; however, there were modifications to the collection basket (Figure 63) and changes in experimental procedures.

King et al. (1978) categorized age-0 fish by size only; we advanced age-0 fish to age 1 on January 1 regardless of size. This practice was used throughout the remainder of the study. At the Bowline Point plant in December 1976, white perch and striped bass were randomly subsampled from large collections for 96–108-h observations of survival. Only the subsampled group is reported here because its initial survival did not differ significantly from that of the whole group.

Analytical Methods

We estimated postimpingement survival after traveling screens were washed and fish were sluiced into the collection basket. Fish were categorized immediately after collection from the basket (initial survival) as live (able to maintain equilibrium and swim normally), stunned (unable to maintain equilibrium), or dead. Live and stunned fish were considered to be survivors and

TABLE 61.—Chronological development of methods for studies of postimpingement fish survival at the Danskammer Point plant intake.

Fish	Time	Fish collection and handling
Impinged	1975–1977	Transport from screens: fish washed off screens were collected at the point where the sluiceway discharged into the river Collection period: 60 min Collection facility: 6-mm-mesh basket (150 cm long × 150 cm wide × 61 cm deep) floating in the river under the sluiceway discharge; basket was immersed to minimize stress on fish; after removal of the collection basket from the river, fish were concentrated in a 106-L collection box in the bottom of the basket Monitoring: usually 6 h, 12 h, 18 h, 36 h, 60 h, and 84–96 h after collection; fish were held in rectangular 1,400-L tanks
Control		Seined or trap-netted from the Hudson River, held 48–96 h before tests, and not fed; controls in 1975 were exposed to the holding tanks but not to the collection basket; therefore, unadjusted extended survival was used as the best survival estimate; controls in 1976 and 1977 were exposed to the same collection period and gear as impinged fish
Impinged	1979–1980	Transport from screens: same as for 1976–1977 Collection period: two sequential 30-min periods (reduced to minimize stress on fish) Collection facility: basket size was changed to 120 cm long × 240 cm wide × 120 cm deep, and two removable trays were installed in the basket bottom Monitoring: 12 h, 18 h, 36 h, and 84 h after collection; fish were held in cylindrical 193-L tanks
Control		Same as for 1975–1977, except that fish were introduced to the collection basket at 1-min intervals during the 30-min collection period; estimated survival represented the average exposure of 15 min

were immediately transferred to a holding facility supplied with ambient flow-through river water; initial survival estimates were based on surviving and dead fish at this time. The holding facility consisted of an enclosed building with large tanks to permit visual observation and monitoring of fish survival (Tables 60, 61). Fish were monitored usually twice per day for the first day following collection, then once per day thereafter until monitoring was terminated (after 96–108 h at Bowline Point; after 84–96 h at Danskammer Point). The survival estimates at the end of the monitoring period (extended survival) took into account the fish that were dead on arrival at the collection baskets, but did not allow for the possibility that live fish stressed by impingement might have been more susceptible than usual to predation or disease had they returned to the river.

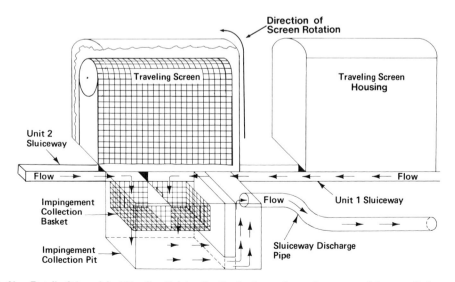

FIGURE 61.—Detail of the original Bowline Point collection basket and traveling screen sluiceway discharge system.

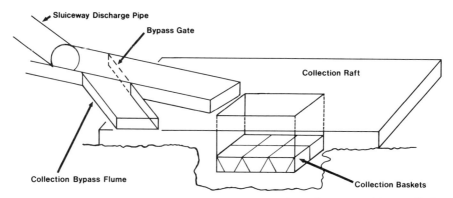

FIGURE 62.—Detail of the modified Bowline Point collection basket and traveling screen sluiceway discharge system.

The time interval between monitoring checks varied within and among the plants due to availability of personnel; however, this did not effect the data discussed in this study. King et al. (1978) showed that survival, as a function of time for postimpinged and control fish held during extended survival monitoring, usually decreased exponentially (a typical survivorship curve) for the first 1–1.5 d and thereafter showed no change in survival rate until monitoring was terminated 3.5–4.5 d later. The survivorship curves in our study showed no slope change within the monitoring interval of 84–108 h, so any extended survival measurement within this period provided the same survival value.

Initial and extended survival estimates, as well as estimates of impinged fish survival adjusted by controls, were calculated according to the formulae in Table 62. Survival was calculated for data pooled across the years of similar experimental methodol-

ogy. Survival of controls exposed to similar collection, handling, and holding procedures as impinged fish were used to estimate the effects of experimental manipulation, and survival estimates for impinged fish were adjusted for these effects. Standard errors were calculated for all data with the exception of white perch and striped bass controls at Bowline Point in 1975–1978. For the latter data, standard errors could not be calculated because the linear regression analysis method used to estimate survival did not provide control sample size (N_C). When we could not collect control fish from the river in winter, or caught too few of them, we took the unadjusted extended survival to be the best estimate of postimpingement survival.

The data for white perch and striped bass at Bowline Point allowed us to examine the relationship between extended postimpingement survival and both water temperature and conductivity by linear regression analysis. Impingement, collection,

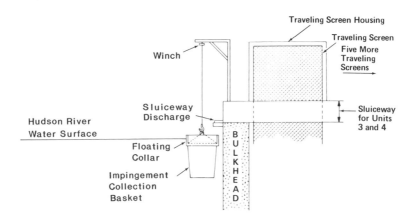

FIGURE 63.—Detail of the Danskammer Point collection basket and traveling screen sluiceway discharge system.

TABLE 62.—Formulae used to calculate survival of impinged fish.

Fish	Survival estimate time	Calculation formula	Definition of terms
Impinged	Initial ($t = 0$ min after collection)	$$P_{Ii} = \frac{L_{Ii} + S_{Ii}}{N_{Ii}}(100)$$	P_{Ii} = initial fish survival proportion L_{Ii} = number of fish alive S_{Ii} = number stunned N_{Ii} = number alive, stunned and dead
		$$SE = \left[\frac{P_{Ii}(1-P_{Ii})}{N_{Ii}}\right]^{1/2}(100)$$	SE = Standard error of P_{Ii} (Snedecor and Cochran 1967)
	Extended ($t = 96$–108 h after collection at Bowline or 84–96 h after collection at Danskammer)	$$P_{Ie} = \frac{L_{Ie} + S_{Ie}}{N_{Ie}}(100)$$	P_{Ie} = Extended fish survival proportion L_{Ie} and S_{Ie} as above N_{Ie} = number alive, stunned, initial dead and extended dead
		$$SE = \left[\frac{P_{Ie}(1-P_{Ie})}{N_{Ie}}\right]^{1/2}(100)$$	SE = standard error of P_{Ie} (Snedecor and Cochran 1967)
Control	Initial (as above)	$$P_{Ci} = \frac{L_{Ci} + S_{Ci}}{N_{Ci}}(100)$$	P_{Ci} = initial fish survival proportion L_{Ci}, S_{Ci}, and N_{Ci} as above for initial
		SE as above for impinged initial survival	
	Extended (as above)	$$P_{Ce} = \frac{L_{Ce} + S_{Ce}}{N_{Ce}}(100)$$	P_{Ce} = extended fish survival proportion L_{Ce}, S_{Ce}, and N_{Ce} as above for extended
		SE as above for impinged extended survival	
Impinged adjusted by controls	Initial (as above)	$$P_{Ai} = \frac{P_{Ii}}{P_{Ci}}(100)$$	P_{Ai} = initial impinged fish survival proportion adjusted by controls
		$$SE = \frac{1}{P_{Ci}}\left[\frac{P_{Ii}(1-P_{Ii})}{N_{Ii}} + P_{Ai}{}^2\left(\frac{P_{Ci}(1-P_{Ci})}{N_{Ci}}\right)\right]^{1/2}(100)$$	SE = standard error of P_{Ai} (Fleiss 1973; Zar 1974)
	Extended (as above)	$$P_{Ae} = \frac{P_{Ie}}{P_{Ce}}(100)$$	P_{Ae} = extended fish survival proportion adjusted by controls
		$$SE = \frac{1}{P_{Ce}}\left[\frac{P_{Ie}(1-P_{Ie})}{N_{Ie}} + P_{Ae}{}^2\left(\frac{P_{Ce}(1-P_{Ce})}{N_{Ce}}\right)\right]^{1/2}(100)$$	SE = standard error of P_{Ae} (Fleiss 1973; Zar 1974)

and holding effects were constant in these experiments. Controls were not available for striped bass and some white perch data; therefore, the unadjusted survival estimates were used. These data probably overestimated mortality because they included the sum effect of impingement, postimpingement collection, and holding. Specific conductance and water temperature were recorded at 24-h intervals during the extended survival holding period and these values were averaged for each experiment.

Conductivity and percent survival were transformed ($\log_{10}$ [conductivity $+1$]; arcsin $\%^{1/2}$) to improve linear fit and to normalize the data. Samples with less than 10 fish were not used so that variability caused by small sample size would be reduced.

Survival of Impinged Fishes

White perch and striped bass were the major species impinged at the Bowline Point and Dans-

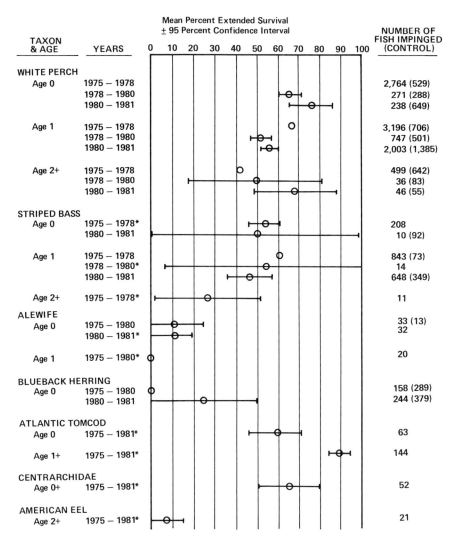

FIGURE 64.—Extended survival of impinged fish 96–108 h after collection at the Bowline Point generating plant, 1975–1981. Survival percentages have been adjusted for control fish survival except for years marked by an asterisk (∗). Confidence intervals could not be calculated for all white perch life stages, or for age-1 striped bass, in 1975–1978. Breaks in the year sequences correspond to changes in sampling methodology.

kammer Point plants (Figures 64, 65). Alewives, blueback herring, American shad, gizzard shad, and Atlantic tomcod were also commonly impinged during high-impingement seasons (fall and spring migratory periods, and winter). Survivals of species collected coincidentally with the main species indicate the range of sensitivity of impingement, but most species were too rare to warrant further statistical evaluation.

Initial postimpingement survival adjusted for control survival exceeded 80% for most species and life stages (EA 1983). Extended survival, the

most relevant index of impingement stress, varied much more widely among species. In general, white perch, striped bass, Atlantic tomcod, gizzard shad, and minnows had better than 50% adjusted survival, among the species that were abundant at one or both power plants. Very few alewives or blueback herring survived 4 d after impingement. For several species, survival estimates improved conspicuously over the years as steps were taken to reduce the experimental stresses of collection, handling, and holding; white perch at Danskammer Point showed this

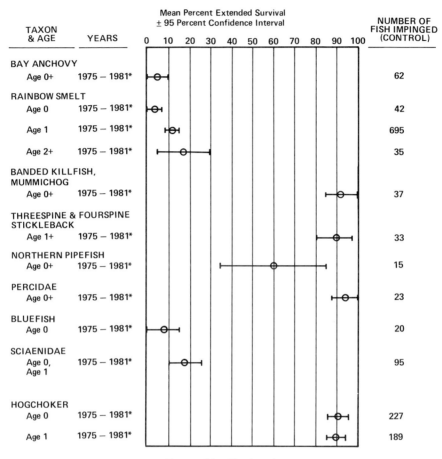

FIGURE 64. Continued.

most conspicuously. There also were variations in extended survival among life stages (age groups), but these were not consistent within or among species nor among years.

Screen Operation and Fish Survival

The results reported above were obtained when the traveling screens were rotated and washed continually. In normal power plant operations, however, screens are only washed at 2-h, 4-h, or longer intervals. King et al. (1978) compared postimpingement survival of white perch and Atlantic tomcod under several operational modes at three Hudson River power plants. In general, survival was greatest where screens were continually washed and declined as the interval between washes increased (Table 63). This is equivalent to saying that survival was inversely related to the time fish were impinged on the screens before they were washed off. Although some species

such as clupeids have low postimpingement survival even when the screens are continually washed, it is likely that the overall impingement impact on fishes may be mitigated if screens are continually rotated during peak impingement periods. Conventional traveling screens are not designed to operate continually year around, but well-maintained screens should be able to function steadily for several days or weeks at key times.

Environmental Correlates of Fish Survival

Studies conducted at Bowline Point demonstrated that conductivity and temperature influenced extended survival of white perch and striped bass. At temperatures above 4.5°C, survival exhibited a logarithmic relationship with conductivity (Figure 66). Water temperature exercised an overriding influence on survival at temperatures below 4.5°C (Figure 67), exhibiting a direct linear relationship with survival. Similar

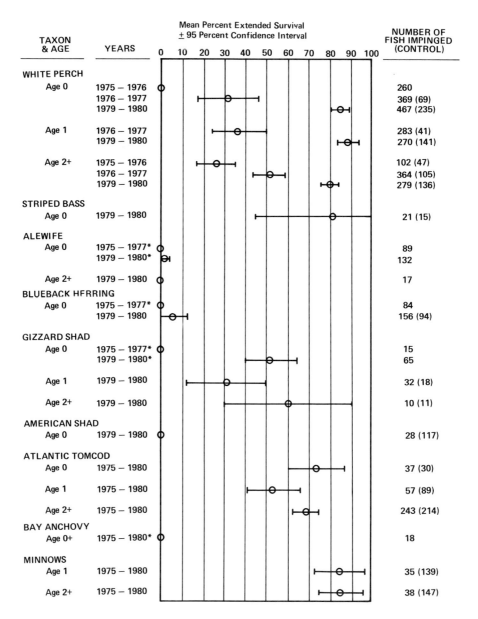

FIGURE 65.—Extended survival of impinged fish 84–96 h after collection at the Danskammer generating plant, 1975–1980. Survival percentages have been adjusted for control fish survival except for years marked by an asterisk (*). Breaks in the year sequences correspond to changes in sampling methodology.

patterns were observed for white perch used as collection controls but not for holding controls; however, the number of control tests conducted was insufficient to warrant more rigorous analysis.

Specific conductance fluctuates in the lower Hudson River estuary with freshwater runoff and tidal cycles (Cooper et al. 1988, this volume). During winter and spring, monthly peaks in conductivity associated with the upriver intrusion of the salt front occur in conjunction with periods of maximum tidal amplitude and low runoff. A direct relationship has been observed at lower Hudson River power plants (TI 1974, 1975) between the rate of impingement and the passage of this salt front. The salt front generally does not extend into the vicinity of Danskammer Point at this time of year.

TABLE 63.—Percent survival of impinged age-0 and age-1 white perch and age-1+ Atlantic tomcod washed from traveling screens at three Hudson River power plants, adapted from King et al. (1978). Data for initial and extended survival have been adjusted for survival of control fish subjected only to postimpingement sampling and holding procedures.[a] Extended survival was measured 96–108 h after impingement at the Bowline Point plant and 84 h after impingement elsewhere. At the time these studies were conducted, no statistical test had been developed to test adjusted survival proportions between screen-wash intervals.

Sampling time	Species (age)	Wash-water pressure (kg/cm^2)	Initial survival (%)			Extended survival (%)		
			Continuous wash	2 h between washes	4 h between washes	Continuous wash	2 h between washes	4 h between washes
Bowline Point plant								
Nov–Dec 1976	White perch (age 0)		100	93	74	100	53	36
Roseton plant								
Nov–Dec 1976	White perch (age 0)	3.5	100	98	98	60	11	23
		7.0	81	94	72	8	7	0
Apr–May 1977	White perch (age 1)	3.5	95	100	92	29	29	36
		7.0	89	97	39	73	33	14
Nov 1976–Mar 1977	Atlantic tomcod (age 1+)		96	93		81	72	
Danskammer Point plant								
Nov–Dec 1976	White perch (age 0)		84	63		40	21	
Apr–May 1977	White perch (age 1)		88	97	39	61	36	9
Nov 1976–May 1977	Atlantic tomcod (age 1+)		89	96		83	87	

[a] See Table 3 for calculation of adjusted initial (P_{Ai}) and extended (P_{Ae}) survival proportions.

Our studies showed that increased conductivity was associated with increased survival of impinged fish. Collins and Hulsey (1963), Miles et al. (1974), and Hattingh et al. (1975) also observed the beneficial effect of saline solutions on fish survival. Such enhanced survival in the holding facility may be related to the alleviation of osmoregulatory dysfunction or hyperglycemia associated with stress (Wedemeyer 1972; Hattingh and Van Peltzen 1974; Miles et al. 1974). We also found that extended survival could be expected to decrease sharply at water temperatures equal to or less than 4.5°C. Increased mortality due to stress (increased osmoregulatory dysfunction) at low temperatures (3°C or less) has been noted in other studies (Stanley and Colby 1971; Umminger 1971; Colby 1973; Umminger and Gist 1973; Otto et al. 1976). Lower lethal limits for white perch and striped bass have not been determined in thermal tolerance studies.

Technical Lessons Learned

There were many changes in collection methodology between and within the plants during this study; the changes were made to improve experimental methodology and to reduce the mortality of experimental fish due to collection and holding. The impingement collection basket was enlarged to reduce sluiceway water turbulence and velocity effects on fish at Bowline Point when the intake was modified. Removable trays were installed in the basket bottom to keep fish immersed in water at both the Bowline Point and Danskammer Point plants. Increased survival was observed after these changes.

Sample collection time was reduced from the 15–30-min period of earlier Bowline Point collections to 5 min in more recent years; collection time at Danskammer was reduced from 60 min to 30 min. Increased white perch survival was generally noted following these changes, although this effect may have been confounded with effects of concurrent basket design alterations. It was important to balance the minimization of collection time with the need to sample over sufficient time to maintain statistically adequate sample sizes.

Control fish should be introduced to the collection basket at regular intervals during impingement collections to insure that stress effects due to collection, handling, and holding can be fac-

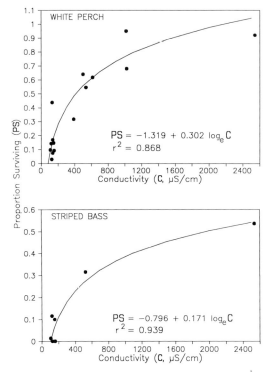

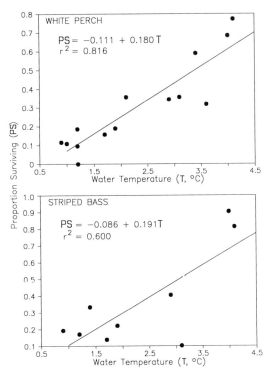

FIGURE 66.—Extended survival of impinged white perch and striped bass related to specific conductance, for temperatures above 4.5°C.

FIGURE 67.—Extended survival of impinged white perch and striped bass related to water temperature, for water temperatures less than 4.5°C.

tored out of postimpingement survival estimates. It required special efforts to collect adequate numbers of control fish for winter studies and to maintain them for several weeks or months. Ice conditions precluded use of conventional collection gear (seine, trawl, or box trap), and controls had to be obtained in fall for use during winter sampling. Extended holding of control fish required periodic treatments with saline water or potassium permanganate for disease control.

We conclude that survival data for species collected at both plants were generally similar and characterize the range of values expected for fishes in the lower Hudson River estuary. Our studies showed that fish species varied greatly with respect to susceptibility to impingement mortality; such mortality was influenced by environmental conditions during holding, design of the collection basket, and operating mode of plant traveling screens. Postimpingement survival of fish can be enhanced by continuous rotation of conventional traveling screens for several days or weeks during peak impingement periods. Impingement survival data can be improved by minimizing experimental stress on impinged and control fish.

Acknowledgments

We express our appreciation to personnel of EA Science and Technology (formerly Ecological Analysts, Incorporated) for assisting in various aspects of specimen collection and data processing. Karen D. Galeazzi of Orange and Rockland Utilities, Incorporated, typed the manuscript. Ronald J. Klauda, Larry W. Barnthouse, and Douglas S. Vaughan provided helpful suggestions and critical reviews of the manuscript. The studies were conducted under contract with Orange and Rockland Utilities, Incorporated, and Central Hudson Gas & Electric Corporation. Portions of the studies were jointly funded by Central Hudson Gas & Electric Corporation, Consolidated Edison Company of New York, Incorporated, Niagara Mohawk Power Corporation, New York Power Authority, and Orange and Rockland Utilities, Incorporated.

References[1]

Barnthouse, L. W., and W. Van Winkle. 1988. Analysis of impingement impacts on Hudson River fish populations. American Fisheries Society Monograph 4: 182–190.

Christensen, S. W., W. Van Winkle, L. W. Barnthouse, and D. S. Vaughan. 1981. Science and the law: confluence and conflict on the Hudson River. Environmental Impact Assessment Review 2:63–88.

Colby, P. G., 1973. Response of the alewives, *Alosa pseudoharengus*, to environmental change. Pages 163–198 *in* W. Chavin, editor. Responses of fish to environmental changes. Thomas, Springfield, Illinois.

Collins, J. L., and A. H. Hulsey. 1963. Hauling mortality of threadfin shad reduced with M.S. 222 and salt. Progressive Fish-Culturist 25:105–106.

Cooper, J. C., F. R. Cantelmo, and C. E. Newton. 1988. Overview of the Hudson River estuary. American Fisheries Society Monograph 4:11–24.

EA (Ecological Analysts). 1983. Impingement survival data for the Bowline Point and Danskammer Point generating stations. Report to Central Hudson Gas and Electric Corporation and Orange and Rockland Utilities.

Fleiss, J. L. 1973. Statistical methods for rates and proportions. Wiley, New York.

Hansen, C. H., J. R. White, and H. W. Li. 1977. Entrapment and impingement of fishes by power plant cooling water intakes: an overview. U.S. National Marine Fisheries Service Marine Fisheries Review 39(10):7–17.

Hattingh, J., F. L. Fourie, and J. H. J. Van Vuren. 1975. The transport of freshwater fish. Journal of Fish Biology 7:447–449.

Hattingh, J., and A. J. J. Van Pletzen. 1974. The influence of capture and transportation on some blood parameters of fresh water fish. Comparative Biochemistry and Physiology A, Comparative Physiology 49:607–609.

Hutchison, J. B. 1988. Technical descriptions of Hudson River electricity generating stations. American Fisheries Society Monograph 4:113–120.

Jensen, L. D., editor. 1977. Third national workshop on entrainment and impingement, section 316(b)—research and compliance. EA Communications, Melville, New York.

Jensen, L. D., editor. 1978. Fourth national workshop on entrainment and impingement. EA Communications, Melville, New York.

King, L. R., J. B. Hutchison, Jr., and T. G. Huggins. 1978. Impingement survival studies on white perch, striped bass, and Atlantic tomcod at three Hudson River power plants. Pages 217–233 *in* L. D. Jensen, editor. Fourth national workshop on entrainment and impingement. EA Communications, Melville, New York.

Miles, H. M., S. M. Loehner, D. T. Michaud, and S. L. Salivar. 1974. Physiological responses of hatchery reared muskellunge (*Esox masquinongy*) to handling. Transactions of the American Fisheries Society 103:336–342.

Otto, R. G., M. A. Kitchel, and J. O. Rice. 1976. Lethal and preferred temperatures of the alewife (*Alosa pseudoharengus*) in Lake Michigan. Transactions of the American Fisheries Society 105:96–106.

Snedecor, G. W., and W. G. Cochran. 1967. Statistical methods. Iowa State University Press, Ames.

Stanley, J. G., and P. G. Colby. 1971. Effects of temperature on electrolyte balance and osmoregulation in alewife (*Alosa pseudoharengus*) in fresh and sea water. Transactions of the American Fisheries Society 100:624–638.

TI (Texas Instruments). 1974. Indian Point impingement study report for the period 15 June 1972 through 31 December 1973. Report to Consolidated Edison Company of New York.

TI (Texas Instruments). 1975. Indian Point impingement study report for the period 1 January 1974 through 31 December 1974. Report to Consolidated Edison Company of New York.

Umminger, B. L. 1971. Osmoregulatory role of serum glucose in freshwater adapted killifish (*Fundulus heteroclitus*) at temperatures near freezing. Comparative Biochemistry and Physiology A, Comparative Physiology 38:141–145.

Umminger, B. L., and D. H. Gist. 1973. Effects of thermal acclimation on physiological responses to handling stress, cortisol and aldosterone injections in the goldfish, *Carassius auratus*. Comparative Biochemistry and Physiology A, Comparative Physiology 44:967–977.

Wedemeyer, G. 1972. Some physiological consequences of handling stress in the juvenile coho salmon (*Oncorhynchus kisutch*) and steelhead trout (*Salmo gairdneri*). Journal of the Fisheries Research Board of Canada 29:1780–1783.

Zar, J. H. 1974. Biostatistical analysis. Prentice Hall, Englewood Cliffs, New Jersey.

[1]See Table 1 for sources of legal documents and unpublished reports pertaining to the Hudson River.

American Fisheries Society Monograph 4:182–190, 1988

Analysis of Impingement Impacts on Hudson River Fish Populations

LAWRENCE W. BARNTHOUSE AND WEBSTER VAN WINKLE

Environmental Sciences Division, Oak Ridge National Laboratory
Post Office Box 2008, Oak Ridge, Tennessee 37831-6036, USA

Abstract.—Impacts of impingement, expressed as reductions in year-class abundance, were calculated for six Hudson River fish populations. Estimates were made for the 1974 and 1975 year classes of white perch, striped bass, Atlantic tomcod, and American shad, and the 1974 year classes of alewife and blueback herring. The maximum estimated reductions in year-class abundance were less than 5% for all year classes except the 1974 and 1975 white perch year classes and the 1974 striped bass year class. Only for white perch were the estimates greater than 10% per year. For striped bass, the 146,000 fish from the 1974 year class that were killed by impingement could have produced 12,000–16,000 5-year-old fish or 270–300 10-year-olds. We also estimated the reductions in mortality that could have been achieved had closed-cycle cooling systems been installed at one or more of three power plants (Bowline Point, Indian Point, and Roseton) and had the screen-wash systems at Bowline Point and Indian Point been modified to improve the survival of impinged fish. Closed-cycle cooling at all three plants would have reduced impingement impacts on white perch, striped bass, and Atlantic tomcod by 75% or more; installation of closed-cycle cooling at Indian Point alone would have reduced impingement impacts on white perch and Atlantic tomcod by 50%–80%. Modified traveling screens would have been less effective than closed-cycle cooling, but still would have reduced impingement impacts on white perch by roughly 20%.

This paper presents quantitative estimates of the impacts of impingement at Hudson River power plants on populations of white perch, striped bass, Atlantic tomcod, American shad, alewife, and blueback herring. These analyses, performed for the U.S. Environmental Protection Agency (EPA), include estimation of the impacts actually imposed on the 1974 or 1975 year classes (or both) of each population, and calculation of the reductions in impact that could have been achieved had cooling towers or modified traveling screens been installed at one or more power plants.

Our measure is the conditional impingement mortality rate (Vaughan 1988, this volume). This measure is equivalent to the fractional reduction in year-class abundance due to impingement, provided that density-dependent mortality is low during the period in which impingement occurs. Conditional impingement mortality rates were calculated for the 1974 year classes of all six populations and for the 1975 year classes of all except alewife and blueback herring. Similar analyses could not be performed for the vulnerable and ecologically important bay anchovy because available data on the distribution, abundance, and mortality of this species were insufficient.

The model used for these analyses and the derivation of constituent equations have been described in detail by Barnthouse et al. (1979) and by Barnthouse and Van Winkle (1981). Like the model used by Texas Instruments (TI) (Englert and Boreman 1988, this volume), it is derived from Ricker's type-II fishery model (Ricker 1975). The conditional impingement mortality rate, computed for an arbitrary time interval, is

$$m = 1 - (1 - A)\exp(u/A); \quad (1)$$

m = conditional impingement mortality rate;
u = impingement exploitation rate;
A = fraction of the initial population dying from all causes during the time interval.

In applying equation (1) to our impact assessment, we (a) decomposed A into components due to impingement mortality and natural mortality and (b) set the time interval for calculation at 1 month rather than 1 year. Separating natural mortality (n) from impingement mortality (m) involved substituting $[1 - (1 - m)(1 - n) = m + n - mn]$ for A in equation (1) and then solving the equation iteratively. This procedure enabled us to assess the potential effectiveness of mitigating measures that would reduce the numbers of fish impinged or increase the survival of impinged fish but would not affect natural mortality rates. The monthly time interval was employed to allow for seasonal variations in natural and impingement mortality.

Data Source and Uncertainties

Impingement

The impingement estimates used in these analyses were obtained from sampling programs conducted at the Bowline Point, Lovett, Indian Point, Roseton, Danskammer, and Albany generating stations. During 1973–1977, impinged fish were collected and enumerated regularly at all six power plants. At Indian Point, all screen washes were monitored and attempts were made to collect, identify, and count all impinged fish. At the other plants, screen washes were monitored for 24 h one or more days per week. At all plants, length-frequency data were obtained, making it possible to calculate approximate age distributions of impinged fish.

Barnthouse (1982) identified two important sources of bias that affect estimates of numbers of fish impinged and killed at power plants: low collection efficiency and high (at least for some species at some plants) survival of impinged fish. For reasons that are not completely understood, not all fish that are impinged and killed are collected and counted during screen-wash monitoring. Experiments with marked fish showed that collection efficiencies at the major plants range from less than 20% at Indian Point unit 2 to nearly 80% at Indian Point unit 3 and Bowline Point. For some species at some plants, the bias due to low collection efficiency appears to be partly or completely offset by the survival of fish impinged, washed off the screens, and returned to the river on days when screen washes are not monitored (Muessig et al. 1988, this volume). Barnthouse (1982) developed a table of adjustment factors to account for the likely biases in impingement estimates for each species at each plant. The impingement estimates employed in the assessments presented here were adjusted by these factors.

Abundance and Mortality

Estimates of the abundance of the 1974 and 1975 year classes of the species of interest, at the time they first became large enough to be impinged, were obtained from the TI field sampling programs (Young et al. 1988, this volume). For white perch, mark–recapture population estimates were available. For the other species, abundance estimates had to be extrapolated from catch–effort data. The uncertainties associated with all of these estimates were large. To account for these uncertainties, upper and lower bounds on the abundance of each year class were esti-

mated. For white perch, these were taken to be the upper and lower 95% confidence limits around the mark–recapture population estimates. For the other species, bounds were calculated from maximum and minimum estimates of sampling gear efficiency, assumed to be 100% (lower population bound) or 20% (upper bound). Estimates of the efficiency of TI's 30-m beach seine (TI 1978) and of other similar gear (Kjelson 1977) are substantially above 20%.

Estimates of mortality rates for impingeable juvenile fish were calculated from TI's weekly (longitudinal river survey) or biweekly (fall shoals survey) abundance estimates for the years 1974 and 1975. The time series for each year class was fitted, by least squares regression, to the equation

$$\log_e P_t = \log_e P_0 - Dt; \qquad (2)$$

P_t = population size on day t;
P_0 = population size on day 0 (the first day of the period of vulnerability to impingement);
D = daily instantaneous mortality rate.

Gear selectivity and migration in and out of the study area bias estimates of D obtained from equation (2). Time-dependent increases in gear avoidance and emigration of fall juveniles would cause equation (2) to overestimate the true mortality rate. Therefore, the values obtained from the regressions were assumed to be upper limits on the rate of natural mortality. There was no straightforward way to calculate lower bounds on D. It seemed reasonable, however, to assume that mortality among young-of-the-year fish of all species should be at least as high as the observed mortality of yearling and older white perch. Data presented by Wallace (1971) had indicated that mortality among yearling and older white perch is probably about 50% per year.

Table 64 shows that the abundances of the populations examined varied over approximately a factor of 50, the Atlantic tomcod being by far the most abundant and the striped bass and alewife the least abundant. Table 64 also shows a rough correspondence between abundance and numbers impinged; however, both the 1974 and 1975 year classes of white perch were impinged in high numbers relative to their estimated abundance. Mortality rates for most of the species were similar, with the notable exception of Atlantic tomcod (Table 64). The very high natural mortality rate estimated for this species is consistent with the observation that the Atlantic tomcod

TABLE 64.—Impingement, abundance, and natural mortality estimates used in impingement impact assessments (from Barnthouse and Van Winkle 1982).

Species	Year class	Total impingement (10^6 fish)[a]	Initial abundance (10^6 fish)[b]	Natural mortality rate	
				Age-0 fish[c]	Age-1+ fish[d]
White perch	1974	2.8–2.9	14–55	0.5–0.8	0.5
White perch	1975	2.4	24–63	0.5–0.8	0.5
Striped bass	1974	0.15	4–20	0.5–0.8	0.5
Striped bass	1975	0.08	5–28	0.5–0.8	0.5
Atlantic tomcod	1974	2.5	200–999	0.98	
Atlantic tomcod	1975	0.5	87–434	0.98	
American shad	1974	0.04	16–78	0.9	0.5
American shad	1975	0.06	16–80	0.9	0.5
Alewife	1974	0.16	4–20	0.5–0.9	0.5
Blueback herring	1975	0.46	29–145	0.5–0.9	0.5

[a]Total number of fish over all years during which members of that year class were impinged.
[b]Estimated abundance of year class at the beginning of its period of vulnerability to impingement.
[c]Expressed as annual mortality, except for Atlantic tomcod. For Atlantic tomcod, the estimate presented is for a 9-month period of vulnerability.
[d]Expressed as annual mortality.

population in the Hudson is composed almost exclusively of young of the year (McLaren et al. 1988, this volume).

Estimates of Impingement Mortality

Equation (1) calculates the magnitude of impingement mortality required to account for the observed number of impinged fish, given the number of fish initially available for impingement, the prevailing rate of natural mortality, and the age of the fish (in months) at the time they are impinged. Clearly, the impact of impinging a given number of fish is inversely related to the size of the year class from which they are removed. More counterintuitively, the impact of impinging a given number of fish of a given age is directly related to the prevailing rate of natural mortality. This is true because natural mortality and impingement "compete" for fish, in the sense that any particular fish can die only once and from only one cause. For any initial population size, a higher impingement mortality rate is required to account for the observed number of impinged fish if natural mortality is high than if it is low. For related reasons, the impact of impinging any particular fish increases with its age because the year class from which it is removed is continuously decreasing in abundance.

To set probable upper and lower bounds on the impact of impingement on the species of interest, we estimated ranges of conditional impingement mortality rates for each species from all possible combinations of initial abundance and natural mortality for the 1974 and 1975 year classes (Table 65). We also estimated conditional impingement mortality rates for white perch under alternative

assumptions of 2- and 3-year vulnerability (Barnthouse and Van Winkle 1981). The two assumptions about the age distribution of impinged white perch constitute a third source of uncertainty affecting impact estimates for this species. Consequently, Table 65 presents two ranges for white perch: a "maximum range" (the highest and lowest conditional impingement mortality rates computed from the eight possible combinations of assumptions) and a "probable range" obtained by excluding the highest and lowest values. Because more and better field data were available for white perch and striped bass than for the other species, impact estimates for these two are more certain than are those for the other four species considered. The least adequate data, and consequently the least certain impact estimates, pertain to alewife and blueback herring.

Conditional impingement mortality rates calculated by McFadden and Lawler (1977) for the utility companies fell within our ranges for striped bass and Atlantic tomcod, but fell outside our ranges for American shad and white perch.

Several unique aspects of the life history of white perch in the Hudson River are responsible for the comparatively high impact of impingement on this species. During the winter, a major fraction of the population resides in the lower and middle estuary, in the vicinity of the Bowline Point, Lovett, and Indian Point plants. Although substantial winter impingement of white perch occurs at all three plants, the numbers impinged at Indian Point exceed by far the combined totals for all other Hudson River power plants (Barnthouse and Van Winkle 1981). This phenomenon appears to be related to the concentration of fish in deep

TABLE 65.—Ranges of estimates of conditional impingement mortality rates for six Hudson River fish species.

| Species (year class) | Oak Ridge estimates | | Utilities' estimate (McFadden and Lawler 1977) |
	Low estimate	High estimate	
White perch (1974)			
Maximum range	0.095	0.588	
Probable range	0.119	0.446	0.113
White perch (1975)			
Maximum range	0.077	0.245	
Probable range	0.115	0.245	
Striped bass (1974)	0.011	0.092	0.042
Striped bass (1975)	0.004	0.035	0.023
Alewife (1974)	0.014	0.043	
Blueback herring (1974)	0.005	0.025	
American shad (1974)	0.001	0.005	0.012
American shad (1975)	0.002	0.011	
Atlantic tomcod (1974)	0.010	0.049	0.015
Atlantic tomcod (1975)	0.006	0.030	

areas of the Hudson River channel near the Indian Point intakes, and in the vicinity of the salt front, which fluctuates above and below Indian Point during the winter (TI 1974, 1975). The mobility of these overwintering fish is greatly reduced by near-freezing water temperatures, increasing their vulnerability to impingement.

The vulnerability of yearling and older white perch contributes significantly to the impact of impingement. Yearling and older fish account for roughly 10% of the number of white perch impinged. In computing conditional mortality rates, a yearling white perch is "worth" 2–5 young of the year (depending on the mortality rate assumed), and a 2-year-old white perch is worth 4–10 young of the year. A major reason for the discrepancy between our conditional mortality rates for the 1974 white perch year class and the corresponding rate (Table 65) calculated by McFadden and Lawler (1977) is that the latter quantified the impact on young of the year only.

Although striped bass, like white perch, are most vulnerable to impingement during the winter, their distribution is centered well downriver from Indian Point (McFadden 1977). Consequently, the impacts of winter impingement on the 1974 and 1975 year classes of striped bass were much lower than the impacts on white perch. Bowline Point, rather than Indian Point, was the primary source of impact.

The extremely low impingement impacts on alewife, blueback herring, and American shad are related to the brief period that these species are concentrated in the vicinity of major power plants during their emigration from the estuary in autumn.

Evaluation of Mitigating Measures

In addition to estimating the impacts actually imposed on the 1974 and 1975 year classes of Hudson River fish populations, we estimated the reductions in impact that could have occurred had mitigation been attempted. The purpose of these analyses was to provide guidance to the EPA as to the biological effectiveness of mitigating technologies being proposed in the hearings and in the settlement negotiations. Two types of mitigation were investigated: installation of closed-cycle cooling systems (cooling towers) to reduce the numbers of fish impinged, and installation of modified traveling screens to increase the survival of impinged fish.

Closed-Cycle Cooling

We considered three closed-cycle cooling configurations: (1) cooling towers at the Roseton, Bowline Point, and Indian Point plants; (2) cooling towers at Bowline Point and Indian Point; and (3) cooling towers at Indian Point only. To calculate the numbers of white perch, striped bass, and

TABLE 66.—Estimates of conditional impingement mortality rates for the 1974 and 1975 Hudson River year classes of white perch, striped bass, and Atlantic tomcod, for three alternative closed-cycle cooling configurations.

| | Closed-cycle cooling assumed at | | | | | |
| | Indian Point, Bowline Point, and Roseton | | Indian Point and Bowline Point[a] | | Indian Point only[a] | |
Species (year class)	Low estimate	High estimate	Low estimate	High estimate	Low estimate	High estimate
White perch (1974)						
Maximum range	0.027	0.150	0.030	0.177	0.042	0.237
Probable range	0.031	0.128	0.036	0.143	0.049	0.195
White perch (1975)						
Maximum range	0.013	0.042	0.019	0.061	0.024	0.078
Probable range	0.020	0.042	0.029	0.061	0.036	0.078
Striped bass (1974)	0.003	0.023	0.003	0.024	0.010	0.081
Striped bass (1975)	0.001	0.013	0.001	0.013	0.003	0.024
Atlantic tomcod (1974)	0.004	0.018	0.004	0.019	0.004	0.019
Atlantic tomcod (1975)	0.001	0.003	0.001	0.004	0.001	0.004

[a]Once-through cooling assumed elsewhere.

Atlantic tomcod that would have been impinged had closed-cycle cooling systems been in operation during the years 1974–1977, we assumed that the number of fish impinged at a particular plant is directly proportional to the volume of water withdrawn by that plant. Thus, the reduction in impingement at each generating unit assumed to have a cooling tower was calculated from the estimated reduction in cooling water withdrawal for that unit (see Barnthouse and Van Winkle 1982 for detailed methods). Under this assumption, the numbers of fish impinged at the three plants would be reduced by 89% (Indian Point Unit 3, winter) to 98% (Bowline Point, all seasons).

Impacts associated with this reduced impingement (Table 66) are based on the same estimates of abundance and mortality used to generate the values in Table 65. Comparison of these two tables shows that the installation of closed-cycle cooling would have greatly reduced the impacts of impingement on all three species. If cooling towers had been built at all three plants, the maximum conditional impingement mortality rates for white perch would have been reduced by about 75% for the 1974 year class and by about 80% for the 1975 year class. Similar reductions could have been achieved for striped bass and Atlantic tomcod. Nearly equal mitigation could have been achieved by closed-cycle cooling at Bowline Point and Indian Point only, the 1975 white perch year class being the only appreciable exception. Closed-cycle cooling at Indian Point units 2 and 3 alone would have reduced the impact of impingement

on white perch and Atlantic tomcod by 50%–80%.

Modified Traveling Screens

During the settlement negotiations, the utilities suggested that impingement impacts could be reduced by installing traveling screens equipped with fish buckets and special screen-wash systems (sometimes called "Ristroph" screens) at Bowline Point and Indian Point. It was claimed that 60% survival of impinged white perch and striped bass could be obtained by use of these screens.

We assisted EPA in evaluating this proposal by estimating the reduction in impact on the 1975 year class of white perch that would have occurred had these screens been in place during 1975–1977. Two cases were examined. In case 1, it was assumed that 60% survival of white perch could be achieved at both Bowline Point and Indian Point. Evidence available at the time suggested that 60% survival was overly optimistic. Cannon et al. (1979) had found no evidence that fish-bucket-type traveling screens were more effective at reducing impingement mortality than were the continuously rotating conventional traveling screens already employed at Bowline Point and Roseton. Therefore, in case 2, it was assumed that impingement survival with the modified traveling screens would be equal to that observed for the existing screens at Bowline Point and Roseton. We had previously estimated (Barnthouse 1982) impingement survival at these two plants to be about 40% for white perch for the screen-wash

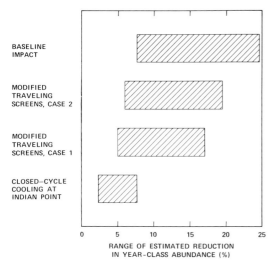

BASELINE
IMPACT

MODIFIED
TRAVELING
SCREENS, CASE 2

MODIFIED
TRAVELING
SCREENS, CASE 1

CLOSED—CYCLE
COOLING AT
INDIAN POINT

0   5   10   15   20   25

RANGE OF ESTIMATED REDUCTION
IN YEAR-CLASS ABUNDANCE (%)

FIGURE 68.—Evaluation of modified traveling screens as a means of reducing the impact of impingement on the Hudson River white perch population, with the 1975 year class as a reference. It is assumed that that continuously rotating traveling screens with fish buckets are installed at all generating units at Bowline Point, Indian Point, and Roseton. In case 1, 60% of the white perch impinged on these screens are assumed to return to the river alive. In case 2, only 40% survive. The top bar shows the range of baseline impact estimates for the 1975 year class (from Table 65). The other bars show corresponding ranges for modified traveling screens (cases 1 and 2) and for closed-cycle cooling at Indian Point units 1 and 2 (from Table 66).

systems and operating modes employed during 1974–1979.

Our results suggested that modified traveling screens would be much less effective than closed-cycle cooling (Figure 68). However, a moderate degree of mitigation could be achieved. Regulations then in force required that all fish impinged at Indian Point be collected and counted (Mattson et al. 1988, this volume). If these regulations were relaxed then, even if the modified screens were no more effective than continuously rotating conventional screens, a 20% reduction in impact could be achieved by increasing survival at Indian Point from 0% to 40%.

Discussion

The maximum reductions in year-class abundance due to impingement at Hudson River power plants were estimated to be less than 5% for all year classes except the 1974 and 1975 white perch year classes and the 1974 striped bass year class. Thus, our results suggest that, for most of the

species examined, impingement is probably not a biologically important source of mortality except, perhaps, when added to other, more serious stresses.

Only for white perch are impingement losses high enough to be a major source of mortality. Our conditional impingement mortality rates are equivalent to reductions in year-class abundance on the order of 10–60%. When combined with entrainment losses, estimated at roughly 10% per year (Boreman and Goodyear 1988, this volume), the total impact of once-through cooling water withdrawal on this population appeared to be in excess of 20% per year class for the years examined. Our understanding of the white perch population and its interactions with other components of the Hudson River ecosystem is insufficient for predicting the long-term effects of these losses.

The estimated reductions in striped bass year-class abundance (up to 10% per year) probably do not, by themselves, constitute a threat to the population as a whole. However, the loss of these fish may have socioeconomic importance. About 146,000 striped bass of the 1974 year class and 80,000 of the 1975 year class were killed by impingement at Hudson River power plants (Table 64). We used the theory of conditional mortality rates to estimate the number of these fish that could have survived to enter the sport or commercial fisheries. Barnthouse and Van Winkle (1982) presented initial population sizes, population sizes at age 2, and conditional impingement mortality rates for the 1974 and 1975 striped bass year classes. These values were used to calculate the number of "equivalent 2-year-olds" impinged from each year class: 42,000–57,000 for the 1974 year class and 23,000–30,000 for the 1975 year class (Appendix).

We used the life table for striped bass developed by Dew (1981) to extrapolate these estimates to numbers of 5- and 10-year-olds. The results of this exercise indicate that the impinged members of the 1974 year class could have produced 12,000–16,000 5-year-old striped bass (the median age for commercially caught striped bass in the Hudson) or 270–370 10-year-old sport fish. Impingement losses from the 1975 year class were equivalent to 6,400–8,400 5-year-olds or 150–190 10-year-olds. Hoff et al. (1988, this volume) developed estimates of annual survival of 5- to 10-year-old striped bass (0.45 for males and 0.60 for females) that are somewhat higher than Dew's estimate (0.47 for both sexes). Using the values from Hoff et al. roughly doubles our estimates of the numbers of equivalent 10-year-olds impinged.

Whether or not the biological impact of white perch impingement or the socioeconomic impact of striped bass impingement is important enough to warrant mitigation is, in our opinion, a socio-political question rather than a scientific one. The parties to the settlement negotiations did consider impingement of these species to be important. Mitigation of impingement at Indian Point was explicitly included in all of the settlement proposals considered.

Although our analysis (Table 66) showed the potential effectiveness of closed-cycle cooling at Indian Point, this solution was considered too costly by the utilities. Due to lack of applicable data, we could not evaluate the potential effectiveness of the angled intake screens that were ultimately agreed on as a mitigating measure. However, our evaluation of the fish-bucket-type traveling screens may have been a factor in the subsequent abandonment of these devices by the negotiators.

There is still considerable doubt as to the feasibility of angled screens as a mitigating measure for large power plants such as Indian Point. However, other means of reducing impingement mortality have been developed and implemented at Hudson River power plants since the period covered in this paper. Impingement of all species at Bowline Point has been substantially reduced following the installation of a barrier net (Hutchison and Matousek 1988, this volume). Experiments at Bowline Point and Danskammer, completed subsequent to our analyses, have shown that the survival of impinged white perch can exceed 60% when traveling screens are operated in the continuous mode (Muessig et al. 1988). Although routine operation in continuous mode is not feasible for existing traveling screen designs, continuous rotation is possible for short periods when impingement is high. The intake structure at Bowline Point was rebuilt in 1979–1980, in part to permit extended operation of the traveling screens in continuous mode. Thus, it appears that relatively simple devices and operational changes may have succeeded where expensive technologies proved impractical.

In conclusion, we note that, given the expense of collecting the data necessary for performing assessments of the kind described here, it is desirable to identify in advance the circumstances that may lead to large (10% or greater) reductions in year-class abundance. Two such circumstances can be identified from the Hudson River studies:

(1) the presence of a major fraction of the population in close proximity to power plants at a time when the fish are stressed and susceptible to being impinged, and
(2) the vulnerability of fish through a major portion of their life-cycle.

When population-level assessment is necessary, information concerning the abundance and life history of the species involved is essential if biological or socioeconomic importance is to be inferred. It is not currently possible to estimate a level of impingement mortality above which population collapse or other clearly adverse long-term impacts may occur. It is possible, however, to use the measure of impact employed in this paper (i.e., the conditional impingement mortality rate) to distinguish between losses that may be important and losses that are clearly trivial. It is also possible to estimate the socioeconomic importance of impinging a given number of fish and to evaluate the reduction in impact that might result from implementing mitigating measures designed to reduce impingement or to increase the survival of impinged fish. We believe it is more fruitful to focus assessment studies on these achievable objectives than on the appealing, but unattainable, objective of long-term impact assessment.

Acknowledgments

We thank J. E. Breck, G. F. Cada, W. P. Dey, R. J. Klauda, D. B. Odenweller, P. Rago, and D. S. Vaughan for their thorough reviews of this manuscript. Research was supported by the U.S. Environmental Protection Agency under interagency agreement (IAG) DOE 40-740-78 (EPA 79-D-X0533) with the U.S. Department of Energy (DOE), by the U.S. Nuclear Regulatory Commission under IAG DOE 40-550-75 with DOE, and by the Office of Health and Environmental Research, DOE, under contract DE-AC05-84OR21400 with the Martin Marietta Energy Systems, Incorporated. Although the research described in this article has been funded wholly or in part by the U.S. Environmental Protection Agency, it has not been subjected to EPA peer review and, therefore, does not necessarily reflect the views of EPA and no official endorsement should be inferred. This is publication 2760 of the Environmental Sciences Division, Oak Ridge National Laboratory.

References[1]

Barnthouse, L. W. 1982. An analysis of factors that influence impingement estimates at Hudson River power plants. Pages I-1–I-49 in L. W. Barnthouse and seven coauthors. The impact of entrainment and impingement on fish populations in the Hudson River estuary, volume 2. Oak Ridge National Laboratory, ORNL/NUREG/TM-385/V2, Oak Ridge, Tennessee.

Barnthouse, L. W., D. L. DeAngelis, and S. W. Christensen. 1979. An empirical model of impingement impact. Oak Ridge National Laboratory, ORNL/NUREG/TM-290, Oak Ridge, Tennessee.

Barnthouse, L. W., and W. Van Winkle. 1981. The impact of impingement on the Hudson River white perch population. Pages 199–205 in L. D. Jensen, editor. Issues associated with impact assessment: proceedings of the fifth national workshop on entrainment and impingement. EA Publications, Sparks, Maryland.

Barnthouse, L. W., and W. Van Winkle. 1982. Impingement impact estimates for seven Hudson River fish species. Pages III-1–III-91 in L. W. Barnthouse and seven coauthors. The impact of entrainment and impingement on fish populations in the Hudson River estuary, volume 2. Oak Ridge National Laboratory, ORNL/NUREG/TM-385/V2, Oak Ridge, Tennessee.

Boreman, J., and C. P. Goodyear. 1988. Estimates of entrainment mortality for striped bass and other fish species inhabiting the Hudson River estuary. American Fisheries Society Monograph 4:152–160.

Cannon, J. B., G. F. Cada, K. K. Campbell, D. W. Lee, and A. T. Szluha. 1979. Fish protection at steam-electric power plants: alternative screening devices. Oak Ridge National Laboratory, ORNL/TM-6472, Oak Ridge, Tennessee.

Dew, D. B. 1981. Impact perspective based on reproductive value. Pages 251–256 in L. D. Jensen, editor. Issues associated with impact assessment: proceedings of the fifth national workshop on entrainment and impingement. EA Publications, Sparks, Maryland.

Englert, T. L., and J. G. Boreman. 1988. Historical review of entrainment impact estimates and the factors influencing them. American Fisheries Society Monograph 4:143–151.

Hoff, T. B., J. B. McLaren, and J. C. Cooper. 1988. Stock characteristics of Hudson River striped bass. American Fisheries Society Monograph 4:59–68.

Hutchison, J. B., and J. A. Matousek. 1988. Evaluation of a barrier net used to mitigate fish impingement at a Hudson River power plant intake. American Fisheries Society Monograph 4:280–285.

Kjelson, M. A. 1977. Estimating the size of juvenile fish populations in southeastern coastal-plain estuaries. Pages 71–90 in W. Van Winkle, editor. Proceedings of the conference on assessing power-plant-induced mortality of fish populations. Pergamon, New York.

Mattson, M. T., J. B. Waxman, and D. A. Watson. 1988. Reliability of impingement sampling designs: an example from the Indian Point station. American Fisheries Society Monograph 4:161–169.

McFadden, J. T., editor. 1977. Influence of Indian Point unit 2 and other steam-electric generating plants on the Hudson River estuary, with emphasis on striped bass and other fish populations. Report to Consolidated Edison Company of New York.

McFadden, J. T., and J. P. Lawler, editors. 1977. Influence of Indian Point unit 2 and other steam-electric generating plants on the Hudson River estuary, with emphasis on striped bass and other fish populations, supplement 1. Report to Consolidated Edison Company.

McLaren, J. B., J. R. Young, T. B. Hoff, I. R. Savidge, and W. L. Kirk. 1988. Feasibility of supplementary stocking of age-0 striped bass in the Hudson River. American Fisheries Society Monograph 4:286–292.

Muessig, P. H., J. B. Hutchison, Jr., L. R. King, R. B. Ligotino, and M. Daley. 1988. Survival of fishes after impingement on traveling screens at Hudson River power plants. American Fisheries Society Monograph 4:170–181.

Ricker, W. E. 1975. Computation and interpretation of biological statistics of fish populations. Fisheries Research Board of Canada Bulletin 191.

TI (Texas Instruments). 1974. Hudson River ecological survey in the area of Indian Point. Annual Report (1973) to Consolidated Edison Company of New York.

TI (Texas Instruments). 1975. Indian Point impingement study report for the period 1 January 1974 through 31 December 1974. Report to Consolidated Edison Company of New York.

TI (Texas Instruments). 1978. Catch efficiency of 100-ft (30-m) beach seines for estimating density of young-of-the-year striped bass and white perch in the shore zone of the Hudson River estuary. Report to Consolidated Edison Company of New York.

Vaughan, D. S. 1988. Introduction [to entrainment and impingement impacts]. American Fisheries Society Monograph 4:121–124.

Wallace, D. C. 1971. Age, growth, year class strength, and survival rates of the white perch, *Morone americana* (Gmelin), in the Delaware River in the vicinity of Artificial Island. Chesapeake Science 12:205–218.

Young, J. R., R. J. Klauda, and W. P. Dey. 1988. Population estimates of juvenile striped bass and white perch in the Hudson River estuary. American Fisheries Society Monograph 4:89–101.

---

[1]See Table 1 for sources of legal documents and unpublished reports pertaining to the Hudson River.

Appendix

If impingement mortality and natural mortality are independent, the number of striped bass surviving to age 2 can be estimated from

$$N_2 = N_0 S_2 (1 - m); \qquad \text{(A1)}$$

N_2 = number of surviving 2-year-olds;
N_0 = number of young of the year (age 0);
S_2 = natural survival rate from age 0 to age 2;
m = conditional impingement mortality rate.

If there had been no impingement, the number of surviving 2-year-olds would have been

$$N_2' = N_0 S_2 = N_2/(1 - m). \qquad \text{(A2)}$$

The number of age-0 fish that would have survived to age 2 had they not been impinged (i.e., the number of "equivalent 2-year-olds" killed by impingement) is

$$N_{E2} = N_2' - N_2 = mN_2'. \qquad \text{(A3)}$$

Combination of equations (A2) and (A3) gives

$$N_{E2} = mN_2/(1 - m). \qquad \text{(A4)}$$

The ranges of equivalent 2-year-olds presented in the Discussion were obtained by applying equation (A4) to the values of N_2 and m presented in Table 16 of Barnthouse and Van Winkle (1982). These values were then extrapolated to equivalent 5- and 10-year-olds by means of the age-specific survival rates in Table 1 of Dew (1981).

American Fisheries Society Monograph 4:191–203, 1988

Development and Sensitivity Analysis of Impact Assessment Equations Based on Stock–Recruitment Theory

IRVIN R. SAVIDGE,[1] JOHN B. GLADDEN,[2] K. PERRY CAMPBELL,[3] AND J. SCOTT ZIESENIS[4]

Texas Instruments Incorporated, Ecological Services Group
Buchanan, New York 10511, USA

Abstract.—A central and unresolved problem in assessing the impact of power plant operation on Hudson River fish populations was the prediction of long-term population changes resulting from impingement and entrainment mortality. A series of equations was developed from the Ricker and Beverton–Holt stock–recruitment models to address this issue. It was assumed that compensation, a response by the population to partially offset added mortality, occurs during a brief period early in the life cycle rather than throughout the life cycle. Equations were developed for the added power plant-induced mortality occurring before or after compensation in each model. Mortality added after compensation resulted in larger estimates of population reduction than mortality added before compensation. The simple deterministic and more complex age-structured models provided identical predictions of long-term changes in population size in most cases. Similarly, the simple models predicted population changes that were nearly identical to the average age-structured model with variable survival except when levels of variation were quite high and the population modeled was semelparous (reproducing at a single age). Advantages of the equations developed herein are their simplicity, generality, and minimal data requirements. Information concerning population rates of increase, magnitudes of impingement and entrainment mortality, and the form of density-dependent relationships are required for both simple and complex models. The simple models, as formulated, cannot incorporate multiple sources of mortality or multiple sources of density dependence in differing life stages, particularly when the rates or functional relationships of these processes are likely to vary from year to year. The predictions of long-term population changes resulting from impingement and entrainment mortality were not attained. The magnitude and form of the compensatory responses of Hudson River fish populations remained undetermined, so that an agreement on this critical component of prediction models could not be reached.

A primary objective of the Hudson River fisheries studies was to estimate the magnitude of impact that new power plants (Indian Point, Roseton, and Bowline Point) would have on the fish populations in the river. Impact prediction dealt with two components: (1) the increase in mortality of young fish as the result of entrainment and impingement, and (2) the long-term consequences of this added mortality for the populations. This paper will discuss the use of models based on simple stock–recruitment relationships that were developed for the prediction of long-term changes in populations subjected to entrainment and impingement.

Although the magnitude of mortality resulting from entrainment and impingement is not disputed (Englert and Boreman 1988, this volume), there has been no general agreement on the long-term consequences of this mortality source for the populations. The reason for this was disagreement between consultants for the U.S. Environmental Protection Agency (EPA) and those for the utilities on the magnitude and functional form of density-dependent relationships used in the predictions.

Any model represents a compromise between reality and generality. A realistic model portrays a system exactly and can generate precise and testable predictions specific to the system. However, systems being modeled usually are known imprecisely, so a complex model usually requires many assumptions about parameter values and functional relationships. The uncertainty associated with predictions increases exponentially with complexity due to the increasing number of unknown or poorly estimated variables (Jester et al.

---

[1] Present address: Department of Chemistry and Biochemistry, University of Colorado, Boulder, Colorado 80309, USA.

[2] Present address: Savannah River Ecology Laboratory, E. I. Dupont de Nemours, Aiken, South Carolina 29808, USA.

[3] Present address: Beak Consultants, Incorporated, 317 Southwest Alder, Portland, Oregon 97204, USA.

[4] Present address: Personnel Research Department, Southland Corporation, Dallas, Texas 75221, USA.

1977). General models such as those derived from stock–recruitment models require fewer detailed assumptions and can yield useful results. However, these models are less realistic and do not necessarily reflect the mechanisms involved in the system being modeled. Rather, the stock–recruitment models require many more general assumptions concerning population growth rates and types of regulation processes. It must be assumed that the multiple parameters and complex relationships involved in population growth and regulation are adequately "averaged" in the few parameters of the stock–recruitment models.

Data on several aspects of the life history and ecology of striped bass were felt to be inadequate for the development of a complex realistic model, particularly in the early stages of the Hudson River study. A realistic model requires data on (1) the effects of environmental variables on spawning success and ichthyoplankton survival, (2) the probability distributions and serial correlations of environmental variables, (3) mortality rates and the variables affecting these rates at intermediate ages (between the ages when juveniles and yearlings leave the estuary until they return as spawners), and (4) the mechanisms and functional forms of compensatory response operative within the population at various densities. Similar problems existed for the other species that were considered. The mechanisms for compensation are generally unspecified but include a host of potential density-dependent processes (Backiel and Le Cren 1967).

Because data needed for a mechanistic model were unavailable, and because we and other utility consultants wished to incorporate the effects of compensation into impact predictions, we attempted to predict the long-term impacts on the basis of stock–recruitment equations available in the literature. This effort was directed primarily at the impacts on striped bass, but attempts were also made to apply the technique to white perch and Atlantic tomcod (McFadden 1977). Only tests of the striped bass model will be reported here.

Simple single-equation models for the estimation of long-term changes in average population size were developed from the stock–recruitment equations of Ricker (1954, 1975) and Beverton and Holt (1957). These equations are referred to as the equilibrium reduction equations (EREs). The EREs require only estimates of conditional mortality rates from entrainment and impingement and an estimate of the "compensatory reserve" under existing environmental conditions. The conditional mortality rate is the proportion of a

cohort killed, and it would be equal to the proportional reduction of a year class if no compensation were to occur. Compensatory reserve is the remaining ability of the population to offset (e.g., through reduced density-dependent mortality or increased fecundity) additional density-independent mortality. The form of the compensation function in these equations is derived from the underlying stock–recruitment equation.

Both the Ricker and the Beverton–Holt stock–recruitment equations are related to the logistic growth equation. The logistic growth equation, when written as a difference equation for one unit of time, is equivalent to the Beverton–Holt equation, and it can be obtained as the limiting form of the Ricker equation (Eberhardt 1977). The formulation of the Ricker stock–recruitment equation permits the number of recruits to vary more greatly than does the Beverton–Holt model. It also permits the recruits of a generation to exceed the long-term carrying capacity, whereas the Beverton–Holt formulation keeps the number of recruits in any generation at or below the carrying capacity.

Various attempts were made to estimate the compensatory reserve required to apply the EREs. The difficulties inherent in fitting stock–recruitment equations to data are discussed by Christensen and Goodyear (1988, this volume). The compensatory reserve of the population to be evaluated is a key parameter in the application of the EREs. It includes both the potential rate of increase of the species and the population-specific, density-independent influences of the environment, and it may be viewed as a population-specific intrinsic rate of increase expressed on a per-generation basis. This parameter is needed for predictions of population response to additional mortality in both complex and simple models.

This paper describes the derivation and testing of simple deterministic models to predict the long-term population consequences of entrainment and impingement. These models, based on conventional stock–recruitment relationships, provide alternatives to complex simulation models and can be used to illustrate the effects of alternative forms and magnitudes of compensation. However, without a reliable estimate of the population-specific compensatory reserve of the Hudson River fish, long-term effects of the power plant could not be predicted.

Methods

Derivation of the equations.—Four models (EREs) were developed to permit a choice of the functional form of compensation (Ricker and Beverton–Holt) and of the timing of compensation (before or after impingement and entrainment). When compensation during a generation occurs before entrainment or impingement, the compensation is acting on the conditional mortality imposed on the previous generation. Only the first ERE described, ERE-1, was developed in time for extensive use in the adjudicatory proceedings held by EPA.

The Ricker stock–recruitment equation represents the theoretical relationship between parental stock size and recruitment stock size at the same age. Discrete generations and a constant environment are assumed. Recruitment (R) is the product of a linear function (α) of the parental stock size (P) and a density-dependent feedback function ($e^{-\beta P}$):

$$R = \alpha P e^{-\beta P}. \tag{1}$$

In this equation, α is the compensatory reserve of the stock, in that it determines the rate of potential population growth per generation in the average existing environment and in the absence of density-dependent feedback, and β is a measure of density-dependent feedback, and so determines the population equilibrium level that will result from the compensatory reserve of the stock.

Ricker (1975) derived a series of equations from equation (1) for the management of fish populations. The equilibrium population size, at which parents (P_r) are just replaced by recruits (R_r), is

$$P_r = R_r = \frac{\log_e \alpha}{\beta}. \tag{2}$$

Model ERE-1 is derived analytically from the Ricker stock–recruitment equation; it is assumed that the added conditional mortality of entrainment and impingement occurs after the period of compensation during each generation. Because the added conditional mortality rate (m = percentage of cohort killed) due to entrainment and impingement is assumed to be density-independent, the effect of this mortality is to proportionately reduce the compensatory reserve of the population. The new equilibrium stock size (P_E) and the new equilibrium recruit stock size (R_E) in the presence of this new mortality is

$$P_E = R_E = \frac{\log_e [\alpha (1-m)]}{\beta}. \tag{3}$$

In this case, it is assumed that the added mortality occurs after the period of compensation and that the density-dependent feedback function is not affected.

The percentage change (PC) in size of the equilibrium population undergoing additional entrainment and impingement mortality is

$$PC = \frac{P_E - P_r}{P_r} \times 100. \tag{4}$$

Model ERE-1 is derived by substituting equations (2) and (3) into equation (4) and simplifying:

$$PC = \left(\frac{\log_e [\alpha (1-m)]}{\log_e \alpha} - 1 \right) \times 100$$
$$= \frac{\log_e (1-m)}{\log_e \alpha} \times 100. \tag{5}$$

Another equation (ERE-2) was derived for the case when the added mortality occurs prior to the period of compensation. In this case, the magnitude of density-dependent feedback within each generation is also proportionally reduced. The density-dependent feedback term in the stock–recruitment equation thus becomes $e^{-\beta(1-m)P}$, and the new equilibrium stock size is

$$P_E = R_E = \frac{\log_e [\alpha (1-m)]}{(1-m)\beta}. \tag{6}$$

The predictive equation for this case (ERE-2) is derived by substituting equations (6) and (2) into equation (4) and simplifying:

$$PC = \left(\frac{\log_e [\alpha (1-m)]}{(1-m)\log_e \alpha} - 1 \right) \times 100. \tag{7}$$

Similar predictive equations were derived from the Beverton–Holt stock–recruitment equation,

$$R = \frac{1}{a + B/P}; \tag{8}$$

R and P are the recruits and parents as before; a is an index of the density-dependent feedback, and B is an index of the compensatory reserve such that $B = 1/\alpha$ of the Ricker equation. In the Beverton–Holt equation, the replacement stock size at which parents (P_r) equal recruits (R_r) is

$$P_r = R_r = \frac{1-B}{a}. \tag{9}$$

For the case of additional entrainment and impingement mortality (m) imposed after the period of compensation, the new equilibrium stock size is

$$P_E = R_E = \frac{1-m-B}{a}. \tag{10}$$

The percent change from the unimpacted population size (ERE-3) is estimated by substituting equations (9) and (10) into equation (4) and simplifying:

$$PC = \frac{-m}{1-B} \times 100. \tag{11}$$

If the impact of entrainment and impingement occurs prior to the period of compensation in the life cycle of the species, the new equilibrium stock size is

$$P_E = R_E = \frac{1-m-B}{a(1-m)}. \tag{12}$$

By substitution of equations (12) and (9) into equation (4), ERE-4 is obtained:

$$PC = \frac{-Bm}{(1-m)(1-B)} \times 100. \tag{13}$$

Assumptions required.—Several assumptions are required in the development of any mathematical model because of inadequate data for the basic parameters and functional relationships. In the development of models describing biological systems, the driving factors and feedback functions are of primary importance. A key assumption in the application of these EREs is the relevancy of the Ricker or Beverton–Holt formulation of the feedback mechanisms. Other assumptions include (1) the applicability of a deterministic model to populations inhabiting a highly variable environment and exhibiting large fluctuations in year-class strengths, (2) the applicability of the stock–recruitment relationships developed for semelparous species (one-time spawner such as Pacific salmon) to iteroparous species (repeat spawner such as striped bass), (3) the timing of entrainment and impingement relative to dominant density-dependent processes (compensation) in the life cycle, and (4) the constancy of other human impacts and background environmental conditions. Of these assumptions, those relating to a variable environment and reproductive strategy were amenable to testing by simulation.

Tests of assumptions.—In order to test the robustness of the ERE models with respect to the assumptions relating to environmental constancy and reproductive strategy, modified Leslie matrix models were constructed that contained several modifications not found in Leslie's (1945) original formulation. The first year of life was incorporated as a submodel; it was divided into three

distinct stages and provision was made for survival and year-to-year variation in density-independent survival in each of the three stages. In these simulations, all density-dependent effects occurred in the second stage following random density-independent variation. Additional density-independent mortality was applied to either the first or third life stage. The density-dependent survival function in the second life stage was derived from the Ricker or Beverton–Holt stock–recruitment equation appropriate to the ERE being evaluated (TI 1980b). Explicit delineation of a submodel for the first year of life by use of Leslie matrix procedures to simulate adult population dynamics has been used by Van Winkle et al. (1974), Hess et al. (1975), Eraslan et al. (1976), Christensen et al. (1977), and others.

The simulation model used to evaluate the EREs was developed and run on the Statistical Analysis System (SAS: Barr et al. 1976). The matrix model was developed with 16 age-classes and advanced in yearly time increments by standard matrix multiplication procedures. Average first-year survivorship for the steady-state population was calculated by the technique of Vaughan and Saila (1976). In the simulations, the total of density-dependent and stochastic density-independent effects on first-year survival was calculated in the first-year submodel, and the single first-year survival value was entered into the population projection (A) matrix for each model iteration.

Testing of the EREs involved comparison of the ERE predictions with predictions from simulations with the modified Leslie matrix model, in which parameters relating to the assumptions of age structure and deterministic survival were manipulated. Four levels of compensatory reserve (α or $1/B = 2, 4, 8,$ or 16) and three levels of conditional mortality from entrainment and impingement ($m = 0.0, 0.1,$ or 0.2) were used in the simulations. The added mortality was imposed either before or after compensation to correspond to the ERE being evaluated. The effect of iteroparity versus semelparity on the predictions was evaluated by running simulations with the model calibrated to age structures and fecundity schedules approximating those observed for Hudson River striped bass, which spawn at multiple ages, and comparing the predictions with the predictions of simulations run with the model calibrated to the same survivorship through age 8 (the approximate mean generation time of the multiple-age model) but with all spawning (equal to the

TABLE 67.—Initial population parameters for single-age (SA) and multiple-age (MA) spawning matrix models. Parameters were derived for equilibrium deterministic population models with no density-dependent regulations.

Age-class	Number of individuals in age-class[a]		Probability of survival to next age-class		Number of "female eggs" produced by females in age-class[b]	
	SA	MA	SA	MA	SA	MA
0	3,075,868,158	3,075,868,158	1.53566×10^{-5}	1.53566×10^{-5}	0	0
1	47,216	47,216	0.4	0.4	0	0
2	18,886	18,886	0.6	0.6	0	0
3	11,332	11,332	0.6	0.6	0	0
4	6,799	6,799	0.6	0.6	0	23,030
5	4,079	4,079	0.6	0.6	0	60,690
6	2,448	2,448	0.6	0.6	0	167,790
7	1,469	1,469	0.6	0.6	0	403,680
8	881	881	0.0	0.6	3,491,939	587,700
9	0	529	0.0	0.6	0	782,500
10	0	317	0.0	0.6	0	881,500
11	0	190	0.0	0.6	0	988,000
12	0	114	0.0	0.6	0	1,044,000
13	0	69	0.0	0.6	0	1,063,000
14	0	41	0.0	0.6	0	1,095,000
15	0	25			0	1,295,500

[a]Individuals in age-class 0 represent the total number of female eggs spawned.
[b]Age-specific fecundities presented for striped bass in TI (1980a) were adjusted to reflect only female eggs and the percentage of females that was mature in each age-class.

multiple-age model) occurring at that age (single-age model). Initial parameter values for these two models are presented in Table 67.

To evaluate robustness of the EREs in the face of environmentally induced variations in density-independent mortality, we imposed the background density-independent mortality as a random variable so that the mean density-independent survival corresponded to the level required by the compensatory reserve being simulated. The standard deviation of the normal probability distribution from which the pseudorandom survivals were drawn was set to provide coefficients of variation (CV = $100 \cdot$ SD/mean) of the mean survival rate of 0, 20, and 50%. In the case of simulations with the 50% CV, the probability distributions of the survivals were truncated at zero. It can be shown that the CV for abundance of juvenile striped bass probably exceeds 50% (TI 1980b).

Six replicate simulations of 100 years were run for each combination of variables in simulations involving environmental variability, and the means of the last 50 years were used for comparison with the ERE predictions. No replicates were necessary for deterministic simulations involving constant survival rates. The total numbers of individuals in yearling and older age-classes were monitored to determine changes in population size.

Results

ERE Predictions

The predicted percentage change (PC) of a population exposed to an additional conditional mortality rate (m) depended on the ERE chosen and the compensatory reserve (α or $1/B$: Figures 69, 70). The magnitude of m above which the population is predicted to become extinct is the same for all four EREs. The critical level of additional density-independent mortality (m_c), above which extinction is predicted by all four equations, is

$$m_c = 1 - (1/\alpha) = 1 - B. \quad (14)$$

This has also been demonstrated by Christensen et al. (1977), who used a more complex modeling approach.

At levels of m less than m_c, the predictions of PC depend on the ERE chosen as well as the compensatory reserve. The predictions of PC, in order of decreasing population size, are ERE-2 (equation 7), ERE-4 (equation 13), ERE-1 (equation 5), and ERE-3 (equation 11). Only in ERE-2 (equation 7), in which impact occurs prior to compensation in the Ricker-based equation, did any predictions of an increased population occur. These predictions of an increased population occurred when

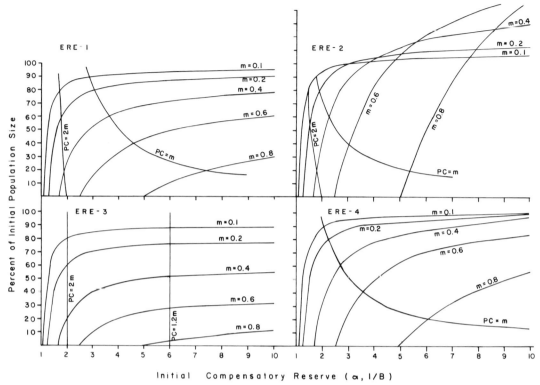

Initial   Compensatory   Reserve $(\alpha, 1/B)$

FIGURE 69.—Relationships between percent population change and the initial compensatory reserve (α or $1/B$ for the Ricker or Beverton–Holt formulations, respectively) at various levels of conditional mortality (m) for four equilibrium reduction equations (ERE). Lines labeled PC = m indicate the levels of compensatory reserve at which the conditional (m) rate expressed as a percentage produces an equal percentage change in population size (PC).

$$\frac{\log_e(1-m)}{-m} > \log_e\alpha, \qquad \text{for } m > 0, \quad (15)$$

so that

$$m < 1 - \frac{\log_e[\alpha(1-m)]}{\log_e\alpha}, \qquad \text{for } m > 0. \quad (16)$$

This is also the value of m below which PC is less than m in ERE-1 (equation 5). However, the possibility that such increases could realistically occur has been strongly challenged by DeAngelis et al. (1977). The decrease in the population predicted by ERE-3 when m is between 0 and m_c is a straight-line relationship with the compensatory reserve (Figures 69, 70).

Robustness to Assumptions

Changes in average population size predicted by the modified Leslie matrix model with a multiple-age spawning structure (MA model) and a single-age spawning structure (SA model) were similar to those predicted by the corresponding EREs when density-independent survival was held constant (Table 68). In addition, the SA and MA models give nearly identical predictions for a given compensation function, α, and magnitude and time of impact. Notable disagreement between the ERE and the deterministic matrix model predictions occurred only for the largest α (16) in the Ricker formulations (ERE-1, 2) and was a result of the oscillatory behavior of Ricker-type models at high alphas (May 1977). It is clear from Table 68 that additional mortality applied after the period of compensation resulted in larger population reductions than mortality applied before, and that populations regulated by a Beverton–Holt type of stock–recruitment function are more severely affected by additional mortality than populations regulated by a Ricker-type function.

The predictions of the MA model with a 20 or 50% coefficient of variation (CV) in density-independent mortality generally agreed well with the predictions made under the condition of no variability (Figures 71, 72). The predictions of the SA model with a 50% CV in density-independent

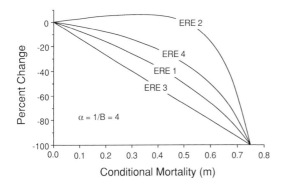

FIGURE 70.—Predictions of percent change in equilibrium population size at various levels of additional mortality applied to the population based on four equilibrium reduction equations (ERE). All predictions are based on an assumed population compensatory reserve of 4 (α for the Ricker equation, $1/B$ for the Beverton–Holt model).

mortality, however, were consistently lower than those of the corresponding EREs because the SA model lost year classes and had no provision for their reinstatement in later generations. This may be an artifact of the truncated normal probability distribution used to generate random density-independent survival rates, or it may indicate a higher sensitivity to environmental variation by less strongly iteroparous species (O'Neill et al. 1981). In simulations with the MA model, lost year classes were reconstituted in subsequent generations by reproduction from existing year classes. With a 20% CV, the SA model generally produced predictions in good agreement with the corresponding ERE predictions.

The only MA-model simulations not in good agreement with ERE predictions were those with $m = 0.10$, α ($=1/B$) $= 4$ and 8, and CV $= 20\%$ in density-independent survival (Figure 71). The ERE-4 simulations produced larger population sizes than expected, whereas ERE-2 simulations, under these conditions, produced smaller population sizes than expected. The reason for these differences is unknown. The MA-model predictions agreed with the corresponding ERE predictions when α was 2 or 16 and CVs were similar, as they did when α was 4 or 8 and the CV was 50% for density-independent survival.

Under the conditions of variable density-independent mortality, the ERE predictions are those of long-term average population size, but they

TABLE 68.—Comparison of equilibrium reduction equations (ERE) and matrix model predictions of population change with constant survival in matrix model simulations. Conditional mortality was applied before and after the population's compensatory reserve was expressed,[a] and matrix models were run for single-age (SA) and multiple-age (MA) populations.[b]

Compensa-tory reserve (alpha)	Conditional mortality		Percent population change					
			Ricker model (ERE-1,2)			Beverton–Holt model (ERE-3,4)		
				Matrix model			Matrix model	
	Before	After	ERE	SA	MA	ERE	SA	MA
2	10	0	−5.78	−5.78	−5.77	−11.11	−11.02	−11.01
	20	0	−15.24	−15.18	−15.15	−25.00	−24.58	−24.53
	0	10	−15.20	−15.20	−15.18	−20.00	−19.85	−19.83
	0	20	−32.19	−32.08	−32.05	−40.00	−39.44	−39.38
4	10	0	2.67	2.66	2.67	−3.70	−3.71	−3.70
	20	0	4.88	4.87	4.88	−8.33	−8.34	−8.33
	0	10	−7.60	−7.60	−7.60	−13.33	−13.33	−13.33
	0	20	−16.10	−16.10	−16.10	−26.67	−26.67	−26.66
8	10	0	5.48	5.61	5.45	−1.59	−1.59	−1.59
	20	0	11.59	11.55	11.55	−3.57	−3.57	−3.57
	0	10	−5.07	−5.18	−5.04	−11.43	−11.43	−11.43
	0	20	−10.73	−10.70	−10.70	−22.86	−22.86	−22.86
16	10	0	6.89	12.32	10.43	−0.74	−0.74	−0.74
	20	0	14.94	18.18	18.14	−1.67	−1.67	−1.67
	0	10	−3.80	−3.63	−4.58	−10.67	−10.67	−10.67
	0	20	−8.05	−7.72	−8.88	−21.33	−21.33	−21.33

[a]The compensatory reserve of the population defined as the rate of population increase per generation under the existing environmental condition prior to the addition of the conditional mortality, if no density dependent feedback were operating.

[b]Matrix models using single-age (SA) and multiple-age (MA) formulations of reproduction corresponding to semelparous and iteroparous life history strategies.

may not correspond well to the population size in any specific year (Figure 72).

Discussion

ERE Predictions

Four variables affect the ERE predictions of long-term population consequences of entrainment and impingement: (1) the compensatory reserve of the population (α or $1/B$), (2) the shape of the applicable stock–recruitment function (Ricker or Beverton–Holt), (3) the timing of the additional mortality resulting from entrainment and impingement relative to the timing of compensation, and (4) the magnitude of the conditional mortality rate (m) associated with entrainment and impingement.

Estimation of the compensatory reserve is the most difficult aspect of the application of the EREs, but a similar problem exists in the application of any population model that incorporates the potential (intrinsic) rate of increase and density-dependent feedback. The compensatory reserve can be thought of as the rate at which the population would increase per generation under existing environmental conditions, but without the source of mortality being evaluated, in the absence of density-dependent feedback. In the application of the EREs, it is assumed that all sources of density-independent mortality remain constant during the time period of interest except for the source of mortality being evaluated. Additional sources, or removal of existing sources, will proportionately change the compensatory reserve of the population. For example, a population with a compensatory reserve (α) of 10 will have a remaining compensatory reserve of 8 after the imposition of an additional density-independent conditional mortality rate of 0.2 (20%). The addition of a second impact with a conditional mortality rate of 0.1 would further reduce the compensatory reserve to 7.2, because the combined added mortality rate is 0.28.

Selection of the stock–recruitment model appropriate to a species is equally difficult. Although the level of compensatory reserve alone determines the magnitude of additional density-independent mortality that will cause extinction, the underlying stock–recruitment relationship, together with the compensatory reserve, determines the predicted change in average population size for less severe (i.e., $m < m_c$) increases in mortality rate (Christensen et al. 1977; DeAngelis et al. 1977). Knowledge of a species' biology guides the

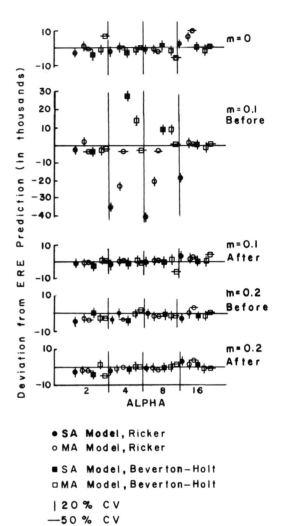

● SA Model, Ricker
○ MA Model, Ricker

■ SA Model, Beverton–Holt
□ MA Model, Beverton–Holt

| 20% CV
— 50% CV

FIGURE 71.—Average deviations of matrix model simulations with variable young-of-the-year survival from equilibrium reduction equation (ERE) predictions (individual data points are means of six simulations). The coefficient of variation (CV) is expressed as a percentage ($100 \cdot$ SD/mean) and m is the added conditional mortality rate applied before or after the period of population compensation. Average numbers of yearlings and older females during 50–100 years of simulation were used to calculate means; initial population size was 93,110 for the single-age (SA) spawner model and 94,395 for the multiple-age (MA) spawner model. The results of simulation with the single-age spawner model and 50% CV are not included.

selection of the most appropriate stock–recruitment relationship. The derivation of the Ricker curve assumes feedback from an initial density with a generation $dN/dt = -bN_0$, for which N_0 is the initial density, whereas the Beverton–Holt

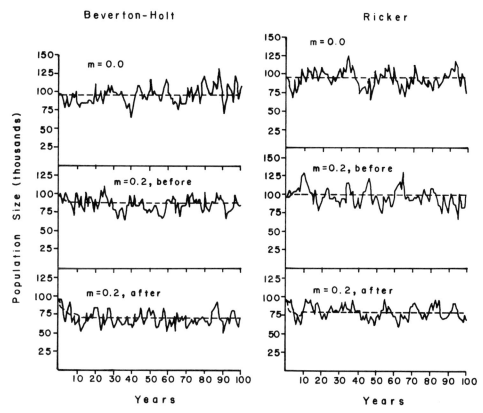

FIGURE 72.—Representative simulation results showing the behavior of models with density-independent first-year survival that was constant (broken lines) or varying (continuous lines of 20% coefficient of variation: 100 · SD/mean). These multiple-age spawner models incorporated compensatory reserve based either Ricker or Beverton–Holt functions ($\alpha = 1/B = 4$) acting during the first year of survival, with an additional conditional mortality (m) of 20% applied either before or after the period of compensation. The results presented are the total population sizes of yearling and older female fish.

derivation assumes feedback from the density at the time of the feedback: $dN/dt = -bN$. For example, a Ricker-type curve can result when cannibalism of one age-class on another is important or when there is a substantial time lag in the response of a predator or parasite to changes in abundance of the species being considered. Populations that are likely to be regulated by available food or habitat and those for which predators respond quickly to changes in abundance are more likely to be characterized by a Beverton–Holt-type curve. Chapman (1973) and Cushing and Harris (1973) provided further descriptions of the potential biological mechanisms underlying the two stock–recruitment relationships.

The prediction of population size at a life stage within the life cycle of a species will reflect the time of the added mortality relative to the time of density-dependent feedback (Christensen et al.

1977). The population size predictions of ERE-1 and ERE-3 are made for a life stage after the added conditional mortality but before occurrence of compensation for that mortality within the life cycle. The population size predictions of ERE-2 and ERE-4 are made for a life stage after compensation has followed the added conditional mortality in the life cycle. The differences in population size projections between ERE-1 and ERE-2 and between ERE-3 and ERE-4, therefore, reflect only differences in the timing of the population size projections relative to the timing of added mortality and compensation. These differences could be important to a fishery if the projections were for harvestable life stages.

In the application of the EREs, it is assumed that the density-dependent population size adjustment occurs either wholly before or wholly after the period of added mortality. If, for example, the life stages subject to entrainment and impinge-ment occur both before and after the period of compensation, equations (5), (7), (11), and (13) would not provide an accurate projection even if the assumed stock–recruitment relationship and the compensatory reserve parameter are correct. This would be the case for combined entrainment and impingement mortality of striped bass if com-pensation occurs in the post–yolk-sac or early juvenile life stages. In that case, the true popula-tion effects would be bracketed by the predictions made by the EREs for mortality before and after compensation.

For species in which compensation occurs over a relatively short interval either temporally or developmentally, accurate projections can still be made if the other assumptions are valid. This can be done by subdividing the imposed conditional mortality rate into mortality before compensation (m_1) and mortality after compensation (m_2),

$$m = 1-(1-m_1)(1-m_2), \qquad (17)$$

and combining the appropriate EREs. The Ricker-based EREs (equations 5 and 7) can be combined to give

$$PC = \left(\frac{\log_e[\alpha(1-m_1)]}{(1-m_2)\log_e\alpha} - 1\right) \times 100, \qquad (18)$$

and the Beverton–Holt-based EREs (equations 11 and 13) can be combined to give

$$PC = \frac{-m_2+m_1m_2-Bm_1}{(1-m_1)(1-B)} \times 100. \qquad (19)$$

If the density-dependent population adjustment extends over a relatively long period and overlaps the added mortality, the EREs can still be used for either bracketing or rough approximation. An extended or multiple-period, density-dependent feedback system is more complex, but may occur naturally and provide fine tuning of the population size and a Beverton–Holt type of stock–recruit-ment relationship. This condition can better be handled by models such as the modified Leslie matrix model used to evaluate the robustness of the EREs; however, the EREs can still provide upper and lower bounds on, or rough approxima-tions of, impact projections.

Population Fluctuations

Fluctuations of natural populations from gener-ation to generation may result from two causes: environmental variability and overcompensation (a density-dependent response greater than needed to offset a changed population density). Fluctuations caused by overcompensation can be readily illustrated by simulating a population with the Ricker model under the condition of a large compensatory reserve (α). Oscillations in the pre-dictions of a Ricker model occur with an α above 7.34 (Ricker 1954). Although an unexploited striped bass population may have a compensatory reserve exceeding 10 (McFadden and Lawler 1977), as indicated by the rapid increase of the commercial catch from the California population during the period following the species' introduc-tion there (Sommani 1972), it is unlikely that the heavily exploited Atlantic coast populations have sufficiently high compensatory reserves that over-compensation can account for the fluctuations in year-class strength. A working estimate of $\alpha = 4$ was used for striped bass in most of the ERE calculations in the Hudson River studies.

The most likely cause of fluctuating year-class strengths in Atlantic coast populations of striped bass is environmental variability. Estuarine envi-ronments are characterized by pronounced envi-ronmental fluctuations and high levels of uncer-tainty (Copeland et al. 1974; Gladden et al. 1988, this volume). These environmental fluctuations are extremely important for striped bass spawning success and ichthyoplankton survival (Polgar 1982). The Hudson River is the least stable (in terms of fluctuations in flow and salinity) of all the estuaries along the Atlantic coast having striped bass populations (Simpson et al. 1973). The de-gree of iteroparity for American shad populations has been associated with environmental fluctua-tions along a latitudinal gradient (Leggett and Carscadden 1978), and this may be relevant to the development of iteroparity in striped bass. An iteroparous life history pattern enables a popula-tion to sustain a high level of egg production in a variable environment and to take advantage of favorable environmental conditions whenever they occur. This persistence resulting from itero-parity was illustrated in the differing responses of the SA and MA models to high levels of variation in survival; year classes of subsequent genera-tions remained missing in SA models after unsuc-cessful reproduction.

Limitations of the EREs

Certain aspects of the EREs may render them inappropriate in some applications. The models are inherently deterministic and ignore certain key attributes of fish populations such as age and fecundity distributions, which may be important in assessment of responses to age-specific mortalities (Murphy 1968; Schaffer 1974). Two shortcomings of ERE models for impact prediction are unavoidable: their inability to handle populations with multiple, nonconcurrent, density-dependent feedback mechanisms; and their inability to handle added mortality sources that are density dependent.

The MA model used to test the predictions of the EREs under conditions of a variable environment and iteroparity appears to be a favorable compromise between complexity and simplicity for evaluation of impacts on populations not amenable to the EREs. This model can be run with the same data and assumptions required for the EREs. Other variables can be determined through calibration, and additional information and functional relationships can be incorporated into the model as they become available. The magnitude and functional form of the commercial and sport fisheries are examples of the type of information that may be incorporated into the MA model but is not amenable to ERE treatment. In the EREs, the fishery is ignored and, thus, is assumed to be a density-independent source of mortality in the existing environment. If the intensity of the fishery changes as a result of changes in population size, the ERE predictions would not reflect this change. Van Winkle et al. (1978) presented a density-dependent function for fishing mortality for an age-structured model. At present, the relationship between stock size and fishing pressure for Hudson River striped bass is unknown.

The EREs derived from Ricker or Beverton–Holt stock–recruitment relationships provide a simple deterministic means of approximating the impacts of imposed density-independent mortality sources such as power plant impingement and entrainment. The EREs have been subject to criticism because they ignore such complexities as iteroparity and environmental variability. However, simulations incorporating these complexities, along with appropriate stock–recruitment-based density-dependent survival functions, showed very good agreement with the results predicted by the EREs. Although more complex population models may be desirable, the EREs are robust models that can offer an alternative to data-intensive mechanistic models for estimating the long-term impacts of entrainment and impingement mortality on fish population.

Acknowledgments

This work was supported by the Consolidated Edison Company of New York, Incorporated; the Power Authority of the State of New York; Central Hudson Gas and Electric Company; and Orange and Rockland Utilities, Incorporated, through contracts with Texas Instruments, Incorporated. Final preparation of this manuscript was supported, in part, by the U.S. Department of Energy through a contract with the University of Georgia's Institute of Ecology and Savannah River Ecology Laboratory under contract DE-AC09-765R00-819 and, in part, by the Department of Chemistry and Biochemistry, University of Colorado, Boulder, Colorado.

References[5]

Backiel, T., and E. D. Le Cren. 1967. Some density relationships for fish population parameters. Pages 261–293 in S. E. Gerking, editor. The biological basis of freshwater fish production. Blackwell Scientific Publications, Oxford, England.

Barr, A. J., J. H. Goodnight, J. P. Sall, and J. T. Helwig. 1976. A user's guide to SAS 76. SAS Institute, Cary, North Carolina.

Beverton, R. J. H., and S. J. Holt. 1957. On the dynamics of exploited fish populations. Fishery Investigations, Series II, Marine Fisheries, Great Britain Ministry of Agriculture Fisheries and Food 19.

Chapman, D. G. 1973. Spawner–recruitment models and estimation of the level of maximum sustainable catch. Rapports et Procès-Verbaux des Réunions, Conseil International pour l'Exploration de la Mer 164:325–332.

Christensen, S. W., D. L. DeAngelis, and A. G. Clark. 1977. Development of a stock–progeny model for assessing power plant effects on fish populations. Pages 196–226 in W. Van Winkle, editor. Assessing the effects of power-plant-induced mortality on fish populations. Pergamon, New York.

Christensen, S. W., and C. P. Goodyear. 1988. Testing the validity of stock–recruitment curve fits. American Fisheries Society Monograph 4:219–231.

Copeland, B. J., H. T. Odum, and F. N. Moseley. 1974. Migrating subsystems. Pages 422–453 in H. T. Odum, B. J. Copeland, and E. A. McMahan, editors. Coastal ecosystems of the United States, volume 3. Conservation Foundation, Washington, D.C.

Cushing, D. H., and J. G. K. Harris. 1973. Stock and recruitment and the problem of density dependence. Rapports et Procès-Verbaux des Réunions, Conseil International pour l'Exploration de la Mer 164:142–155.

DeAngelis, D. L., S. W. Christensen, and A. G. Clark. 1977. Responses of a fish population to young-of-the-year mortality. Journal of the Fisheries Research Board of Canada 34:2124–2132.

Eberhardt, L. L. 1977. Relationship between two stock–recruitment curves. Journal of the Fisheries Research Board of Canada 34:425–428.

---

[5]See Table 1 for sources of legal documents and unpublished reports pertaining to the Hudson River.

Englert, T. L., and J. Boreman. 1988. Historical review of entrainment impact estimates and the factors influencing them. American Fisheries Society Monograph 4:143–151.

Eraslan, A. H., and six coauthors. 1976. A computer simulation model for the striped bass young-of-the-year population in the Hudson River. Oak Ridge National Laboratory, ORNL/NUREG-8, EDS-766, Oak Ridge, Tennessee.

Gladden, J. B., F. R. Cantelmo, J. M. Croom, and R. Shapot. 1988. Evaluation of the Hudson River ecosystem in relation to the dynamics of fish populations. American Fisheries Society Monograph 4: 37–52.

Hess, K. W., M. P. Sissenwine, and S. B. Saila. 1975. Simulating the impact of the entrainment of winter flounder larvae. Pages 1–29 in S. B. Saila, editor. Fisheries and energy production: a symposium. Lexington Books, Lexington, Massachusetts.

Jester, D. B., Jr., D. L. Garling, Jr., A. R. Tipton, and R. T. Lackey. 1977. A general population dynamics theory for largemouth bass. Virginia Polytechnical Institute and State University, Department of Fisheries and Wildlife Science, Blacksburg.

Leggett, W. C., and J. E. Carscadden. 1978. Latitudinal variation in reproductive characteristics of American shad (Alosa sapidissima): evidence for population specific life history strategies in fish. Journal of the Fisheries Research Board of Canada 35:1469–1478.

Leslie, P. H. 1945. On the use of matrices in certain population mathematics. Biometrika 33:183–212.

May, R. M. 1977. Thresholds and breakpoints in ecosystems with a multiplicity of stable states. Nature (London) 269:471–477.

McFadden, J. T., editor. 1977. Influence of Indian Point unit 2 and other steam electric generating plants on the Hudson River estuary, with emphasis on striped bass and other fish populations. Report to Consolidated Edison Company of New York.

McFadden, J. T., and J. P. Lawler, editors. 1977. Supplement to influence of Indian Point unit 2 and other steam electric generating plants on the Hudson River estuary, with emphasis on striped bass and other fish populations, supplement 1. Report to Consolidated Edison Company of New York.

Murphy, G. I. 1968. Patterns in life history and the environment. American Naturalist 102:391–403.

O'Neill, R. V., R. H. Gardner, S. W. Christensen, W. Van Winkle, J. H. Carney, and J. B. Mankin. 1981. Some effects on parameter uncertainty in density-independent and density-dependent Leslie models for fish populations. Canadian Journal of Fisheries and Aquatic Sciences 38:91–100.

Polgar, T. T. 1982. Factors affecting recruitment of Potomac River striped bass and resulting implications for management. Pages 427–442 in V. S. Kennedy, editor. Estuarine comparisons. Academic Press, New York.

Ricker, W. E. 1954. Stock and recruitment. Journal of the Fisheries Research Board of Canada 11:559–623.

Ricker, W. E. 1975. Computation and interpretation of biological statistics of fish populations. Fisheries Research Board of Canada Bulletin 191.

Schaffer, W. M. 1974. Optimal reproductive effort in fluctuating environments. American Naturalist 108: 783–790.

Simpson, J. J., R. Bopp, and D. Thurber. 1973. Salt movement patterns in the lower Hudson. Paper 9 in Hudson River ecology, 3rd symposium. Hudson River Environmental Society, Bronx, New York.

Sommani, P. 1972. A study on the population dynamics of striped bass in the San Francisco estuary. Doctoral dissertation. University of Washington, Seattle.

TI (Texas Instruments). 1980a. 1977 year class report for the multiplant impact study of the Hudson River estuary. Report to Consolidated Edison Company of New York.

TI (Texas Instruments). 1980b. 1978 year class report for the multiplant impact study of the Hudson River estuary. Report to Consolidated Edison Company of New York.

Van Winkle, W., D. L. DeAngelis, and S. R. Blumm. 1978. A density-dependent function for fishing mortality rate and a method for determining elements of a Leslie matrix with density-dependent parameters. Transactions of the American Fisheries Society 107:395–401.

Van Winkle, W., B. W. Rust, C. P. Goodyear, S. R. Blum, and P. Thall. 1974. A striped bass population model and computer programs. Oak Ridge National Laboratory, ORNL/TM-4578, Oak Ridge, Tennessee.

Vaughan, D. S., and S. B. Saila. 1976. A method for determining mortality rates using the Leslie matrix. Transactions of the American Fisheries Society 105:380–383.

Comments[6]

Savidge et al. (1988, this volume) acknowledge the absence of a reliable estimate of the compensatory capacity of Hudson River fishes. My reasons for agreeing with this view are given in Christensen and Goodyear (1988, this volume) and in my comments on Lawler (1988, this volume). One trusts that the "working estimate of α = 4" (page 200) is to be taken merely as an example, and not as a meaningful estimate.

The utility of the equilibrium reduction equation (ERE) approach is limited by two factors. First, the impossibility of estimating the compensatory capacity of Hudson River striped bass with currently available data has been discussed at length; it remains to be demonstrated whether or not it will be any more possible for other species in other locations. Second, knowing what kind of parent–progeny relationship is appropriate to describe a particular population usually will not be easy to determine. This is a prerequisite for applying an ERE.

SIGURD W. CHRISTENSEN
Environmental Sciences Division
Oak Ridge National Laboratory
Post Office Box 2008
Oak Ridge, Tennessee 37831-6036, USA

---

[6]The Hudson River studies produced no agreement on the long-term impact of power plants on the river's fish fauna, in large part because the issue of biological compensation by the fish stocks for plant-induced mortality could not be resolved. The four contiguous papers in Section 3 by Savidge et al., Lawler, Christensen and Goodyear, and Fletcher and Deriso address these problems. Because this topic remains controversial, each of these authors was invited to comment on papers by the others. Not all chose to respond; the remarks of those who did are reproduced here.

References

Christensen, S. W., and C. P. Goodyear. 1988. Testing the validity of stock–recruitment curve fits. American Fisheries Monograph 4:219–231.

Lawler, J. P. 1988. Some considerations in applying stock–recruitment models to multiple-age spawning populations. American Fisheries Society Monograph 4:204–218.

Savidge, I. R., J. B. Gladden, K. P. Campbell, and J. S. Ziesenis. 1988. Development and sensitivity analysis of impact assessment equations based on stock–recruitment theory. American Fisheries Society Monograph 4:191–203.

American Fisheries Society Monograph 4:204–218, 1988

Some Considerations in Applying Stock–Recruitment Models to Multiple-Age Spawning Populations

John P. Lawler

Lawler, Matusky & Skelly Engineers, One Blue Hill Plaza
Pearl River, New York 10965, USA

Abstract.—Several approaches used during the Hudson River power plant hearings (1977–1980) to permit quantitative estimation of the impact of power plant operation on the river's striped bass population are presented. The Ricker stock–recruitment model was adopted as a starting point to provide a reasonable working conceptualization of the stock–recruitment processes operating in this river. This model was then modified in a variety of ways to reflect multiple-age spawning by striped bass, and the results were fit to data on the river's commercial striped bass catch and effort, organized to reflect certain average age distribution parameters. Environmental variation in the system, represented in an overall fashion by the spring and summer variation in the river's freshwater flow, and the effect of certain assumed modes of cannibalism were also included in these analyses. Quantitative estimates of the Ricker parameters α and β, representing "compensatory reserve" and density-dependent mortality, respectively, in the system were extracted from these fits. Finally, an approach to modeling the influence of certain density-dependent and density-independent growth factors on the various modes of mortality is presented. All of the above procedures are directed toward estimating the change a given impact may induce on the stock–recruitment model parameters α and β. These changes in α and β are used in various equilibrium reduction models to estimate the percentage change in the equilibrium spawning population that, all other things remaining equal, can be expected in the presence of the impact under study.

This paper presents approaches that were developed during the Hudson River power plant hearing for the purpose of estimating the impact of human removal of a given fraction of each year's fish production. These are illustrated with available data on striped bass; the methodology, however, should be useful in evaluating impacts of a similar nature on other fish populations.

The existence of some form of density-dependent population control of striped bass is assumed in this presentation and the data base is accepted as the best available. The methodology involves a mathematically simple model with sufficient biological features to provide a reasonably accurate depiction of the expected impact. Where the data fall short of what might be desirable, care is taken to bound the interpretation; for example, in the absence of data on year-to-year variation in the age-distributed parameters (percentage composition, fecundity, etc.), the available data are assumed to reflect the average behavior of each parameter, and the implications of so doing are discussed. The methodology employed at the time of the Hudson River hearings has not been updated via recourse to the published literature since the close of those hearings. Note is made of several more recent papers on the subjects of this presentation that have come out of the Fisheries

Laboratory, Lowestoft, England (Shepherd and Cushing 1980; Horwood and Shepherd 1981; Shepherd 1982).

Basic Concepts

For this work, Ricker's stock density-dependent stock–recruitment model was adopted because it is a simple overall description of a population process subject to density-independent and density-dependent mortality and incorporates intrinsic oscillation into the population. It can be readily modified to reflect variation in either or both the density-independent or density-dependent factors at any life stage in the recruits from birth to spawning and, though originally applied to single-year spawners (Pacific salmon), can be modified, as shown in this paper, to reflect the behavior of a multiple-age spawning population.

The basic derivation of the Ricker equation is developed below to set the stage for the mathematical steps that were necessary in making the modifications employed in the Hudson River hearing. The elements of these modifications are then described. The next section presents the actual multiple-year-spawner modification, including application of the available data base to extract estimates of the density-independent and density-dependent mortality parameters. These results are followed by a summary of some of the methodologies developed and used to estimate the power plant impact.

Begin by recognizing the differential equation of fish survival behavior that underlies Ricker's analysis for a single-year spawner. Ricker (1973) gives

$$\frac{d\,p(t)}{dt} = -(k_1 + k_0 P_0)\,p(t); \qquad (1)$$

$p(t)$ = time- (stage-) variable density or number per water body of the fish species being analyzed, from egg stage up to recruitment into the spawning population;

P_0 = population of spawners responsible for p;

k_1 = sum of all unit mortality rates for density-independent mortality in p;

k_0 = unit mortality rate associated with stock density-dependent mortality in p.

Integration of equation (1) between the limits of $p = P_0\overline{E}$ at $t = 0$, the moment of spawning, and $p = R$ at $t = T$, the time of recruitment to the spawning population, yields

$$R = P_0\overline{E} \exp(-k_1 T) \exp(-k_0 P_0 T); \qquad (2)$$

R = population of fish recruited to the spawning population;

$\overline{E}$ = mean egg production per spawning adult.

For any year, the spawning population is given by the general notation P, and the subscript can be dropped. The quantity $\overline{E} \exp(-k_1 T)$ represents the reproductive reserve, or the net effect of birth and density-independent mortality only, and is designated alpha (α). The quantity $\exp(-k_0 P\,T)$ represents the effect of stock density on survival, an effect that probably operates only during the first year of life; this term is written simply $\exp(-\beta P)$.[1] Equation (2) is rewritten

$$R = \alpha P \exp(-\beta P), \qquad (3)$$

in which form it is the Ricker stock–recruitment model. Beta (β) represents the existence of compensation within a population (i.e., the tendency of the population to adjust to changes in its own size, and so to oscillate about a long-term equilibrium). The overall second-order (i.e., density-

dependent) mortality rate ($k_0 P$) will increase when P, the parent population, is high and will decrease when P is low, causing the population to oscillate about an equilibrium population P_E. Thus, β or $k_0 t$ also represents the unit level of stock density-dependent mortality in the system. For any given α, when β is large, the equilibrium population and the population size in general tend to be small; when β is small, the reverse is true.

Two differences between the α and β parameters are worth noting, particularly in view of a common tendency to use the term "compensatory" in describing each of these two parameters. The first difference is that α is a measure of reproductive reserve; that is, it is a positive factor: the larger it is, the larger will be both the equilibrium level and the population in general (Goodyear 1977). Beta, on the other hand, is a negative factor because it measures only mortality; population sizes increase in inverse proportion to it. The second difference is that α, the reproductive reserve, considers density-independent mortality only: it measures the ability of the population to reproduce in the face of forces that operate without regard to the size of that population. Beta, conversely, represents density-dependent mortality only: it measures the factors of mortality (e.g., cannibalism, lack of food) that are directly related to the size of the population.

Alpha measures the number of fish that would survive to adulthood per adult spawner if density-independent mortality alone were operating. Standing alone, α is quite hypothetical; so, far from being by itself a direct measure of compensation, it ignores compensation by pretending that all mortality is independent of density. But for this very reason, it becomes a measure of compensatory reserve, because what it represents, by implication, is the amount of compensatory or density-dependent mortality (β) required to bring the actual survival ratio of recruits to spawners to unity on a long-term basis. Thus, if it is assumed that the population remains at equilibrium over the long term, and such an equilibrium cannot be explained by a long-term fortuitous balance in density-independent factors only (see, for example, Royama 1977), the system is perceived to require the concepts embodied in both α and β. As illustrated above for the Ricker model, these concepts of compensatory reserve and density-dependent mortality are interrelated. For a given model, this interrelationship will influence the unit population parameters.

Beverton and Holt (1957) showed what they

---

[1]The choice of T in this quantity is immaterial. Had k_0 been operative only over some $t_c < T$, the integration would yield a multiplier $\exp(-k_0 P t_c)$ in equation (2). The quantity $k_0 t$ is simply β, which is determined empirically.

perceived as the necessity for the existence of density-independent and density-dependent factors, operating conceptually as described above, by noting that a noncompensatory stock-recruitment model in which α, or $\overline{E} \exp(-k_1 T)$, is exactly unity is precisely in equilibrium at all values of P; it would be colinear with the 45° equilibrium line in a stock-versus-recruitment graph. Obviously, such a model is highly unstable, any perturbation pushing it into either unlimited population growth or total decline. This argument, admittedly, is oversimplified. It is possible to develop density-independent stochastic models in which populations persist for very long periods without exploding or collapsing. Royama (1977), however, treated this subject in depth and showed that density-independent "regulation" is fragile, at best, and unable to maintain persistence over the long run. He stated, referring to persistence in the wide sense, "A density-independent regulation will not practically occur in nature."

Modification of equation (3) to permit its application to a multiple-age spawning system, as well as to one in which year-to-year environmental variation is perceived to influence unit mortality rates, proceeds by rewriting equation (2) for the usual situation wherein k_1 and k_0 vary with the life history stage of the recruits; that is, as the age of the recruit population increases from eggs to larvae to juveniles and so on, both unit rates can be expected to change, usually in a decreasing manner. The k_0, for example, is generally viewed as zero, or nonoperative, after the recruits pass the yearling stage. Integration of equation (1) over a continuous series of piecewise smooth intervals, within each one of which k_1 and k_0 both remain constant, yields

$$R(t) = P_0 \overline{E} \exp\left(\sum_{i=1}^{n} -k_{1i} t_i\right)$$
$$\cdot \exp\left(\sum_{i=1}^{n} -k_{0i} P_0 t_i\right). \quad (4)$$

In equation (4), the index i represents each specific life history stage within which both the density-independent and density-dependent mortality rate processes are assumed to be reasonably represented by a constant unit mortality rate. Equation (4) still represents the case of single-age spawning given by equations (2) and (3); the mortality terms have simply been written on a life stage basis. The variable t_i is the duration of time for each stage.

For the multiple-age spawning, equation (4) must be modified to recognize that the progeny contribution by recruits (now parents) of the original parent population P_0 must reflect that some of these recruits spawn more than once. Hence, it is insufficient to simply account for survival up to the first time these recruits spawn, because some but not all of them will survive another year, and so on, to spawn again and again.

The mathematical details of this modification were developed in careful detail by McFadden and Lawler (1977: Part 2-IV-C-4,5). It was shown therein that α can be expressed as a multifaceted function of stage-dependent, density-independent unit mortality rates and associated stage durations and age-dependent reproduction parameters. It was further shown how, given estimates of these parameters for all relevant life history stages, this elaborate function can be evaluated to estimate α. A consideration of these details is given in the next section of this paper.

In a similar vein, equation (4) can be modified to include the influence of environmental variation on the density-independent unit mortality rate parameter, k_1, and the influence of density-dependent factors, such as cannibalism, on the density-dependent unit mortality rate parameter, k_0. For example, suppose, in the absence of specific data on year-to-year variation in the density-independent mortality rate, a long-term average value, $\overline{k}_1$, is postulated, along with the assumption that the actual rate, k_1, varies linearly with some measure of the river's freshwater flow, Q (mean, $\overline{Q}$), during each spawning season; that is,

$$k_1 = k_{1_0} + \gamma' Q,$$

or

$$k_1 = \overline{k}_1 + k_1'$$
$$= \overline{k}_1 + \gamma' [Q - \overline{Q}].$$

Then equation (2) would be rewritten as

$$R = P_0 \overline{E} \exp(-\overline{k}_1 T) \exp[-\gamma' (Q - \overline{Q}) T]$$
$$\cdot \exp(-k_0 P_0 T),$$

or

$$R = \alpha P \, e^{-\gamma(Q - \overline{Q})} e^{-\beta P_0},$$

in which $\alpha = \overline{E} \exp(-\overline{k}_1 T)$ and $\gamma = \gamma' T$.

Examination of equations (1) and (4) also suggests that, just as k_1 and k_0 can vary from stage to stage, P_0 can also be redefined from stage to stage without interference with the basic piecewise integration of equation (1). Thus, if a primary density-dependent influence on the recruits, when they reach the yearling stage, is viewed as cannibalism by that year's adults, that stage of the integration of equation (1) might include P_{0+1} rather than P_0 as the stock-density parameter. Similarly, P_{0-1} might be employed to represent the influence of cannibalism by yearlings on young-of-year recruits (age-0 juveniles).

Application of these conceptual and mathematical modifications of the basic Ricker model were made during the Hudson River power plant hearings. These applications are illustrated in the following sections.

Multiple-Age Spawning Approach

Extension of the Ricker model to multiple-age spawning is achieved by estimating the actual recruit population expected to be associated with a given year's parent population. This requires information on the age-distributed parameters of percentage composition in each age-class (all fish), sex ratio, percentage of mature fish, and fecundity. Let P_k, the parent population in year k, be

$$P_k = \sum_{i=y_1}^{y_n} F_{gi,k} \, (Y/f)_k; \tag{5}$$

$(Y/f)_k$ = catch/effort index of population in calendar year k;

$F_{gi,k}$ = age-distributed specific percentage composition parameter (defined below for various alternative approaches) for age-group i;

y_1 = earliest age of adult spawning;

y_n = latest age of adult spawning.

The corresponding recruit population is

$$R_k = \sum_{i=y_1}^{y_n} F_{gi,k+i} \, (Y/f)_{k+i}. \tag{6}$$

Equation (6) indicates that the recruits or progeny of parent year k become parents themselves over a span of years $k+i$; for striped bass i may span 3–18 years.

Because $(Y/f)_k$ in equation (5) is independent of i, it can be removed from the summation. This permits normalization of equations (5) and (6) so that the normalized parent population, $\bar{P}_k$, is always given by $(Y/f)_k$, the actual catch per effort in year k.[2] The corresponding normalized recruit population, $\bar{R}_k$, is

$$\bar{R}_k = \frac{\sum\limits_{i=y_1}^{y_n} F_{gi,k+i} \, (Y/f)_{k+i}}{\sum\limits_{i=y_1}^{y_n} F_{gi,k}}. \tag{7}$$

The age-distributed specific percentage composition parameter, F_{gi}, can represent several measures of spawning stock; these include, for a given age, all fish, all mature fish, all female fish, all mature female fish, and all eggs. For this analysis, we have used all female fish and all eggs as two measures of parents and associated progeny.

For all female fish,

$F_{gi} = F_{fi}$, the age distribution of all females in the catch;

for all eggs,

$$F_{gi} = F_{fi} \times f_{fmi} \times E_i;$$

F_{fi} = age distribution of females;

f_{fmi} = fraction of mature females, each age-group;

E_i = average fecundity of mature females, each age-group.

For the case of all female fish, which has been designated "multiple-age spawning" analysis (MAS), the progeny contribution is

$$R_k = \frac{\sum\limits_{i=y_1}^{y_n} F_{fi,k+i} \, (Y/f)_{k+i}}{\sum\limits_{i=y_1}^{y_n} F_{fi,k}}. \tag{8}$$

Though the overbar in equation (7) has been

---

[2]Note that $(Y/f)_k$ will normally be reported as biomass, rather than as numbers, whereas the derivations of the basic Ricker model and all of the modifications proposed herein are presented on a numbers basis. It can be shown that the basic model and the modifications can be derived on, or converted to, a weight basis, with no change in the functional form of the results. Alternatively, catch-per-effort biomass data can be converted to numbers, provided that the F_{gi} distribution includes unit weight data on each age-class, as well as to the age-distributed percentage composition by either weight or numbers in the catch. In this study, due to insufficient years of data for some parameters, a stable age distribution is assumed, or better, an average age distribution is assumed to be reasonably representative of the stock–recruitment behavior. The age-distributed parameters are all on a numbers basis. For a stable age distribution, the relationship between total weight and total numbers of each catch is constant so, for this case, catch-per-effort by weight data can be used with the age-distributed parameters obtained from fish numbers data.

dropped in equation (8), R_k is understood to be the normalized recruit population.

For the stock–recruitment model in which progeny egg production is regressed on parent egg production, the normalized recruit egg production (REP) is

$$\text{REP}_k = \frac{\sum_{i=y_1}^{y_n} F_{fi,k+i} \cdot (f_{fmi,k+i} \cdot E_{i,k+i} \, (Y/f)_{k+i}}{\sum_{i=y_1}^{y_n} F_{fi,k} \cdot f_{fmi,k} \cdot E_{i,k}}. \quad (9)$$

Because data are not available on the calendar-year variation of each of the age-distributed parameters in equations (8) and (9), the $k+i$ and k indices of those parameters are dropped and estimates of R_k and REP_k are obtained under the assumption that the existing information on the age-distributed parameters represent average values of these parameters. This will result in additional error in the error term of the Ricker stock–recruitment regression model. This model (without error terms) is

$$R = \alpha P \exp(-\beta P). \quad (10)$$

Equation (10) is symbolically identical to equation (2), the single-age-spawner Ricker model. If the stock density-dependent term $\exp(-\beta P)$ operates only on immature fish (before the first year of recruit spawning), this term is conceptually identical to its counterpart in equation (2). However, as noted earlier, α is now a multifaceted combination of stage-dependent, density-independent mortality rates and associated stage-duration and age-dependent reproduction parameters. The precise functional dependence of α on these parameters was given by McFadden and Lawler (1977: Section IV-C-4,5).

The parameters α and β in equation (10) are evaluated by nonlinear least-squares fitting of time-series data on R and P to equation (10). Any year's P is given by Y/f for that year; R is obtained from equation (8) for regressions involving fish and from equation (9) for regressions involving eggs.

Equation (10) can be expanded to evaluate the influence of environmental variables on the stock–recruitment relationship. In recognition that many facets of the early life history behavior are directly or indirectly influenced by river freshwater flow (e.g., spawning above the salt front; temperature-dependent growth and survival), flow was incorporated as a linear influence on the density-independent unit mortality rate (k_i) in equation

(11) in the early life history stages. As illustrated in greater mathematical detail under "Basic Concepts," this yields

$$R_k = \alpha \, \{\exp[\gamma(Q_{jk} - \overline{Q}_j)]\} \, P_k \exp(-\beta P_k); \quad (11)$$

Q_{jk} = average river freshwater flow in month j of year k;

$\overline{Q}_j$ = long-term average freshwater flow in month j;

γ = $\gamma' t$;

γ' = proportionality constant between incremental flow $(Q_{jk} - \overline{Q}_j)$ and unit mortality rate increment over average unit mortality rate;

t = the duration of the early life history state over which flow influences the unit mortality rate.

Similar modification of equation (10) can be made to reflect cannibalism by older fish on younger ones of the same species. Data gathered in the Hudson River during the spring of 1974 showed yearling striped bass in the stomachs of adults. Data gathered in late summer and early fall of 1973 through 1975 showed young of the year in the stomachs of yearlings (McFadden 1978). Data gathered in Haverstraw Bay in December 1980 show similar evidence of cannibalism of young of the year by yearling and perhaps 2-year-old striped bass.

These observations suggest the inclusion of intraspecific predation (cannibalism) as a factor influencing the stock density-dependent mortality rate term $\exp(-\beta P)$ in equation (10). This modification is

$$R_k = \alpha P_k \exp[\gamma(Q_{jk} - \overline{Q}_j)]\exp$$
$$-[\beta_0 P_k + \beta_1 P_{k+1} + \beta_{-1} P_{k-1}]; \quad (12)$$

$\beta_1 P_{k+1}$ = density-dependent mortality due to predation by adults on yearlings, occurring in year $k+1$ for which the adult population is P_{k+1};

$\beta_{-1} P_{k-1}$ = density-dependent mortality due to predation by yearlings on young of year, occurring in year k; the yearling population in that year is assumed to be roughly proportional to its parent population, which is P_{k-1}.

The foregoing models (equations 10–12) were fit to the Hudson River striped bass catch-per-effort data of 1950 through 1975, with the definitions of progeny population given in equations (8) and (9), and with some adjustments in the

TABLE 69.—Commercial landings of striped bass in New York water of the Hudson River, 1950–1975, indices of fishing effort (stake and anchor gill nets), stock abundance indices (catch per effort), and numbers of recruits used in spawner–recruit models.

Year	Landings, Y (kg) Total[a]	Landings, Y (kg) Gill nets[b]	Gill-net area, A (m²)	Fishing hours allowed weekly, H	Index of fishing effort, f ($10^{-6} AH$)	Stock abundance index, P (gill-net Y/f)	Number of recruits, R[c] Egg model	Number of recruits, R[c] Multiple-age spawning model
1950	4,327	10,082[d,e]	55,813[f]	132	7.37	1,368	3,131	2,811
1951	7,865	16,125[d,e]	40,409[f]	96	3.88	4,156	3,439	3,114
1952	13,539	23,974[d,e]	41,161[f]	108	4.45	5,387	3,521	3,051
1953	8,778	13,090[d,e]	41,417[f]	108	4.47	2,928	3,697	3,474
1954	25,402[g]	30,777[d,e]	68,898[f]	108	7.44	4,137	3,545	4,322
1955	33,294	31,030[d]	113,716[f]	108	12.28	2,527	2,691	3,694
1956	42,105	39,242[d]	114,906[f]	108	12.41	3,162	2,204	2,296
1957	38,329	35,723[d]	113,792[f]	108	12.29	2,907	2,297	2,043
1958	34,973	32,594[d]	112,758[f]	108	12.18	2,676	2,686	2,328
1959	60,374	56,269[d]	101,696[f]	120	12.20	4,612	3,051	2,330
1960	60,283	56,184[d]	93,269[f]	120	11.19	5,021	3,830	3,060
1961	32,070	29,889[d]	92,947[f]	120	11.15	2,681	4,012	3,735
1962	21,818	20,335[d]	96,649[f]	120	11.60	1,753	4,036	4,246
1963	21,183	19,743[d]	66,655[f]	120	8.00	2,468	3,597	4,110
1964	13,381	12,471[d]	57,614[f]	120	6.91	1,805	3,495	4,097
1965	16,647	15,876	52,178	120	6.26	2,536	2,705	2,673
1966	20,114	19,414	50,720	120	6.09	3,188	2,801	2,620
1967	24,786	24,397	44,766	120	5.37	4,543		
1968	27,579	27,352	58,706	120	7.04	3,885		
1969	34,998	34,998	53,808	120	6.46	5,418		
1970	20,820	19,382	59,744	120	7.17	2,703		
1971	11,225	11,204	34,247	120	4.11	2,726		
1972	8,140	8,140	36,783	120	4.41	1,846		
1973	30,407	22,596	32,360	120	3.88	5,824		
1974	13,758	11,740	92,456	120	11.09	1,059		
1975	20,947	18,688	106,385	120	12.77	1,463		

[a]From "Fishery Statistics of the United States," published by the U.S. Department of Commerce. Data have been converted from pounds.

[b]Data for 1965–1975 were provided by Fred Blossum, National Marine Fisheries Service, and converted from pounds.

[c]Values were calculated by the procedure in Table 70.

[d]Total landings multiplied by 0.932, the average fraction of total landings that was gill-net landings during 1965–1975, when there were direct estimates of gill-net catches.

[e]Gill-net landings were adjusted (upward) for the fishery's conversion from multifilament to monofilament nylon nets. A conversion rate of 20% per year was assumed (0% in 1950, 100% in 1955). Monofilament nets were assumed to be 2.5 times more efficient than multifilament nets, a value midway in the range of 2 to 3 reported by McCombie and Fry (1960), Muncy (1960), and Hamley (1975).

[f]Values were increased 1.2 times from recorded estimates to account for nets that caught primarily striped bass but were not covered during surveys (National Marine Fisheries Service formula).

[g]Approximate estimate.

Y/f values of 1950 through 1954 (Tables 69, 70). The adjustments in the 1950–1954 Y/f values were made to permit inclusion of five additional data points by comparison to the data of 1955 to 1975 used earlier (McFadden 1977; McFadden et al. 1978) in the proxy and other analyses. This was done because of the intention to include several independent variables in the multiple-regression analysis, making an expanded data base highly desirable. This in no way suggests this data set is better or poorer than the earlier one; it simply added precision to the results to be obtained through multiple regression though, perhaps, at the expense of biases caused by the adjustments.

Data on freshwater flows (Table 71) originally covered 7 months, February–August, the period covering all stages in the early life history of striped bass up to the point year-class strength begins to be determined. At this point, the developing recruit population is expected to be less subject to environmental fluctuation than it is in its earliest stages. February was chosen as the starting period because upriver runoff in this month can influence downriver flow and salinity distributions in March, unless spawners first move upriver.

Selection of a 7-month period for the flow parameter has the advantage of covering any flow-related (or flow-correlated) environmental factors that influence survival throughout the early life history. It has the disadvantage of smoothing point effects, so that if the real control were, for example, salinities during April, rather

TABLE 70.—Computational procedures employed to obtain the population *(P)* and recruitment *(R)* values of Table 69.

P, all models

Values of *P,* the index of parent population are identical to the adjusted Y/f (catch per effort) values, for all models evaluated.

R, all models

For both the multiple-age and eggs-on-eggs models, recruit values are composed of weighted fractions of percent values lagged 5 through 9 years, reducing the available parent data to 1950 through 1966.

R, multiple-age spawning model

Recruits for the multiple-age model, by equation (8) in which k denotes year, are

$$R_k = \sum_{i=5}^{9} F_{fi} P_{k+i} ;$$

F_{fi} = average fraction of all females of age-group i in the population.

Values used for F_{fi} are the arithmetic averages of the spawning stock and commercial catch fractions of females given in Table 2-VIII-7 of McFadden and Lawler (1977), normalized[a] to yield

$$\sum_{i=5}^{9} f_{fi} = 1.0.$$

These are

Age:	5	6	7	8	9
F_{fi}:	0.50	0.32	0.10	0.04	0.03.

R, egg production model

Recruit egg production (REP, normalized) for the eggs-on-eggs model is

$$REP_k = \frac{1}{\sum_{i=5}^{9} f_{fi} f_{fmi} E_i} \sum_{i=5}^{9} F_{fi} f_{fmi} E_i P_{k+i};$$

F_{fi} = defined and valued above;
f_{fmi} = fraction of mature females in the female population according to values given in Table 2-VIII-5 of McFadden and Lawler (1977);
E_i = fecundity per female according to values given in Table 2-VIII-5 of McFadden and Lawler (1977).

$$\sum_{i=5}^{9} F_{fi} f_{fmi} E_i$$

represents the actual recruit egg production: the total number of eggs produced by progeny of parent population P_k throughout their productive spawning lives. This was used as the period age 5 through age 9. A comparison analysis for ages 5–12 showed no significant difference in the values of REP, and required three additional years of P_k values, reducing the available data set to 14 years instead of 17 years.

Actual parent egg production (PEP) is equal to P_k times

$$\sum_{i=5}^{9} F_{fi} f_{fmi} E_i.$$

This was normalized by dividing by the constant

$$\sum_{i=5}^{9} F_{fi} f_{fmi} E_i$$

to make PEP_k equivalent to P_k (i.e., to $(Y/f)k$). The quantity REP_k was divided by the same constant to obtain the above statement for REP_k. Use of these secondary normalized values, described above, is still valid because numerator and denominator are multiplied by the same single normalizing constant.

The value of

$$\sum_{i=5}^{9} F_{fi} f_{fmi} E_i$$

is 3.259×10^5.

[a]This secondary normalization (i.e., the normalization process that leads to equations 7–9 is not what is taking place here) is appropriate when an average age structure is used.

TABLE 71.—Average 3-month (Q_3) and 7-month (Q_7) freshwater flows in the Hudson River, 1950–1975. Data are based on U.S. Geological Survey records of Green Island near Albany, New York.

Year	Q_3 Feb–Apr (m³/s)	Q_7 Feb–Aug (m³/s)
1950	623	399
1951	827	520
1952	758	523
1953	729	508
1954	679	491
1955	728	430
1956	666	479
1957	403	280
1958	631	417
1959	636	379
1960	782	486
1961	526	405
1962	561	352
1963	532	347
1964	567	322
1965	354	224
1966	475	344
1967	471	340
1968	497	409
1969	670	459
1970	658	407
1971	657	515
1972	715	695
1973	763	556
1974	656	483
1975	648	478
1950–1966	406[a,b]	614[a,b]
1951–1966	407[a,c]	616[a,c]

[a]Mean value for time period.
[b]Values of $\bar{Q}_7$ and $\bar{Q}_3$ used in equations (11) and (12).
[c]Used in equation (12) when P_{k-1} is included as a parameter.

than a multiplicity of factors throughout the 7 months, it would not show up as strongly in the multiple regression. Accordingly, some regressions of equation (12) were made with 3-month flow averages, February–April. This period was chosen because earlier analyses suggested that several early life stage responses correlated with April flows.

When environmental variation (flow) and an additional stock density-dependent mortality factor (cannibalism) were included in the model, more of the variance in recruitment could be explained (Table 72). Alpha levels rose with inclusion of each additional parameter, especially once cannibalism was included, and values of this parameter became significantly different from zero at low risk of type-I error.

These results were not unexpected. When cannibalism was ignored, in the face of known substantial rates of such predation (McFadden et al. 1978), density-dependent behavior was limited to only one population variable: stock density in the year of recruit birth. When the unmodified Ricker model was fit to the data, what was actually density-dependent behavior was then lumped into the density-independent factor (α), resulting in a biased-low value of α. When the model given by equation (12) was fit to the multiple-age spawning data on R and P by non-linear multiple-regression analysis, positive values of β_{-1} and β_{+1} were obtained. This resulted in more of the variance being explained by density-dependent factors, which in turn resulted in higher levels of α.

Alpha also increased with the inclusion of the flow variable only (cannibalism ignored). The β_0 values (not shown) also increased over those in the unmodified model fits, suggesting that inclusion of the flow variable in the Hudson River data base resulted in a greater fraction of the mortality being explained by the stock density-dependent factor. This reduced the level of density-independent mortality and increased α.

Lower values of α were obtained for the egg production analysis than for the female fish analysis. This can be explained in terms of the considerations set forth under the proxy analysis (McFadden et al. 1978) and by recognizing that, on a relative basis, the age-distributed weighting factors (the F_{gi} of equations 5 and 6 and their counterparts in later equations) shift the contribution to the older age-groups as one moves from the multiple-age spawning (female fish) model (equation 8) to the egg production model (equation 9). The latter occurs because fecundity and percentage maturation, the two additional factors introduced in the calculation of the egg production model F_{gi}, both increase with age.

The proxy analysis showed that the best indicator of the progeny of a given parent year is the catch 5 years later, because the largest fraction of the catch consists of 5-year-old fish. The catch 6 years later will contain some 6-year-olds and so on but, with each additional year, the relationship of catch to the parent year in question is progressively more blurred by the relatively large numbers of fish whose birth occurred in other years.

Similar reasoning applies to the comparison between the results of fitting equations (8) and (9) to the basic yield-per-effort data. The recruit egg production values are more heavily weighted toward the older age-groups than are the recruit female fish values, and thus can be expected to

TABLE 72.—Summaries of nonlinear least-square fits of Ricker models for Hudson River striped bass when multiple-age spawning, freshwater river flows, and cannibalism by adults on yearlings and by yearlings on young of the year are incorporated in the models.

Flow variable[a]	Cannibalism variable[b]	Alpha[c]			(R^{*2})[d]	Significance[e]			
		Mean	95% confidence interval			$\gamma \neq 0$	$\beta_{-1} \neq 0$	$\beta_0 \neq 0$	$\beta_1 \neq 0$
			One-tailed	Two-tailed					
Multiple-age spawning model (R_k defined by equation 8)[f]									
None	None	3.5	>2.2	2.0–5.0	0.70			<0.01	
Q_7	None	4.0	>2.4	2.1–5.8	0.72	NS		<0.01	
Q_7	P_{k+1}	5.8	>3.3	2.8–8.8	0.81	NS		<0.01	<0.05
Q_7	P_{k-1}, P_{k+1}	6.5	>2.8	1.9–11.0	0.80	NS	NS	<0.01	<0.05
Q_3	None	4.0	>2.6	2.3–5.8	0.74	NS		<0.01	
Q_3	P_{k+1}	6.5	>4.0	3.5–9.6	0.85	<0.05		<0.01	<0.01
Q_3	P_{k-1}, P_{k+1}	7.3	>3.8	3.0–11.7	0.84	<0.05	NS	<0.01	<0.01
Egg production model (R_k defined by equation 9)[f]									
None	None	2.7	>1.9	1.7–3.7	0.67			<0.01	
Q_7	None	3.0	>2.0	1.8–4.3	0.69	NS		<0.01	
Q_7	P_{k+1}	3.7	>2.2	1.8–5.5	0.72	NS		<0.01	NS
Q_7	P_{k-1}, P_{k+1}	3.1	>1.4	1.0–5.3	0.71	NS	NS	<0.01	NS
Q_3	None	3.1	>2.1	1.9–4.3	0.70	NS		<0.01	
Q_3	P_{k+1}	3.9	>2.4	2.0–5.9	0.75	NS		<0 01	NS
Q_3	P_{k-1}, P_{k+1}	3.5	>1.7	1.3–5.7	0.73	NS	NS	<0.01	NS

[a] Q_7 = February–August average flow;
Q_3 = February–April average flow.
[b] P_{k+1} = the adult population in year $k+1$ responsible for cannibalism on the recruits of year k when they are yearlings;
P_{k-1} = a measure of the yearling population in year k responsible for cannibalism on the recruits of year k when they are young of the year. This yearly population is assumed to be roughly proportional to its parent population, which is P_{k-1}.
[c] Density-dependent mortality rate.
[d] "R^{*2}" is the percentage reduction in the error sum of squares obtained by using the given model as opposed to $R = \alpha P$. It is precisely equal to $R_R^2 - R_L^2/1 - R_L^2$, wherein R_R^2 and R_L^2 are, respectively, the multiple correlation coefficients for the two no-intercept nonlinear models, $R = \alpha P$ and $R = \alpha P \exp(-\beta P) f(Q, P_k, P_{k-1})$. It is a relative measure of the improvement in the explanation of variance measured around the abscissa, relative in the sense that it measures the reduction in unexplained variance ($SSE_L - SSE_R$) as a fraction of the variance that remained unexplained after the model $R = \alpha P$ is applied.
[e] $\gamma = \gamma' t$; γ' = proportionality constant between incremental flow ($Q_{jk} - \overline{Q}_j$) and unit mortality rate increment over average unit mortality rate;
t = the duration of the early life history state over which flow influences the unit mortality rate.
β_{-1} = unit density-dependent mortality rate due to predation of yearlings on young of the year.
β_0 = unit density-dependent mortality rate due to parent stock density.
β_{+1} = unit density-dependent mortality rate due to predation of adults in yearlings.
NS = not statistically significant at the 0.05 level of risk of type-I error. Empty cells correspond to absence of either a flow or a cannibalism variable, or both.
[f] Recruit population in year k.

give different results.[3] These results are considered to be poorer, for the reasons given above.

Regression analysis of the Hudson River striped bass data by the proxy and multiple-age spawning approaches consistently showed poorer correlation and lower values of α as the mean time spread between parents and progeny, however defined, increased (Table 72). In every case, the statistical parameters chosen as indicators of sig-

nificance showed, for the female fish regressions, significance levels equal to or better than those obtained for the comparable egg production regression. No significant improvement was found when additional variables were included in this latter model.

For all of these reasons, the α levels obtained through the egg production model are probably biased low and do not represent the actual α level for this population. The α levels obtained by the multiple-age spawning model are considered to be reasonable estimates for Hudson River striped bass.

Effect of Changes in Beta

During the Hudson River power plant proceedings, Ricker's equation was used to make quanti-

---

[3] Further, serial correlation probably exists in the catch/effort data and, therefore, in both the P_k and R_k variables in the regression analysis. This, of course, is a natural result of the multiple-year contribution of each year class to both the catch and the spawning stock. The effect of this on the assumed independence of the independent variable P_k in the regression analysis undoubtedly has some influence on the significance of the results.

tative estimates of the impact of plant operations on the Hudson River striped bass population. In the early stages of these proceedings, it was assumed that β, the coefficient of the density-dependent term in Ricker's equation, is an unchanging constant and that additional density-independent stresses, such as power plant entrainment and impingement, change only α. Under this assumption, β does not appear in the equilibrium reduction equation used to compute reduction in the equilibrium spawning stock (see, for example, McFadden and Lawler 1977 or Savidge et al. 1988, this volume). As shown above under "Basic Concepts," β is composed of the second-order unit mortality rate, k_0, and the time period, t_c, over which the stock density-dependent mortality actually occurs. As detailed by McFadden and Lawler (1977) and summarized here, t_c may change with the density of young-of-the-year striped bass and with ambient temperature in the Hudson River. Increases in density-independent mortality and river temperature, each occasioned by power plant operation, may reduce t_c, which, in turn, will reduce β. Should this occur, further compensatory response will result. Predictions of long-range impact at a given conditional mortality rate will be lower than they will if β is assumed to remain constant.

Attempts were made during the power plant proceedings to document density-dependent growth by Hudson River striped bass. Among these attempts were efforts to correlate length-frequency data with estimates of standing crop size at various life history stages, with and without the simultaneous influence on growth of such environmental variables as temperature and freshwater flow. So far these attempts have been inconclusive, in large measure because there are insufficient years of data to permit adequate statistical treatments. Whether or not density-dependent growth can be demonstrated for Hudson River striped bass is immaterial to this presentation, however. My purpose is to show how density-dependent growth, if it exists or is assumed to exist in the early life history stages of fish, will influence the modeling of reduction in the equilibrium spawning stock via the use of traditional stock–recruitment models. Certainly, the concept of density-dependent growth in the early life history stages of fish is sufficiently grounded to merit investigation of how such growth will influence quantitative procedures for estimating human impacts on fish populations.

Faster growth tends to reduce the period of time (t_c in the Ricker model) over which the young of the year (age-0 fish) are susceptible to density-dependent causes of mortality such as predation and cannibalism because the larger larvae are generally stronger and more mobile.

Density of age-0 fish is assumed to influence growth and mortality as follows: at higher densities, competition increases, growth is slower, t_c is longer, and mortality during the early life history stage is greater.[4] Because Ricker's β is directly proportional to t_c, β will also tend to increase at higher densities. Analyses by McFadden and Lawler (1977) showed that when β is density dependent, predicted reductions in the equilibrium spawning stock are lower at a given level of m (human-caused conditional mortality) than those obtained with an equilibrium reduction equation (ERE), in which β is assumed constant. For example, in the case of moderate[5] density-dependent growth, which results in equation (2-IV-33) of McFadden and Lawler (1977), plant-induced conditional mortality rates, ranging between 5 and 25%, produced equilibrium spawning stock reductions between 2 and 15%. By comparison, the comparable classical Ricker result ranges between 4 and 21%.

In their treatise on the development of stock–recruitment models, Beverton and Holt (1957) stated that "factors which are not themselves influenced by the density of the population can nevertheless have an indirectly density-dependent effect. For example, since growth is considerably influenced by temperature, the latter will be partly responsible for determining the length of time during which the larvae are exposed to predation." In other words, t_c and thus β may be influenced by temperature as well as by density (see also footnote 4).

---

[4]Actually, reduction in the duration of time it takes for a larva to move through certain early life history stages will influence the density-independent mortality as well as the density-dependent mortality that takes place during those stages. This means that, if such reduction in t_c actually occurs, α will increase as a result of any reduction in density-independent mortality. The possibility that an increase in density-independent mortality in the system, occasioned by certain human activities such as power plant entrainment, will be simultaneously accompanied by some reduction in the preexistent level of density-independent mortality is not accounted for in present versions of the equilibrium reduction equations (see, for example, Savidge et al. 1988).

[5]A linear dependence of t_c on stock density was characterized as "moderate" in the original work to distinguish it from two other scenarios of growth (i.e., "weak" and "strong").

This suggests that the temperature effect of the power plants might be examined as a possible offset to the plant cropping effect. Regardless of year-to-year variation in the naturally occurring ambient temperature, plants operating in a once-through mode will increase the temperature of receiving waters. Therefore, the plants considered alone represent a density-independent factor that can influence density-dependent mortality.

McFadden and Lawler (1977) incorporated Beverton and Holt's suggested analysis of a temperature influence on β into the ERE to examine the effect of increases in ambient temperature on the predicted reductions in the equilibrium spawning stock. Plant-induced conditional mortality rates, ranging between 5 and 25% with α at 4.0, produced equilibrium spawning stock reductions between -3% (population increase) and 20% for plant-induced average river temperature rises between 0.5 and 1.5°C. Again, by comparison, a range of 4 to 21% was obtained by use of the classical Ricker model.

For example, with the temperature coefficient of growth equal to 1.05, plant-induced conditional mortality equal to 5%, and the change in river temperature equal to 1.0 and 1.5°C, the plant-induced average river temperature results in growth rates that could more than offset power plant cropping, so that the population equilibrates at sizes higher than those prevailing prior to plant operations (shown above as a negative reduction in the equilibrium spawning stock). These results indicate that, all other things remaining equal, the heat discharged from once-through cooling systems has the potential to offset the cropping effect of the plants. Of course, this projection rests on assumptions that temperatures do not exceed the optimum for the species and life stage, that additional food is available for the fish growth and for the increased energy expenditure from greater activity and other metabolic processes, and that, although the striped bass are stimulated to greater activity and foraging success, the same is not happening to their predators.

Modification for Dual Density Dependence

All of the foregoing models consider the unit mortality rate throughout each stage of the life history to be independent of the changing density of any cohort of fish. The term "stock density dependent" is used simply to recognize that the unit mortality rate, though independent of the changing density of the cohort, $p(t)$, does depend on the stock density, P_0, responsible for that cohort.

At about the same time Ricker introduced his classical model (equation 2), Beverton and Holt (1957) presented a similar analysis, except that they considered early life stage mortality in a given year to partially depend on the continuously changing early life stage density in that year rather than on the given stock density in that year. Based on this assumed mechanism for the operation of compensation, Beverton and Holt developed an equation relating recruits to stock.

McFadden and Lawler (1977) fit the first-year striped bass survival data to the Beverton and Holt model to obtain rate constants for the single-stage Beverton and Holt equation. So far as the assumption is valid that this model reasonably represents the extant processes, the resulting rate constants suggested that a substantial portion of the mortality during the first year of life is attributable to density-dependent processes. Further, application of the rate constants in the Beverton–Holt stock–recruitment equation gave lower predictions of long-range reductions in the equilibrium spawning stock (15% reduction) than the classical Ricker equation did (21% reduction) with α equal to 4.0 and the same conditional mortality rate ($m = 0.25$).

An alternative Beverton–Holt analysis of the 1975 survival curve, based on the assumption that survival during the egg and yolk-sac stages was due only to density-independent causes, gave a set of mortality rates that attributed nearly all of the natural mortality during the first year of life to density-independent causes. In spite of the small portion of total mortality that was attributed to density-dependent causes, the resulting stock-recruitment equation predicted reductions in the equilibrium spawning stock (19% at $m = 0.25$) lower than those obtained from the classical Ricker equation with α equal to 4.0.

Christensen et al. (1977) and McFadden and Lawler (1977) have shown independently that when the Beverton–Holt equation is extended to include stock density-dependent mortality, the resulting stock–recruitment equation has many of the properties of the Ricker equation, including the ability to reproduce the oscillatory patterns apparent in the abundance of many fish stocks. As expected, the addition of the stock density-dependent term to the Beverton–Holt formulation provides additional compensatory response to new sources of density-independent mortality such as power plants. McFadden and Lawler (1977) have shown that if 50% of the density-independent mortality rate in the Beverton–Holt equation is attributed to stock density-dependent sources of mortality, then the predicted reduction in the equilibrium spawning stock is reduced from 15% to 12% at $m = 0.25$.

References[6]

Beverton, R. J. H., and S. H. Holt. 1957. On the dynamics of exploited fish populations. Fishery Investigations, Series II, Marine Fisheries, Great Britain Ministry of Agriculture Fisheries and Food 19.

Christensen, S. W., D. L. DeAngelis, and A. G. Clark. 1977. Development of a stock–recruitment progeny model for assessing power plant effects on fish population. Pages 196–226 in W. Van Winkle, editor. Assessing the effects of power-plant-induced mortality on fish populations. Pergamon, New York.

Goodyear, C. P. 1977. Assessing the impact of power plant mortality on the compensatory reserve of fish populations. Pages 186–194 in W. Van Winkle, editor. Assessing the effects of power-plant-induced mortality on fish populations. Pergamon, New York.

Hamley, J. M. 1975. Review of gillnet selectivity. Journal of the Fisheries Research Board of Canada 32:1943–1969.

Horwood, J. W., and J. G. Shepherd. 1981. The sensitivity of age-structured populations to environmental variability. Mathematical Biosciences 57:59–82.

McCombie, A. M., and F. E. J. Fry. 1960. Selectivity of gill nets for lake whitefish, Coregonus clupeaformis. Transactions of the American Fisheries Society 89:176–184.

McFadden, J. T., editor. 1977. Influence of Indian Point unit 2 and other steam electric generating plants on the Hudson River estuary, with emphasis on striped bass and other fish populations. Report to Consolidated Edison Company of New York.

McFadden, J. T., and J. P. Lawler, editors. 1977. Influence of Indian Point unit 2 and other steam electric generating plants on the Hudson River estuary, with emphasis on striped bass and other fish populations, supplement 1. Report to Consolidated Edison Company of New York.

McFadden, J. T., Texas Instruments, and Lawler, Matusky & Skelly Engineers. 1978. Influences of the proposed Cornwall pumped storage project and steam electric generating plants on the Hudson River estuary, with emphasis on striped bass and other fish populations. (Revised.) Report to Consolidated Edison Company of New York.

Muncy, R. J. 1960. A study of the comparative efficiency between nylon and linen gill nets. Chesapeake Science 1:96–102.

Ricker, W. E. 1973. Critical statistics from two reproduction curves. Rapports et Procès-Verbaux des Réunions, Conseil International pour l'Exploration de la Mer 174:333–346.

Royama, T. 1977. Population persistence and density dependence. Ecological Monographs 47:1–35.

Savidge, I. R., J. B. Gladden, K. P. Campbell, and J. S. Ziesenis. 1988. Development and sensitivity analysis of impact assessment equations based on stock–recruitment theory. American Fisheries Society Monograph 4:191–203.

Shepherd, J. G. 1982. A versatile new stock–recruitment relationship for fisheries, and the construction of sustainable yield curves. Journal du Conseil, Conseil International pour l'Exploration de la Mer 40:67–75.

Shepherd, J. G., and D. H. Cushing. 1980. A mechanism for density-dependent survival of larval fish as the basis of a stock–recruitment relationship. Journal du Conseil, Conseil International pour l'Exploration de la Mer 39:160–167.

---

[6]See Table 1 for sources of legal documents and unpublished reports pertaining to the Hudson River.

Comments[7]

During the Hudson River studies and power plant hearings, various attempts were made to incorporate components of environmental variations into the impact prediction efforts. These attempts ranged from analysis of the effects of freshwater flow and water temperature on year-class strength and growth to incorporation of environmental variables and random variations in the stock–recruitment equations. The Ricker equation (Ricker 1954) was considered applicable (McFadden and Lawler 1977) and was used as a starting point by investigators at Lawler, Matusky & Skelly Engineers and at Texas Instruments. The interpretation of the Ricker model and of the appropriateness of various modifications differed among the various investigators. These differences between the consultants to the utilities are apparent in the emphasis on α by Savidge et al. (1988, this volume) in contrast to the emphasis on β by Lawler (1988, this volume).

The Ricker stock–recruitment equation for a specific population is defined by two of the three parameters α, β, and P_0. The equilibrium population size or "carrying capacity," P_0, relative to the actual population size at a given time determines or is determined by β. Each of the parameters is affected by numerous environmental (and genetic) components of the population under study. Either parameter can be subdivided into components to reduce the generality of the model and adapt it to a specific system. Savidge and Ziesenis (1980) described the hierarchy of influences acting within α which transform the "intrinsic rate of increase" of a species into the "compensatory reserve," or α, specific to a population under specified environmental conditions. Subdivision of β may also be appropriate when the functional relationships of specific density-dependent mechanisms can be determined. Substantial improvement in the fit of the stock–recruitment model for American shad in the Con-

necticut River was achieved by removing the influence of June flows (Lorda and Crecco 1987). Incorporation of specific terms for growth, mortality, and recruitment was achieved by Jensen (1984) in a logistic surplus-production model for American lobster *Homarus americana* and spiny dogfish *Squalus acanthias*.

Fundamental relationships of temperature-induced changes are not sufficiently understood to instill confidence in an incorporation of such changes into the stock–recruitment equation for Hudson River striped bass via either density-dependent (β) or density-independent (α) mechanisms. Exploration of these potential mechanisms with models is interesting, but our understanding of the relationships is not yet at the stage at which these mathematical refinements are of value in impact prediction.

IRVIN R. SAVIDGE
Department of Chemistry and Biochemistry
University of Colorado
Boulder, Colorado 80309, USA

References[6]

Jensen, A. L. 1984. Logistic surplus-production model with explicit terms for growth, mortality and recruitment. Transactions of the American Fisheries Society 113:617–626.

Lawler, J. P. 1988. Some considerations in applying stock–recruitment models to multiple-age spawning populations. American Fisheries Society Monograph 4:204–218.

Lorda, E., and V. A. Crecco. 1987. Stock–recruitment relationship and compensatory mortality of American shad in the Connecticut River. American Fisheries Society Symposium 1:469–482.

McFadden, J. T., and J. P. Lawler, editors. 1977. Influence of Indian Point unit 2 and other steam electric generating plants on the Hudson River estuary, with emphasis on striped bass and other fish populations, supplement 1. Report to Consolidated Edison Company of New York.

Ricker, W. E. 1954. Stock and recruitment. Journal of the Fisheries Research Board of Canada 11:559–623.

Savidge, I. R., J. B. Gladden, K. P. Campbell, and J. S. Ziesenis. 1988. Development and sensitivity analysis of impact assessment equations based on stock–recruitment theory. American Fisheries Society Monograph 4:191–203.

Savidge, I. R., and J. S. Ziesenis. 1980. Sustained yield management. Pages 405–409 in S. D. Schemnitz, editor. Wildlife management techniques. The Wildlife Society, Washington, D.C.

---

[7]The Hudson River studies produced no agreement on the long-term impact of power plants on the river's fish fauna, in large part because the issue of biological compensation by the fish stocks for plant-induced mortality could not be resolved. The four contiguous papers in Section 3 by Savidge et al., Lawler, Christensen and Goodyear, and Fletcher and Deriso address these problems. Because this topic remains controversial, each of these authors was invited to comment on papers by the others. Not all chose to respond; the remarks of those who did are reproduced here.

In April 1984, at Lawler's indirect request, I sent him a draft copy of Christensen and Goodyear (1988, this volume; hereafter referred to as "our paper"). Its contents should have come as no surprise to him because the full analysis on which our paper is based was filed as testimony in 1979. Basically, our analysis demonstrates that estimates of α (alpha) from the curve-fitting exercise (CFE) Lawler conducted for the Environmental Protection Agency's (EPA) hearings are unreliable to the point of being useless. Lawler's (1988, this volume) current results are new, but they are subject to most of the same causes for doubt as were the original results. Before they are given any more credence than the original results, they should be tested as described in our paper. I am disappointed to see these results characterized as "reasonable estimates for Hudson River striped bass," and even to see discussion of likely directions of biases for certain of the models, without even acknowledgment, much less discussion, of the issues we have openly raised.

Lawler is now using nonlinear regression to fit his models (see his Table 72). As we discuss briefly in our paper, the limited work we have done with nonlinear regression gave very similar results to those obtained with linear regression. I would not expect the switch to nonlinear regression to make the CFE work substantially better.

Most of Lawler's models are new variants of the original approach, and most of these new variants have higher R^2 values than the original models. Values of R^2 in the original CFE were not good indicators of the performance of the models in the testing procedure we used. In addition, our procedure (with linear regression) shows that the conventional tests for statistical "significance" of the estimated parameters are unreliable. Incidentally, for the four cases in Lawler's Table 72 that are directly comparable with the analysis performed for the hearings, the current values of α are similar to the original values (two are higher and two are lower), but the current R^2 values are all slightly lower than the original R^2 values.

Our testing protocol should be applied to Lawler's analyses as a necessary, but not sufficient, condition for accepting his estimates as meaningful or reasonable. If, unexpectedly, they should "pass" the procedure, other questions (e.g., validity of the underlying model) would still need to be addressed.

Knowing how the "multiple-age spawning model" and the "egg production model" performed in their previous incarnation (i.e., when fitted with linear-regression techniques) is of interest. These models are identified in our paper as the "multiple-age" and the "eggs-on-eggs" models, respectively. In our paper, there is no "cannibalism variable," and the Q_7 flow variable is used. Neither of these two models, nor any of the other models tested, provided reliable estimates of the α values that were specified in runs of our simulation model.

Readers may well wonder what could be "wrong" with the approach Lawler has taken. The mathematics itself, while rooted in classical stock-recruitment theory, extends into new areas. The basic derivations are carefully, even elegantly, explained. Where, then, is the problem?

In introducing his equation (5), Lawler says "Let P_k, the parent population in year k, be . . .," and gives an expression that involves a catch-per-effort index of the population. The particular sets of assumptions under which his equation (5) would actually provide the needed index of parental egg production are not spelled out, but they could have been. How closely the needed assumptions are met for either the "all female fish" or "all eggs" analyses were lively topics of dispute at the EPA hearings. This question, however, pales next to the question of the assumptions needed for his equations (7), (8), or (9) to properly represent "recruits." Only under steady-state conditions could such equations be accurate, and under steady-state conditions the parameters of a stock-recruitment model cannot be estimated. Therefore, even if the Ricker model applies to the Hudson River striped bass population and if the catch-per-effort index were error-free (which it is not), the necessary indices of stock and particularly of recruits cannot be extracted from the catch-per-effort index.

I have one additional thought on a tangentially related topic. Lawler invokes the "proxy analysis" as a basis for preferring the results from his "female fish analysis" over those from the "egg production analysis." An empirical comparison of the proxy analysis with the "generation time" approach is given in our paper. Both are found wanting for the kinds of data available for the Hudson River striped bass, but the generation time approach, besides performing better on balance, has a logical basis that would permit it to work well for species with the right kind of life history, given the right kind of data. I know of no such basis for the proxy analysis. I am coming to feel more and more strongly that the less that is said about the proxy analysis the better!

Two summary thoughts, taken from the discussion section of our paper, bear repeating here. The first is a quote, taken from a different study, that summarizes well the main lesson learned from the exercise described in our paper: "A regression fit of the model to real but 'inappropriate' data may yield a very 'good' statistical fit but unrealistic estimates" (Gallucci and Quinn 1979). The second is our strong feeling that it is better to admit our inability to make a useful estimate than to rely on estimates that only by happenstance might be close to being correct.

SIGURD W. CHRISTENSEN
Environmental Sciences Division
Oak Ridge National Laboratory
Post Office Box 2008
Oak Ridge, Tennessee 37831-6036, USA

References

Christensen, S. W., and C. P. Goodyear. 1988. Testing the validity of stock–recruitment curve fits. American Fisheries Society Monograph 4:219–231.

Gallucci, V. F., and T. J. Quinn II. 1979. Reparameterization, fitting, and testing a simple growth model. Transactions of the American Fisheries Society 108:14–25.

Lawler, J. P. 1988. Some considerations in applying stock–recruitment models to multiple-age spawning populations. American Fisheries Society Monograph 4:204–218.

American Fisheries Society Monograph 4:219–231, 1988

Testing the Validity of Stock–Recruitment Curve Fits

Sigurd W. Christensen

Environmental Sciences Division, Oak Ridge National Laboratory
Post Office Box 2008, Oak Ridge, Tennessee 37831-6036, USA

C. Phillip Goodyear[1]

U.S. Fish and Wildlife Service, Office of Biological Services, National Power Plant Team
Ann Arbor, Michigan 48105, USA

Abstract.—The utilities relied heavily on the Ricker stock–recruitment model as the basis for quantifying biological compensation in the Hudson River power case. They presented many fits of the Ricker model to data derived from striped bass catch and effort records compiled by the National Marine Fisheries Service. Based on this curve-fitting exercise, a value of 4 was chosen for the parameter alpha in the Ricker model, and this value was used to derive the utilities' estimates of the long-term impact of power plants on striped bass populations. We developed and applied a technique to address a single fundamental question: If the Ricker model were applicable to the Hudson River striped bass population, could the estimates of alpha from the curve-fitting exercise be considered reliable? The technique involved constructing a simulation model that incorporated the essential biological features of the population and simulated the characteristics of the available actual catch-per-unit-effort data through time. The ability or failure to retrieve the known parameter values underlying the simulation model via the curve-fitting exercise was a direct test of the reliability of the results of fitting stock–recruitment curves to the real data. The results demonstrated that estimates of alpha from the curve-fitting exercise were not reliable. The simulation-modeling technique provides an effective way to identify whether or not particular data are appropriate for use in fitting such models.

The importance of biological compensation—stabilizing density-dependent mortality that enables populations to withstand some added stress—as an issue in the Hudson River power case was recognized by the scientists and lawyers on both sides long before the U.S. Environmental Protection Agency (EPA) hearings began. This topic had been a source of substantial contention beginning in 1972 during the U.S. Atomic Energy Commission hearings relating to licensing of unit 2 at the Indian Point nuclear generating station. In these hearings, compensatory functions were used either in the young-of-year models of entrainment and impingement (Lawler 1972a, 1972b; USNRC 1975; Eraslan et al. 1976) or in the life cycle models used in conjunction with the young-of-year models (Van Winkle et al. 1974). A serious drawback to this approach to compensation was the lack of data either to guide the selection of the functions or to estimate their parameters (Van Winkle et al. 1976; Christensen et al. 1981).

In 1975, Texas Instruments presented an analysis based on catch statistics for Hudson River striped bass (TI 1975). Catch and effort records compiled by the National Marine Fisheries Service (NMFS) from the commercial fishery operating in the river during the spring were combined to form estimates of catch per unit effort (CPUE). These data were lagged by 5 years to form estimates of stock (parents) and recruits (progeny). A straight line fitted to these stock–recruitment estimates had a significant negative slope. This analysis was recognized at Oak Ridge National Laboratory (ORNL) as being a departure from the previous approaches to evaluating compensation in the Hudson River and one deserving of investigation. It stimulated research into the application of stock–recruitment theory to power plant impacts on age-0 fish (Christensen et al. 1977).

The first major report by Texas Instruments on its Hudson River research, released early in 1977 (McFadden 1977), gave fits of the Ricker stock–recruitment model (Ricker 1954, 1958, 1975) to lagged CPUE data in an attempt to quantify the compensatory reserve of the striped bass (i.e., the amount of additional density-independent mortality the population can sustain without collapsing). The utilities' main testimony, presented in July 1977, relied on a greatly expanded curve-fitting

[1]Present address: National Marine Fisheries Service, Southeast Fisheries Center, Miami Laboratory, 75 Virginia Beach Drive, Miami, Florida 33149, USA.

exercise (CFE) as the basis for quantifying compensation (McFadden and Lawler 1977). The CPUE data were manipulated, through a number of processing approaches, to form estimates of stock and recruits, and the Ricker stock–recruitment model (usually in linearized form) was fitted to each of these. The fundamental equation for the Ricker model (modified from Ricker 1975) is

$$R = \alpha P_A \exp(-\beta P_B); \qquad (1)$$

R denotes an index of recruits, P_A denotes an index of parental egg production ("stock"), and P_B denotes some index of stock or parental egg production that adversely influences young-of-year mortality. The parameter beta (β), the nonlinear term in the Ricker model, represents a negative influence of stock, or parents, on survival of age-0 fish. It is the "compensatory parameter" in the Ricker model because, as the population size decreases due to stress (e.g., from power plant mortality), mortality due to beta decreases because of the decrease in stock size. This enables the population to have a new, although lower, equilibrium size, provided the amount of added stress is not too great. The parameter alpha (α) represents the balance between fecundity and density-independent mortality in the population. It determines the slope of the Ricker curve at the origin (Figure 73). Because the slope at the origin of a stock–recruitment curve is a direct measure of the excess reproductive capacity under "ideal" conditions for individual reproduction (i.e., when population size is so low that compensatory mortality does not operate), alpha is an index of the compensatory reserve of the population (Christensen et al. 1977; Goodyear 1977, 1980). In the context of the Hudson River power case, it was the key parameter to be estimated, because high values of alpha imply high compensatory reserve and low power plant impact, and vice versa. This is so because, in the Ricker model, the percentage change in equilibrium population size due to entrainment impacts can be calculated from the "equilibrium reduction equation" (Savidge et al. 1988, this volume). This equation involves only alpha and the conditional mortality rate for the entrainment impact; beta cancels out.

Although estimates of alpha obtained from these different fits often varied considerably, the fits appeared to be statistically significant (i.e., estimates of beta always appeared to differ significantly from zero with reference to conventional tabular statistics). The utilities concluded that assuming an alpha value of 4 for striped bass was reasonable, and they used this value in projecting

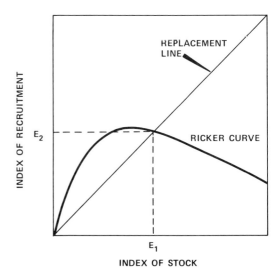

FIGURE 73.—A generalized Ricker curve, defining the relationship between stock (parents) and recruitment (progeny) in a fish population. The slope of the curve at the origin of the graph is equal to the value of alpha, the key parameter in predicting the response of the population to stress. Equilibrium values for stock (E_1) and for recruitment (E_2) are given by the intersection of the Ricker curve with the replacement line.

the population-level response of striped bass to power plant impacts. A subsequent utility exhibit, introduced during EPA's cross-examination of the utilities' case, used a slightly expanded CPUE data set, introduced river flow as a variable, and added some new processing approaches (Anonymous 1978). The CFE was an intricate and extensive effort on the part of the utilities, and it was defended continually by their consultants during cross-examination spanning a 4-month period in late 1977 and early 1978.

Causes for Doubt

The utilities' treatment of compensation was taken seriously by the government's technical consultants. If correct, it would represent a breakthrough in a critical area that had previously defied consensus. If not correct, it would be important to determine and demonstrate this convincingly, lest the same approach be used by others.

Study of and reflection on the utilities' case for compensation led to the identification of the following concerns: (1) the applicability of the Ricker model to the Hudson River striped bass population was questionable; (2) some of the effort data and all of the catch data compiled by NMFS and used to calculate the CPUE values were based on recall surveys (interviews with fishermen conducted once per year), and hence were questionable; (3) in the absence of age-specific data, the CPUE values, even if reliable, were conceptually unsuitable for deriving indices of either stock or recruits; (4) the scatter in the data about the fitted curve, although in most cases not unlike plots that have been published in the fisheries literature, was disturbing; and (5) the transformation involved in linearizing the model introduced potential bias in the fits (although a limited number of nonlinear fits presented by McFadden and Lawler 1977 yielded very similar results). These concerns were intensified when it was found that some of the utilities' curve-fitting techniques could be applied to randomly generated data, by both linear and nonlinear regression, and still yield significant fits of the model; if the utilities' logic were followed, this would lead to inferences of compensatory reserve (Goodyear 1979). Thus, even if the Ricker model were an appropriate model for the Hudson River striped bass population, it appeared doubtful that fits of the model to the available data could produce reliable estimates of alpha. Therefore, a means of testing the validity of the CFE was needed.

We developed a validation method, based on simulation modeling, to address this problem. This simulation technique is similar to one proposed years ago by McFadden (1963), the utilities' primary biological consultant. It is also conceptually related to the use of neutral models in hypothesis testing (Caswell 1976), and to applications of simulation techniques to stock–recruitment estimation procedures (Ludwig and Walters 1981) and to other problems in fisheries (e.g., Rivard and Bledsoe 1978).

Basis of the Validation Method

If the proposed CFE produces reliable (i.e., accurate and precise) estimates of alpha from Hudson River data, it must also produce reliable estimates from data generated by an appropriate simulation model. Conversely, if the CFE, when applied to the simulated data, produces unreliable (i.e., inaccurate or imprecise) estimates of the known alpha in the simulation model, then estimates of alpha based on the Hudson River data will at best be similarly unreliable. The phrase "appropriate simulation model" is important. To be appropriate, the model must meet certain criteria. First, in order for the test to be fair, the data generated by the model (simulated CPUE indices) must resemble the natural CPUE data with re-

TABLE 73.—Comparison of the utilities' curve-fitting exercise (CFE) with our validation procedure (VP); CPUE is catch per unit effort.

Step	CFE	VP
1	Assume Ricker model	Assume Ricker model
2	Parameters unknown	Specify parameters
3	Calculate CPUE values	Generate CPUE values
4	Process CPUE values	Process CPUE values
5	Fit Ricker model	Fit Ricker model; recycle to step 3 as needed
6	Accept estimates as valid	Compare estimates with the "truth"

spect to characteristics that could influence the results (e.g., variability and periodicity). Second, if the CFE itself, not the underlying assumption that the Ricker model applies to the population, is to be tested, the simulation model must be based on the Ricker model. A third criterion was used to facilitate use of the results as testimony: when parameter values were needed, they were based on the utilities' data when possible. When direct data were not available, the opinions of the utilities' consultants, if known, were used in most cases to avoid needless contention.

Our simulation model directly incorporates the mechanisms underlying the Ricker model. We generalized it for multiple age-classes using the techniques described by Christensen et al. (1977). It is conceptually a Leslie matrix model, with survival rates during the first year of life fixed within a given year but varying between years depending either on the number of age-1 fish or on the number of eggs spawned in the year. Construction of this model makes the formal (mathematical) assumption (for purposes of this validation procedure only) that the Ricker model applies to the Hudson River striped bass population. This is the first step in the validation procedure (VP), which is compared in Table 73 to the utilities' CFE. Both the CFE and the VP in Table 73 begin with the assumption that the Ricker model is valid. The second step differs. Alpha (equation 1), the key parameter, is not known for the CFE; indeed, the purpose of performing the CFE is to estimate alpha. In the VP, the value of alpha is specified as part of the procedure; it becomes the "truth" for that model run. The purpose of using the VP is to compare estimates of alpha with this "truth" as the basis for inferences about the reliability of the CFE. Other model parameters are needed in step 2 of the VP; a subset of these is needed in the CFE for some ways of processing CPUE data (step 4), but otherwise the CFE does not require parameter estimates. In the CFE, data on landings and effort from the Hudson River are used to calculate CPUE values (step 3), whereas, in the VP, analogous CPUE values from the simulated multiple-age-class population are generated. The CPUE values are processed in step 4 of both procedures in exactly the same manner to obtain estimates of stock and recruits. In the CFE, approximately 11 different processing approaches were used to estimate stock–recruit indices from the single 26-year CPUE time series; these same processing approaches were incorporated into the simulation model program. In step 5 of both procedures, the log-transformed Ricker model is repeatedly fitted, by means of linear least squares, to stock–recruit data obtained from each of the utilities' approaches. As indicated in Table 73, steps 3 through 5 in the VP can be repeated with different sets of random elements as many times as desired for a given set of population parameters to obtain numerous estimates for comparison. In the final step, estimates of alpha obtained from the CFE are accepted as valid. In contrast, estimates of alpha from the VP can be compared with the true value of alpha ("truth" as specified in step 2), as a test of the validity of using the CFE to estimate the (assumed) actual but unknown alpha in the real world.

The actual validation model used in the VP, named SRVAL (Christensen et al. 1979, 1982a, 1982b), can be described by three main equations:

$$N_0(t-1) = \sum_{i=1}^{15} a_i N_i(t-1); \qquad (2)$$

$$N_i(t) = N_{i-1}(t-1)S_i(t-1), \text{ for } i = 1 \text{ to } 14; \quad (3)$$

$$N_{15}(t) = [N_{14}(t-1) + N_{15}(t-1)]S_{15}(t - 1); \quad (4)$$

N is the number of fish; the subscript on N indicates the age of the fish at the time of spawning; t is time in years; a_i represents the average fecundity (egg production) of a fish of the subscripted age; and S_i represents the probability of survival from age $i - 1$ to age i. Equation (4) represents an aggregation of 15-year-old and older fish into a single "bulked" age-group. Although the S terms in equations (3) and (4) can, in general, be time-varying, we have varied only S_1 in this work. Simulated CPUE was generated from the model as a weighted sum of numbers-at-age, with a random term representing uncertainty in the real estimates:

$$CPUE(t) = \sum_{i=1}^{15} Q_i N_i R(t); \qquad (5)$$

the Q_i are factors incorporating (a) age-specific weight, and (b) an age-specific index of vulnerability to the fishery developed from the catch of the four fishermen in the utilities' 1976 "simulated commercial fishery" (see Christensen et al. 1982a); $R(t)$ is a normal or lognormal random variable (obtained from independently, identically distributed random variates) with a mean of one and adjustable variance.

All the parameters in equations (2) to (5) are specified by the modeler prior to a run except for $R(t)$ (specification described later) and S_1, the probability of surviving from age 0 (an egg) to age 1. The term S_1 is related to the Ricker model as follows:

$$S_1(t) = \frac{\alpha}{K} \exp[G(t) - \beta P_B(t)]; \qquad (6)$$

alpha and beta are parameters in the Ricker model; K is a constant based on life-table parameters (see Appendix); $G(t)$ is a random variable (see Appendix), including (a) a component related empirically to simulated river flow (following Anonymous 1978), and (b) a correction factor to remove from the population a tendency for random variation of this form to cause a population to increase (see Appendix A of Christensen et al. 1982a, or Goodyear and Christensen 1984); P_B represents either the number of eggs or the number of fish age 1 and older. The feedback term, $\exp[-\beta P_B(t)]$, constitutes the stock-dependent mortality (Harris 1975) that causes the Ricker model to have an (nontrivial) equilibrium point. Additional details about the model and its application were given by Christensen et al. (1982a).

Recall that step 2 in the VP (Table 73) involves specifying parameter values for a particular run. First, age-specific values for striped bass survival and fecundity in equations (2) through (4) and the Q_i factors for equation (5) were selected. Next, alpha in equation (6) was assigned a value (typically, 1.5, 5, or 20 for the exploited stock, to represent very weak, intermediate, and strong compensation, respectively). The particular value used for beta was unimportant for purposes of this study; a value that produced an arbitrary and constant equilibrium population size was calculated for each run. The type of stock-dependent mortality (feedback) in the Ricker model—$P_B(t)$ in equation (6)—was linked either to yearly egg production or to population size. Finally, the magnitude of the variation term—$R(t)$ in equation (5)—and of the random component of $G(t)$ in equation (6) (with its related correction factor) were varied in iterative trial runs until the median variance in model-simulated CPUE values matched the variance in the Hudson River striped bass CPUE values.

Having thus determined appropriate parameter values, a model run was made. In each run, the computer program generated 120 sets of simulated CPUE time series, processing each set by 12 approaches to generate simulated stock–recruit data assemblages. Two curve fits were performed on each of these 12 assemblages. The first type of curve fit involved a linearized form of the Ricker model (equation 1), identical to that used by the utilities in their main curve-fitting exercise (McFadden and Lawler 1977):

$$\log_e(R/P) = \widehat{\log_e\alpha} - \hat{\beta}P; \qquad (7)$$

the circumflex (ˆ) indicates that the parameter is now as estimated from regression, rather than as specified in equation (6). The second type of curve fit involved a linearized modification of the Ricker model that included flow (see equation 6 and Appendix), as used by the utilities (Anonymous 1978):

$$\log_e(R/P) = \widehat{\log_e\alpha} + \hat{\gamma}D - \hat{\beta}P; \qquad (8)$$

D is, for each spawning year, the difference between the simulated river flow in that year and the mean simulated flow for the spawning years included in the particular simulated data set; gamma—a component of $G(t)$ in equation (6)—is a parameter relating river flow to age-0 survival. The results of the 2,880 curve fits from each run were obtained as printed and magnetic tape output. The Statistical Analysis System (Barr et al. 1976) was used to analyze these results. Twenty-eight runs were performed with different values of alpha and other model parameters.

Code Verification

Several procedures were used to verify the computer code. First, a parallel model was developed independently by the second author, incorporating some but not all of the processing approaches. With identical starting conditions and random numbers, identical results were obtained for each of the several lag approaches tested. Second, the Hudson River CPUE and flow time series were substituted for the first model-generated series. This enabled comparison of the model-generated curve fits with the utilities' fits, and verified the processing approach and curve-fitting parts of the model. Third, an artificial set of parameter values was inferred, which should enable perfect estimates to be obtained from two of the processing approaches (matrix models "8" and "13" developed in equations 2-IV-8 through 1-IV-13 of McFadden and Lawler 1977). These processing approaches involve a matrix technique to try to establish age structure in the population from the annual CPUE indices so that indices of stock and recruits can be obtained. The artificial parameters enabling retrieval of perfect estimates are: perfect knowledge of flow provided to the matrix model, flow being the only random variable; perfect knowledge of fecundity and survival provided to the matrix model; all reproductive ages included in the matrix; only one age of fish represented in the CPUE index (i.e., only one age of fish caught); and the appropriate type of stock-dependent mortality used in the underlying Ricker model. Special test cases meeting these conditions were set up; as expected, the source value of alpha (the value of alpha used in the particular run) was retrieved from the curve fits.

Results

The stock–recruitment validation model was used in two investigations: (1) two methods of applying simple lags to CPUE data to generate stock–recruit indices were compared, and (2) the validity of the fits of the Ricker model to the Hudson River data (i.e., of the CFE) was tested.

The first investigation arose because one of the techniques used to process CPUE data to generate stock–recruit indices for the CFE involved the application of simple lags to the CPUE data. The CPUE was taken as an index of stock in a given year; then, one must determine the best lag to use in selecting a CPUE value to represent recruits. One answer proposed during the hearings was to use a lag time of 5 years, corresponding to the age of fish considered by the utilities to dominate the CPUE index. Thus, the CPUE value for year $t+5$ would represent recruits to be paired with the CPUE value for year t, representing stock size in year t. This came to be known as the "proxy" approach. An alternative answer was to use a lag time of 7 years, based on the mean generation time of striped bass; this came to be known as the "generation-time" approach. Attempts to reach agreement on the superiority of one of these two approaches during cross-examination failed.

TABLE 74.—Partial results of applying the validation methodology to compare the efficacies of the "proxy" approach (5-year lag) and the "generation-time" approach (7-year lag) to processing catch-per-unit-effort data. Assumptions used in parameterizing the model were those a utility witness felt would support the proxy approach.

Case	Approach used	Lag (years)	Random variation?	Source (true) alpha	Fitted alpha		Mean square error	Mean bias (fitted minus source alpha)
					Mean±SD	Range		
P1	Proxy	5	Yes	1.25	2.96±0.88	0.92–4.93	3.73	1.71
	Generation-time	7	Yes	1.25	2.13±0.76	0.82–4.54	1.35	0.88
P2	Proxy	5	Yes	10	5.27±1.35	2.51–9.27	24.33	−4.73
	Generation-time	7	Yes	10	7.52±2.14	2.90–12.87	10.78	−2.48
P3	Proxy	5	No	10	5.97±0.40	5.14–6.38	16.52	−4.03
	Generation-time	7	No	10	9.05±0.06	8.94–9.10	0.92	−0.95

The validation model proved an ideal tool for investigating this question. Model runs were made for a hypothetical case that a proponent of the proxy approach had suggested would be appropriate for that approach. The purpose of this exercise was to give the proxy approach every opportunity to succeed by constructing an idealized case exactly as suggested by its supporters; the realism of the case was irrelevant for this purpose. The hypothetical case was constructed as follows. On the average, 5-year-olds make up 40% of the catch. The 6-, 7-, and 8-year-olds each make up 20% of the catch. The 6-, 7-, and 8-year-old fish do all of the spawning, in equal proportions (i.e., each of these ages contributes 33.3% of the eggs, on the average). Then, according to the proxy approach, the 5-year lag would be appropriate to use in constructing stock–recruit pairs. According to the generation-time approach, 7 years would be the most appropriate lag.

When alpha in the model was specified as 1.25, the proxy approach yielded estimates of this alpha ranging, in 120 runs, from 0.92 to 4.93 and averaging 2.96 (Table 74). The difference between this mean estimate (2.96) and the true value of alpha (1.25) yielded the mean bias (1.71). The generation-time approach in this case yielded a substantially lower mean bias of 0.88. The generation-time approach also usually resulted in a lower standard deviation of alpha estimates than did the proxy approach.

The mean square error (MSE) shown in Table 74 is calculated as follows:

$$\text{MSE} = \text{variance} + (\overline{\text{bias}})^2. \qquad (9)$$

The MSE is thus a linear combination of the variance and the square of the mean bias. A statistician would prefer to minimize the variance and then estimate and correct for the bias. Our attempts to develop corrections for the bias, however, were unsuccessful. In such a case, if one needed to choose between two estimators, it would be reasonable to prefer the estimator with the lower MSE. For all three cases in Table 74, the generation-time approach resulted in a substantially lower MSE in alpha estimates than did the proxy approach. Thus, in all three cases examined, the generation-time approach (use of a 7-year lag) performed better on balance than the proxy approach (use of a 5-year lag), based on the bias, variance, and MSE of the estimates. This result notwithstanding, neither approach was very reliable in estimating alpha for cases incorporating random variation.

The second investigation—testing the validity of the fits of the Ricker model to the Hudson River data (the CFE)—involved the same model and procedures, but the various sets of parameter values were chosen to be consistent with what is known about the actual striped bass population. Twelve processing approaches were applied to the model-generated data, representing all those that had been applied to the Hudson River data in the CFE. The focus was not directly on comparing the performance of specific processing approaches. Rather, we wished to determine whether any or all of the processing approaches resulted in reliable estimates of the known alpha in the model, for any or all cases representing a range of reasonable assumptions about the actual Hudson River striped bass population.

Seven cases were constructed for analysis (Christensen et al. 1979). These cases differed in the magnitude of assumed adult mortality, the

TABLE 75.—Comparison of source (true) apha values specified in the simulation model with fitted alpha values obtained from model output for case 6. Results are shown for 6 of the 12 processing approaches tested. Simulated river flow was included in the fitted model. Each line in the table is based on 120 simulated catch-per-unit-effort (CPUE) time series. The conditions for case 6 are: annual adult survival = 0.43; stock-dependent mortality based on number of eggs; lognormal random variation in young-of-year survival; normal random error in CPUE index; and controlled variation in yearling population. See Christensen et al. (1979) for additional biological parameters.

| Run number | Processing approach[a] | Source (true) apha | Fitted alpha | | Mean square error | Mean bias (fitted minus source apha) |
			Mean±SD	Range		
55	5-year lag	1.25	2.99±1.18	1.16–10.44	4.44	1.74
	7-year lag	1.25	3.13±1.55	1.24–12.67	5.94	1.88
	Multiple-age model	1.25	3.22±1.15	0.95–8.41	5.24	1.97
	Eggs-on-eggs model	1.25	3.33±1.28	1.33–10.26	6.02	2.08
	Matrix model "8"	1.25	1.57±0.76	0.32–4.07	0.68	0.32
	Matrix model "13"	1.25	4.16±5.62	0.24–35.81	40.16	2.91
56	5-year lag	5	3.35±1.08	1.60–6.26	3.92	−1.65
	7-year lag	5	3.59±1.39	1.83–11.09	3.93	−1.41
	Multiple-age model	5	3.47±1.03	1.64–9.68	3.41	−1.53
	Eggs-on-eggs model	5	3.57±1.10	2.03–11.65	3.28	−1.43
	Matrix model "8"	5	1.61±0.64	0.22–4.27	11.97	−3.39
	Matrix model "13"	5	8.27±15.13	0.15–123.90	239.57	3.27
57	5-year lag	20	3.92±1.50	1.98–14.87	263.07	−16.08
	7-year lag	20	3.76±1.17	1.88–8.07	267.24	−16.24
	Multiple-age model	20	3.98±1.09	2.37–9.79	259.98	−16.02
	Eggs-on-eggs model	20	3.83±0.96	2.32–8.21	264.52	−16.17
	Matrix model "8"	20	1.70±0.54	0.59–3.83	338.01	−18.30
	Matrix model "13"	20	10.35±7.17	0.38–46.07	145.34	−9.65

[a] The 5- and 7-year lags are described in the text. The multiple-age and the eggs-on-eggs models are more complex methods of processing CPUE data; see Anonymous (1978) for a description. The matrix models "8" and "13" are described in equations 2-IV-4 through 2-IV-13 of McFadden and Lawler (1977); see especially equations 2-IV-8 and 2-IV-13.

type of stock-dependent mortality, and the location, form, and magnitude of random variation or error in the model and in the simulated CPUE index. Each of the seven cases contained at least three runs, each incorporating one of a fixed set of true alpha values (1.25, 5, or 20, representing very weak, intermediate, and strong compensation, respectively). Results were presented in 140 tables (Christensen et al. 1979). Table 75 contains data from three of these tables that relate to case 6 (one of the latest and most finely tuned applications of the model) for six of the 12 processing approaches.

The results in Table 75 demonstrate the main conclusion reached in the testimony presenting the full analysis: none of the fitting techniques provided reliable estimates of alpha for any of the cases. For low true alpha values (1.25) in the model simulating the Hudson River striped bass, the CFE consistently tended to overestimate the true value (compare the "mean fitted alpha" values in Table 75 with the corresponding "source [true] alpha" values). Figure 74 graphically compares one of the 120 curve fits summarized in the first line of Table 75 with the shape of the true underlying curve. The poor correspondence between the two curves, the overestimation of the

true alpha value (slope at the origin), and the extreme scatter in the data (typical of both the Hudson River data and the simulated data) are evident. In contrast, alpha values of 5 and higher usually were underestimated. Overall, estimates of alpha from the CFE were very unresponsive to changes in the underlying true alpha specified in the model.

The test for significance of beta (i.e., the test for a non-zero slope of the fitted linearized model in equation 7 or 8), used extensively by McFadden and Lawler (1977), was also found not to be reliable. In two runs involving a true value of 0.0 for beta (corresponding to no compensation and, therefore, an alpha value of 1.0), this test judged a very substantial majority of the thousands of estimates of beta to be "significantly" higher than 0.0 ($P < 0.05$). This spurious statistical result is consistent with the findings from other studies (Goodyear 1979; Robson 1979; Kenney 1982).

The relative efficiency of the various processing methods, as judged by the MSE criterion, varied among true alpha values and among cases. This variation was evident even in the subset of results from case 6 shown in Table 75. The MSE criterion did not show any of the processing methods to be either noticeably bet-

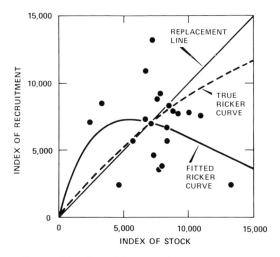

FIGURE 74.—One of the curve fits summarized in the first line of Table 75. The dashed curve indicates the true Ricker model underlying the simulation. The data points were obtained by applying the 5-year-lag approach to the simulated catch-per-unit-effort values. The solid curve is the result of fitting the Ricker model to the data points. The fitted curve is clearly not a good representation of the underlying model.

ter or noticeably worse than the others when all of the cases were considered.

Discussion

The inevitable conclusion to be drawn from our findings is that reliable estimates of alpha cannot be obtained from CPUE estimates for the Hudson River striped bass population, regardless of the fitting techniques employed. Our assumption that only the Ricker function regulates the model population should increase the chances of successful curve fitting of that function. However, the curve-fitting techniques examined produce unreliable estimates of alpha when applied to model-generated data. The estimates from the CFE applied to the actual Hudson River CPUE time series can be expected to be, if anything, still less reliable—to the point of being useless. These estimates cannot form the basis for a sound decision about the Hudson River striped bass population, even if the Ricker model were known accurately to describe the sole mechanism regulating the population. There is no known way to estimate the compensatory reserve of the population from available data. We strongly feel that it is better to admit our inability to make a useful estimate than to rely on estimates that only by happenstance might be close to being correct.

Other investigators of these Hudson River curve fits, using different approaches, have identified additional bases for doubt as well (Bossert 1979; Fletcher and Deriso 1979; Goodyear 1979; Levin 1979; Robson 1979; Rohlf 1979). By 1980, key personnel at Texas Instruments considered the issue of the stock–recruitment relationship for Hudson River striped bass to be unresolved (Klauda et al. 1980). A conclusion reached in a different study (an analysis of the von Bertalanffy growth model) serves well to summarize the main lesson learned from this exercise: "A regression fit of the model to real but 'inappropriate' data may yield a very 'good' statistical fit but unrealistic estimates" (Gallucci and Quinn 1979).

The main reason for our result lies in the inability of the CPUE index to form the basis of good indices of both stock and recruits. There would be greater hope of obtaining reliable estimates if a sufficiently long time series of age-structure data for Hudson River striped bass were available, or if the CPUE values actually represented fish of only one age. Without such age-structure information to enable derivation of reliable indices of both stock and recruits, it is futile to attempt to fit the Ricker model to a multiple-age spawner such as the striped bass. The data are simply not appropriate for this purpose. The processed indices used to fit the model likely contain so much error that they more nearly represent random numbers than indices of stock and recruits. It has been shown (Goodyear 1979) that either linear or nonlinear fits of the Ricker model to random data with variation characteristic of that of the CPUE indices consistently yield positive estimates of alpha, similar to those obtained by the utilities.

Age-structure information, particularly if it were known to be reliable, would provide much greater hope of obtaining reliable results. Even if such information were available, however, investigation of the other concerns listed in the earlier section entitled "Causes for Doubt" would still be needed. Ludwig and Walters (1981) also recognized the need for caution when stock–recruitment models are fitted to data, and Walters (1985) emphasized the value of Monte Carlo simulation for investigating possible biases.

The application of simulation methods for purposes of validating parameter estimates does not always lead to rejection of the estimates. For example, Rivard and Bledsoe (1978), performing an analogous simulation study relating to the Pella–Tomlinson stock production model, were able to confirm their parameter estimation procedures. However, curve fits of stock–recruitment models seem to be subject to greater difficulties, some of which have been investigated by Ludwig and Walters (1981) and Walters and Ludwig (1981).

We feel that the validation approach we have applied here will be most useful in a research setting. It can be used before field data are collected to predict whether a particular sampling protocol would be likely to enable reliable inferences to be drawn about the population, given assumptions about sampling error. In the context of stock–recruitment analyses, the question is usually whether or not existing historical data on stock parameters can be used to draw inferences about mechanisms regulating the population. We feel that, although it is not a simple and easy procedure, the general concept of validation that we have developed here should be applied to any stock–recruitment exercise for which there is reasonable question about the appropriateness of the data or methodology and on which important decisions could depend.

Acknowledgments

L. W. Barnthouse, R. J. Klauda, W. Van Winkle, and D. S. Vaughan critically reviewed the manuscript and provided helpful suggestions. Special thanks are extended to B. L. Kirk, who wrote much of the computer code and provided substantial technical assistance. Sigurd W. Christensen's contribution was supported in part by the Office of Nuclear Reactor Regulation, U.S. Nuclear Regulatory Commission, under interagency agreement DOE 40-544-75 with the U.S. Department of Energy (DOE); in part by the Division of Biomedical and Environmental Research, DOE, under contract DE-AC05-84OR21400 with Martin Marietta Energy Systems, Incorporated; and in part by the Office of Research and Development, U.S. Environmental Protection Agency, under interagency agreement DOE 40-740-78 (EPA 79-D-X0533) with DOE. This is publication 2876 of the Environmental Sciences Division, Oak Ridge National Laboratory.

References[2]

Anonymous (attributed to J. P. Lawler). 1978. Recent stock–recruitment analyses. Testimony (Exhibit UT-58) before the U.S. Environmental Protection Agency, Region II, in the matter of Adjudicatory Hearing Docket C/II-WP-77-01.

---

[2]See Table 1 for sources of legal documents and unpublished reports pertaining to the Hudson River.

Barr, A. J., J. H. Goodnight, J. P. Sall, and J. T. Helwig. 1976. A user's guide to SAS 76. SAS Institute, Cary, North Carolina.

Bossert, W. H. 1979. Empirical estimation of compensatory mortality in the striped bass population of the Hudson River. Testimony (Exhibit EPA-211) before the U.S. Environmental Protection Agency, Region II, in the matter of Adjudicatory Hearing Docket C/II-WP-77-01.

Caswell, H. 1976. Community structure: a neutral model analysis. Ecological Monographs 46:327–352.

Christensen, S. W., D. L. DeAngelis, and A. G. Clark. 1977. Development of a stock–progeny model for assessing power plant effects on fish populations. Pages 196–226 in W. Van Winkle, editor. Assessing the effects of power-plant-induced mortality on fish populations. Pergamon, New York.

Christensen, S. W., C. P. Goodyear, and B. L. Kirk. 1979. An analysis of the validity of the utilities' stock–recruitment curve-fitting exercise. Testimony (Exhibit EPA-208) before the U.S. Environmental Protection Agency, Region II, in the matter of Adjudicatory Hearing Docket C/II-WP-77-01.

Christensen, S. W., C. P. Goodyear, and B. L. Kirk. 1982a. An analysis of the validity of the utilities' stock–recruitment curve-fitting exercise and "prior estimation of beta" technique. The impact of entrainment and impingement on fish populations in the Hudson River Estuary, volume 3. Oak Ridge National Laboratory, ORNL/NUREG/TM-385/V3 (NUREG/CR-2220/V3), Oak Ridge, Tennessee.

Christensen, S. W., B. L. Kirk, and C. P. Goodyear. 1982b. A user's guide for the stock–recruitment model validation program. Oak Ridge National Laboratory, ORNL/TM-8216 (NUREG/CR-2562), Oak Ridge, Tennessee.

Christensen, S. W., W. Van Winkle, L. W. Barnthouse, and D. S, Vaughan. 1981. Science and the law: confluence and conflict on the Hudson River. Environmental Impact Assessment Review 2:63–88.

Eraslan, A. H., and six coauthors. 1976. A computer simulation model for the striped bass young-of-the-year population in the Hudson River. Oak Ridge National Laboratory, ORNL/NUREG-8, Oak Ridge, Tennessee.

Fletcher, R. I., and R. B. Deriso. 1979. Appraisal of certain arguments, analyses, forecasts, and precedents contained in the utilities' evidentiary studies on power plant insult to fish stocks of the Hudson River estuary, volume 1. Testimony (Exhibit EPA-218) before the U.S. Environmental Protection Agency, Region II, in the matter of Adjudicatory Hearing Docket C/II-WP-77-01.

Gallucci, V. F., and T. J. Quinn II. 1979. Reparameterizing, fitting, and testing a simple growth model. Transactions of the American Fisheries Society 108:14–25.

Goodyear, C. P. 1977. Assessing the impact of power plant mortality on the compensatory reserve of fish populations. Page 186–195 in W. Van Winkle, editor. Assessing the effects of power-plant-induced mortality on fish populations. Pergamon, New York.

Goodyear, C. P. 1979. The reliability of the utilities' regression estimates of parameters of the Ricker model. Testimony (Exhibit EPA-214) before the U.S. Environmental Protection Agency, Region II, in the matter of Adjudicatory Hearing Docket C/II-WP-77-01.

Goodyear, C. P. 1980. Compensation in fish populations. Pages 253–280 in C. H. Hocutt and J. R. Stauffer, Jr., editors. Biological monitoring of fish. Lexington Books, Lexington, Massachusetts.

Goodyear, C. P., and S. W. Christensen. 1984. Bias-elimination in fish population models with stochastic variation in survival of the young. Transactions of the American Fisheries Society 113:627–632.

Harris, J. G. K. 1975. The effect of density-dependent mortality on the shape of the stock and recruitment curve. Journal du Conseil, Conseil International pour l'Exploration de la Mer 36:144–149.

Kenney, B. C. 1982. Beware of spurious self-correlations! Water Resources Research 18:1041–1048.

Klauda, R. J., W. P. Dey, T. B. Hoff, J. B. McLaren, and Q. E. Ross. 1980. Biology of Hudson River juvenile striped bass. Marine Recreational Fisheries 5:101–123.

Lawler, J. P. 1972a. Effect of entrainment and impingement at Indian Point on the population of the Hudson River striped bass. Modifications and additions to testimony of April 5, 1972. Testimony (October 30, 1972) before the U.S. Atomic Energy Commission in the matter of Consolidated Edison Company of New York, Inc. (Indian Point station, unit no. 2). Docket 50-247.

Lawler, J. P. 1972b. The effect of entrainment at Indian Point on the population of the Hudson River striped bass. Testimony (April 5, 1972) before the U.S. Atomic Energy Commission in the matter of Consolidated Edison Company of New York, Inc. (Indian Point station, unit no. 2). Docket 50-247.

Levin, S. A. 1979. The concept of compensatory mortality in relation to impacts of power plants on fish populations. Testimony (Exhibit EPA-212) before the U.S. Environmental Protection Agency, Region II, in the matter of Adjudicatory Hearing Docket C/II-WP-77-01.

Ludwig, D., and C. J. Walters. 1981. Measurement errors and uncertainty in parameter estimates for stock and recruitment. Canadian Journal of Fisheries and Aquatic Sciences 38:711–720.

McFadden, J. T. 1963. An example of inaccuracies inherent in interpretation of ecological field data. American Naturalist 97:99–116.

McFadden, J. T., editor. 1977. Influence of Indian Point unit 2 and other steam electric generating plants on the Hudson River estuary, with emphasis on striped bass and other fish populations. Report to Consolidated Edison Company of New York.

McFadden, J. T., and J. P. Lawler, editors. 1977. Influence of Indian Point unit 2 and other steam electric generating plants on the Hudson River estuary, with emphasis on striped bass and other fish populations, supplement 1. Report to Consolidated Edison Company of New York.

Ricker, W. E. 1954. Stock and recruitment. Journal of the Fisheries Research Board of Canada 11:559–623.

Ricker, W. E. 1958. Handbook of computations for biological statistics of fish populations. Fisheries Research Board of Canada Bulletin 119.

Ricker, W. E. 1975. Computation and interpretation of biological statistics of fish populations. Fisheries Research Board of Canada Bulletin 191.

Rivard, D., and L. J. Bledsoe. 1978. Parameter estimation for the Pella-Tomlinson stock production model under nonequilibrium conditions. U.S. National Marine Fisheries Service Fishery Bulletin 76:523–534.

Robson, D. S. 1979. Critique of some statistical analyses of Hudson River striped bass data. Testimony (Exhibit EPA-210) before the U.S. Environmental Protection Agency, Region II, in the matter of Adjudicatory Hearing Docket C/II-WP-77-01.

Rohlf, F. J. 1979. Analysis of (1) population density and growth and (2) striped bass stock recruitment models. Testimony (Exhibit EPA-208) before the U.S. Environmental Protection Agency, Region II, in the matter of Adjudicatory Hearing Docket C/II-WP-77-01.

Savidge, I. R., J. B. Gladden, K. P. Campbell, and J. S. Ziesenis. 1988. Development and sensitivity analysis of impact assessment equations based on stock–recruitment theory. American Fisheries Society Monograph 4:191–203.

TI (Texas Instruments). 1975. First annual report for the multiplant impact study of the Hudson River estuary, volumes 1 and 2. Report to Consolidated Edison Company of New York.

USNRC (U.S. Nuclear Regulatory Commission). 1975. Final environmental statement related to operation of Indian Point nuclear generating plant, unit no. 3, volumes 1 (NUREG-75/002) and 2 (NUREG-75/003). Docket 50-286.

Van Winkle, W., S. W. Christensen, and G. Kauffman. 1976. Critique and sensitivity analysis of the compensation function used in the LMS Hudson River striped bass models. Oak Ridge National Laboratory, ORNL/TM-5437, Oak Ridge, Tennessee.

Van Winkle, W., B. W. Rust, C. P. Goodyear, S. R. Blum, and P. Thall. 1974. A striped-bass population model and computer programs. Oak Ridge National Laboratory, ORNL/TM-4578, Oak Ridge, Tennessee.

Walters, C. J. 1985. Bias in the estimation of functional relationships from time series data. Canadian Journal of Fisheries and Aquatic Sciences 42:147–149.

Walters, C. J., and D. Ludwig. 1981. Effects of measurement errors on the assessment of stock–recruitment relationships. Canadian Journal of Fisheries and Aquatic Sciences 38:704–710.

Appendix

Equation (6) in the text, which is derived in this appendix, gives the probability of survival through the first year of life for a multiple-age-class (matrix) fish population model. It also contains the parameters α (alpha) and β (beta) used in the Ricker stock–recruitment model. Deriving equation (6) in the text requires extending stock–recruitment theory, which was developed for non-iteroparous fish (i.e., fish that spawn only once before dying), to iteroparous (repeat-spawning) fish populations. The approach is similar in concept to that presented by Christensen et al. (1977) and McFadden and Lawler (1977). Because of the desire to retain alpha and beta as parameters in the matrix model (McFadden and Lawler 1977), it is necessary to put the matrix model into the same form as the Ricker model. The purpose behind some of the manipulations may, therefore, not be immediately evident.

The Ricker stock–recruitment model (or, more generally, parent–progeny model) (Ricker 1954, 1958, and 1975) may be written as

$$R = \alpha P \exp(-\beta P); \qquad \text{(A-1)}$$

R is a measure of recruitment or progeny, and P is a measure of the parental stock that produced the recruits. Alpha (α) and beta (β) are parameters, as explained in the text. A minor but important generalization of the Ricker model can be written as

$$R = \alpha P_A \exp(-\beta P_B); \qquad \text{(A-2)}$$

P_A is the parental contribution to the population's self-regeneration term (αP_A), and P_B is the parental contribution to the feedback term [$\exp(-\beta P_B)$]. This convention allows the two components of the equation to be based on alternative measures of parental stock. Whether the Ricker model is applied to noniteroparous or to iteroparous populations, P_A must logically be an index of (i.e., directly proportional to) egg production. If "progeny" (R) are defined in the same units as "parents" (P_A), alpha can be easily interpreted as an index of compensatory reserve (i.e., the amount of additional density-independent mortality the population can sustain without collapsing). This convention is necessary if the "equilibrium reduction equation" (Savidge et al. 1988) is to be applied.

With the noniteroparous fish species for which the Ricker model was developed, the number of spawning fish could be used for P_A under the reasonable assumption that all spawners produce an equal number of eggs. With iteroparous species, this assumption is not appropriate unless the population's age structure and size is at a steady state. Only under these conditions is stock size (number of fish) proportional to egg production (i.e., "mean fecundity" actually represents the fecundity of the mean member of the stock). Steady-state conditions will, therefore, be imposed during the derivation. In addition, because the matrix model requires an expression for the probability of survival through the first year of life, it is necessary to develop the equation in terms of parental egg production, rather than simply in terms of number of parents or stock.

We begin by defining two useful terms. The "stock value" of a recruit, $\overline{V}$, is the total lifetime expected contribution of a yearling recruit to current and future "stocks" (defined as the number of fish age 1 or older in the population). The stock value of a recruit is defined as

$$\overline{V} = 1 + \sum_{i=2}^{15} \prod_{j=2}^{i} \overline{S}_j; \qquad \text{(A-3)}$$

$\overline{S}_j$ is the (time-invariant) probability of survival from age j-1 to age j. The second term is the fecundity per unit stock, $\overline{E}$, defined for the population at a steady state as

$$\overline{E} = \frac{a_1 + \sum_{i=2}^{15} \left(a_i \prod_{j=2}^{i} \overline{S}_j \right)}{\overline{V}}. \qquad \text{(A-4)}$$

The term a_i in equation (A-4) is the mean fecundity (egg production) of a fish of the subscripted age, calculated as

$$a_i = (ff_i) \, (fm_i) \, (emf_i); \qquad \text{(A-5)}$$

ff_i is the female fraction of age-i fish; fm_i is the mature fraction of ff_i; emf_i is the number of eggs produced by each mature female.

For the population at a steady state, an equation for the number of recruits in the matrix model can be written

$$R_1 = P_A \overline{E} \exp(-d) \exp(-\beta P_B); \qquad \text{(A-6)}$$

R_1 is the number of 1-year-old recruits at time $t+1$; P_A is the number of fish (age 1 and older) in the spawning stock at time t; d is a density-independent mortality rate constant for the first year of life.

With consideration still restricted to the steady-state situation, both sides of equation (A-6) can be multiplied by $\overline{V}$ to yield

$$R_1\overline{V} = \overline{V}\,\overline{E}\,\exp(-d)P_A\,\exp(-\beta P_B).\quad\text{(A-7)}$$

Examination of equation (A-7) shows it to be in the same form as the generalized Ricker model (equation A-2), whereby recruits (the left-hand side of equation A-7) are defined in units of stock (i.e., P_A). The convenience of this convention has already been mentioned. The parameter alpha in equation (A-2) is given in equation (A-7) by

$$\alpha = \overline{V}\,\overline{E}\,\exp(-d),\quad\text{(A-8)}$$

which can be rearranged to give

$$\exp(-d) = \frac{\alpha}{\overline{V}\,\overline{E}}.\quad\text{(A-9)}$$

We can now derive an expression for the steady-state probability of surviving the first year of life ($\overline{S}_1$) that will involve the parameters of the Ricker model. By definition, $\overline{S}_1$ is the number of 1-year-old recruits divided by the number of eggs spawned, all at steady-state conditions:

$$\overline{S}_1 = \frac{R_1}{P_A\,\overline{E}}.\quad\text{(A-10)}$$

Substitution of equation (A-9) into equation (A-7), and rearrangement, provides an expression for the numerator of equation (A-10):

$$R_1 = \frac{\alpha P_A}{\overline{V}}\,\exp(-\beta P_B).\quad\text{(A-11)}$$

Substitution of equation (A-11) into equation (A-10) yields the survival rate from eggs to age-1 fish at a steady state:

$$\overline{S}_1 = \frac{\alpha}{\overline{V}\,\overline{E}}\,\exp(-\beta P_B).\quad\text{(A-12)}$$

Equation (A-12) relates the parameters alpha and beta in the Ricker stock–recruitment model to a multiple-age-class fish population model. Because progeny ($R_1\overline{V}$) and parents (P_A) were measured in the same units (number of fish) beginning with equation (A-7), the "equilibrium reduction equation" (Savidge et al. 1988) is applicable.

We now relax the constraint of steady-state conditions to introduce variation into the model:

$$S_1(t) = \frac{\alpha}{K}\,\exp[r(t)-C+\gamma H(t)-\beta P_B(t)];\quad\text{(A-13)}$$

K is a constant equal to $\overline{V}\,\overline{E}$; $r(t)$ is a normal random variable (obtained from independently, identically distributed random variates) with mean of zero and adjustable variance; C is a correction factor to remove from the population a tendency for random variation of this form to cause a population to increase (see Appendix A of Christensen et al. 1982a, or Goodyear and Christensen 1984); gamma (γ) is a specified parameter relating river flow to young-of-year mortality (see Anonymous 1978); H represents a normal random variable with mean of zero and variance of 3,471, simulating the variance in Hudson River flow over the period 1950–1975 (Anonymous 1978); and $P_B(t)$ represents (in our application) either the number of eggs or the number of fish age 1 and older.

For simplicity, a single variable $G(t)$ was defined as

$$G(t) = r(t) - C + \gamma H(t).\quad\text{(A-14)}$$

Substitution of $G(t)$ into equation (A-13) gives equation (6) in the text.

Comments[3]

The problems of estimating the parameters of the Ricker (1954) stock–recruitment equation from

curve fitting in the presence of other sources of variation in the data are well illustrated by Christensen and Goodyear (1988, this volume). Estimation of alpha (α) is critical in modeling populations in which density-dependent feedback occurs. The "compensatory reserve" or α is the rate at which a population would increase in a defined environment in the absence of density-dependent feedback and is a necessary component of population modeling regardless of the functional formulation of the feedback mechanisms.

The problems in estimating the compensatory reserve of populations, either through advanced

---

[3]The Hudson River studies produced no agreement on the long-term impact of power plants on the river's fish fauna, in large part because the issue of biological compensation by the fish stocks for plant-induced mortality could not be resolved. The four contiguous papers in Section 3 by Savidge et al., Lawler, Christensen and Goodyear, and Fletcher and Deriso address these problems. Because this topic remains controversial, each of these authors was invited to comment on papers by the others. Not all chose to respond; the remarks of those who did are reproduced here.

curve-fitting procedures or through the use of alternative methods, deserve further attention. The results of Christensen and Goodyear (1988) serve as a dramatic illustration of the unreliability of estimates of alpha from curve-fitting exercises when the data have large unexplained variability resulting from sampling error or fluctuating environmental conditions. The compensatory reserve of a population is a function of the "innate capacity for increase" of the population, as determined by the gene pool and modified by the prevailing environmental conditions (Savidge and Zienesis 1980). Because environmental conditions may fluctuate from year to year and generation to generation, the "effective alpha" at a specific time may fluctuate from year to year. Lorda and Crecco (1987) identified an environmental component associated with river flow that affects the variability of data on Connecticut River American shad relative to the fit of a modified Ricker stock–recruitment curve. By removing this component, they markedly improved the fit of the curve to their data. Partitioning α into component parts may provide a means of identifying year-to-year variation in components of the environment that affect reproduction and density-independent sources of mortality. Adequate details relative to extant data may not be available, however, or the data may not be tractable to partitioning. Even in the absence of sampling error and annual variation, catch–effort data on multiple-age stocks without details about age and sex composition could result in biased estimates because of the averaging across year classes that would result from differential parental stock size.

The "equilibrium reduction equation" described by Savidge et al. (1988, this volume) cannot be applied to the estimation of impact without reliable estimates of α. Approximations of α may be obtained from data not requiring curve fitting. Data on rates of increase of populations of the same species newly introduced into a similar habitat or recovering from reduced population levels may be used to estimate α when available. Because the compensatory reserve is determined both by the genetics of the stock and the specific environment of the population, care must be taken when estimates of α are extrapolated from other stocks.

The problems identified by Christensen and Goodyear (1988) should stimulate development of methods to identify and remove extraneous (non–stock-dependent) variations from data prior to stock–recruitment curve fitting and to develop alternative procedures of estimating the compensatory reserve of populations. The identification of criteria to determine the presence of extraneous nonextractable variation that will render a curve-fitting exercise unreliable would be useful, as would a procedure to estimate the relationship of small amounts of extraneous variation to the reduction of reliability.

IRVIN R. SAVIDGE
Department of Chemistry and Biochemistry
University of Colorado
Boulder, Colorado 80309, USA

References

Christensen, S. W. and C. P. Goodyear. 1988. Testing the validity of stock–recruitment curve fits. American Fisheries Society Monograph 4:219–231.

Lorda, E. and V. A. Crecco. 1987. Stock–recruitment relationship and compensatory mortality of American shad in the Connecticut River. American Fisheries Society Symposium 1:469–482.

Ricker, W. E. 1954. Stock and recruitment. Journal of the Fisheries Research Board of Canada 11:559–623.

Savidge, I. R., J. B. Gladden, K. P. Campbell, and J. S. Ziesenis. 1988. Development and sensitivity analysis of impact assessment equations based on stock–recruitment theory. American Fisheries Society Monograph 4:191–203.

Savidge, I. R. and J. S. Ziesenis. 1980. Sustained yield management. Pages 405–409 *in* S. D. Schemnitz, editor. Wildlife management techniques. The Wildlife Society, Washington, D. C.

American Fisheries Society Monograph 4:232–244, 1988

Fishing in Dangerous Waters: Remarks on a Controversial Appeal to Spawner–Recruit Theory for Long-Term Impact Assessment[1]

R. Ian Fletcher[2]

Ecosystems Research Center, Cornell University, Ithaca, New York 14853, USA

Richard B. Deriso

International Pacific Halibut Commission, Seattle, Washington 98195, USA

Abstract.—In the Hudson River power plant case, the defending utility companies appealed to the (perceived) practices of commercial fishery managers as precedent for their own uses of spawner–recruit models in forecasting the effects on stock abundances of long-term water withdrawals by the power plants. In contrast to this perception of spawner–recruit models as commonly accepted instruments of fishery management, our survey of managers and regulatory agencies revealed instead a universal rejection of such models. The inadequacies of spawner–recruit (or parent–progeny) models were attributed variously to the effects of environmental uncertainty, to imperfect knowledge of population regulation, and to the want of sufficient biological meaning in the models themselves. In general, impact assessment little resembles fishery regulation, as the options and privileges commonly exercised by a fishery manager are closed to impact management. In either case, long-term forecasting remains unreliable owing to the problem, as yet unsolved, of predicting reproduction and recruitment in natural populations.

During the course of the U.S. Environmental Protection Agency hearings known as the Hudson River power plant case, the defending utility companies advanced a thesis of natural mitigation that eventually became the major sticking point of the case. In brief, the utilities argued that however severe on fish life the immediate effects of water withdrawals by the power plants might seem to be, in terms of the numbers of fish entrapped and killed, the ultimate effects of those killed were not likely to reduce the fish stocks by very much owing to the offsetting benefits of density-dependent compensation. According to the notion behind this self-regulating process, any significant reduction in fish abundance (in particular, the early life stages in the case of the power plant kills) would accrue as a benefit to the surviving young by reducing competition amongst themselves, thereby enhancing their own changes for continued survival. This hypothesized increase in survival probability was believed by the utilities to be sufficiently great as to safely moderate the direct mortalities imposed by the power plants.

The consultants for the utility companies selected the Ricker (1954) spawner–recruit curve (in equilibrium form) as the model for demonstrating compensation, and added several variations of the Ricker curve along the way. The arguments supporting the choice of spawner–recruit theory ultimately turned on two propositions, each of which was to elicit controversy that filled thousands of pages of testimony by the time the hearings ended.

The utilities argued, as their first proposition, that, because the past effects of fishing on any stock could be represented satisfactorily with density-dependent spawner–recruit curves (especially the Ricker curve), the same theory and curves could be employed, in like fashion, as a reliable means for predicting the long-term consequences of juvenile losses to power plants. The second, and supporting, proposition (which is the proposition we examine in this paper) dealt with precedent: the utilities argued that such proposed uses of spawner–recruit curves were preceded by a long history of successful and parallel applica-

[1]Publication reference ERC–046 of the Ecosystems Research Center, Cornell University. Partial funding for this research was provided by the Office of Research and Development, U.S. Environmental Protection Agency, under cooperative agreement CR-811060. The work and conclusions published herein represent the views of the authors and do not necessarily represent the opinions, policies, or recommendations of the Office of Research and Development or of Cornell University. The Environmental Protection Agency and Cornell do not endorse any commercial products used in the study.

[2]Current address: Great Salt Bay Experimental Station, Damariscotta, Maine 04543, USA.

tions of the curves in fishery management (e.g., that spawner–recruit curves enjoyed a "widespread use and acceptance" in the management of commercial fish stocks). On being challenged in court to support this claimed use and acceptance, the utilities responded with a list of supporting citations, taken mainly from the fisheries literature, that was entered into the hearing records as Utilities Exhibit UT–59 (reproduced here in the Appendix).

The Question of Precedent

Precedent was obviously a crucial issue for the utility companies. If they could show that spawner–recruit models *are* commonly used for assessing and managing fisheries, then the arguments supporting similar uses for impact management were likely to be viewed by the court as altogether plausible. This confidence in the management values of spawner–recruit models was expressed with unquestioned conviction throughout the utilities' testimony. In Utilities Exhibit UT–3, for example, they had advanced the belief that "The concepts upon which this impact prediction is based were developed through experience in managing commercial fisheries," and the chief scientific witness for the utilities affirmed during cross-examination that with the use of the spawner–recruit models and data fittings contained in their exhibits, the utilities were in fact "attempting to manage the Hudson River striped bass stock on the basis of forecasting the future state of that stock and other stocks of the Hudson River."

This appeal by utilities to the practices of fishery managers was probably made in good faith; the perception of spawner–recruit models as widely accepted instruments of stock management is not uncommon. But, in our view, that perception was incorrect. We asserted that no fishery management agency would be so imprudent as to regulate any stock of consequence on the basis of data fits to spawner–recruit models, and we summarize here the basis for that view.

Management Uses of Spawner–Recruit Models

The utilities cited 20 reports and communications in UT–59 as evidence of the claimed regulatory applications and spawner–recruit models. We, in turn, responded with an item-by-item review of UT–59, which included a survey of the cited authors (or agencies). A detailed report of that examination was given by Fletcher and

Deriso (1979) and will not be repeated at length here. We believe, however, the questions and uncertainties surrounding the management uses of spawner–recruit models to be of such general interest that we give here a number of excerpts from the statements of agencies and authors on their views and experiences with such models.

Pacific Salmon

The first item of Utilities Exhibit UT–59 refers to an evidentiary assertion by utilities to the effect that "the management of all Pacific salmon stocks is based on a Ricker stock/recruitment model, and has been for several years." Aside from the unequivocal ring of the statement, the assumption behind it is not unfamiliar. Unless one were closely acquainted with assessment practices in northwestern North America, one might easily suppose that if the Ricker spawner–recruit curve had any management validity at all, it would surely make an appearance in the regulatory practices in Pacific salmon fisheries (the theoretical workings of the Ricker model, in particular, being suited more to semelparous breeders of a common age than to iteroparous fishes of mixed ages).

The view from afar is a misleading view, however, and the several agencies that manage Pacific salmon are just as autonomous in their practices as fishery agencies anywhere else. The Pacific salmon fisheries consist of hundreds of distinct stocks of five distinct species (coho, sockeye, chum, pink, and chinook salmon), and these stocks are managed, to greater and lesser extents, by the agencies of at least four political entities (Oregon, Washington, Alaska, and British Columbia). Cooperative management among these agencies is limited to very few fisheries. Moreover, the habits and population dynamics of the five salmon species are far from identical; their fisheries are extremely diverse, and the management problems reflect that diversity. Some salmon fisheries are oceanic (on prespawning fish); others are estuarine and riverine (on spawning runs); some are troll fisheries, other are trawl, gill-net, purse-seine, dip-net, or angling fisheries. And, although all fives species are semelparous (they all spawn once and die), some species mature and spawn at 4 years of age (spawning runs being composed of single year classes), whereas others mature over ages that range from 2 to 6 years or more (their runs being composed, therefore, of mixed year classes). Owing to these differences in the fisheries and habits of the five salmon species, no regulatory model of any kind really emerged from our survey as the one preferred means for managing Pacific salmon.

We solicited information from managers of salmon fisheries in Oregon, Washington, Alaska, and Canada on their regulatory practices and experiences with spawner–recruit curves. One of us surveyed the managers with the following inquiry:

> I am doing some research into the management uses of spawner-recruit models that is very difficult to glean from the scientific literature. In the journals I find theoretical treatments, speculation, author's opinions, and various syntheses of historical data on recruitment, but virtually no commentary or documentation on the actual managing of fish stocks by direct forecasting from spawner-recruit models. Would you be so kind as to tell me of *any* stocks (whether demersal, pelagic, estuarine, or anadromous) that [your agency] might actually manage, or might have managed, over any length of time at all, on the basis of data fits to spawner-recruit curves (such as Ricker's, or any other parent-progeny model)? And if so, how good or bad have the long-range forecasts been? That is, have subsequent observations agreed with management forecasts to any significance?
>
> And finally, what is your view, as a fishery scientist, on the value in direct regulation of current spawner-recruit models and theory? That is, do you think the general stock-recruitment problem of fishery science has been resolved to any extent sufficient to the risking of long-term (or even short-term) management strategies on current models of spawner-recruit theory?

The full texts of the replies are reproduced in Fletcher and Deriso (1979). The reply from Washington, on salmon regulation there, was written by Samuel Wright, Chief of the Harvest Management Division, Department of Fisheries, who emphasized that "All important fisheries management decisions are made on the basis of in-season run size updates . . . none of these is based on spawner–recruit models."

The reply from Robert Gonsolus, Anadromous Fish Supervisor of the Oregon Department of Fish and Wildlife, contained this passage: "Oregon does not manage any of our salmon stocks based on spawner–recruit curves. We did examine production–escapement data in the early 1960s, but the fit to the Ricker curve was not too promising, even at that time" (i.e., before the Columbia and Snake River dams were built and the hatcheries started, which has made the management problem even more complicated). On the question of the general reliability of spawner–recruit models, Gunsolus added: "I do not believe the use of spawner–recruit models is presently of much value . . . I believe there are too many unpredictable factors affecting the survivals of juveniles and adults which would cloud the relationship."

Our Alaska inquiry was answered by Kenneth Parker, Deputy Director of Commercial Fisheries, Alaska Department of Fish and Game. Parker provided information and data on the Bristol Bay sockeye salmon fishery (the largest of its kind in the world). With the use of a complex simulator, the research group at Bristol Bay prepares a tentative, preseason forecast of the upcoming run size of sockeye salmon. That information then goes to the Bristol Bay management group, and they use it, along with other assessment information, for preseason planning. But, as Parker indicated in his reply, the actual setting of a given season's regulations is more a strategy of plan-as-you-go, as the manager "depends more on other run indicators such as test-fishing results, escapement curves, and catch per unit effort in the commercial fleet." Although the preliminary forecasts of the research group seem to be based on accurate data and fairly sophisticated methods of simulation, Parker said that "the track record for the Bristol Bay forecasts has been variable with errors ranging from 12% to 200%. On the average, the preseason forecast has been within 40% of the actual run . . . at first inspection these errors may seem too great to be of any value to either the fishery manager or to the industry, and in some years this is true." (The forecasts exceeded the actual size in 11 of 19 years, by as much as 240%.) Parker did not know of any other managers of Alaskan fisheries who employed spawner–recruit models, and or own inquiries produced no additional leads to management applications of spawner–recruit models in any Alaskan fishery.

Our inquiry to W. E. Johnson, Director-General of Fisheries, Pacific Region, Canada, was answered by F. C. Withler, Acting Director of the Pacific Biological Station at Nanaimo. The Withler reply was especially pertinent to another citation of UT–59 (that of Shepard and Withler 1958, on the spawning stocks and production of salmon in the Skeena River system). Like many other scientific inquiries of its kind, the Shepard and Withler work is an assemblage and examination of historical data (in this case, data on spawners and associated returns), accompanied by the authors' attempts at discerning some natural pattern between the variables of interest. As the authors said, "The purpose of the present paper is to examine the available information in more detail [than past studies of the Skeena system] and to characterize the relationship, if any, between the abundance of spawners and the size of the resulting stock."

Like every citation of exhibit UT–59, as it turned out, the Shepard and Withler paper contained no claim of regulatory applications of spawner–recruit models. Despite some 30 years of study by the regulatory agency and various fishery scientists, the relationships between recruitment and the spawning stocks of the Skeena River system are not understood. The Skeena fisheries (18 major stocks and 20 or more minor stocks of sockeye salmon alone) are not managed in any way from spawner–recruit models, and the one attempt by the managers to employ information from a spawner–recruit relationship ended in failure. On the question of forecasting the future abundance of Hudson River fish stocks, Withler wrote in his reply: "In the matter of forecasting abundance, I doubt that a spawner–recruit curve would be very useful—certainly this was the case for the Skeena sockeye."

Because of the salmon commentary attributed to Donald Cushing, we also sent a letter of inquiry directly to Cushing, similar to the ones we directed to the other respondants, but also appealing to his broader experience with the world's fisheries. In his reply, Cushing was silent on salmon but cited two considerations of stock–recruitment relationships by management, both of which apply to herring stocks: "In answer to your question, the British Columbian herring stock was recovered by an intuitive use of a stock–recruitment relationship, rather than an explicit one. At the present time a stock–recruitment relationship for North Sea herring is being used explicitly to try to stage a recovery. Fishing is banned completely. . . . Beyond these two examples I know of no real use of such relationships."

Atlantic Fisheries

In exhibit UT–59, the utilities cited the International Commission for the Northwest Atlantic Fisheries (ICNAF) as managing groundfish, Atlantic herring, Atlantic cod, and shrimp with spawner–recruit models (including Ricker's). We directed our letter of inquiry to L. R. Day, Executive Secretary of ICNAF, who said in his reply that ". . . although I know of innumerable papers on the stock and recruitment problem, I know of

no management agency, including ICNAF, that has regulated a fishery from spawner–recruit curves. Nor do I think the problem has been resolved to the extent sufficient to reaching even short-term management strategies on current models of spawner–recruit theory."

In exhibit UT–59, management of the Atlantic menhaden fishery is attributed to the National Marine Fisheries Service (NMFS) and the Ricker curve as the management model. A paper by Nelson et al. (1977) on the dependence of menhaden stock strength on larval transport, is cited as the supporting reference, but the paper is a purely scientific work, and its authors did not suggest any anticipated application of the work to Atlantic menhaden management. Nevertheless, we sent off our standard inquiry to the Northeast Fisheries Center of NMFS at Woods Hole and received responses from Richard Hennemuth, Acting Center Director, and from Michael Sissenwine, Deputy Division Director of Resource Assessments. With respect to the regulation of Atlantic fisheries in general, Hennemuth wrote: "First off, outside of ICNAF, there has been very little direct control of catch or effort in the Atlantic. Thus, the application of any model is rare. In those cases where catch limits have actually been imposed, I know of no cases which were based on spawner–recruit curves." As to the possible management value of current spawner–recruit theories, Hennemuth said that "Your question on the value of S/R curves in management is rather more difficult. The data to which they are fitted seldom give any confidence that the result is meaningful."

The reply from Sissenwine contained a wealth of information on the management practices of the National Marine Fisheries Service, but on the question of the Atlantic menhaden fishery, Sissenwine replied: "There is no catch or effort management of the population at present [1979]." On the specific issue of spawner–recruit models, he added, "I know of no fish stocks of the Atlantic Ocean (or its estuaries) for which management is based on a spawner–recruit model."

Gulf of St. Lawrence

Two papers cited in exhibit UT–59 are reports on digital simulations of fisheries in the Gulf of St. Lawrence, namely Lett and Kohler (1976) on Atlantic herring, and Lett and Doubleday (1976) on Atlantic cod. Spawner–recruit models are cited in UT–59 as being the claimed models of management for those stocks, but again, neither reference contains any evidence that a spawner–recruit model has ever been employed in the regulation of either fishery. In regards to the simulation of the Atlantic herring fishery, the authors noted: "The uncertainty in the relationship between stock biomass and larval production, coupled with the variance in the year class size vs. larval relationship, make the scatter in the general Ricker stock–recruitment relationship no surprise. It indicates that recruitment is independent of stock size over a fairly wide range. Furthermore, the simulated results indicate that the only time any pattern in a stock diagram emerged was when the [simulated] fishery was collapsing."

We directed our letter of inquiry to Patrick Lett, then head of the modeling group of the Canadian Department of Fisheries and Oceans at Dartmouth, Nova Scotia, asking for a clarification on the intent of his papers, and also making the usual request for information on the possible uses of spawner–recruit models in fishery management. Lett's reply contained this passage: "Neither of the papers . . . were meant to be taken as expositions of the successful managing or regulating of a fishery with a spawner–recruit curve. Indeed, I believe they demonstrate just the opposite." Further, "The Department of Fisheries and Oceans has not managed any fishery, over any length of time, based upon the fit of a spawner–recruit curve." Lett concludes by saying that "No fishery manager in his right mind would manage a fishery based on a spawner–recruit relationship unless it was for fun, or had some experimental basis. A fish stock of commercial value would definitely not be managed in this fashion."

Pacific Oceanic Fisheries

Two items of exhibit UT–59, Thompson and Bell (1934) and Southward (1968), are worth mentioning because of their historical importance and their relationship to the management practices of the International Pacific Halibut Commission. Both references are cited in UT–59 as evidence of management by spawner–recruit models, but the Thompson and Bell paper appeared long before the first works on spawner–recruit theory. The work of Thompson and Bell was an early attempt at constructing a yield model for the Pacific halibut fishery. Unlike the spawner–recruit models in question, recruitment strength is not a density-dependent variable in the Thompson and Bell model, which is an attribute (or deficiency) shared by such formulations as the Ricker equilibrium yield model (Ricker 1975) and the Beverton–Holt

yield-per-recruit model (Beverton and Holt 1957), collectively designated (especially in the older literature) as dynamic pool models.

The Southward work on Pacific halibut was a mathematical simulation in which alternative strategies were pursued on a digital computer. Of the various regulatory strategies simulated by Southward, those constructed on spawner–recruit models were the least satisfactory. Southward introduced the Ricker spawner–recruit model into the simulation (his equation 7) and said, "It will be shown that unrealistic results are obtained when this relationship is used to estimate recruitment." Southward also tried and rejected the Beverton–Holt recruitment model, and noted that "over the range of observed stock size, the [actual] recruitment was constant" and provided no evidence of density-dependence in either recruitment or survival.

We directed our letter of inquiry to Bernard Skud, then Director of the International Pacific Halibut Commission, asking for information on the Commission's possible uses of spawner–recruit models. Skud replied: "The Commission has not based any of its management decisions on spawner–recruit models." Referring to the lack of a discernable relationship between spawner and recruitment abundances, Skud said that "On numerous occasions the Commission has used yield-per-recruit models, which is tantamount to assuming constant recruitment . . . That is not to say that the Commission believes recruitment is constant, but rather that annual variations in recruitment could not be related to the abundance of spawners or to any other stock parameter."

The Pacific yellowfin tuna fishery was also listed in exhibit UT–59 as one presumably managed by a stock–recruitment model; the Inter-American Tropical Tuna Commission (IATTC) was given as the management agency. Two papers by Schaefer (1957, 1967) were cited as supporting references. We directed our letter of inquiry to Robin Allen, Senior Scientist at the Commission. Allen replied: "My answer to your letter of 27 October is very brief. The IATTC has never used spawner–recruit models as a basis for management."

Fur Seals

The North Pacific Fur Seal Commission was listed in exhibit UT–59 as managing the fur seal *Callorhinus ursinus* herd by means of a "stock–recruitment" model, with Cushing (1974) as the supporting citation. But contrary to that supposition, all efforts at establishing a stock–recruitment (a parent–progeny) relationship for the Pribilof Islands fur seal herd have been unsuccessful, and the nature of the density-dependent control in the population is unknown. In 1956, for planning purposes only, the Commission tried to apply a parent–progeny model to the fur seal data, but the predictions proved to be so erroneous that further uses of such models were abandoned. A history of the management practices of the Fur Seal Commission, and the general failures of population models, has been given by Chapman (1964, 1979). Nevertheless, we sent a letter of inquiry to George Harry, the U.S. member of the Commission. Harry's reply contained this opening passage: "In your letter dated March 13 you asked whether the North Pacific Fur Seal Commission bases its management decisions on a formulated parent–progeny model. Management of the Pribilof Islands fur seal herd is not based on such models." With regard to the Commission's use of forecasting or planning models, Harry replied that "Dr. Douglas Chapman does make a forecast of the harvest each year, but even this is not used for management purposes."

Detailed discussions of the remaining items of exhibit UT–59 and complete copies of the correspondence excerpted here are contained in Fletcher and Deriso (1979).

The Population Renewal Problem

Irrespective of the pros and cons of the utilities' assessment methods, the foregoing examination of exhibit UT–59 points up the striking disparity between the activities of fishery scientists and the actual practices of fishery managers in dealing with population renewal, the single most important problem of fishery science. The utilities made the obvious mistake of not distinguishing between scientific inquiry and fishery regulation, which is not an uncommon mistake in matters of impact assessment. But why does renewal theory (or parent–progeny mechanics), which gets so much scientific press, get so little notice by fishery managers? For all our papers and learned discussions and data fittings over the 30 years since the work by Ricker (1954) on spawner–recruit theory, we seem to be little farther along with practical solutions than we were when Ricker started us out on the problem. We are still obliged to view every known spawner–recruit formulation as an unsubstantiated hypothesis; none has yet gained the authority we ascribe to the laws of nature. What are we missing?

The failures of spawner–recruit models are often attributed to the stochastic influences of the environment. Larkin (1978), for example, wrote: "Because of environmental vicissitudes, there is a wide gap between the theory of stock and recruitment and what seems to happen naturally. Whenever there are sufficient data to test theory, actuality departs significantly from the deterministic expectation."

But do the deterministic truths we think we are seeking merely lurk beneath the environmental noise? Are they that close to the surface? In Chapman (1973), following a theoretical examination of spawner–recruit formulations, we find this dispirited conclusion: "It would appear that if we pick a particular spawner-recruit model . . . to fit real data, we are merely picking a convenient empirical curve which may have, but probably does not have biological meaning."

Ricker (1975) made a similar admission: "Unfortunately, our knowledge of population regulatory mechanisms in nature is so slight that it is usually difficult to choose among different curves on [the basis of biological arguments], so we usually fit the simple curve that looks most reasonable."

Among the many postulated spawner–recruit formulations in the literature, two curve geometries have received the most attention, one the general dome-shaped curve of Ricker (1954) and the other the general asymptotic or first curve of Beverton and Holt (1957). In a discussion on the derivations of the two curves, Gulland (1974) observed: "The difficulty has been that because of the scatter of points, both types of curves could be fitted, and, in the case of the Ricker type, could be fitted with a wide range of alternatives."

Is the scatter of points the difficulty? If the scatter is real, that is where the truth must lie. Perhaps we are asking too much of determinism in the first place. Is density-dependent compensation really a *deterministic* law of nature? Why is it so easy to imagine but so hard to find? Probably the best theoretical argument favoring the deterministic persistence of populations is given in a splendid paper by Royama (1977). But often, exploited fish populations seem to wax and wane without apparent cause, and sometimes, with no signs of forewarning, they collapse to extinction or nearly so. Some recover after a removal of exploitation; others do not. That side of the argument is convincingly put forth in an equally valuable (but often overlooked) monograph by Reddingius (1971). Perhaps chance and the extrava-

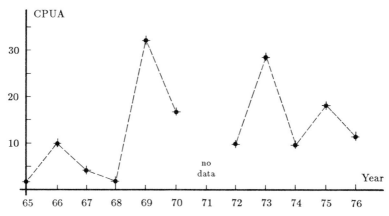

FIGURE 75.—Density indices (catch per unit area, CPUA) of young-of-the-year striped bass, Hudson River stock. (From testimonial transcript Tr 1092-1093, R. J. Klauda, 16 December 1977.)

gant waste by Nature of her progeny play a greater role in population renewal than we have been inclined to believe.

Fishery Regulation and Impact Assessment Contrasted

A fishery manager (who in reality is a manger of fishermen) must deal with animal populations in the wild, over which she seldom has any measure of direct control. As a consequence, much of commercial fishery management is management by guess and by error. Regulations are imposed and quotas are set on the basis of imperfect information and uncertain assessments, often in response to fluctuating catches, and sometimes with unexpected consequences.

In contrast to the variant behavior of natural stocks, the assessing of an industrial impact is often pursued in a steady-state (or equilibrium) setting, as it was by the utility companies in the Hudson River power plant case. But in view of the highly unsteady conditions of the estuarine environment, the expectation of finding equilibia in recruitment or stock abundances is vanishingly small. In the case of the striped bass stock, the utilities own measures of young-of-the-year abundances (Figure 75) give us some notion of the departures from equilibrium in recruitment. The abundance indices of Figure 75 do not demonstrate steady-state conditions, but apparent random fluctuations instead. Between 1968 and 1969, for example, the index of striped bass recruitment exhibited a 25-fold fluctuation in magnitude. Over the years for which the data are reported, the mean deviation of the abundance indices (7.95) is

60% of the mean of the indices (12.99), which is not unexpected in commercial fish stocks.

In addition to the extent that a steady-state assumption might invalidate an adapted fishery model, the *mode* of exploitation associated with a fishery differs from that imposed by a power plant. A fishery takes adults (or older juveniles); a power plant takes eggs, larvae, and very young juveniles for the most part. That difference did not escape the notice of the utilities. In support of a thesis defending the "ecological soundness" of killing juvenile fish, they aver (in Utilities Exhibit UT-4) that

Killing some fish during the egg, larval, and juvenile stages (which is what power plants do) is equivalent to killing the parents that would have produced these young (which is what fisheries do).

From that hypothesis, the utilities proceeded on the following corollary:

The entire foundation of fishery management experience and principles, therefore, can be applied with confidence to problems of power-plant impact.

As to the experience and principles connected with fishery management, what we have learned through *experience* in managing commercial fisheries we have learned largely through experience in harvesting postjuvenile fishes. Such knowledge does not extent with any confidence to the systematic killing of eggs, larvae, and juveniles. In cases of exploitation or industrial impacts where the very young suffer the heavier losses, we move into a little-explored area of stock assessment. There is at least one important and not well understood difference between killing the young and killing adults. The removal of premature young from an iteroparous stock eliminates their breeding opportunities altogether, which eventually reduces the long-term reproductive potential of the stock to a greater extent than removing post-reproductive adults. As to the *principles* of fishery management (the first of which is the preservation of the stock), a fishery manager usually has three basic privileges that he may exercise in controlling a fishery:

(1) through gear regulations, the manager can adjust the lower limit on the size (hence the age) of the exploited portion of the stock;

(2) through trial and error (or, more euphemistically, through adaptive management strategies), the manager can bring a stock into a general stage of abundance and reproduction where overexploitation remains unlikely;

(3) and, most commonly, a fishery manager can react to an unexpected stock decline through a quick (and selective) curtailment of exploitation.

These privileges are essentially closed to impact management.

Another Dispirited Conclusion

One obvious purpose of this paper has been the demonstration of our own conviction that the answer to long-term impact forecasting does not lie with routine (or even creative) uses of existing spawner–recruit models. We also hope we have shown that impact management scarcely resembles fishery management, as most of the regulatory options commonly exercised by fishery managers are not open to the regulators of power plant operations. At the very least, the techniques and methods of analysis that fail in fishery assessments are just as likely to fail in impact assessments.

We see no clear path to reliable, long-term predictions of stock behavior, whether the agent of mortality is a fishery or a power plant. But as decisions must be made and impact statements written, we conclude with a suggestion. Some notion of impact risks can be constructed, provided enough of the right kinds of data have been collected (which is seldom the case). Given such information as age-specific fecundities, percent of animals mature by age, natural adult mortality and fishing morality (if a fishery on the stock of interest exists), together with the estimated crop-

ping fraction of the young by the power plant, then a year-class model can be constructed for estimating such critical impacts as the probable losses from expected egg production and the probable loss of harvest from the fishery. This sort of analysis was employed by Deriso and Quinn (1983) for estimating the *current* loss in stock production owing to a change in exploitation of immature Pacific halibut.

What we have suggested in the foregoing paragraph is still frought with uncertainty. Environmental conditions have a severe effect on spawning success, and the mortalities in a population of year classes will vary from year to year (thus creating a suite of age- and year-specific probabilities in survival, which are usually impossible to separate). To do much more, or to expect long-term certitude from any assessment procedure is unreasonable, and the ultimate decision makers are owed such advice. One cannot, with any reliability, expect to forecast the consequences to future generations of current juvenile losses. Once again, we have run square against the renewal problem—the problem of reproduction and recruitment in natural populations. For the time being, it remains the central problem of impact assessment just as much as it remains the central problem of fishery science.

References[3]

Beverton, R. J. H., and S. J. Holt. 1957. On the dynamics of exploited fish populations. Fishery Investigations, Series II, Marine Fisheries, Great Britian Ministry of Agriculture Fisheries and Food 19.

Chapman, D. G. 1964. A critical study of the Pribilof fur seal population. U.S. Fish and Wildlife Service Fishery Bulletin 63:657–669.

Chapman, D. G. 1973. Spawner–recruit models and the estimation of the level of maximum sustainable catch. Rapports et Procès-Verbaux des Réunions, Conseil International pour l'Exploration de la Mer 164:325–332.

Chapman, D. G. 1979. Forecast of the kill of male seals, St. Paul Island. Report to the North Pacific Fur Seal Commission, Seattle, Washington.

Cushing, D. H. 1974. A link between science and management in fisheries. U.S. National Marine Fisheries Service Fishery Bulletin 72:859–864.

Deriso, R. B., and T. J. Quinn. 1983. The Pacific halibut resource and fishery in regulatory area 2: II. Estimates of biomass, surplus production, and reproductive value. International Pacific Halibut Commission Scientific Report 67.

Fletcher, R. I., and R. B. Deriso. 1979. Appraisal of

---

[3]See Table 1 for sources of legal documents and unpublished reports pertaining to the Hudson River.

certain arguments, analyses, forecasts, and precedents contained in the utilities' evidentiary studies on power plant insult to fish stocks of the Hudson River estuary, volume 1. Testamony (Exhibit EPA-218) before the U.S. Environmental Protection Agency, Region II. Adjudicatory Hearings Docket C/II-WP-77-01.

Gulland, J. A. 1974. Fishery science, and the problems of management. Pages 413–431 *in* F. H. Jones, editor. Sea fisheries research. Wiley, New York.

Larkin, P. A. 1978. Fisheries management: an essay for ecologists. Annual Review of Ecology and Systematics 9:57–73.

Lett, P. F., and W. G. Doubleday. 1976. The influence of fluctuations in recruitment on fisheries management strategy, with special reference to Gulf of St. Lawrence cod. International Commission for the Northwest Atlantic Fisheries Selected Papers 1: 171–192.

Lett, P. F., and A. C. Kohler. 1976. Recruitment: a problem of multispecies interaction and environmental perturbations, with special reference to Gulf of St. Lawrence herring. Journal of the Fisheries Research Board of Canada 33:1353–1371.

Nelson, W. R., M. C. Ingram, and W. E. Schaaf. 1977. Larval transport and year-class strength of Atlantic menhaden, *Brevoortia tyrannus*. U.S. National Marine Fisheries Service Fishery Bulletin 75: 23–41.

Reddingius, J. 1971. Gambling for existence. Bibliotheca Biotheoretica 12.

Ricker, W. E. 1954. Stock and recruitment. Journal of the Fisheries Research Board of Canada 11:559–623.

Ricker, W. E. 1975. Computation and interpretation of biological statistics of fish populations. Fisheries Research Board of Canada Bulletin 181.

Royama, T. 1977. Population persistence and density dependence. Ecological Monographs 47:1–35.

Schaefer, M. B. 1957. A study of the dynamics of the fishery for yellowfin tuna in the eastern tropical Pacific Ocean. Inter-American Tropical Tuna Commission Bulletin 2:247–268.

Schaefer, M. B. 1967. Fishery dynamics and the present status of the yellowfin tuna population of the eastern Pacific Ocean. Inter-American Tropical Tuna Commission Bulletin 12:87–136.

Shepard, M. P., and F. C. Withler. 1958. Spawning stock size and resultant production for Skeena sockeye. Journal of the Fisheries Research Board of Canada 15:1007–1025.

Southward, G. M. 1968. A simulation of management strategies in the Pacific halibut fishery. Report of the International Pacific Halibut Commission 47.

Thompson, W. F., and F. H. Bell. 1934. Biological statistics of the Pacific Halibut fishery. (2) Effects of changes in fishing intensity upon total yield and yield per unit of gear. International Fisheries Commission Report 8.

Appendix: Photoreproduction of Utilities Exhibit UT–59.

EXHIBIT UT *59*

Managerial Applications of Models

Model	Species	Agency	Reference
Ricker	Pacific salmon	Dept.Env. Canada	Cushing 1978
Schaefer	Whales	IWC	"
Schaefer; Ricker; Beverton & Holt	herring; groundfish cod; shrimp	ICNAF	Day 1978
Ricker	Atlantic menhaden	NMFS	Nelson, et al. 1977
Ricker	Atlantic herring	Dept. Env. Canada	Lett & Kohler 1976
complex stock recruit	cod	ICNAF	Lett & Doubleday 1976
Schaefer	Blue whales	IWC	Cushing 1974
stock-recruitment	Pacific fur seal	NPFSC	Cushing 1974
stock-recruitment	Alaska sockeye salmon	Alaska?	Dahlberg 1973
Ricker	Pink salmon		Cushing & Harris 1973
"	Red salmon		"
"	Chum salmon		"
"	Atlantic herring		"
"	Pacific herring		"
"	California sardine		"
"	halibut		"
"	Petrale Sole		"
"	Plaice		"
"	Cod		"
"	Haddock		"
Ricker	Chinook salmon		VanHyning 1973
stock-recruitment	Antarctic fin whales	IWC	Allen 1971

Appendix (continued).

-2-

Model	Species	Agency	Reference
Ricker	American plaice		Powles 1969
Ricker;Beverton & Holt	Pacific halibut	IPHC	Southword 1968
Schaefer	Pacific yellowfin tuna	IATTC	Schaefer 1967
Schaefer	Redfish	ICNAF	Parsons et al, 1966
stock-recruitment	Pacific sardine		Radovich 1962
Ricker	Sockeye salmon		Shepherd & Withler 1958
Schaefer	Pacific yellowfin tuna	IATTC	Schaefer 1957
Schaefer	Pacific halibut; California sardine		Schaefer 1954
logistic	Flatfish		Graham 1935
logistic	Pacific halibut		Thompson & Bell 1934

Appendix (continued).

References on Stock-Recruitment Models

Thompson, W.F. & F.H. Bell. 1934. Rep. Int. Fish Comm. 8, 49 pp.

Graham, M. 1935. Cons. Expl. Mer. 10: 264-274.

Nelson, W.R., M.C. Ingham, & W. E. Schaaf. 1977. Fish Bull.:
Vol. 75, No. 1, pp. 23-41.

Schaefer, M.B. 1954. Inter-Amer. Trop. Tuna Comm., Bull. 1(2): 27-56.

Schaefer, M.B. 1957. Inter-Amer. Trop. Tuna Comm., Bull. 2(6): 245-285.

Schaefer, M.B. 1967. Inter-Amer. Trop. Tuna Comm., Bull. 12(3): 87-136.

Powles, P.M. 1969. J. Fish. Res. Bd. Can. 26: 1205-1236.

Southward, G.M. 1968. Int. Pac. Halibut Comm. Rept. 47.

Allen, K.R. 1971. Annex 6. 21'st Ann. Rept., Internat. Whaling
Comm., pp. 58-63.

Radovich, J. 1962. Calif. Fish and Game 48: 123-140.

Lett, P.F. & A.C. Kohler. 1976. J. Fish. Res. Bd. Can. 33: 1353-1371.

Lett, P.F. & W.G. Doubleday. 1976. Int. Comm. Northwest Atl. Fish.
Sel. Pap. 1.1: 171-193.

Cushing, D.H. 1974. Fish. Bull. Vol. 72, No. 4, pp. 859-864.

Shepherd, M.P. & F.C. Withler. 1958. J. Fish. Res. Bd. Can 15: 1007-1025.

Dahlberg, M.L. 1973. Cons. Int. Expl. Mer. Rapp. 164: 98-105.

Cushing, D.H. & J.G.K. Harris. 1973. Rapp. P.V. Reun., Cons. Int.
Explor. Mer. 164: 142-155.

Van Hyning 1973. Rapp. P.V. Reun., Cons. Int. Explor. Mer 164:

Parsons, L.S., A.T. Pinhorn, and D.G. Parsons. 1966. ICNAF Res.
Bull. No. 12: 37-48

Cushing, D.H. 1978. Personal Communications to Texas Instruments.

Day, L.R. 1978. Personal Communications to Texas Instruments.

Comments[4]

This paper was fascinating. However, had I been a manager of a salmon fishery asked to respond to the quoted inquiry, my response could potentially have been colored by the wording of the second paragraph of the inquiry. This para-

---

[4]The Hudson River studies produced no agreement on the long-term impact of power plants on the river's fish fauna, in large part because the issue of biological compensation by the fish stocks for plant-induced mortality could not be resolved. The four contiguous papers in Section 3 by Savidge et al., Lawler, Christensen and Goodyear, and Fletcher and Deriso address these problems. Because this topic remains controversial, each of these authors was invited to comment on papers by the others. Not all chose to respond; the remarks of those who did are reproduced here.

graph made clear to me what answer was anticipated, but the substance of the answers by the respondents probably was not biased by such a perception. I am more optimistic than the authors that the renewal problem will be better understood, if not solved, for some species in some places. How much data will be required to accomplish this (and how practical it may be to collect such data more than a few times) is an open question indeed.

SIGURD W. CHRISTENSEN
Environmental Sciences Division
Oak Ridge National Laboratory
Post Office Box 2008
Oak Ridge, Tennessee 37831-6036, USA

American Fisheries Society Monograph 4:245–254, 1988

Implications of Power Plant Mortality for Management of the Hudson River Striped Bass Fishery

C. Phillip Goodyear[1]

U.S. Fish and Wildlife Service, Office of Biological Services, National Power Plant Team
Ann Arbor, Michigan 48105, USA

Abstract.—The Atlantic coastal stock of striped bass apparently declined from colonial times to the early 1930s and subsequently recovered. The reasons for the decline and recovery are not known, but fishing remains a possible explanation, which would suggest population sensitivity to increased mortality. Evidence suggests that fishing mortality has been increasing in recent years and will continue to increase in the absence of management intervention. The consequence of increased fishing mortality is an increase in the marginal effect of the power plant mortality which, based on the utilities' models and parameter fits, could result in important reductions in the Hudson River striped bass population. Any management actions imposed to arrest population decline or to increase yield per effort in the fishery would be required to mitigate the impact of the power plants by reducing fishing mortality. It is estimated that a 20% conditional power plant mortality is equivalent to a 14% increase in the number of average fishermen using the stock. Consequently, should any management intervention be required on behalf of the population, managers would be required to reduce fishing mortality by about 14% just to account for the power plant mortality.

Young striped bass are killed by entrainment and impingement at the Bowline Point, Indian Point, and Roseton power plants along the Hudson River. The obvious problem faced by those responsible for regulatory and management decisions is to interpret what these losses represent. Because of this problem, the scientists who were responsible for the analysis of the impact during the Hudson River power plant case (both for the government and for the utilities) were under considerable pressure to provide quantitative predictions of the effect of the power plant losses both on the future size of the Hudson River striped bass population and on the fishery associated with this population. Such forecasts were uncertain at best because of two primary difficulties.

The major problem with forecasting effects was (and remains) our inability to adequately describe, much less quantify, the specific mechanisms that regulate density-dependent survival of striped bass. Such mechanisms are termed compensatory and are in part responsible for the relationship that exists between parental stock and recruitment of progeny. Few, if any, would contest the general notion that compensatory phenomena exist in nature; the question is how much mortality a population can sustain before the compensatory responses are overwhelmed to the point that the

population would become economically or biologically extinct without management intervention.

The second difficulty lies in our limited knowledge of the historical fishing mortality associated with the Hudson River stock and our inability to predict accurately the levels of fishing mortality that the stock will experience during the next several decades. Accurate knowledge of fishing mortality is required because the response of the population to the power plant mortality depends on the combined effects of other existing sources of mortality on the population (Goodyear 1977, 1980a).

I address the problem in this paper by first presenting evidence that the Atlantic coastal striped bass resource, to which the Hudson stock contributes, has not been a stable one clearly unaffected by human population. I then provide evidence that fishing mortality has probably been increasing, estimate its 1977 level, and project future growth. I use the projected growth to illustrate the sensitivity of the utilities' small impact estimates for the Hudson River stock (McFadden and Lawler 1977) to the assumption of constant fishing mortality, and conclude by casting the power plant mortality in terms of an equivalent increase in fishermen.

The body of this paper is derived from analyses performed in 1978 and incorporated into testimony presented in May 1979. Some of the analyses proceeded on an assumption of increasing fishing mortality, which the population eventually

---

[1]Present address: National Marine Fisheries Service, Southeast Fisheries Center, Miami Laboratory, 75 Virginia Beach Drive, Miami, Florida 33149, USA.

would not be able to sustain. In retrospect, the Chesapeake stocks were the first of the Atlantic coastal stocks to decline. This event precipitated the current management intervention to reduce fishing mortality coastwide. Excessive fishing mortality is a leading candidate explanation for the decline in this stock, perhaps in concert with other sources of human-induced mortality such as environmental contaminants (USDOI and USDOC 1986). In addition to the management actions, the closures due to polychlorinated biphenyl (PCB) contamination have been extended beyond the Hudson River. These closures and accompanying efforts to minimize illegal sale of contaminated fish also have contributed to reduced fishing on the Hudson stock.

There is a direct analogy between power plant and other mortality agents in terms of their interactions with fishing to influence striped bass populations (Goodyear 1980a, 1983, 1985). The methods applied in this paper are equally applicable to the analysis of other sources of mortality when conditional survival probabilities can be estimated.

Trends in Abundance

The degree to which human activities have already depressed the Hudson River striped bass stock from its primitive (unexploited) level is an indicator of the potential importance of power plant mortality to the striped bass population. A stable population near its primitive level would argue for small impacts. In contrast, a severely depressed population would suggest sensitivity to exogenous mortality and argue for large impacts.

Tagging studies indicate that the major portion of the fishery for the Hudson stock of striped bass is not in the Hudson River itself but in the coastal waters from New Jersey to Massachusetts, where more than 70% of the fishermen recaptures of striped bass tagged in the Hudson have been made (Young 1976; McFadden and Lawler 1977). Given that the number of tags lost through tagging mortality and tag shedding increases with time, and that most of the fish recaptured by fishermen in the Hudson were taken shortly after the fish were released, the actual portion of the fishery for striped bass that occurs beyond the Hudson is likely to be higher than these data indicate. It also appears that the fishery that takes the Hudson stock in the coastal waters is largely a hook-and-line fishery, because 96–97% of the tag returns from these waters were from anglers.

These data indicate that the Hudson stock is exploited primarily in coastal waters from northern New Jersey northward, where it is mixed with southern stocks. Trends in exploitation rates in this area are important to both the Hudson and southern stocks. However, the landings reflect the relative magnitudes of the contributing stocks and (at least in recent years) are normally dominated by contributions from the Chesapeake Bay. Consequently, the early records of abundance in the area inhabited by the Hudson stock represent a combination of the Hudson and Chesapeake stocks. Thus, inferences on relative stability of stock size in this area only serve as a guideline to the sensitivity of the species to exogenous mortality.

Koo (1970) summarized historical information on the abundance of striped bass, which suggested a substantial depletion of the combined Atlantic coastal striped bass stocks between colonial times and the early 1930s. He attributed the reduction to overfishing, dam construction, and pollution. He also concluded that the stock was recovering from its depleted state in the early 1930s and was in a comparatively healthy state by the mid-1960s. It is possible that this increase in the stock was the result of management decisions that reduced fishing mortality. For example, Pearson (1938) attributed the threefold increase in Maryland striped bass landings between 1934 and 1935 to the prohibition of purse-seine fishing after 1932.

Size limits were also imposed on the fishery for striped bass in New York and New Jersey in 1939 (Higgins 1940), in Massachusetts in 1947 (Pfuderer et al. 1975), and in Connecticut in 1949 (Pfuderer et al. 1975). These size limits were imposed to increase yield per recruit based on Merriman's (1941) analysis. However, they also would have increased the reproductive potential of the recruits and would have favored population growth.

It is not possible to actually know the causes of the past depletion and subsequent recovery of the Atlantic coastal population of striped bass. However, one interpretation of the available information is that the stock declined in response to overfishing and subsequently recovered in response to management actions that reduced fishing mortality.

This interpretation, if valid, suggests population sensitivity to exogenous mortality, including that imposed by power plants. Increases in the prereproductive mortality rate from any source could have a catastrophic effect on the stock. The extent to which this conclusion may be applicable to the Hudson stock depends on both the validity of the interpretation and the extrapolation from the behavior of the mixed coastal stock to the Hudson itself. Both steps are uncertain. However, if each is valid, increased fishing mortality would lead to important reductions in the striped bass population, and the marginal impact of the power plant mortality could cause important reductions in the striped bass population.

Trends in Fishing

It is likely that the hook-and-line component of fishing mortality has increased in the recent past and, in the absence of management intervention, it will continue to increase in the foreseeable future. Alperin (1966) commented that the number of saltwater anglers in New York was increasing at a rapid pace during the period from 1955 to 1965. Using the same data, Schaefer (1968) projected that the number of saltwater anglers would continue to grow at an annual rate of increase of 6.7%. These and more recent data indicate a substantial increase in returns by anglers since 1936. The data in Table 76 are for striped bass that were either tagged in New York waters or in the Chesapeake Bay and recaptured along the Atlantic coast outside the Chesapeake. For both groups, most of the recaptures were made from New Jersey northward, which is the area frequented by the Hudson stock. The data indicate an increasing fraction of fishing mortality due to angling over the period 1936–1976. Although, as Schaefer (1968) noted, part of the shift likely has been due to a decline in commercial fishing effort,

the increase in numbers of anglers reported by the U.S. Department of Commerce (USDOC 1976) indicates that a part of the increased proportion of tags taken by the hook-and-line fishery has been due to increased numbers of anglers.

The comparatively low fraction (55%) of hook-and-line recaptures in the data presented by Young (1976) for striped bass tagged at Long Island in the years 1973–1975 (Table 76) appears to conflict with the high proportion of hook-and-line recaptures (96–97%) reported by McFadden and Lawler (1977) and B. M. Florence (Maryland Department of Natural Resources, personal communication). However, the fish that were tagged in the study reported by Young were purchased from commercial haul seiners and were released in the same area. Furthermore, for economic reasons, only small striped bass were purchased. Because smaller fish tend to be recaptured near the release areas (McFadden and Lawler 1977), it is not surprising that a higher proportion of returns came from the commercial fishery than was indicated in the data reported by McFadden and Lawler and by Florence.

In comparing the percentage of returns from anglers and commercial net fishermen to obtain the distribution of fishing mortality between these two groups, one must assume both groups are returning the tags with equal frequency. However, unless the return rates for each category of fishermen have changed over the years, the rela-

TABLE 76.—Recaptures of striped bass by anglers and commercial net fishermen. Only striped bass tagged in New York waters or tagged in the Chesapeake Bay and recaptured in Atlantic coastal waters are included.

Year of tagging	Location of tagging	Number of recaptures	% recaptured by		Comments	Source
			Commercial fishermen	Anglers		
1936–1938	Chesapeake Bay	31	83	17		Valdykov and Wallace (1938)
1954–1956	Westhampton Beach, New York	50	68	32		Schaefer (1968)
1956–1961	Great South Bay, New York	242	25	75		Alperin (1966)
1961	Great South Beach, New York	27	44	56	<600 mm	Schaefer (1968)
1961–1963	Great South Beach, New York	62	32	68	≥600 mm	Schaefer (1968)
1972–1973	Chesapeake Bay	112	4	96	≥5.5 kg	Florence (personal communication)
1972–1974	Hudson River	38	11	89	All recaptures	McFadden and Lawler (1977)
		28	4	96	Coastal recaptures	McFadden and Lawler (1977)
1973–1975	Amagansett, New York	279	45	55		Young (1976)
1976	Hudson River	146	16	84	All recaptures	Texas Instruments (unpublished)
		108	3	97	Coastal recaptures	Texas Instruments (unpublished)

tive trends in returns from each group should still accurately reflect changes in the pattern of exploitation between the two groups. Adequate data to examine the historical trends in frequency of returns for the two groups probably do not exist. However, the results presented in Table 76 agree with data presented by Merriner (1976), which indicate that angling in the 1970s accounted for more than 80% of the coastwide annual landings of striped bass.

Projected Fishing Mortality

Growth in Effort

The preceding discussion indicates that the mortality of striped bass attributable to the hook-and-line fishery has been increasing. If past indications are a guide, the mortality due to anglers may be expected to continue to increase. At some time, this increasing effort will cause a stock decline and force management measures to either halt the increase or to reduce the level of fishing.

Accurate forecasts of the future growth in hook-and-line fishing effort probably are not possible. However, for the purpose of this analysis, I will employ the U.S. Department of Commerce estimates (USDOC 1976) on growth in the marine recreational fishery. These data indicate a linear increase in the number of fishermen participating, as opposed to a compounded rate of increase. The rate of increase in 1955 was about 7% but had dropped to about 3% by 1970. If the same trend continues, the rate of increase will drop to about 1.7% by the year 2000. According to these data, a 68% increase in the number of anglers can be expected between 1976 and 2000. An associated increase in fishing mortality will make the Hudson stock more sensitive to the mortalities of young at the Hudson River power plants. The actual increase in the number of fishermen and associated change in fishing mortality is, of course, unknown and may be higher or lower than indicated by the Department of Commerce data.

Estimate of Existing Fishing Mortality Rates

If the instantaneous rates of fishing mortality and tag loss are each relatively constant during the several months following the application of the tags, it is possible to estimate the fishing mortality rate from the tag return data by the method outlined by Ricker (1975). Let

$$Z' = F + M + U; \qquad (1)$$

Z' = instantaneous rate of disappearance of tags from the marked population;
F = instantaneous rate of fishing mortality;
M = instantaneous rate of natural mortality;
U = instantaneous rate of loss of tags not attributable to fishing or natural mortality.

Further, let

$$A' = (1 - S') = (1 - e^{-Z'}); \qquad (2)$$

A' = fraction of marks that disappear;
S' = $1 - A'$ = "survival" probability of tags.

The apparent exploitation fraction, u', during a time i is then the fraction of tagged fish recovered with tags still attached and can be estimated with equation (4.4) of Ricker (1975):

$$u' = \frac{R_1 + R_2 + \ldots + R_n}{Y[1 + S' + (S')^2 + \ldots + (S')^{n-1}]}; \qquad (3)$$

R_i = tag returns during interval i;
Y = initial number of marks.

The fraction of tags actually reported was divided by 0.62 to correct for nonreporting of recaptures (Chadwick 1968). The extent of tagging mortality was not investigated during the study and was assumed to be nonexistent in the utilities' calculations. Clearly, the assumption of no tagging mortality underestimates the actual loss. For this reason, I used both 0 and 20% reductions of the numbers tagged to adjust the initial numbers of marks at large (Y in equation 3) for mortalities resulting from the tagging experience.

The data employed in the estimates were obtained from the utilities in the form of a listing of a computer file that contained all of the tag and recapture data for their marking experiments with older striped bass. These studies were described by McLaren et al. (1981). The analysis was restricted to legal-sized fish (i.e., 434 mm total length or longer) and only recaptures by fishermen (not research teams) were included in the analysis. The recaptures were grouped into 30-d intervals (Table 77). The 30-d apparent survival (S') for this data set is 0.74, which corresponds to a Z' of 0.30 and an A' of 0.26. Substitution of these values into equation (3), with a correction for nonresponses, for assumed tagging mortalities of 0 and 20% provides estimates of u' (30 d) of 0.033 and 0.041, respectively. The corresponding values of instantaneous fishing mortality can be calculated from

$$F = \frac{u'Z'}{A'}; \qquad (4)$$

and are 0.038 and 0.048 respectively, where F is

TABLE 77.—Summary of recaptures by commercial and recreational fishermen of striped bass tagged in the Hudson River, by time at large.

Days at large	Number of recaptures
0–29	33
30–59	22
60–89	24
90–119	16
120–149	7
150–179	8
180–209	6
210–329	3

defined by a 30-d interval. Landings of striped bass north of New Jersey are much reduced during the winter (Koo 1970). Consequently, these rates are assumed to apply for 9 monthly intervals (270 d) during the year, which provides estimates of the conditional annual fishing mortality between 0.29 ($F = 0.34$) and 0.35 ($F = 0.43$) or approximately 32%. If the natural survival for the Hudson stock of striped bass is in the same range as that reported by Miller (1974) for the California stock (i.e., about 0.8), the total annual survival is approximately 0.54. This value of total survival is slightly lower than the value of 60% reported by Hoff et al. (1988, this volume).

It should be kept in mind that the fishing mortality rates derived in the preceding paragraphs are based on data collected when the striped bass fishery in the Hudson was closed because of PCB pollution. The fishing mortality rate might have been greater if the commercial fishery had been open. This may be an important consideration if the commercial landings in the Hudson represent an important part of the total exploitation of the stock, and fishing mortality actually declined as the result of the closure. Striped bass will continue to be taken as a bycatch in the Hudson fishery for American shad; moreover, the highest reported commercial landings in the Hudson occurred in 1976, the first year of the PCB closure. Consequently, it is assumed herein that the closure did not significantly change fishing mortality.

Projected Fishing Mortality

The effect of an increasing hook-and-line fishery is likely to be an increase in the total fishing mortality regardless of the trend in commercial fishing. The total instantaneous fishing mortality rate, F, is the sum of that due to commercial net fishermen (F_c) and to anglers (F_a). Specifically,

$$F = F_c + F_a. \qquad (5)$$

If we assume that the commercial net-fishing effort will remain constant and there will be no change in catchability, the value of F_c will also remain constant in the future. The conditional fishing mortality rate that was previously derived (0.32) corresponds to a calculated value for F of 0.389. Because 16% of the tag returns from the utilities' 1976 tagging experiment were from the commercial net fishery, the value of F_c would be $0.16 \cdot F$ or approximately 0.062. Similarly, the value of the instantaneous rate of angling mortality (F_a) would be $0.84 \cdot F$ or approximately 0.327. The value of F_a is also given by

$$F_a = q_a E_a; \qquad (6)$$

q_a = catchability coefficient for the hook-and-line fishery;
E_a = effort of the hook-and-line fishery.

If the trends in hook-and-line fishing effort reported by the USDOC (1976) hold for the remainder of this century, by the year 2000 the number of anglers will increase about 68% over the 1976 level. If catchability remains constant, the instantaneous fishing mortality due to the hook-and-line fishery in the year 2000 would be 1.68 times the 1976 mortality rate, or about 0.55. The total rate of fishing mortality would be about 0.61.

Constant catchability means that the fraction of the available stock captured by the average fisherman (a very small number) will not change over the range of stock sizes that will exist during the

TABLE 78.—Life history data used in the simulations of the impact of power plant mortality on the size of the Hudson River population of striped bass. The fecundity and maturity data were taken from McFadden and Lawler (1977).

Age	X[a]	R[b]	L[c]	v[d]
1	0	0	0.5	0
2	0	0	0.5	0.2
3	658	0.04	0.5	0.5
4	658	0.07	0.5	1.0
5	578	0.19	0.5	1.0
6	714	0.43	0.5	1.0
7	928	0.86	0.5	1.0
8	1,310	0.89	0.5	1.0
9	1,570	1.00	0.5	1.0
10	1,760	1.00	0.5	1.0
11	1,980	1.00	0.5	1.0
12	2,090	1.00	0.5	1.0
13	2,130	1.00	0.5	1.0
14	2,190	1.00	0.5	1.0
15–25	2,590	1.00	0.5	1.0

[a] X = age-specific fecundity (thousands of eggs).
[b] R = fraction of females that are mature.
[c] L = age-specific fraction that are female.
[d] v = vulnerability to the fishery.

forecast period. However, if effort continues to grow, catch per effort and total catch will decline, even without the losses due to the power plants. A consequence of this observation is that management intervention is likely sometime during the period to arrest declining catch rates. Any effective management action will reduce the fraction of the stock killed by the average fisherman, and thereby violate the assumption of constant catchability. Because of this possibility, assumptions were made for subsequent analyses that catchability would be abruptly decreased in either 1980 or 1990 by management actions that halt the increase in fishing mortality. (Substantial restrictive changes in fishing regulations for striped bass in this area actually were imposed beginning in the mid-1980s: USDOI and USDOC 1986.)

Effect of Growth in Fishing on the Robustness of the Utilities' Impact Estimates

The anticipated growth in angling will reduce the capacity of the stock to sustain additional mortality beyond the level that may have existed before power plant operations began. Thus, impact estimates based on stock–recruitment relationships fitted to landings data might substantially underestimate the actual plant impact, even if the selected model was appropriate and the

parameters were well estimated. Difficulties with the parameter estimates are discussed elsewhere (Christensen and Goodyear 1988, this volume).

The effect of the anticipated growth in the fishery on the robustness of the utilities' estimates of impact was examined by simulation. I incorporated the Ricker stock–recruitment relation in a simulation model (Goodyear 1980b) with estimates of Ricker's alpha in the range 3–5 to bracket the utilities' estimates based on landings data from 1955–1975. Population size was simulated with the life history data in Table 78. Natural mortality was set at 0.2 and fishing mortality at 0.25 in 1965. A linear increase was assumed for hook-and-line fishing effort from 1965 to either 1980 or 1990 (when management measures were assumed to halt the increase) and commercial fishing effort was assumed to be constant. The value of Ricker's alpha was set at 3 (Figure 76), 4 (Figure 77), or 5 (Figure 78) for the year 1965 by the method of Goodyear (1980b).

The population was initialized with the stable age distribution and trajectories were computed for the conditions of no power plant mortality and a 15% and 25% conditional power plant mortality. The power plant mortality was assumed to begin in 1975 and to continue until 2015.

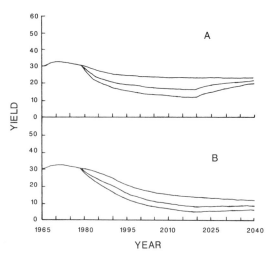

FIGURE 76.—Simulated yields (in arbitrary units) of the Hudson River striped bass population with and without power plant mortality of young fish and an assumed Ricker function regulating the survival of young with an alpha of 3.0. The angling effort is assumed to increase at estimated historical rates until 1980 (A) or 1990 (B) and remain constant thereafter. In each panel, the upper, middle, and lower curves represent power plant mortalities of 0%, 15%, and 25%, respectively.

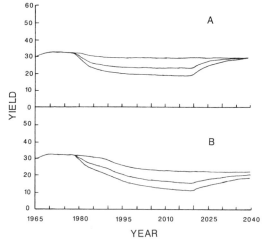

FIGURE 77.—Simulated yields (in arbitrary units) of the Hudson River striped bass population with and without power plant mortality of young fish and an assumed Ricker function regulating the survival of young with an alpha of 4.0. The angling effort is assumed to increase at estimated historical rates until 1980 (A) or 1990 (B) and remain constant thereafter. In each panel, the upper, middle, and lower curves represent power plant mortalities of 0%, 15%, and 25%, respectively.

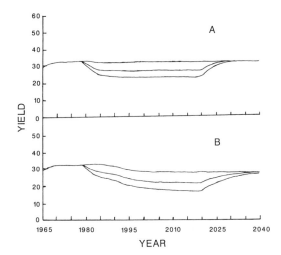

TABLE 79.—Calculated percentage reductions in Hudson River striped bass yields resulting from a conditional power plant mortality *(m)* of 15% or 25% for several assumed values of Ricker's alpha and a linear increase in hook-and-line effort from 1965 to either 1980 or 1990, after which fishing mortality remains constant. Corresponding population trajectories are presented in Figures 76–78.

Alpha (1965)	Last year of increasing fishing	Calculated percentage reduction	
		$m = 0.15$	$m = 0.25$
3	1980	29.2	48.7
3	1990	40.0	61.7
4	1980	20.5	35.8
4	1990	30.0	49.7
5	1980	16.0	28.4
5	1990	23.1	39.9

FIGURE 78.—Simulated yields (in arbitrary units) of the Hudson River striped bass population with and without power plant mortality of young fish and an assumed Ricker function regulating the survival of young with an alpha of 5.0. The angling effort is assumed to increase at estimated historical rates until 1980 (A) or 1990 (B) and remain constant thereafter. In each panel, the upper, middle, and lower curves represent power plant mortalities of 0%, 15%, and 25%, respectively.

The percentage reductions in yield attributable to the power plant mortality in the year 2015 are presented in Table 79. The simulated reductions by 2015 range from 16 to 40% for a conditional power plant mortality of 15%, and from 28 to 62% for a conditional power plant mortality of 25%. Thus, even if the utilities consultants' appeal for the Ricker model and their parameter estimates are accepted, it is apparent that serious stock reductions attributable to the power plants could occur during the plants' anticipated operational lifetimes, depending on what occurs in the fishery.

Because no accurate data exist for the past and none exist for the future, accurate forecasts of plant impact would be impossible even if the Ricker model were appropriate and its parameter values had been well estimated. Of the many life history assumptions used to illustrate this point, the only one required is that fishing mortality does not remain constant through time.

Power Plant Impacts on Management Options

An alternative to attempting to forecast the impact of power plant mortality on the population size or its associated yield is to evaluate the impact in terms related to the opportunity for fishermen to use the resource. The goal here is to reduce the impacts of power plant and fishing mortality to a common scale. The power plant impact can then be cast in terms of a comparable increase in the total number of (average) fishermen participating in the fishery.

The analysis is based on the compensation ratio, CR (Goodyear 1977, 1980a), which is a measure of the overall change in the survival and fecundity parameters that must occur for a population undergoing exploitation to stabilize at a new equilibrium. It is defined as

$$CR = \frac{v_e}{v_0}; \qquad (7)$$

v_e = viability index of the exploited stock;
v_0 = viability index of the virgin stock.

The viability index (v) represents all of the compensatory density-dependent factors that operate throughout the life history of a population, given that the population is able to maintain, or fluctuate about, an equilibrium. It is given by

$$v = \frac{1}{Ps_0}; \qquad (8)$$

P = potential fecundity per recruit;
s_0 = probability of survival from density-independent sources of mortality between the deposition of eggs and recruitment.

It is assumed that an individual is recruited to the progeny population at the time it becomes 1 year of age. Thus, the power plant mortality resulting from entrainment and first-year impingement is included in the parameter s_0 such that

$$s_0 = s_n s_p; \qquad (9)$$

s_n = survival from natural causes of mortality for age-class 0;
s_p = survival from power plant mortality for age-class 0.

Potential fecundity per recruit (P) is the average lifetime production of eggs per recruit at equilibrium population densities, plus those eggs that would have been produced under conditions of optimum growth and natural mortality. It is determined as

$$P = \sum_{i=1}^{n} X_i R_i L_i \prod_{j=0}^{i-1} S_j; \qquad (10)$$

S_j = density-independent annual survival probabilities by age;
X_i = maximum mean fecundity of mature fe males at age i;
R_i = maximum fraction of age-i females that are mature;
L_i = fraction of age-class i that is female;
n = number of age-classes in the population.

The parameter values needed to solve for P were taken from Table 78 under the assumption that existing exploitation was sufficient to fully release any potential density-dependent growth and survival of adults that might occur. Specifically, no further reduction in abundance would cause the recruits to be subject to less natural mortality, to be more fecund upon maturity, or to mature earlier.

The probability for a recruit to survive to age i is simply the product of the annual total survival probabilities to which the recruit is exposed prior to entering the ith age-class. The total survival probabilities are higher in the case of an unexploited population than for an exploited population. Thus, the value of P is higher for the unexploited population, and the corresponding viability index, v, is lower for the unexploited population (v_0) than for the exploited population (v_e). The relationship simply reflects that the probability of an egg surviving to produce a recruit egg has to be higher in the exploited population if the stock is not tending toward extinction.

Substitution of equations (8) and (9) into equation (7), under the assumption that the density-independent natural survival rates for the exploited and unexploited populations are the same, yields

$$CR = \frac{P_0}{P_e s_p}; \qquad (11)$$

P_0 = potential recruit fecundity in the absence of fishing mortality;
P_e = potential recruit fecundity in the exploited stock;
s_p = power plant survival probability.

The estimated values of CR equate the impact of fishing and power plant mortality based on their impacts on the intrinsic growth rate of the population (Goodyear 1980a). The difference between CR determined for fishing alone and that determined for fishing and power plant mortality represents the marginal impact of the power plant (Figure 79).

Several consequences of the power plant mortality can be deduced from these curves. Stock size and catch per effort are inversely related to CR (Goodyear 1980a). Thus, if fishing remains constant, the imposition of power plant mortality will decrease the stock size and the catch per unit effort, thereby denying the opportunity for fishermen to take striped bass. Furthermore, the imposition of power plant mortality reduces the threshold above which fishing mortality will induce undesirable reductions in abundance (Goodyear 1977, 1980a).

Against a background of gradually increasing fishing mortality, as seems to be the case for striped bass, the addition of power plant mortality would favor premature population collapse. Management intervention to arrest a declining stock (or declining catch per unit effort) would attempt

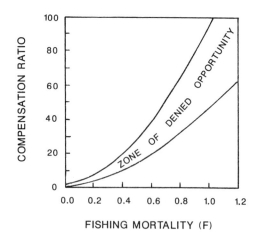

FIGURE 79.—Plot of the compensation ratio as a function of fishing mortality with (upper curve) and without (lower curve) a conditional power plant mortality of 0.5.

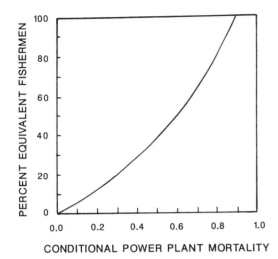

CONDITIONAL POWER PLANT MORTALITY

FIGURE 80.—Relationship between the conditional power plant mortality rate of juvenile striped bass and the equivalent percentage increase in the number of striped bass fishermen of average efficiency.

to restore population growth by reducing fishing mortality, and the required reduction would be greater by the change caused by the power plant mortality. It is perhaps noteworthy that it is very unlikely that managers could distinguish the influence of the power plants, but the regulations they would be required to impose would have to offset the power plant losses.

Goodyear (1980a) termed the area between such curves (Figure 79) the "zone of denied opportunity" because the difference between the two curves represents the substitution of power plant mortality for fishing mortality. For any fishery management strategy based on maintenance of adequate spawning stock, the opportunity for fishermen to take striped bass is reduced by the difference between the two curves. This reduction can be accomplished either by reducing the number of fishermen or by reducing their ability to catch fish (e.g., with length limits). In either case, fishermen must be denied some of their opportunity to take striped bass because of the power plant mortality.

These observations suggest that one useful way to place the power plant mortality into a management perspective is to state the impact in terms of an equivalent fractional increase in fishermen of average efficiency. A plot of the equivalent percentage change in striped bass fishermen as a function of the conditional power plant mortality rate is presented in Figure 80. It is based on an assumed instantaneous fishing mortality rate of

0.389 and an assumed natural mortality rate of 0.2. Under these assumptions, a total conditional power plant mortality of 0.2 (Boreman and Goodyear 1988, this volume) is equivalent to an increase of 14% in the number of striped bass sport and commercial fishermen of average efficiency. In a management setting, this would require a 14% reduction in allowable fishing mortality just to offset the power plant losses.

It is not presently possible to separately manage the Hudson stock of striped bass because it mixes with southern stocks in coastal waters (Goodyear 1978). Any management scheme based on reduced fishing to mitigate the power plant mortality suffered by the Hudson stock would require similar reductions in fishing for the stocks from other spawning grounds that frequent the area used by Hudson fish during their summer migrations. As a consequence, the required reduction in effort would result in a greater loss in yield to fishermen than would be attributable to the Hudson stock alone.

An important assumption in the foregoing analysis is that the power plant mortality would affect the population in the same manner as fishing mortality. It must be remembered that the entire aquatic community in the nursery area for striped bass in the Hudson may be modified by the power plants. Neither the magnitude nor the ultimate consequence of any possible perturbation is known. The importance of this observation is that the entire nature of the population regulatory process may be altered, thereby distorting the adequacy of the comparison between power plant and fishing mortality.

References[2]

Alperin, I. M. 1966. Dispersal, migration and origins of striped bass from Great South Bay, Long Island. New York Fish and Game Journal 13:79–112.

Boreman, J., and C. P. Goodyear. 1988. Estimates of entrainment mortality for striped bass and other fish species inhabiting the Hudson River estuary. American Fisheries Society Monograph 4:152–160.

Chadwick, H. K. 1968. Mortality rates in the California striped bass population. California Fish and Game 54:228–246.

Christensen, S. W., and C. P. Goodyear. 1988. Testing the validity of stock–recruitment curve fits. American Fisheries Society Monograph 4:219–231.

Goodyear, C. P. 1977. Assessing the impact of power plant mortality on the compensatory reserve of fish populations. Pages 186–195 in W. Van Winkle,

[2]See Table 1 for sources of legal documents and unpublished reports pertaining to the Hudson River.

editor. Assessing the effects of power-plant-induced mortality on fish populations. Pergamon, New York.

Goodyear, C. P. 1978. Management problems of migratory stocks of striped bass. Marine Recreational Fisheries 2:75–94.

Goodyear, C. P. 1980a. Compensation in fish populations. Pages 253–280 in C. H. Hocutt and J. R. Stauffer, Jr., editors. Biological monitoring of fish. Lexington Books, Lexington, Massachusetts.

Goodyear, C. P. 1980b. Oscillatory behavior of a striped bass population model controlled by a Ricker function. Transactions of the American Fisheries Society 109:511–516.

Goodyear, C. P. 1983. Measuring effects of contaminant stress on fish populations. American Society for Testing and Materials Special Technical Publication 802:414–424.

Goodyear, C. P. 1985. Toxic materials, fishing, and environmental variation: simulated effects on striped bass population trends. Transactions of the American Fisheries Society 114:107–113.

Higgins, E. 1940. Progress in biological inquiries 1939. U.S. Bureau of Fisheries, Administrative Report 39, Washington, D.C.

Hoff, T. B., J. B. McLaren, and J. C. Cooper. 1988. Stock characteristics of Hudson River striped bass. American Fisheries Society Monograph 4:59–68.

Koo, T. S. Y. 1970. The striped bass fishery in the Atlantic states. Chesapeake Science 11:73–93.

McFadden, J. T., and J. P. Lawler, editors. 1977. Influence of Indian Point unit 2 and other steam electric generating plants on the Hudson River estuary, with emphasis on striped bass and other fish populations, supplement 1. Report to Consolidated Edison Company of New York.

McLaren, J. B., J. C. Cooper, T. B. Hoff, and V. Lander. 1981. Movements of Hudson River striped bass. Transactions of the American Fisheries Society 110:158–167.

Merriman, D. 1941. Studies of the striped bass (*Roccus saxatilis*) of the Atlantic coast. U.S. Fish and Wildlife Service Fishery Bulletin 50(35).

Merriner, J. V. 1976. Differences in management of marine recreational fisheries. Pages 123–131 in H. Clepper, editor. Marine recreational fisheries. Sport Fishing Institute, Washington, D.C.

Miller, L. W. 1974. Mortality rates for California striped bass (*Morone saxatilis*) from 1965–1971. California Fish and Game 60:157–171.

Pearson, J. C. 1938. The life history of the striped bass, or rockfish, *Roccus saxatilis* (Walbaum). U.S. Bureau of Fisheries Bulletin 49(28):825–851.

Pfuderer, H. A., S. S. Talmage, B. N. Collier, W. Van Winkle, Jr., and C. P. Goodyear. 1975. Striped bass—a selected, annotated bibliography. Oak Ridge National Laboratory, ORNL–EIS 75-73, Oak Ridge, Tennessee.

Ricker, W. E. 1975. Computation and interpretation of biological statistics of fish populations. Fisheries Research Board of Canada Bulletin 191.

Schaefer, R. H. 1968. Size, age composition and migration of striped bass from the Long Island surf. New York Fish and Game Journal 15:117–118.

USDOC (U.S. Department of Commerce). 1976. A marine fisheries program for the nation. U.S. National Marine Fisheries Service, Washington, D.C.

USDOI (U.S. Department of the Interior) and USDOC (U.S. Department of Commerce). 1986. Emergency striped bass study: report for 1985. U.S. Fish and Wildlife Service, Washington, D.C.

Valdykov, V. D., and D. H. Wallace. 1938. Is the striped bass (*Roccus lineatus*) of Chesapeake Bay a migratory fish? Transactions of the American Fisheries Society 67:67–86.

Young, B. H. 1976. A study of striped bass in the marine district of New York. New York Department of Environmental Conservation, Division of Marine Resources, Completion Report AFC-8, Albany.

American Fisheries Society Monograph 4:255–266, 1988

Relative Contributions of Hudson River and Chesapeake Bay Striped Bass Stocks to the Atlantic Coastal Population

W. Van Winkle, K. D. Kumar,[1] and D. S. Vaughan[2]

Environmental Sciences Division, Oak Ridge National Laboratory
Post Office Box 2008, Oak Ridge, Tennessee 37831-6038, USA

Abstract.—Fourteen variables derived from 13 morphological characters were used in a stepwise discriminant analysis and a maximum-likelihood analysis to estimate the relative contributions of striped bass stocks from the Hudson River and Chesapeake Bay to the Atlantic coastal striped bass population. The analyses made use of the data collected by Texas Instruments in 1975, and were designed to focus on relative contributions by sex and year class to populations north of Chesapeake Bay and north of the Hudson River. The discriminant function method misclassified approximately 20% of the fish sampled on the spawning grounds. When applied to the data set for fish collected in the ocean, the two methods of analysis resulted in estimates of relative contribution of the Hudson stock to the Atlantic coastal population that varied considerably among year classes. In particular, the estimated relative contribution for the 1965 year class was between 40 and 50%, whereas the relative contributions for the 1966, 1968, and 1969 year classes were 10% or less.

One of the major issues in the Nuclear Regulatory Commission (NRC) licensing hearings for operation of Indian Point units 2 and 3 was the relative stock composition of the Atlantic coastal striped bass population. If Hudson River fish were major contributors to the coastal population, and if entrainment and impingement mortality of young-of-the-year striped bass at power plants along the Hudson were high, the Indian Point facility might contribute substantially to a reduction in the commercial and sport fisheries for striped bass along the Atlantic coast. In response to this concern, Consolidated Edison Company of New York funded a study by Texas Instruments (Grove et al. 1976; Berggren and Lieberman 1978).

Conventional wisdom, based on indirect evidence from commercial fisheries catch data and recoveries of tagged fish, has been that 90% or so of the legal-sized striped bass in the population migrating up and down the Atlantic coast from Maine to Cape Hatteras, North Carolina, comes from stocks spawning in the Chesapeake Bay and its tributaries. This wisdom was supported by the results of the Texas Instruments study, in which Berggren and Lieberman (1978) used observations of morphological characters and discriminant analysis to quantitatively estimate the relative contribution of striped bass stocks from various estuaries to the striped bass fishery along the Atlantic coast. These authors estimated relative contributions of 91% from the Chesapeake Bay, 6% from the Hudson River, and 3% from the Roanoke River, North Carolina. These estimates were based on a sample of 2,471 ocean fish that included males and females of all ages from all 10 geographical strata and all 6 temporal strata defined in that study.

Our hypothesis was that the relative contributions are likely to vary substantially from year class to year class. We recognized the problems with limited sample sizes for certain year classes, but still believed that the range of year-class variation in relative contribution could be estimated from the original Texas Instruments data set. In 1977, we obtained on tape the complete data set for this Texas Instruments study. We repeated the Texas Instruments analysis and obtained identical results. Then we developed and applied alternative methods of analysis and focused on estimating relative contribution by sex and year class. We argue that the time is propitious to repeat this study.

Methods

Spawning-stock data.—The collection and processing of spawning-stock specimens was described by Berggren and Lieberman (1978). Briefly, mature striped bass were collected from the natal rivers (Figure 81) of major stocks along

---

[1]Present address: SAS Institute Incorporated, Box 8000, Cary, North Carolina 27511, USA.

[2]Present address: National Marine Fisheries Service, Southeast Fisheries Center, Beaufort Laboratory, Beaufort, North Carolina 28516, USA.

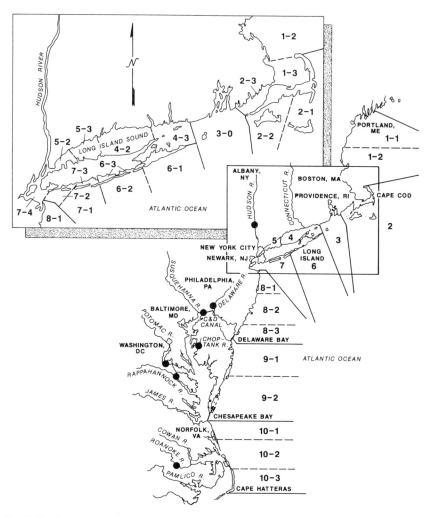

FIGURE 81.—Collection regions for the Atlantic coastal populations of striped bass, showing geographical stratification and substratification. Collection sites for spawning-stock specimens are indicated by dots on source rivers (Berggren and Lieberman 1978).

the Atlantic coast during the spawning season of 1975. Sampling was designed to obtain nearly equal numbers of male and female striped bass and a minimum of 10 individuals in each of several length categories in order to assure an adequate representation of the sexes and multiple year classes in the spawning-stock collections. The following 13 counts and measurements were made for each fish: number of lateral line scales, number of left pectoral rays, number of right pectoral rays, number of second dorsal rays, number of anal rays, number of upper-arm gill rakers, fork length, snout length, head length, internostril width, distance from focus to first annulus of

scales, distance from focus to second annulus of scales, and age as determined from scale annuli.

For our analysis, we deleted the Roanoke fish from the spawning-stock data set. Our reasoning was as follows. Tag–recapture studies do not indicate appreciable migration of Roanoke fish north of the entrance to Chesapeake Bay or appreciable migration of Hudson fish south of the entrance to Chesapeake Bay. The controversy concerning the relative stock composition of the ocean population concerns primarily the area north of Chesapeake Bay. By deleting the Roanoke fish, we assume that we are simplifying the data set in a manner that will more accurately allow us to estimate what we are primarily interested in, without the complicating and confounding effects of including a third stock with its own differences among ages and between sexes. Consistent with our deletion of Roanoke fish from the spawning-stock data set is our deletion of all fish from stratum 10 (south of the entrance to Chesapeake Bay) from the ocean data set (Figure 81).

Berggren and Lieberman (1978) reported that the relative contributions of the Roanoke stock to stratum 1 (Pemaquid Neck Light on the coast of Maine south to Race Point Light at the tip of Cape Cod and including all of Cape Cod Bay) were 11.5% (9 of 82 fish) during May–June and 4.6% (3 of 58 fish) during July–August. The relative contribution of Roanoke stock to stratum 2 (Race Point Light south along the outer coast of Cape Cod to the Massachusetts–Rhode Island border) was 24.0% (20 of 82 fish) in September–October 1975. These results may reflect extensive migrations by the Roanoke stock, but they are at odds with the extensive tag-recapture data (Clark and Baldrige 1984) currently available for the Roanoke stock. An alternative, and we feel more likely, interpretation is that these results are artifacts of the discriminant analysis procedure.

To minimize bias due to sex and year-class differences in the characters, we deleted all sex and year-class combinations of fish sampled on spawning grounds if there were fewer than two fish for either the Hudson or Chesapeake stock (Table 80). We repeated the analysis with a criterion of fewer than one fish for either spawning stock, and there were no pronounced differences in the results. The resulting data set includes 4–8-year-old and 10–11-year-old males and 6–7-year-old and 9–11-year-old females; 28 of 164 fish were deleted from the Hudson stock and 78 of 231 fish from the Chesapeake stock.

Ocean data.—Collection and processing of the ocean specimens were described by Berggren and Lieberman (1978). The same counts and measurements were made on the oceanic striped bass as on the spawning-stock striped bass. Geographic stratification consisted of dividing the Atlantic coast from Maine to Cape Hatteras into 10 collection regions, some with substratification (Figure 81); temporal stratification consisted of dividing the calendar year into six 2-month periods. As indicated above, we deleted from the ocean data set all fish caught in stratum 10 (Table 80). We also deleted from the ocean data set all fish that did not belong to one of the sex and year-class combinations retained in the spawning-stock data set (Table 80). This left us with 798 ocean-caught striped bass for an analysis that we call case A.

Because there was only one 5-year-old female in the Hudson spawning-stock sample, this sex–age combination was not retained in the ocean data set, resulting in the deletion of 1,123 fish or 44% of the total number of fish in the ocean sample. The low abundance of 5-year-old females in the Hudson spawning stock is not surprising because Hudson female striped bass become sexually mature at age 6 and older, whereas Chesapeake females become sexually mature at age 4 and older (Hoff et al. 1988, this volume). However, because of the dominance of 5-year-old females in the ocean sample, a separate analysis was done that included this sex–age combination in the spawning-stock and ocean data sets. The dominance in the ocean sample of 5-year-old striped bass, both female and male (Table 80), reflects the 1970 dominant year class produced in the Chesapeake.

Three other analyses (cases B, C, and D) were performed with the discriminant function method but not with the maximum-likelihood method (see next section). The objective of these analyses was to test hypotheses for specific geographic strata. These three analyses involved deleting fish from strata 8 and 9 (case B), from strata 5 and 7-4 (case C), and from strata 5, 7-4, 8, and 9 (case D).

Statistical methods.—The primary goal of the statistical analysis was to estimate the relative contribution (p_1) of the Hudson River striped bass stock to the Atlantic coastal striped bass population. We estimated p_1 by two independent methods, the discriminant function method and the maximum-likelihood method.

Using equations estimated from the spawning-stock data set for striped bass known to be from either the Hudson or the Chesapeake, the discriminant function method attempts to classify each of the ocean fish of unknown origin as belonging to one group or the other. The methods of estimating the discriminant function and the "confusion matrix" and the derivation of the equation for calculating the relative contribution from the confusion matrix are provided in the Appendix. The estimates of contribution obtained from the discriminant analysis were corrected for bias due to misclassification by using percentages of specimens from each spawning stock that were misclassified into the other spawning stock. This is the same procedure used by Berggren and Lieberman (1978).

TABLE 80.—Sex, year class, and age composition of striped bass collected by Texas Instruments in 1975 in the Hudson River and Chesapeake Bay (spawning-stock data set) and in the ocean (ocean data set).

Year class	Age (years)	Number of legal-sized fish[a]		Number of ocean fish[a,b]
		Hudson	Chesapeake	
Males				
1973	2	0	(3)	(4)
1972	3	0	(26)	(16)
1971	4	11	20	87
1970	5	17	64	196
1969	6	13	5	25
1968	7	2	2	6
1967	8	5	3	7
1966	9	(8)	0	(13)
1965	10	13	2	3
1964	11	7	3	5
1963	12	(2)	0	(1)
1962	13	(2)	0	(1)
Total		68 (12)	99 (29)	329 (35)
Females				
1973	2	0	0	(4)
1972	3	0	0	(100)
1971	4	0	(4)	(234)
1970	5	(1)	(29)	(1,123)
1969	6	9	10	166
1968	7	7	6	48
1967	8	(1)	(11)	(78)
1966	9	18	24	151
1965	10	17	5	38
1964	11	17	9	66
1963	12	(11)	(1)	(76)
1962	13	(1)	(1)	(7)
1961	14	0	0	(6)
1960	15	(1)	0	(2)
1959	16	0	0	(4)
1958	17	(1)	(3)	(3)
1955	20	0	0	(1)
Total		68 (16)	54 (49)	469 (1,638)
Both sexes				
Total		136 (28)[c]	153 (78)[c]	798 (1,673)

[a]Numbers in parentheses indicate fish in that sex and year class combination that were deleted from the spawning-stock data set and the ocean data set for our analysis. Criterion for deletion was fewer than two fish from either the Hudson River or the Chesapeake Bay. Legal-sized striped bass are fish greater than or equal to 406.5 mm fork length.

[b]All 51 ocean fish in stratum 10 (south of the entrance to Chesapeake Bay) were deleted from the ocean data set. Five fish in strata other than stratum 10 were deleted because sex was undetermined. One fish in stratum 9 was deleted because it was collected in Chesapeake Bay rather than the ocean. The 2,471 (798 + 1,673) fish in this table plus the 57 deleted fish just listed sum to 2,528 fish, which is the total sample size in Table 5 of Berggren and Lieberman (1978).

[c]These totals do not include four fish from the Hudson and one from the Chesapeake, which accounts for the discrepancy with the 168 Hudson fish and 232 Chesapeake fish reported by Berggren and Lieberman (1978). We did not include these five fish because they were not assigned an age by Texas Instruments due to conflicting age estimates based on scale annuli (J. T. Lieberman, Texas Instruments Incorporated, personal communication).

The maximum-likelihood method treats the task of estimating p_1 as a problem in estimating the parameters of two or more normal distributions from a mixture of these normal distributions; p_1 is estimated directly without classification of individual fish (Odell and Basu 1976). As a result, there is no equivalent of the bias problem encountered with the discriminant function method, and thus there is no confusion matrix or need to adjust maximum-likelihood estimates of p_1. Details are given in the Appendix.

We decided not to transform any of the basic data on the 13 morphological characters because we could not test statistically whether or not any transformations gave a better or poorer fit to a multivariate normal distribution; such a distribution is assumed for both the discriminant function and maximum-likelihood methods. This agrees with procedures followed by Berggren and Lieberman (1978).

Under ideal circumstances, one would like to conduct the analysis for each sex and year-class combination separately so that one could obtain a clearer picture of the contribution pattern. However, because the sample sizes were not sufficiently large to allow such an analysis, we took the alternative route of attempting to "correct" the data for these effects.

Consider a specific sex (S) and year class (Y). For this sex and year-class combination (SY), we obtain the mean of the jth morphological character, $\bar{v}_j^{(SY)}$, over both spawning-stock samples. Then, for each fish in this sex and year-class combination in either spawning-stock sample, we define the new character

$$Y_{ijk}^{(SY)} = v_{ijk}^{(SY)} - \bar{v}_j^{(SY)};$$

$v_{ijk}^{(SY)}$ is the original value of the jth character for the kth fish from the ith spawning stock in sex and year-class combination SY. This mode of correction is based on a linear model for the effect of sex and year class on each morphological character. Because the averaging is done over both spawning stocks, it is essential that one must have data from both sources for a given sex and year class. As a consequence, several sex and year-class combinations were dropped from the analysis due to lack of data for both spawning stocks. Table 80 shows the sex and year-class combinations used in this study.

Alternative methods of accounting for sex and year-class effects were explored, involving regression of each morphological character on age or fork length for males and females separately. We preferred the above equation because it was the simplest method and involved the fewest assumptions.

The utility of the discriminant function is maximized when only the most "discriminating" characters are used in the function. We felt that the relationship between the 13 measured characters is complex and not fully understood or known, so we included the squares and cross products of these measures in the discriminant function, resulting in a total of 104 possible independent variables. This procedure is analogous to the second-degree polynomial approach used in empirical response-surface methods (Cochran and Cox 1957). The variables included in the function were determined by the stepwise discriminant function method (SAS 1982). We permitted a square of a measured character or a cross product

of two characters to be in the model even if the character itself did not appear. This allows the discriminant function to be more general than the one permitted by the usual quadratic discriminant function method used by Berggren and Lieberman (1978), which requires that the measured character itself be selected for the model before its squares or cross products are considered. The same character variables selected for the discriminant function method were used for the maximum-likelihood method.

Results and Discussion

Discriminant Function and Confusion Matrix

The stepwise linear discriminant function procedure resulted in selection of 14 (Table 81) of the 104 character variables by criteria described in the Appendix. Berggren and Lieberman (1978) found that five variables best discriminated among Hudson, Chesapeake, and Roanoke stock. In decreasing order of importance they were (1) the ratio of snout length to internostril width; (2) the ratio of the distance between the first and second scale annuli to the distance between the focus and the first annulus; (3) the sum of rays in the left and right pectoral, second dorsal, and anal fins; (4) the number of upper-arm gill rakers (including rudimentary rakers); and (5) the number of scales along the lateral line. Although there is not a direct correspondence between our 14 variables and these 5 variables selected by Berggren and Lieberman, it is apparent that snout length, internostril width, distance between focus and first annulus, distance between first and second annuli, and numbers of rays in the various fins are the most discriminating morphological characters in both analyses.

The confusion matrix for the spawning-stock data showed that of the 136 striped bass from the Hudson River, 28 (21%) were misclassified as Chesapeake fish (Table 82). Of the 153 striped bass from the Chesapeake, 30 (20%) were misclassified as Hudson fish. These misclassification percentages are higher than might be desired, but they are about the same as those of Berggren and Lieberman (1978), who misclassified 23% (39 of 168 fish) of the Hudson fish to the Chesapeake Bay or Roanoke River and 32% (74 of 232 fish) of the Chesapeake Fish to the Hudson or Roanoke.

Some of this misclassification undoubtedly arose because real differences between stocks were confounded with differences due to sex and

TABLE 81.—Variables selected in the discriminant function as determined by stepwise discriminant analysis of Hudson River and Chesapeake Bay striped bass.

Variable number	Description	F^a
1	Snout length	55.0
2	Fork length	49.3
3	Number of rays on left pectoral fin	12.3
4	Distance from focus to first annulus of scale	10.2
5	Distance from first annulus to second annulus of scale	10.5
6	Product of number of soft rays on second dorsal fin and number of scales along lateral line	5.88
7	Product of number of rays on left pectoral fin and number of scales along lateral line	4.98
8	Internostril width	4.86
9	Product of number of rays on right pectoral fin and number of upper-arm gill rakers, including rudimentary rakers	3.61
10	Product of internostril width and number of upper-arm gill rakers, including rudimentary rakers	3.53
11	Product of number of soft rays on anal fin and head length	3.78
12	Square of number of soft rays on anal fin	5.12
13	Square of number of soft rays on second dorsal fin	3.12
14	Product of number of soft rays on anal fin and internostril width	3.87

[a] F = value of the F-statistic to remove a character from the discriminant function. The larger the F value, the more important that character is as a discriminator between the two stocks.

age. For example, when we repeated the analysis using only 5-year-old males from the Hudson and Chesapeake spawning stocks, only 1 of 17 (6%) Hudson fish was misclassified to the Chesapeake and only 2 of 64 (3%) Chesapeake fish were misclassified to the Hudson. The reason for selecting 5-year-old males for this example is that sample size was largest for this sex-age combination (Table 80).

Estimates of Relative Contribution

Errors in estimating the relative contributions of the Hudson stock to the coastal population were judged by the absolute value of the difference between the estimated and true relative contributions in the spawning-stock data by sex-age combination (Table 83). Neither of the two statistical methods we used resulted in a consistently smaller error than the other. Error varied with sample size as expected; the larger the sample, the smaller the error tended to be. Except for 7-, 10-, and 11-year-old males (sample sizes of 4, 15, and 10 fish, respectively), the error was less than 10%; for half of the 12 sex–age combinations, it was less than 5%. If this study is repeated, we recommend that length categories and sample sizes be selected to minimize the chance of an error greater than 10% for each sex–age combination in the spawning-stock data.

Estimates of relative contribution of the Hudson stock to the Atlantic coastal population of striped bass north of Chesapeake Bay ranged from 0 to 79%, depending on year class, sex, and method of estimation (Figure 82). The results indicate marked differences among year classes, and the two methods of estimating relative contribution gave give similar results.

Because 85% of the striped bass in the ocean sample were female (Table 80), it is appropriate to pay particular attention to the p_1 values in Figure 82 for females. The two estimates for 10-year-old Hudson females are between 40 and 50%, suggesting that the 1965 year class from the Hudson was relatively strong. The estimates for 6-, 7-, and 9-year-old females are all less than 10%, and these estimates are based on reasonably large sample sizes. The estimates for 7-year-old males are also low, consistent with the estimates for females. The separate analysis that included 5-year-old (1970 year class) female striped bass in the spawning-stock and ocean data sets resulted in an estimate of relative contribution of 1% from the

TABLE 82.—Confusion matrix for the striped bass spawning-stock data.

Actual	Classified[a]		Total fish
	Hudson ($j = 1$)	Chesapeake ($j = 2$)	
Hudson ($i = 1$)	108 fish $\phi 11 = 0.79$	28 fish $\phi 12 = 0.21$	136
Chesapeake ($i = 2$)	30 fish $\phi 21 = 0.20$	123 fish $\phi 22 = 0.80$	153

[a] ϕij = number in row i and column j divided by the total for row i.

TABLE 83.—Absolute value of the difference between the estimated and true relative contributions (p_1) of Hudson River striped bass in the 1975 spawning-stock data set. Values for this error are given by sex and year-class combination for p_1 values calculated by both methods of analysis.

Year class	Age (years)	Sample size[a]	Discriminant function method			Maximum-likelihood method		
			p_1			p_1		
			Estimate	True	Error[b]	Estimate	True	Error[b]
Males								
1971	4	31	0.433	0.355	0.078	0.375	0.355	0.020
1970	5	81	0.217	0.210	0.007	0.202	0.210	0.008
1969	6	18	0.789	0.722	0.067	0.808	0.722	0.086
1968	7	4	0.513	0.500	0.013	0.326	0.500	0.174
1967	8	8	0.720	0.625	0.095	0.593	0.625	0.032
1965	10	15	0.679	0.867	0.188	0.872	0.867	0.005
1964	11	10	0.844	0.700	0.144	0.858	0.700	0.158
Females								
1969	6	19	0.470	0.474	0.004	0.410	0.474	0.064
1968	7	13	0.450	0.538	0.089	0.500	0.538	0.038
1966	9	42	0.395	0.429	0.034	0.460	0.429	0.031
1965	10	22	0.814	0.773	0.041	0.767	0.773	0.006
1964	11	26	0.641	0.654	0.013	0.661	0.654	0.007

[a]Calculated from Table 1.
[b]Error = estimate − true.

Hudson. This is consistent with the strong 1970 year class from the Chesapeake, the largest on record (Clark and Baldrige 1984).

That the relative contribution from the Hudson varies is to be expected, because the historical record does not indicate a marked tendency for dominant or weak year classes to occur in the Hudson and Chesapeake simultaneously (Klauda et al. 1980). Our analysis indicates that the relative contribution varies from less than 5% to as high as 40 to 50%. The upper bound is less certain than the lower bound because sample sizes were small for most Hudson year classes that seemed to make a large relative contribution.

The estimates of relative contribution in Figure 82 are for the area north of Chesapeake Bay (case A). Because few tagged Hudson striped bass have been recovered south of Sandy Hook, New Jersey (McLaren et al. 1981), we predicted that by deleting the ocean fish from strata 8 and 9 (case B), the resulting relative contribution values (p_1) would be greater than those for case A. Next, we noted that the geographic area encompassed by case A does not include any of the Chesapeake Bay, although case A does include areas analogous to Chesapeake Bay at the mouth of the Hudson River: New York Harbor, Raritan Bay, and the western half of Long Island Sound. Therefore, we predicted that by deleting the fish from strata 5 and 7-4 (case C), the resulting p_1 values would be lower than those for case A. In testing both of

these predictions we included only those sex–age combinations for which sample size was greater than 10 fish and the estimate of relative contribution was greater than 1% for case A.

The analyses supported both predictions (Table 84). Every sex–age combination that showed a substantial decrease in sample size between case A and cases B or C gave p_1 values that changed in the direction expected. For the 1964 year class of female striped bass, as an example, deletion of 12 fish collected between Chesapeake Bay and Sandy Hook (strata 8 and 9) caused the relative contribution of the Hudson stock to increase from 18% (case A) to 23% (case B). Such results have limited value of themselves, but their agreement with our expectations is reassuring.

Our final analysis involved deletion of strata 5, 7-4, 8, and 9 (case D), leaving the area from eastern Long Island up through New England. The purpose of this analysis was to obtain estimates of p_1 for a region where relative contributions are currently a topic of debate. Our findings (Table 84) include examples in which p_1 is lower than that for case A (e.g., 4-year-old males) and higher than that for case A (e.g., 10- and 11-year-old females). We conclude, as we did for case A, that the relative contribution of the Hudson stock to this smaller geographic area varies considerably between sexes and among year classes, ranging from less than 10% to as great as 40 to 50%.

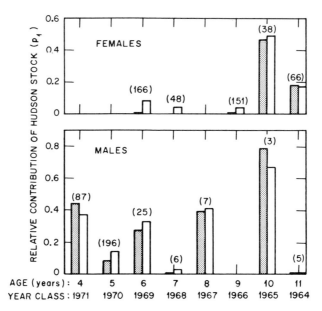

FIGURE 82.—Estimates from the 1975 ocean data of the relative contribution of the Hudson River striped bass to the coastal striped bass population north of Chesapeake Bay, for seven year classes of males and five year classes of females (sample sizes in parentheses), determined by the discriminant function method (stippled bars) and the maximum-likelihood method (clear bars).

Now that the 1970 dominant year class from the Chesapeake is no longer prominent in the ocean population, and no dominant year classes have appeared in the Chesapeake since 1970 (Clark and Baldrige 1984), the time seems propitious to repeat the study. This point is especially valid because the Hudson River population produced a good year class in 1978 and an exceptional year class in 1983 (Clark and Baldrige 1984). If the study is repeated, one implication of our further analysis of the 1975 Texas Instruments data is that the sampling design should be modified to include more length categories, more fish per length category for each sex, or both, so that adequate numbers of striped bass are available to permit estimation of relative contribution for as many sex and year-class combinations as possible. Another implication is that possibly only those character variables found to be the most discriminating in both Berggren and Lieberman's (1978) and our discriminant analyses should be measured. This simplification would save money with little sacrifice in ability to estimate relative contributions.

TABLE 84.—Discriminant function estimates from the 1975 ocean data of the relative contributions (p_1) of the Hudson stock to the coastal striped bass population by sex and year-class combination for four geographic regions (cases A, B, C, and D).[a] N denotes number of fish.

Year class	Age (years)	Case A		Case B		Case C		Case D	
		N	p_1	N	p_1	N	p_1	N	p_1
					Males				
1971	4	87	0.441	82	0.452	57	0.205	52	0.222
1970	5	196	0.082	148	0.124	170	0.026	122	0.034
1969	6	25	0.274	21	0.379	20	0.181	16	0.296
					Females				
1965	10	38	0.464	35	0.511	37	0.468	34	0.535
1964	11	66	0.179	54	0.232	66	0.185	54	0.232

[a]Refer to Figure 81. Case A: region north of Chesapeake Bay; case B: strata 8 and 9 deleted; case C: strata 5 and 7-4 deleted; case D: strata 5, 7-4, 8, and 9 deleted, leaving eastern Long Island and New England.

Acknowledgments

We thank S. W. Christensen and J. A. Solomon for their critical reviews of this paper. The data analyzed were made available to us by Texas Instruments, Incorporated, through Consolidated Edison Company of New York, Incorporated, as part of the operating license proceedings for Consolidated Edison's Indian Point Nuclear Generating Plant, unit 3, before the U.S. Nuclear Regulatory Commission. This paper is based to a large extent on the report by Van Winkle and Kumar (1982). This research was supported by the U.S. Nuclear Regulatory Commission's Office of Nuclear Regulatory Research under interagency agreement DOE 40-550-75 with the U.S. Department of Energy under contract DE-AC05-84OR21400 with Martin Marietta Energy Systems, Incorporated. This is publication 2349 of the Environmental Sciences Division, Oak Ridge National Laboratory, Oak Ridge, Tennessee.

References

Anas, R. E., and S. Murai. 1969. Use of scale characters and a discriminant function for classifying sockeye salmon (*Oncorhynchus nerka*) by continent of origin. International North Pacific Fisheries Commission Bulletin 26:157–192.

Berggren, T. J., and J. T. Lieberman. 1978. Relative contribution of Hudson, Chesapeake, and Roanoke striped bass, *Morone saxatilis*, stocks to the Atlantic coast fishery. U.S. National Marine Fisheries Service Fishery Bulletin 76:335–345.

Clark, W., and M. Baldrige. 1984. Emergency striped bass research study, report for 1982–1983. Report to Congress by the Secretaries of Interior and Commerce, Washington, D.C.

Cochran, W. G., and G. M. Cox. 1957. Experimental designs. Wiley, New York.

Fournier, D. A., T. D. Beacham, B. E. Riddell, and C. A. Busack. 1984. Estimating stock composition in mixed stock fisheries using morphometric, meristic, and electrophoretic characteristics. Canadian Journal of Fisheries and Aquatic Sciences 41:400–408.

Fukuhara, F. M., S. Murai, J. J. LaLanne, and A. Sribhibhadh. 1962. Continental origin of red salmon as determined from morphological characters. International North Pacific Fisheries Commission Bulletin 8:15–109.

Gnanadesikan, R. 1977. Methods for statistical data analysis of multivariate observation. Wiley, New York.

Grove, T. L., T. J. Berggren, and D. A. Powers. 1976. The use of innate tags to segregate spawning stocks of striped bass, *Morone saxatilis*. Pages 166–176 in M. Wiley, editor. Estuarine processes, volume 1. Academic Press, New York.

Hoff, T. B., J. B. McLaren, and J. C. Cooper. 1988. Stock characteristics of Hudson River striped bass. American Fisheries Society Monograph 4:59–68.

Kendall, M. G., and A. Stuart. 1973. The advanced theory of statistics, volume 2. Hafner, New York.

Klauda, R. J., W. P. Dey, T. B. Hoff, J. B. McLaren, and Q. E. Ross. 1980. Biology of Hudson River juvenile striped bass. Marine Recreational Fisheries 5:101–123.

Kshirsagar, A. M. 1978. Multivariate analysis. Dekker, New York.

McLaren, J. B., J. C. Cooper, T. B. Hoff, and V. Lander. 1981. Movements of Hudson River striped bass. Transactions of the American Fisheries Society 110:158–167.

Odell, P. L., and J. P. Basu. 1976. Concerning several methods for estimating crop acreages using remote sensing data. Communications in Statistics. Part A: Theory and Methods 5:1091–1114.

Peters, C., and W. A. Coberly. 1976. The numerical evaluation of the maximum-likelihood estimate of mixture proportions. Communications in Statistics. Part A: Theory and Methods 5:1127–1135.

Rao, C. R. 1952. Advanced statistical methods in biometric research. Wiley, New York.

SAS (Statistical Analysis System). 1982. SAS user's guide: basics. SAS Institute, Cary, North Carolina.

Tubbs, J. D., and W. A. Coberly. 1976. An empirical sensitivity study of mixture proportion estimators. Communications in Statistics. Part A: Theory and Methods 5:1115–1125.

Van Winkle, W., and K. D. Kumar. 1982. Relative stock composition of the Atlantic coast striped bass population: further analysis. U.S. Nuclear Regulatory Commission, NUREG/CR-2563, Washington, D.C.

Appendix: Statistical Methods

Discriminant Function Method

Given a sample of N striped bass collected in the ocean, the discriminant function will classify $\hat{N}_1$ (the "^" denotes an estimate as opposed to the "true" value) as belonging to the Hudson River stock. The relative contribution of the Hudson stock (p_1) is given by

$$p_1 = \alpha \left(\frac{\hat{N}_1}{N} \right) + \beta; \qquad \text{(A-1)}$$

α and β are constants that can be estimated from the spawning-stock data set. In this section we discuss (1) the method of estimating the discriminant function, (2) the method for estimating the "confusion matrix," and (3) the derivation of equation (A-1) from the confusion matrix.

Estimation of the linear discriminant function.— Let

$$y_1 = (y_{11}, y_{12}, ..., y_{1K}) \text{ and } y_2 = (y_{21}, y_{22}, ..., y_{2K})$$

be the vectors of K character variables for each fish (both sexes and all ages) sampled from the Hudson River and Chesapeake Bay, respectively. Let n_1 and n_2 be the numbers of fish sampled in each of the spawning stocks. We further define the following terms:

$\bar{y}_i$ = mean sample vector of character variables for spawning stock i ($i = 1, 2$);
S_i = variance–covariance sample matrix for spawning stock i ($i = 1, 2$);
n = $n_1 + n_2$ = total sample size from the spawning stocks;
$\bar{y}$ = $n_1\bar{y}_1 + n_2\bar{y}_2/n$ = overall mean sample vector of character variables;
W = $1/n-2[(n_1 - 1)S_1 + (n_2 - 1)S_2]$ = within-group, variance–covariance sample matrix;
B = $n_1(\bar{y}_1 - \bar{y})(\bar{y}_1 - \bar{y})' + n_2(\bar{y}_2 - \bar{y})(\bar{y}_2 - \bar{y})'$ = between-group, variance–covariance sample matrix, in which the prime denotes the transpose of the vectors of differences ($\bar{y}_i - \bar{y}$).

It is assumed that the y_i are samples from multivariate normal distributions and that the S_i are estimates of a common variance–covariance matrix. If the vector $z = a'y$ denotes a linear combination of the original character variables, a one-way analysis of variance for the derived variable z

will lead to the following F-ratio of the between-groups mean square to the within-group mean square:

$$F = \frac{a'Ba}{a'Wa}. \qquad \text{(A-2)}$$

If we choose the elements of the coefficient vector a such that this F-ratio is maximized, we are selecting the linear combination of the original character variables that best "discriminates" between the two stocks. The value of a is given by the eigenvector corresponding to the largest eigenvalue of $W^{-1}B$ (see Gnanadesikan 1977). Once the coefficient vector a has been determined, we can then classify the ith fish in the ocean sample as belonging to the Hudson stock if

$$a'x_i > a' (\bar{y}_1 + \bar{y}_2)/2; \qquad \text{(A-3)}$$

x_i is the vector of K character variables for the ith fish in the ocean sample. Otherwise, the ith fish is classified as belonging to the Chesapeake stock. The reader should refer to Rao (1952) and Gnanadesikan (1977) for more details.

The confusion matrix.—Once the discriminant function has been estimated, we can evaluate the effectiveness of the discriminant function by estimating the "confusion matrix." Let ϕ_{ij} represent the proportion of the ith spawning stock that was classified as belonging to the jth spawning stock. Hence, ϕ_{11} and ϕ_{22} represent the proportions correctly classified as Hudson and Chesapeake, respectively, whereas ϕ_{12} and ϕ_{21} represent the proportions misclassified (e.g., ϕ_{12} represents the proportion of the Hudson spawning stock misclassified as Chesapeake spawning stock). The confusion matrix can be estimated by the jackknife method (also called the U-method; Kshirsagar 1978). The confusion matrix is estimated as follows:

Step (1). Compute the discriminant function for the two spawning-stock data sets combined, except for the ith fish.
Step (2). Classify the ith fish by the discriminant function computed in step 1.

We repeat the two steps for all the fish in the combined spawning-stock data set. The reader is referred to Kshirsagar (1978) for a general discussion of different methods of estimating the confusion matrix. The overall effectiveness of the discriminant function is then given by the ratio of the total number of fish misclassified to the total number of fish in the spawning stock data set. The lower this number, the better the discriminant function.

Estimation of relative contribution (p_1).—The number of ocean fish classified as Hudson, N_1, may be expressed as

$$\hat{N}_1 = \text{Prob}(1,1)N_1 + \text{Prob}(2,1)N_2; \quad \text{(A-4)}$$

N_1 and N_2 are the true number of Hudson and Chesapeake fish in the ocean sample, respectively; Prob(1,1) is the probability a Hudson fish is classified as Hudson, and Prob(2,1) is the probability a Chesapeake fish is classified as Hudson. Dividing both sides by $N = N_1 + N_2$ gives

$$\hat{p}_1 = \text{Prob}(1,1)p_1 + \text{Prob}(2,1)(1 - p_1); \quad \text{(A-5)}$$

$\hat{p}_1$ is the proportion of the ocean sample classified as Hudson, and p_1 is the "true" proportion of the ocean sample from the Hudson. Assuming ϕ_{11} equal to Prob(1,1) and ϕ_{21} equal to Prob(2,1) and solving for p_1, we get

$$p_1 = \frac{\hat{p}_1 - \phi_{21}}{\phi_{11} - \phi_{21}}. \quad \text{(A-6)}$$

Then

$$p_1 = \alpha \hat{p}_1 + \beta \quad \text{(i.e., equation A-1);} \quad \text{(A-7)}$$

$$\alpha = 1/(\phi_{11} - \phi_{21}), \quad \text{(A-8)}$$

and

$$\beta = -\phi_{21}/(\phi_{11} - \phi_{21}). \quad \text{(A-9)}$$

This method of adjusting the estimates of contribution obtained directly from the discriminant analysis for bias due to misclassification is akin to moment estimation (Kendall and Stuart 1973). It is the same procedure used by Berggren and Lieberman (1978: their adjusted estimates); also see Fukuhara et al. (1962) and Anas and Murai (1969), who used this procedure to adjust estimates of stock contribution for sockeye salmon. If $p_1 < \phi_1$ or $\phi_{11} < \phi_{21}$ (but not both), the estimate of p_1 will be negative, indicating that the method is unable to estimate the contribution because it is very small or because the sample size is small. Whenever a negative p_1 was obtained, we set it equal to zero, unlike the procedure used by Berggren and Lieberman (1978).

Maximum-Likelihood Method

The maximum-likelihood method treats the task of estimating p_1 as a "mixture-of-normals" prob-

lem, and p_1 is estimated directly without classification of individual fish (Odell and Basu 1976; Peters and Coberly 1976; Tubbs and Coberly 1976; Fournier et al. 1984). As a result, there is no equivalent of the bias problem encountered with the discriminant function method and, thus, there is no need to adjust maximum-likelihood estimates of p_1 as in equation (A-7).

Let $f_1(y_1)$ and $f_2(y_2)$ be the K-dimensional probability density functions of the character variables for the Hudson stock and Chesapeake stock, respectively. The density function of the character variables for the ocean stock then is a mixture of these two spawning-stock density functions. In our case, we define for the ocean sample (denoted by x) a binomial distribution leading to the density function

$$f(x) = \hat{p}_1 f_1(x) + (1 - \hat{p}_1)f_2(x); \quad \text{(A-10)}$$

$\hat{p}_1$ is the estimated contribution of Hudson stock to the Atlantic Ocean population. We further assume that $f_i(x)$ is a K-dimensional, multinormal distribution with mean vector θ_i and variance–covariance matrix Σ_i. Furthermore, to remain consistent with the linear discriminant function method, let $\Sigma_1 = \Sigma_2 = \Sigma$.

Given the spawning-stock data, we can readily obtain the usual estimates of θ_1, θ_2, and Σ. Hence, $f(x)$ can be re-written as

$$\widehat{f(x)} = \hat{p}_1 \widehat{f_1(x)} + (1 - \hat{p}_1)\widehat{f_2(x)}. \quad \text{(A-11)}$$

One can then obtain the maximum-likelihood estimate of p_1 by forming the likelihood function of $f(x)$, differentiating the function with respect to p_1, and setting this derivative equal to zero. After some algebraic manipulations we obtain the equation

$$\hat{p}_1 = \frac{\hat{p}_1}{N} \sum_{i=1}^{N} \frac{\widehat{f_1(x_i)}}{\hat{p}_1 \widehat{f_1(x_i)} + (1 - \hat{p}_1)\widehat{f_2(x_i)}}; \quad \text{(A-12)}$$

N is again the number of fish in the ocean sample. Because $\hat{p}_1$ occurs on both sides of the equation, a fixed-point solution method is used to estimate p_1. The above equation is re-written as

$$\hat{p}_1^{(r)} = \frac{\hat{p}_1^{(r-1)}}{N} \cdot \sum_{i=1}^{N} \frac{\widehat{f_1(x_i)}}{\hat{p}_1^{(r-1)}\widehat{f_1(x_i)} + [1 - \hat{p}_1^{(r-1)}]\widehat{f_2(x_i)}}; \quad \text{(A-13)}$$

$\hat{p}_1^{(r)}$ = estimate of p_1 at the rth iteration.

The estimation algorithm is

Step (0). Let $\hat{p}_1^{(0)} = 0.5$. This is the initial estimate of p_1.

Step (r). Substitute the estimate of p_1 at the $(r-1)$th iteration ($\hat{p}_1^{(r-1)}$ in equation A-13) to obtain $\hat{p}_1^{(r)}$. If $|\hat{p}_1^{(r)} - \hat{p}_1^{(r-1)}|$ is less than some small number ε (0.00001 in our case), the algorithm has converged and $\hat{p}_1^{(r)}$ is our best estimate of p_1. Otherwise we repeat step r.

It can be readily shown that, given $\hat{f}_1(x)$ and $\hat{f}_2(x)$, the algorithm will converge. When $\hat{p}_1^{(r)}$ is close to zero or one, the algorithm is stopped, and $\hat{p}_1^{(r)}$ is set equal to zero or one, respectively.

One of the potential advantages of this maximum-likelihood method is that it is theoretically possible to calculate a confidence interval about $\hat{p}_1$, whereas this is not possible with the discriminant function method. The procedure for doing this, however, would not be straightforward and would require developing additional computer programs.

American Fisheries Society Monograph 4:267–273, 1988

SECTION 4: NEGOTIATED SETTLEMENT AGREEMENT AND BEYOND

Hudson River Settlement Agreement: Technical Rationale and Cost Considerations

LAWRENCE W. BARNTHOUSE

Environmental Sciences Division, Oak Ridge National Laboratory[1]
Post Office Box 2008, Oak Ridge, Tennessee 37831-6036, USA

JOHN BOREMAN

National Marine Fisheries Service, Northeast Fisheries Center, Woods Hole Laboratory
Woods Hole, Massachusetts 02543, USA

THOMAS L. ENGLERT

Lawler, Matusky & Skelly Engineers, One Blue Hill Plaza, Pearl River, New York 10965, USA

WILLIAM L. KIRK

Consolidated Edison Company of New York, 4 Irving Place, New York, New York 10003, USA

EDWARD G. HORN[2]

New York State Department of Environmental Conservation
50 Wolf Road, Albany, New York 12233, USA

Abstract.—In an effort to end litigation over open-cycle cooling at Hudson River power plants, out-of-court negotiations began in August 1979. On December 19, 1980, an agreement that was acceptable to all parties was reached. As an alternative to building cooling towers at the Indian Point, Bowline Point, and Roseton generating stations, the utilities agreed to a variety of technical and operational changes intended to reduce entrainment and impingement. In addition, they agreed to supplement the production of striped bass in the Hudson River estuary by means of a hatchery, to conduct a biological monitoring program, and to fund an independent research foundation for study of Hudson River environmental problems. In this paper, we discuss the role of technical advisors in the settlement negotiations. We then describe the rationale for each of the mitigating measures, including cost considerations, and discuss the objectives and status of the riverwide monitoring program and the Hudson River Foundation.

By the summer of 1979 the U.S. Environmental Protection Agency's (EPA's) hearings had continued for more than a year and a half with no end in sight. Most of the data and analyses presented in the first three sections of this monograph had been submitted in testimony. It was becoming clear to both the scientists and the lawyers involved in the case that the scientific facts would not support an unequivocal decision regarding the need for cooling towers at Indian Point, Bowline Point, and Roseton. In May 1979, EPA's consultants had concluded that, considering the combined effects of entrainment and impingement (see Vaughan 1988, this volume, for definitions of these terms), 10–20% of each year class of striped bass, white perch, Atlantic tomcod, and American shad might be killed at power plant intakes (Barnthouse et al. 1984; Barnthouse and Van Winkle 1988, this volume; Boreman and Goodyear 1988, this volume). Estimates previously developed by the utilities' consultants were similar in magnitude. There was no consensus on how or whether it was possible to estimate the long-term effects of this mortality

[1]Operated by Martin Marietta Energy Systems, Incorporated, under contract DE-AC05–84OR21400 with the U.S. Department of Energy.

[2]Current address: New York State Department of Health, 2 University Plaza, Albany, New York 12237, USA.

on the abundance and persistence of the affected populations (Christensen and Goodyear 1988; Fletcher and Deriso 1988; Lawlcr 1988; Savidge et al. 1988, all this volume).

Much greater mortality than was being imposed by the Hudson River utility industry is often sustained by exploited fish populations, and the utilities felt justified in their contention that cooling towers would be an unnecessary burden on rate-payers (Bergen 1988, this volume). However, from a regulatory perspective, 10–20% annual mortality was too large to be ignored by the EPA and by its state-level counterpart, the New York State Department of Environmental Conservation (NYSDEC). The sport and commercial fisheries for striped bass and American shad might be adversely affected even if the survival of the populations themselves was not threatened (Goodyear 1988, this volume). Moreover, EPA's statcd policy was that power-plant-induced mortality should be minimized, regardless of its biological sustainability (Yost 1988, this volume).

In an effort to end the stalemate, out-of-court negotiations involving EPA, the Hudson River utility companies, NYSDEC, the New York State Attorney General, the Hudson River Fishermen's Association, the Scenic Hudson Preservation Conference, and the Natural Resources Defense Council were begun in August of 1979. The issue of cooling towers was pivotal. As a matter of policy, cooling towers had been endorsed by EPA as the "best available technology" for minimizing entrainment and impingement mortality. This position was controversial and was not fully supported either by the state or by the three citizens' organizations, both because of the enormous expense of the towers and because of the unavoidable impacts of the 61-m-high structures on the scenic beauty of the Hudson Valley.

The negotiations proceeded on two levels. Policy-level discussions involving utility company executives, state and EPA administrators, and lawyers representing all of the parties were mediated by Russell Train, a former Administrator of EPA and the Chairman of the World Wildlife Fund. The details of the policy-level negotiations have been described by Talbot (1983) and need not be repeated here. Simultaneously, scientists working for the contesting parties attempted to develop a sound technical basis for a settlement. At first, the consultants for each agency or organization worked independently as technical advisors to their respective negotiators. As negotiations continued through 1979 and 1980, informal

contacts between EPA and utility consultants were established (encouraged and mediatcd by Albert Butzel, attorney for Scenic Hudson) to speed the development and evaluation of settlement proposals.

The possibility of a settlement hinged on finding effective alternatives to cooling towers for reducing entrainment and impingement mortality. Many potentially effective technological approaches existed for reducing impingement (Cannon et al. 1979). However, no technological solutions were available for reducing entrainment mortality, the primary concern for most of the Hudson River fish populations. The best alternative means of reducing entrainment was alteration of plant operating conditions. Throughout most of the year when river temperatures are low, cooling water pumping rates could be reduced, thus reducing the number of organisms entrained; the entrained water would become hotter at reduced flows, but would not reach lethal temperatures. During the summer, when reduced pumping would be counterproductive because of increased thermal mortality of entrained organisms, maintenance outages at some plants could be scheduled to coincide with the expected peak abundance of vulnerable life stages in the vicinity of each plant.

The first task faced by the technical advisors was to develop a method for evaluating the effectiveness of the various combinations of technologies and operating schemes at reducing entrainment and impingement and comparing them to the expected reduction afforded by EPA's demand for cooling towers at all three plants. It was immediately apparent that the conditional entrainment and impingement mortality rates (Vaughan 1988) used to assess the impacts of existing operation provided a suitable measure for comparing the possible alternatives. The degree of reduction in conditional mortality rates due to cooling towers could readily be approximated by the expected reduction in cooling water withdrawal rates (e.g., Barnthouse and Van Winkle 1988; Boreman and Goodyear 1988). The increase in survival of impinged fish duc to improved intake structure design could be easily expressed in the same terms (Barnthouse and Van Winkle 1988). Estimating the effectiveness of modified operating schedules required the use of an ichthyoplankton transport–entrainment model to calculate the spatiotemporal distribution of life stages. Suitable models had been developed by both utility and EPA consultants. After initial model comparisons confirmed their similarity, given similar inputs (Englert et al.

1988, this volume), the empirical transport model (Boreman and Goodyear 1988) was selected on grounds of ease and speed of use.

During the first 9 months of the negotiations, the task of the technical advisors was greatly complicated by the desire of EPA to include a "mitigation test" as part of any settlement. The test would involve defining a target degree of mitigation, expressed as a fraction of the mitigation afforded by six cooling towers, and then setting up an environmental monitoring program to measure the actual degree of mitigation. If the test were failed, cooling towers would be required. Options for making such a determination included monitoring changes in abundance of the populations and monitoring changes in the key parameters of the entrainment and impingement models. Neither option was found to be feasible. From the beginning, EPA's consultants opposed the idea of basing a test on measurement of population abundance on the grounds that background variability and measurement error would preclude detection of any change in population size due to altered power plant operations. A test based on measurement of key model parameters appeared attractive initially but, ultimately, proved unfeasible as well. Given the many possible alternative interpretations of the inevitably imperfect data that would be collected, it appeared likely that litigation would resume over the question of whether the test had been passed or failed. After an unproductive and sometimes acrimonious meeting in May 1980, the idea of a mitigation test was abandoned.

Policy-level negotiations and informal technical discussions continued throughout the summer and fall of 1980. The negotiating parties exchanged proposals until a package was reached that afforded a degree of mitigation acceptable to EPA and other parties at a cost acceptable to the utilities. On December 19, 1980, the historic settlement agreement was signed by all parties. For the 10-year duration of the settlement, no cooling towers would be required. As an alternative, the utilities agreed to a variety of technical and operational changes intended to reduce entrainment and impingement. In addition, they agreed to supplement the production of striped bass in the Hudson River estuary by means of a hatchery, to conduct a biological monitoring program, and to fund an independent research foundation for study of Hudson River environmental problems. In the remainder of this paper, we describe the rationale for each of the mitigative measures, including cost considerations, and discuss the objectives and status of the riverwide monitoring program and the Hudson River Foundation. Other papers in this section provide detailed discussions of the most important mitigative measures.

Summary of Mitigative Measures

Entrainment Mitigation

Based on the premise that properly timed reductions in cooling water withdrawals would substantially reduce the number of organisms entrained by the generating stations, a detailed scheme of flow reductions and scheduled outages was developed for the Indian Point, Bowline Point, and Roseton plants. The agreement called for all three plants to reduce their cooling water pumping rates during the winter and spring. A plan to reduce cooling water withdrawals throughout the year was considered, but was rejected because the reduction in numbers of organisms entrained would have been more than offset by an increase in thermal mortality of organisms that were entrained (Englert et al. 1988). To facilitate the planned reduction in withdrawals, the utilities agreed to install variable-speed water circulation pumps at the Indian Point facilities.

The annual maintenance shutdowns at Bowline Point and Roseton were scheduled to coincide with expected periods of peak entrainment at each plant (Englert et al. 1988). At the two Bowline Point generating units, the agreement specified 30 unit-days of outages between May 15 and June 30 and an additional 31 unit-days between July 1 and July 31. At the two Roseton generating units, 30 unit-days of outage between May 15 and June 30 were required. The nuclear units at Indian Point are refueled at approximately 16–18 month intervals. Because other routine maintenance is performed during refueling, scheduling of an annual maintenance shutdown during the entrainment season could not be arranged. However, the agreement specified that an average of 42 unit-days of outages should occur at Indian Point units 2 and 3 between May 10 and August 10 of each year.

To provide for electrical system reliability and unanticipated contingencies, especially during the June–July period when power demand in the New York City area is at or near its annual peak, a system of "cross-plant outage credits" was developed (Englert et al. 1988). This credit system permits the utilities to deviate from the specified outage schedule without changing the overall degree of entrainment mitigation. In particular, relief from the outages of Bowline and Roseton is allowed when scheduled or unscheduled shutdowns cause the Indian Point facilities to exceed their outage requirements.

Impingement Mitigation

Settlement conditions intended to reduce fish impingement included the continued seasonal deployment of a fish barrier net at Bowline Point and the installation of a new fish protection device, angled screens, at Indian Point. Experimental use of the barrier net, which takes advantage of the unique intake configuration of Bowline Point (water is withdrawn from a pond adjacent to the Hudson rather than from the river itself), began in 1977. Hutchison and Matousek (1988, this volume) demonstrated that, when this device was deployed during the winter impingement peaks, impingement of striped bass and white perch was significantly reduced.

Angled screens are a novel technology intended to divert fish present in an intake bay away from the traveling screens and to return them safely to the river (Cannon et al. 1979). At the time of the settlement, angled screens were considered to be a promising means for reducing the impingement of white perch, Atlantic tomcod, and striped bass at Indian Point. This belief was based principally on results of utility-sponsored flume studies. However, results from field-scale tests were unavailable and it was uncertain whether an effective system based on the angled screen concept could be built at Indian Point. Therefore, the agreement stipulated that if the parties later agreed that it was "not desirable or possible" to construct angled screens at Indian Point, alternative measures could be substituted up to a total cost of $20 million. Subsequent field-scale studies and reanalysis of the original flume data (Fletcher 1985) showed that, although fish can be diverted by angled screens, the mortality imposed on the diverted fish increases with the size of the diversion system. For a system as large as would be required at Indian Point, Fletcher (1985) concluded that angled screens would probably prove ineffective at reducing mortality due to impingement. Plans to install these screens at Indian Point have now been abandoned. The best alternative method for reducing impingement at Indian Point appears to be continuously rotated vertical traveling screens, modified to effectively collect impinged fish and return them to the river alive. Studies performed at other power plants, including Bowline Point and Roseton, have consistently shown that 40% or greater survival of many fish species can be achieved with these systems (Barnthouse and Van Winkle 1988; Muessig et al. 1988, this volume). A screen was installed at Indian Point in late 1984 for field testing, and parallel work to refine system performance was performed in a flume. Based on results of testing completed in late 1986, the parties to the settlement are expected to agree to installation of modified vertical traveling screens at Indian Point units 2 and 3.

Striped Bass Stocking Program

Although the mitigative value of stocking striped bass could not be firmly documented, the utilities agreed to construct and operate a hatchery to produce 600,000 76-mm-long young-of-the-year striped bass for stocking the Hudson River each year. Studies performed during the 1970s showed that hatchery-reared striped bass survive after release into the Hudson (McLaren et al. 1988, this volume). If the target production were reached and there were no difference in survival between hatchery-reared and wild fish, the annual impingement loss of striped bass would be offset, but the entrainment mortality remaining after mitigation probably would not be. The hatchery began operation in 1983. Production has increased steadily from 63,000 fish during the first year to 564,000 fish in 1986 (EA Engineering, Science, and Technology 1987). Evaluation of the effectiveness of the striped bass stocking program is continuing as part of the biological monitoring program specified in the settlement.

Benefits and Costs of the Settlement

Some of the ecological benefits of the settlement can be estimated from the entrainment mitigation assessment described by Englert et al. (1988). Benefits due to mitigation of impingement impacts from the Bowline barrier net, the angled screens at Indian Point, and the hatchery could not be quantified; only the improvements in entrainment mortality resulting from flow reductions and scheduled outages at the power plants could be estimated. For striped bass and white perch, approximately half of the entrainment mitigation would result from cooling water flow reductions during the cooler months and half from spring and summer outages at Bowline Point and Roseton (Table 85). All mitigation of Atlantic tomcod entrainment would result from flow reductions; the outage schedule specified in the settlement would slightly increase entrainment of this species

TABLE 85.—Percentage reductions[a] in conditional entrainment mortality rate for the flow reductions and scheduled outages at each Hudson River power plant.

Agreement component	Striped bass	White perch	Atlantic tomcod	American shad	Bay anchovy
Flow reduction at Indian Point	9.4	3.7	32.6	1.6	0.4
Flow reductions at Bowline Point and Roseton	7.8	9.4	24.4	7.2	4.5
Subtotal for flow reductions[b]	16.5	12.8	49.0	8.7	4.9
Bowline Point and Roseton outages in May and June	8.6	7.5	−9.3	9.6	4.3
Bowline outage in July	3.9	3.7	0	4.0	3.7
Outage at Indian Point	0	2.8	−2.3	4.0	3.5
Subtotal for outages[b]	12.2	13.4	−11.8	16.7	11.1
Total[b]	26.6	24.5	43.0	23.9	15.4

[a]Percent reduction normalized to the reduction predicted if cooling towers had been constructed at Bowline Point, Indian Point, and Roseton.

[b]The total percent reduction (P_T) is calculated from the individual reductions for each mitigating measure (P_i) as $P_T = 100[1 - \Pi(1 - P_i/100)]$.

relative to presettlement conditions. Outages would be needed to substantially mitigate entrainment of American shad and bay anchovy because entrainable life stages of these species are uncommon during the specified period of flow reductions. The outages also would contribute significantly to reducing the entrainment of striped bass and white perch. Outages at Indian Point, which are required to average 42 unit-days per year between May 10 and August 10, would contribute relatively little to overall entrainment mitigation. The baseline impacts used to develop Table 85 reflect the historical frequency of scheduled and unscheduled outages at Indian Point prior to the settlement; this frequency approximates the 42 unit-days specified in the agreement. Several outage schemes that would have increased the effectiveness of outages at Indian Point were developed (Englert et al. 1988) but were rejected because of cost and system reliability considerations. This seemingly incongruous aspect of the settlement reflects the willingness of the utilities to trade reduced summer power production at Bowline Point and Roseton (which normally would be operating during this period of high power demand) for the opportunity to operate Indian Point in the most economical manner.

Costs to the utilities of implementing the settlement agreement included capital and one-time costs, recurring annual costs, the carrying costs of capital, and other factors such as inflation and the costs of purchasing replacement power during outages. When all of these costs were considered, the revenue requirements for the settlement totaled approximately $180 million (1980 dollars; Table 86). The associated annual costs to customers, levelized over a 20-year recovery period, were approximately $17.7 million. Of the estimated total revenue requirements, slightly more than half ($104 million) were associated with direct mitigation of entrainment losses (flow reductions and outages). The direct cost of mitigating impingement (i.e., the installation of angled screens, or an equivalent alternative) at Indian Point was only about one third of the direct entrainment mitigation cost. This is consistent with the lesser contribution of impingement to total power plant impact on most species of concern (Barnthouse and Van Winkle 1988; Boreman and Goodyear 1988). The striped bass hatchery was relatively inexpensive to build and operate and offered potential mitigation for both entrain-

TABLE 86.—Projected costs (in millions of 1980 U.S. dollars) of implementing the Hudson River settlement agreement.

Agreement component	Capital or one-time costs	Base-year costs[a]	Total revenue requirement[b]
Flow reductions	15.0		27.3
Outage at Indian Point		5.6	65.1
July outage at Bowline Point		1.0	11.6
Angled screens	20.0		35.9
Hatchery		0.7	4.1
Monitoring and research	12.0	2.0	30.0
Other	0.5		8.2
Total	47.5	9.3	182.2
Annual levelized cost[c]			17.7

[a]Base year is 1981; inflation and other factors affect cost in later years.

[b]Includes capital carrying charges, operating costs, cost of replacement power, and inflation.

[c]Annual cost associated with recovery of total revenue requirements for all settlement conditions over a 20-year period.

TABLE 87.—Estimated costs (in millions of 1980 U.S. dollars) of construction and operation of cooling towers at Bowline Point, Roseton, and Indian Point.

Plant	Capital construction	Total revenue requirement[a]
Bowline Point	85.7	246.9
Roseton	71.5	260.2
Indian Point unit 2	150.1	702.8
Indian Point unit 3	170.0	594.0
Total	477.3	1,803.9
Annual levelized cost[b]		175.3

[a]Includes capital carrying charges, operations and maintenance costs, and the replacement cost of power.
[b]Annual cost associated with recovery of total revenue requirements for six cooling towers over a 20-year period.

ment and impingement. It was anticipated that research and monitoring programs would contribute less immediately to impact mitigation, but would increase the likelihood that future mitigation would be cost-effective.

Although these settlement costs were substantial, they were much smaller than the estimated costs of constructing and operating cooling towers. The total revenue requirements for the construction of cooling towers at all six of the generating units at Bowline Point, Indian Point, and Roseton (as specified in EPA's original permits) were estimated to be approximately $1,800 million, implying annual costs to customers, over a 20-year recovery period, of $175.3 million (Table 87). Thus, the settlement was expected to provide 15–43% (depending on species) of the impact reduction that might have been obtained with cooling towers (Table 85) at approximately 10% of the cost.

Biological Monitoring Program

The settlement agreement included provisions for a long-term biological monitoring program to be conducted by the utility companies with a budget of $2 million per year. The program has several objectives.

(1) The effectiveness of mitigation expected from reduced flows and outages is to be evaluated. The utilities are attempting to estimate the degree to which entrainment impacts are reduced each year by the outages and flow reductions imposed during that year. Both in-plant entrainment monitoring and riverwide ichthyoplankton sampling are being conducted to provide the necessary data. The information obtained will be used to determine whether refinement of the flow reduction and outage schedules would be beneficial.

(2) The effectiveness of the striped bass stocking program is to be evaluated. All hatchery-reared striped bass are being marked with coded wire tags. Sampling programs are being conducted to evaluate the survival of hatchery-reared striped bass after release and to estimate their contribution to subsequent spawning runs.

(3) Impingement at Bowline Point, Indian Point, and Roseton is being monitored. These data are being used to estimate the magnitude of impingement mortality at Hudson River power plants. Impingement counts are also being used to evaluate the effectiveness of the barrier net at Bowline Point and will be used to evaluate the effectiveness of new traveling screens at Indian Point. Impingement collections also provide a supplemental source of information on the biological characteristics of Hudson River fish populations.

(4) The status of selected fish populations in the Hudson River is being monitored for indications of power plant impacts. A riverwide survey is being conducted with a variety of sampling gears to monitor trends in the abundance of young-of-the-year striped bass, alosids, and bay anchovies, and of adult white perch and Atlantic tomcod.

Hudson River Foundation

The settlement negotiators recognized the potential value of creating an independent institution to fund studies of Hudson River environmental problems. To this end, the Hudson River Foundation for Science and Environmental Research was established "to sponsor scientific, economic and public policy research on matters of environmental, ecological and public health concern and to publish the results of such research." The Foundation began operation in May 1982 with an initial endowment of $12 million. The long-term goals of the Foundation, as expressed in the most recent program plan (Hudson River Foundation 1986), are to

- explain the dynamic interactions among the physical, chemical, and biological processes that are important to the fishery resources of the ecosystem;

- increase knowledge about the processes that influence spatial and temporal distributions and abundances of major organisms in the Hudson River;

- increase knowledge of the variability of these distributions in relation to natural and anthropogenically induced changes in environmental conditions;

- establish a system to describe comprehensively

all anthropogenic activities affecting the Hudson River;

- develop, through funding support, a community (scientific, governmental, and public) with expertise to make scientifically sound judgments about uses of the river's resources; and
- offer scientific and informational assistance to governmental bodies or the general public within the Foundation's mandate.

The Foundation solicits proposals for scientific and public policy research and for educational programs. Through 1987, 107 grants have been awarded. The current areas of primary interest to the Foundation are the life cycles and population biology of resource organisms; the base of Hudson River food webs; the source, disposition, and role of toxic substances; hydrodynamics and sediment transport; and education, public policy and decision making. The Foundation sponsors an annual Hudson River Symposium at which grantees present and discuss their work. It has also sponsored special symposia on "Critical Data Needs for Shad on the Atlantic Coast," "Issues Related to PCB Dredging in the Hudson River," and "Acidification Events and Anadromous Fishes." In addition, the Foundation offers travel grants and student fellowships, operates a field station at Garrison, New York, and coordinates tag returns for the utilities' biological monitoring program.

Acknowledgments

The authors gratefully acknowledge the helpful comments of S. W. Christensen, R. J. Klauda, D. J. Suszkowski, W. Van Winkle, D. S. Vaughan, and two anonymous reviewers. This is publication 3019 of the Environmental Sciences Division, Oak Ridge National Laboaratory.

References[3]

Barnthouse, L. W., J. Boreman, S. W. Christensen, C. P. Goodyear, W. Van Winkle, and D. S. Vaughan. 1984. Population biology in the courtroom: the Hudson River controversy. BioScience 34:14–19.

Barnthouse, L. W., and W. Van Winkle. 1988. Analysis of impingement impacts on Hudson River fish populations. American Fisheries Society Monograph 4: 182–190.

Bergen, G. S. P. 1988. The Hudson River cooling tower proceeding: interface between science and law. American Fisheries Society Monograph 4:302–306.

Boreman, J., and C. P. Goodyear. 1988. Estimates of entrainment mortality for striped bass and other fish species inhabiting the Hudson River estuary. American Fisheries Society Monograph 4:152–160.

Cannon, J. B., G. F. Cada, K. K. Campbell, D. W. Lee, and A. T. Szluha. 1979. Fish protection at steam-electric power plants: alternative screening devices. Oak Ridge National Laboratory, ORNL/TM-6472, Oak Ridge, Tennessee.

Christensen, S. W., and C. P. Goodyear. 1988. Testing the validity of stock–recruitment curve fits. American Fisheries Society Monograph 4:219–231.

EA Engineering, Science, and Technology. 1987. Hudson River striped bass hatchery: 1986 overview. Report to Consolidated Edison Company of New York.

Englert, T. L., J. Boreman, and H. Y. Chen. 1988. Plant flow reductions and outages as mitigative measures. American Fisheries Society Monograph 4:274–279.

Fletcher, R. I. 1985. Risk analysis for fish diversion experiments: pumped intake systems. Transactions of the American Fisheries Society 114:652–694.

Fletcher, R. I., and R. B. Deriso. 1988. Fishing in dangerous waters: remarks on a controversial appeal to spawner–recruit theory for long-term impact assessment. American Fisheries Society Monograph 4:232–244.

Goodyear, C. P. 1988. Implications of power plant mortality for management of the Hudson River striped bass fishery. American Fisheries Society Monograph 4:245–254.

Hudson River Foundation. 1986. Hudson River fund: goals and call for proposals—1987. Hudson River Foundation for Science and Environmental Research, New York.

Hutchison, J. B., and J. A. Matousek. 1988. Evaluation of a barrier net used to mitigate fish impingement at a Hudson River power plant intake. American Fisheries Society Monograph 4:280–285.

Lawler, J. P. 1988. Some considerations in applying stock–recruitment models to multiple-age spawning populations. American Fisheries Society Monograph 4:204–218.

McLaren, J. B., J. R. Young, T. B. Hoff, I. R. Savidge, and W. L. Kirk. 1988. Feasibility of supplementary stocking of age-0 striped bass in the Hudson River. American Fisheries Society Monograph 4:286–291.

Muessig, P. H., J. B. Hutchison, Jr., L. R. King, R. J. Ligotino, and M. Daley. 1988. Survival of fishes after impingement on traveling screens at Hudson River power plants. American Fisheries Society Monograph 4:170–181.

Savidge, I. R., J. B. Gladden, K. P. Campbell, and J. S. Ziesenis. 1988. Development and sensitivity analysis of impact assessment equations based on stock–recruitment theory. American Fisheries Society Monograph 4:191–203.

Talbot, A. R. 1983. Settling things: six case studies in environmental mediation. The Conservation Foundation, Washington, D.C.

Vaughan, D. S. 1988. Introduction [to entrainment and impingement impacts]. American Fisheries Society Monograph 4:121–123.

Yost, T. B. 1988. Science in the courtroom. American Fisheries Society Monograph 4:294–301.

---

[3]See Table 1 for sources of legal documents and unpublished reports pertaining to the Hudson River.

American Fisheries Society Monograph 4:274–279, 1988

Plant Flow Reductions and Outages as Mitigative Measures

THOMAS L. ENGLERT

Lawler, Matusky & Skelly Engineers, One Blue Hill Plaza, Pearl River, New York 10965, USA

JOHN BOREMAN

National Marine Fisheries Center, Northeast Fisheries Center, Woods Hole Laboratory
Woods Hole, Massachusetts 02543, USA

HWANG Y. CHEN

Lawler, Matusky & Skelly Engineers

Abstract.—The component of the Hudson River settlement agreement that was subjected to the most detailed technical analysis and received central attention during the settlement negotiations was the schedule for reductions in cooling water flow rates and outages for maintenance and refueling at the power plants. Factors considered in the development of such a schedule included possible increases in thermal mortality of ichthyoplankton as a result of flow reductions, plant-by-plant timing of flow reductions and outages to achieve maximum reduction of entrainment mortality, and costs of outages and their effect on system reliability. A mathematical model based on empirical spatiotemporal distributions of ichthyoplankton was used to evaluate the effectiveness of alternative schedules in reducing entrainment mortality of striped bass, white perch, river herrings, Atlantic tomcod, and bay anchovy. In conjunction with the other considerations identified above, the model output was used to develop a set of flow reductions and plant outages that mitigated entrainment impact at acceptable cost and allowed utilities to meet electric power demands. The reductions in entrainment mortality associated with various plant outages were used to develop a system of cross-credits that allowed the utilities as a group to receive credit for unscheduled outages at any of the plants. The model results also provided the basis for a point system to determine whether or not the utilities have met specified goals for outages over the life of the settlement agreement.

Throughout the negotiations that culminated in the Hudson River settlement agreement, considerable discussion and analysis centered on the magnitude of reductions in cooling water flows at the power plants that could be used to reduce the impact of the plants on the river biota. Two approaches to reducing the water withdrawals were considered: reduction of flows by operation of fewer pumps or by throttling pumps; and scheduled outages of units at times when the impact was expected to be largest. It became evident as discussions and the analyses proceeded, however, that there were several biological, economic, and system-reliability factors that needed to be considered to establish a realistic schedule for flow reductions and outages. For instance, during summer when ambient river temperatures are high, flow reductions could result in excessively warm discharge water and heightened mortality of entrained organisms.

It also became evident that the timing of flow reductions and outages at specific plants would greatly affect reductions in entrainment and ther-mal mortality. A further complication was that pumps at the Indian Point plant could not be throttled and the only means of reducing flow was to turn off the pumps entirely.

An important consideration was the cost of flow reductions and, in particular, of outages. During periods of high river temperatures, flow reductions increase back pressure and thereby reduce power output. Outages, of course, are extremely costly during periods of high power demand; the cost of a 1-week outage of a unit at Indian Point was estimated at $7 million (Austin 1981). The ability of the electric power grid to continue to supply electricity in the event of an unexpected failure of some component of the system was also an important consideration.

To provide a quantitative means of evaluating the various costs and benefits of numerous alternatives for flow reductions and outages at the three principal power plants under consideration (Roseton, Indian Point, and Bowline Point, comprising six units), a mathematical model was employed to provide a measure of conditional entrainment mortality, the percentage reduction in the number of young-of-the-year fish in the absence both of other sources of mortality and of density-dependent population responses. The empirical transport model (ETM), developed by the U.S. Environmental Protection Agency (EPA) and Oak Ridge National Laboratory (ORNL), was employed for this purpose (Boreman et al. 1982). Conditional entrainment mortality was selected as the measure of impact because it provided a direct, uncomplicated means of comparing the relative impacts of specific power plants on various life stages of the species under consideration. A model incorporating compensatory responses of the populations to power plant cropping would have provided a more realistic measure of impact; however, there was no agreement on how a compensatory response could be quantified for the species involved. Moreover, the effect of including compensation in a measure of relative impact probably would be small. In the remainder of this paper, we describe how the ETM was chosen, parameterized, and implemented, and how the model output was used to schedule flow reductions and outages of the power plants and to establish standards for utility compliance with the schedule.

Model Selection

During the course of the proceedings to determine whether cooling towers should be built at the Hudson River power plants, two principal models had played central roles in the evaluation of entrainment impact. These were the ETM, used by consultants to the EPA, and the real-time life cycle model (RTLC) used by consultants to the utilities (Christensen and Englert 1988, this volume). The ETM, which had been applied to several species, was based directly on ichthyoplankton data collected as part of the utility studies. It did not simulate hydrodynamic transport or vertical migration. The RTLC simulated organism transport in the estuary with equations describing the hydrodynamics of the Hudson River and the vertical and diurnal migration of striped bass larvae. It was decided for purposes of the settlement agreement to apply the ETM because of its ease of use and its applicability to species other than striped bass. In addition, it was learned through a series of model runs that the RTLC and ETM gave similar predictions of impact for striped bass when the same data were input to the two models (Englert and Boreman 1988, this volume).

Species Considered

Five fish species were selected for evaluation. Striped bass and American shad are economically important to recreational and commercial fisheries (Klauda et al. 1976; Hoff et al. 1988, this volume). White perch and bay anchovy are relatively abundant in entrainment and impingement samples (Barnthouse and Van Winkle 1988; Boreman and Goodyear 1988, both this volume). Atlantic tomcod has a peculiar life cycle that includes maturation within 1 year and a winter spawning season (McLaren et al. 1988, this volume). River herrings (alewife and blueback herring) were initially chosen by EPA consultants for estimation of entrainment and impingement impacts. During settlement negotiations, however, all parties agreed that the potential impacts on river herrings were too low for these species to be considered for scheduling of outages and flow reductions.

Baseline Parameters

Evaluation of various flow reductions and outage schemes required the establishment of baseline biological parameters that would be held constant while plant flows and operating schedules were varied. Baseline flow conditions and outage schedules were also specified to establish a reference impact against which the effectiveness of mitigation could be measured.

Schedules and rates of cooling water withdrawal projected for the five power plants were adapted directly from estimates provided by the utilities (Boreman and Goodyear 1988). The only outages incorporated into the baseline once-through and closed-cycle flow schedules were those expected for routine maintenance. Projected cooling water flows for Lovett and Danskammer were based on an average of the 1974 and 1975 schedules for outages and the assumption that Lovett units 1–3 and Danskammer units 1 and 2 would be used very little, if at all, in the future.

Biological parameter values used in the ETM to analyze the effects of flow reductions and outages were the same values used by EPA consultants in their direct testimony (Boreman et al. 1982; Boreman and Goodyear 1988). The only exception to those values was that spatial and temporal distributions of the entrainable life stages were based on averages of the 1974–1978 values, rather than on the 1974 and 1975 values. Parties in the settlement negotiations agreed that use of a 5-year average for the distributions probably represented future biological conditions better than either the

TABLE 88.—Percent effectiveness of alternative power plant outages and flow reductions in reducing conditional entrainment mortality of young Hudson River fish. Percentages are relative to the effectiveness of cooling towers (no cooling water withdrawals).

Alternative	Percent effectiveness				
	Striped bass	White perch	Atlantic tomcod	American shad	Bay anchovy
Bowline Point and Roseton outages (May–Jun)	8.6	7.5	−9.3	9.6	4.3
Bowline Point outage (Jul)	3.9	3.7	0	4.0	3.7
Indian Point outages[a] (18-month fuel cycle)					
42 d, May 1–Jul 31	0	2.8	−2.3	4.0	3.5
42 d, May 1–Jun 30	5.5	2.8	2.3	2.4	0.4
42 d, May 15–Jun 30	14.0	7.5	2.3	6.4	1.2
49 d, May 1–Jul 31	3.9	5.6	0	5.6	4.3
56 d, May 1–Jul 31	7.8	7.5	1.2	8.0	5.3
Flow reduction at Indian Point (dual-speed pump)	9.4	3.7	32.6	1.6	0.4
60% flow at Indian Point year-round	−33.1	0.9	[b]	13.1	17.3
12-month fuel cycle at Indian Point	7.8	7.5	1.2	8.0	5.3

[a]Outage periods are in unit-days.
[b]Not calculated because a thermal model is not available.

1974 or 1975 values. Furthermore, the distribution data for 1976–1978 were already summarized and easily obtainable during the negotiation period. Data pertaining to the other biological parameters (life stage durations, temporal egg distributions) were not as accessible and were expected to have little, if any, influence on the output of the ETM. Baseline plant impact parameters (organism withdrawal ratios and through-plant mortality) and thermal mortality equations were the same as presented in EPA testimony.

Application of the ETM

After the baseline parameters were developed for each of the species considered, ETM runs were made to compute the conditional mortalities under different scenarios of outage schedules and flow reductions. The percent effectiveness (PE) of a proposed mitigating measure was determined by comparing its estimated conditional mortality rate (CM_{pro}) with those computed for minimum (CM_{min}) and maximum (CM_{max}) mitigating measures:

$$PE = 100 \frac{CM_{min} - CM_{pro}}{CM_{min} - CM_{max}}.$$

The minimum mitigation case is the presettlement condition whereby power plants are run in once-through operating modes. The maximum mitigation is based on the hypothetical case of six cooling towers, two each at Bowline Point, Roseton, and Indian Point, which would require much smaller water withdrawals. The baseline CM_{min} values already included the effects of unscheduled

outages at Indian Point, so the computed percent effectiveness was 0 for striped bass when 42 days of outages were scheduled at that plant during May 1–July 31 (Table 88). However, targeting the Indian Point outage during May 10–August 10 would provide additional protection from entrainment mortality. Routine maintenance outages at Bowline Point and Roseton were included in the minimum mitigation case but were scheduled during April and May. If these outages occurred in May–June, calculated mortality of Atlantic tomcod would increase 9.3%, although other species would be helped by the later scheduling (Table 88). This illustrates the types of trade-offs in mitigation alternatives among species, plant operating schedules, and costs that were necessary in arriving at the final set of outages in the settlement agreement.

One major consideration in the selection of flow reduction was the thermal mortality of entrained organisms. Thermal mortality is determined from ambient (acclimation) temperature, exposure temperature, exposure duration, and, in some cases, length of the organisms. Although fewer organisms are entrained when plant flows are reduced, entrainment mortality generally increases due to higher temperatures in the discharge, especially when the ambient temperature is only slightly below the lethal threshold (Boreman et al. 1982). For example, when the flow at Indian Point was modeled at 60% of full flow year-round, the estimated entrainment loss of striped bass increased by about 33% (Table 88), a result of increased discharge temperatures after May 30. Therefore, flow reductions were limited to those periods of the year when the ambient temperatures are lower than those normally reached by May 30.

Based on these evaluations, the following outage schedules were judged as the "optimal" mitigative measures acceptable to all parties and were, therefore, included in the settlement agreement.

Bowline Point: off line for 30 unit-days between May 15 and June 30 and for 31 unit-days during July each year.

Roseton: off line for 30 unit-days between May 15 and June 30 each year.

Indian Point: off line for an average of 42 unit-days per year between May 10 and August 10 during the 10-year settlement period from 1981 to 1991.

The settlement agreement also contained the following flow reduction requirements.

Indian Point: installation of dual-speed pumps at units 2 and 3 to allow operation at reduced flows when ambient temperatures permit, primarily before May 30.

Bowline Point and Roseton: operation at reduced flows when ambient temperatures permit, primarily before May 30.

These mitigative measures are intended principally to reduce entrainment losses at the existing power plants. The estimated reductions in entrainment losses of striped bass, white perch, and Atlantic tomcod achieved by these measures are 30, 25, and 45%, respectively, of the reduction that would result from construction of cooling towers at Bowline Point, Roseton, and Indian Point. Other mitigative measures, such as the barrier net at Bowline Point (Hutchison and Matousek 1988, this volume), are intended to reduce impingement losses. Plant outages and flow reductions also will reduce impingement losses; however, relatively little impingement occurs during the entrainment period.

Cross-Credit System

As stated in the settlement agreement, the outage obligations at Bowline Point and Roseton must be met each year, but they may be satisfied in alternative ways allowed by a cross-plant outage credit system. The outages at Indian Point are subject to the following compliance schedule.

A minimum of 140 unit-days of outage shall have been taken at the end of the fourth year of the ten year term of this agreement. If this minimum is not met, a minimum of 175 unit-days of outage shall have been taken by the end of the fifth year.

A minimum of 315 unit-days of outages shall have been taken at the end of the eighth year of this agreement. If this minimum is not met, a minimum of 371 unit-days of outage shall have been taken at the end of the ninth year.

At the end of the tenth year of this agreement, 420 unit-days of outage shall have been taken, or 434 unit-days if it is elected to assure the additional 14 unit-days of outage at Indian Point in place of the aggregate 31 unit-days outage required at Bowline in the month of July during the second five-year term of the agreement.

In six of the ten years of this agreement, there shall be a minimum of 14 unit-days of outage between May 15 and July 15.

The cross-credit system was established to allow the utilities greater flexibility in meeting the outage schedule while still achieving the same mitigation. This system permits the utilities to gain credits for the planned or forced outages at each plant beyond those required in the settlement agreement and to apply these credits against the required Bowline Point and Roseton outages. The system is based on the conditional entrainment mortality computed by the ETM for striped bass; a mitigation index expressed in credit points is assigned to each plant unit each week between mid-April and mid-September (Table 89). The base condition is the conditional mortality of striped bass resulting from full-load operation at all plants. The reduction in conditional mortality from the base condition predicted to result when a unit is taken out of service during a given week is multiplied by a factor of 10 for convenience to obtain the cross-credit points.

Whenever a specified outage at either the Bowline Point or the Roseton plant is missed, it can be made up by outages at other plants that achieve a number of credit points equal to that of the missed outage. For purposes of assessing cross-credits, the outages at Bowline Point and Roseton have been computed to have the following credit point (CP) values, which are simply the sums of the weekly values (Table 89) prorated, when necessary, for partial weeks.

Bowline: 30 unit-days, May 15–June 30, 4.4 CP.
Bowline: 31 unit-days, July 1–July 31, 5.5 CP.
Roseton: 30 unit-days, May 15–June 30, 4.0 CP.

TABLE 89.—Cross-credit values of single-unit and double-unit outages for each Hudson River power plant by week.

Week	Roseton		Indian Point		Bowline Point	
	Two units off line	One unit off line	Two units off line	One unit off line	Two units off line	One unit off line
Apr 19–Apr 25	0	0	0	0	0	0
Apr 26–May 2	0	0	0.1	0.1	0	0
May 3–May 9	0.3	0	1.1	0.6	0	0
May 10–May 16	1.3	0.3	3.4	1.7	0.2	0.1
May 17–May 23	3.3	0.8	10.0	5.0	0.7	0.4
May 24–May 30	4.1	0.9	15.7	7.8	1.2	0.6
May 31–Jun 6	4.0	1.2	17.1	8.5	2.1	1.1
Jun 7–Jun 13	3.6	1.2	19.5	9.7	2.4	1.2
Jun 14–Jun 20	3.1	1.0	17.2	8.6	2.5	1.3
Jun 21–Jun 27	2.0	0.7	12.6	6.3	2.9	1.5
Jun 28–Jul 4	1.3	0.5	8.5	4.3	3.3	1.7
Jul 5–Jul 11	0.9	0.4	6.6	3.3	3.4	1.7
Jul 12–Jul 18	0.7	0.3	5.2	2.6	2.9	1.5
Jul 19–Jul 25	0.4	0.2	3.0	1.5	1.7	0.9
Jul 26–Aug 1	0.1	0.1	1.1	0.6	0.7	0.4
Aug 2–Aug 8	0	0	0.4	0.2	0.2	0.1
Aug 9–Aug 15	0	0	0.1	0.1	0.1	0.1
Aug 16–Aug 22	0	0	0	0	0	0
Aug 23–Aug 29	0	0	0	0	0	0
Aug 30–Sep 5	0	0	0	0	0	0
Sep 6–Sep 12	0	0	0	0	0	0
Sep 13–Sep 19	0	0	0	0	0	0

If the May–June outage at Bowline Point is missed entirely, it could be made up by additional outages at other plants beyond those required under the settlement agreement, provided those outages totaled 4.4 credit points. As can be seen from Table 89, only a few additional days of outage at Indian Point would make up the deficiency during most of the May–June period.

Of course, if an Indian Point outage (for example) is used to provide credits for a Bowline Point or Roseton outage obligation, it cannot also be credited against the Indian Point obligation.

Discussion

During the protracted EPA hearings on the Hudson River power plant case, it became apparent that all parties would never agree about the probable long-term impact of power plant operations on fish resources. As a result, settlement discussions were begun and the focus shifted instead to the immediate reductions in conditional entrainment and impingement mortality that would occur with various mitigation alternatives. Acceptance of the conditional mortality generated by the ETM as a measure of impact implied acceptance that biological phenomena associated with density-dependent mortality and other regulatory mechanisms affecting fish populations were not issues in the settlement. It was evident from the hearing record that technical resolution of

those issues could not be achieved without additional long-term studies. However, the conditional mortality computed by the ETM provided a means of comparing the relative effectiveness of various mitigation measures without concern about density-dependent mortality.

With agreed-upon input parameter values for the ETM, evaluation of mitigation schemes was simply a matter of altering the values for power plant flows in the ETM until an "optimal" set of outages and flow reductions acceptable to the utilities, the regulatory agencies, and the intervenor groups was obtained. Placing the negotiations into a purely analytical framework, however, meant that the negotiators had to accept the ETM output on faith. Arguments began to focus on the numbers produced by the runs (estimates of conditional entrainment mortality rates) rather than on their biological importance. At one point during the negotiations, for example, an argument arose concerning whether or not a 5% increase in the conditional mortality of a certain species was an acceptable consequence of a 2-week delay in the shutdown of one of the power plants. At this point the forest had been lost for the trees, because that rise in mortality was unimportant biologically and could not be detected with state-of-the-art techniques for measuring year-class strength in fish populations (Vaughan and Van Winkle 1982). Once the scientists explained to the negotiators that the argument was now in the realm of pure speculation, the negotiations were back on course.

The ETM proved to be a useful device in the negotiation of the settlement agreement. It provided a means of assessing the relative effectiveness of the various combinations of plant outages and flow reductions, and it provided a means by which the difference between fact and speculation could be recognized. Having a quantitative measure of the relative effectiveness of various mitigation measures in reducing entrainment impact also permitted an evaluation of the cost-effectiveness of the various measures.

References[1]

Austin, G. E. 1981. Letter to J. Michael Harrison, Administrative Law Judge, Public Service Commission. Response to staff's interrogatory re cooling tower settlement agreement. Consolidated Edison Company of New York.

Barnthouse, L. W., and W. Van Winkle. 1988. Analysis of impingement impacts on Hudson River fish populations. American Fisheries Society Monograph 4: 182–190.

Boreman, J., and C. P. Goodyear. 1988. Estimates of entrainment mortality for striped bass and other fish species inhabiting the Hudson River estuary. American Fisheries Society Monograph 4:152–160.

Boreman, J. and seven coauthors. 1982. Entrainment impact estimates for six fish populations inhabiting the Hudson River estuary, volume 1. The impact of entrainment and impingement of fish populations in the Hudson River estuary. Oak Ridge National Laboratory, ORNL/NUREG/TM-385/V1, Oak Ridge, Tennessee.

Christensen, S. W., and T. L. Englert. 1988. Historical developemnt of entrainment models for Hudson River striped bass. American Fisheries Society Monograph 4:133–142.

Englert, T. L., and J. Boreman. 1988. Historical review of entrainment impacts and the factors influencing them. American Fisheries Society Monograph 4: 143–151.

Hoff, T. B., J. B. McLaren, R. E. Schmidt, and W. P. Dey. 1988. Stock characteristics of Hudson River striped bass. American Fisheries Society Monograph 4:59–68.

Hutchinson, J. B., and J. A. Matousek. 1988. Evaluation of a barrier net used to mitigate fish impingement at a Hudson River power plant intake. American Fisheries Society Monograph 4:280–285.

Klauda, R. J., K. P. Campbell, and M. Nittel. 1976. Commercial fishery for American shad in the Hudson River: fishing effort and stock abundance trends. Pages 107–134 in Proceedings of an American shad workshop. University of Massachusetts, Amherst.

McLaren, J. B., T. H. Peck, W. P. Dey, and M. Gardinier. 1988. Biology of Atlantic tomcod in the Hudson River estuary. American Fisheries Society Monograph 4:102–112.

Vaughan, D. S., and W. Van Winkle. 1982. Corrected analysis of the ability to detect reduction in year-class strength of the Hudson River white perch (*Morone americana*) population. Canadian Journal of Fisheries and Aquatic Sciences 39:782–785.

---

[1]See Table 1 for sources of legal documents and unpublished reports pertaining to the Hudson River.

American Fisheries Society Monograph 4:280–285, 1988

Evaluation of a Barrier Net Used to Mitigate Fish Impingement at a Hudson River Power Plant Intake

J. B. HUTCHISON, JR.

Orange and Rockland Utilities, Incorporated
One Blue Hill Plaza, Pearl River, New York 10965, USA

J. A. MATOUSEK

Lawler, Matusky & Skelly Engineers
One Blue Hill Plaza, Pearl River, New York 10965, USA

Abstract.—A multifilament nylon net of 0.95-cm bar mesh was deployed as a physical barrier to fish in front of the Bowline Point power plant cooling water intake on the Hudson River from 1976 to 1985. The barrier net was deployed during the historical peak impingement months of October–May. The primary species impinged on the intake screens during this period were young-of-the-year and yearling white perch, striped bass, rainbow smelt, alewife, blueback herring, and American shad, generally ranging from 5 to 10 cm in total length. When the barrier net was deployed, median impingement of all fish was 91% lower than during comparable periods before the net was installed. A mark–recapture population estimate indicated that 230,000 yearling striped bass and white perch were in the embayment outside the net in April 1982; over a 9-d study period, only 1.6% of this estimated population was impinged. Concurrent survival probability studies of fish marked and released at locations inside and outside the barrier net showed that fish released inside had 72% lower survival ($P \le 0.0001$) than those released outside the net. Gill-net catches were significantly lower inside than outside the net ($P \le 0.05$). This study led to improvements in barrier-net deployment, including changes in the anchoring system, use of a debris boom, installation of an air-bubbler system to prevent ice accumulation, and sectioning of the net to facilitate removal and cleaning.

The impingement of large numbers of fish on the cooling water intake screens of power generating stations has led to management concerns over the potential impact of impingement on the aquatic community. Several methods to minimize the biological impacts associated with large volume water withdrawal are currently in use or under investigation. They include special intake screens to minimize mortality, reduction in water withdrawal (cooling towers), behavioral barriers (pneumatic guns, air-bubble curtains), and physical barriers (barrier nets, porous dikes).

The Bowline Point generating station, located on the estuarine lower Hudson River approximately 60 km north of New York City, commenced commercial operation in the fall of 1972. A monitoring program begun at that time indicated increased impingement of fish on the intake traveling screens during cold-water periods. Most impinged fish were young-of-the-year and yearling white perch, but many striped bass, an important commercial and sport species, were among them. Because the plant's cooling water intake is located on a 49-hectare embayment (Hutchison 1988, this volume), the use of a barrier net was considered as a means to reduce impingement.

The initial barrier-net evaluation was conducted during spring 1976 and fall–winter 1976–1977. Initial results indicated that (1) a barrier net could be deployed without affecting plant operation, (2) the net would be able to withstand normal debris loadings (logs, tires, leaves) and algal growth, and (3) 0.95-cm-bar-mesh netting was more effective than 1.27-cm netting at reducing the number of fish entering the plant intake (Edwards and Hutchison 1980).

A seasonal program of barrier-net deployment (October–May) began in 1977. The impingement monitoring program begun in 1972 was continued as a means of monitoring barrier-net effectiveness. In addition, special mark–recapture and fish-sampling studies were conducted inside and outside the barrier net.

The barrier net was adopted as an impingement-reduction technique during the Hudson River settlement negotiations; its continued deployment at the Bowline Point plant was made a condition of the final settlement agreement (Barnthouse et al. 1988, this volume). This paper summarizes data from pre- and postdeployment impingement collections and from postdeployment studies to determine the effectiveness of the barrier net in

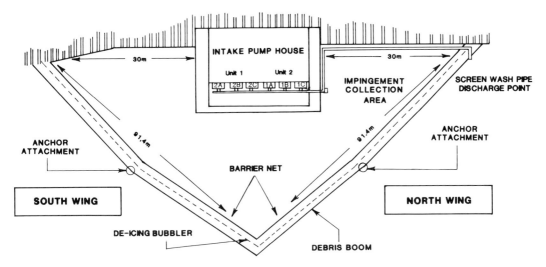

FIGURE 83.—Deployment schematic of the Bowline Point barrier net.

mitigating impingement. A discussion of design and deployment modifications made over the years to further increase the effectiveness of the net is also included.

Methods

Barrier-net design and deployment.—The barrier net, 182 m long and 15 m deep, was set in a V-configuration approximately 15 m from the intake at the closest point (Figure 83). The mean water depth at the net apex was approximately 11.5 m; depths ranged from 11 to 12 m at low and high tide, respectively. The depth decreased gradually along each wing to a mean of 3 m at each bulkhead. Multifilament knotted nylon net material was used because of its availability, ease of handling, low replacement cost, and high tensile strength.

In 1979, a debris boom was placed in front of the barrier net, and an air bubbler was suspended at a depth of 3 m along the outside of the net to reduce ice formation during the winter months. In 1981, net construction was changed from three to seven panels of equal length, laced together during deployment, which facilitated net maintenance and replacement. Danforth anchors were used at the apex and along the wings.

Impingement studies.—Starting in 1973, 24-h impingement collections were made one to three times per week throughout the October–May evaluation period at the intake traveling screens. All fish collected were identified to species, and the daily number per 10^6 m^3 of water withdrawn by the plant (impingement rate) was calculated. Daily

impingement rate data were grouped into pre- and postdeployment periods. A nonparametric median test (Siegel 1956) was used to determine whether or not there were significant differences in median impingement rates between the two periods.

Mark–recapture studies.—The number of striped bass and white perch in the embayment was estimated between 5 April and 13 April 1982. Impingement monitoring conducted during the same time period permitted the calculation of barrier-net effectiveness for the duration of this experiment. Fish movement in or out of the embayment was blocked by a 0.95-cm-bar-mesh net placed across the inlet. Fish to be marked were collected in a 4.9-m otter trawl towed for 5 min and in 30.5-m beach seines. The fish were held onshore in 75-L containers for 1 h before they were marked by fin-clip combinations that designated date and release location; then, they were held for another hour and returned to the embayment. Collection, marking, and release started on 5 April and continued over four consecutive days. On 12–13 April, surface, midwater, and bottom trawls were used to collect fish for use in modified Petersen and Schnabel population estimates (Ricker 1975).

Impingement was monitored on 7 of 9 d of the survey with 24-h collections. The estimated number of white perch and striped bass impinged on the intake screens for the test period was derived by multiplying the mean impingement rate by the daily plant flow.

The proportion of marked fish surviving over time was analyzed by life table techniques (Cutler

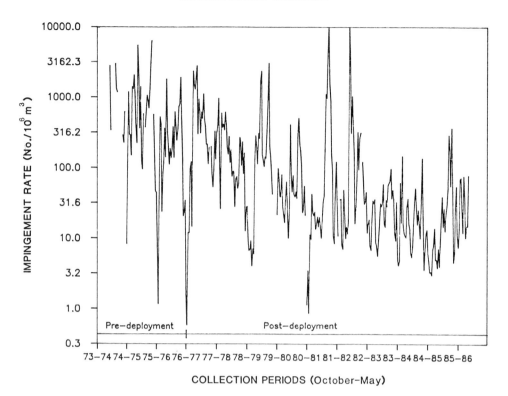

FIGURE 84.—October–May impingement rates (fish/10^6 m^3 of intake water) at the Bowline Point station before and after deployment of a barrier net, 1973–1986.

and Ederer 1958; Miller 1981). A generalized Wilcoxon test was used to determine whether or not there were significant differences in survival between fish released inside and those released outside the barrier net.

Gill-net studies.—Monofilament experimental gill nets with 15.2 × 3.1-m panels of 0.95-, 1.3-, 2.5-, 3.8-, and 5.1-cm square mesh were fished for approximately 24 h on each of 13 dates during 1979–1982 to evaluate the barrier net. On each sampling date (1979–1981), bottom gill nets were anchored at three or four locations outside and at one location inside the barrier net. During 1982, two gill-net stations were sampled inside the barrier net; surface-anchored gill nets were sampled at the same time as bottom nets at all inside and outside net stations. Gill nets were set parallel to the barrier net inside the net and parallel to the shore outside the net in the embayment. Fish abundance was expressed as number caught per fishing hour, and collections from inside and outside the barrier net were compared. A *t*-test was used to compare the total number of fish caught at the inside station with the total caught at each outside station by date.

Results and Discussion

Plant Impingement Studies

Impingement on the intake traveling screens has been monitored at the Bowline Point station since 1972. Species composition over the 13-year period was approximately 75% white perch, 15% striped bass, 5% rainbow smelt, 1% clupeids, and less than 4% other species. Young-of-the-year and yearling fish (5–10 cm in total length) made up 90–95% of the collections. During September 1973–March 1976, before the barrier net was deployed, the median daily impingement rate for the October–May period was 371 fish/10^6 m^3 (Figure 84; Table 90), compared with 33 fish/10^6 m^3 during 1977–1985 when the barrier net was deployed. The difference was significant ($\chi^2 = 48.7$; $P \leq 0.0001$). When the barrier net failed, immediate and extremely high impingement rates were observed; when the net was repaired, the impingement rate decreased dramatically. There has been no evidence of a change in fish distribution or a decline in riverwide fish abundance to account for those lower impingement rates (Battelle 1983;

TABLE 90.—Statistical analysis of impingement data collected during periods of barrier-net deployment from 1973 to 1985 at the Bowline Point plant.

Impingement statistic	Predeployment (1973–1976)	Postdeployment (1977–1985)
Mean rate (fish/10⁶ m³)	821.8	232.3
SE	132.9	59.0
Number of observations	74	262
Median rate (fish/10⁶ m³)	371.3	33.0

Boreman and Klauda 1988, this volume; Klauda et al. 1988, this volume).

Mark–Recapture Studies

Two separate mark–recapture studies incorporating release locations inside and outside the barrier net were conducted to evaluate barrier-net effectiveness. Fish-marking (fin clips, latex injection) and impingement-monitoring studies were conducted on 10 dates between December 1977 and May 1978 (Edwards and Hutchison 1980). The percentage of marked fish released inside the barrier net and subsequently impinged ranged from 8.6 to 60.7% and averaged 41.6%. Corre-

sponding recovery percentages for marked fish from outside the net ranged from 0 to 22.3% and averaged 4.2%. These experiments were repeated during 1979–1980 with similar results. Because the embayment itself was not blocked off, some of the marked fish released outside the barrier net could have left the area.

In the 1982 survey, impingement was monitored for 7 of 9 d following the initial release of marked fish. Of the fish released inside the barrier net and recaptured, 93% were impinged within 2 d and 99% within 6 d; of those released outside and recaptured, 55% were impinged within 2 d and 96% within 6 d.

Fish released outside the barrier net had significantly higher survival rates ($P \leq 0.0001$) than fish released inside the net (Figure 85; Table 91). Survival rates of white perch and striped bass were not significantly different from each other (Table 91).

The estimated embayment population outside the barrier net during the 9-d mark–recapture study was 214,390 fish based on the modified Petersen calculation and 229,555 fish based on the

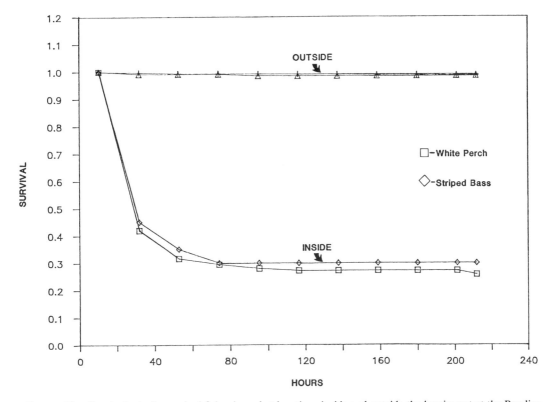

FIGURE 85.—Survival rate for marked fish released at locations inside and outside the barrier net at the Bowline Point plant.

TABLE 91.—Life table analysis of marked striped bass and white perch released at locations inside and outside a barrier net deployed in front of the Bowline Point plant cooling water intake and recaptured on the plant's intake screens (5–13 April 1982).

Species and release location	Initial number	Median survival time, h (SE)	Proportion surviving (SE)	Breslow statistic	P
Striped bass					
Inside	20	19.25 (6.47)	0.3000 (0.1025)		
Outside	317	>211	0.9842 (0.0070)	191.681	<0.0001
White perch					
Inside	136	18.23 (2.22)	0.2577 (0.0387)		
Outside	1,394	>211	0.9871 (0.0031)	1112.336	<0.0001
Inside release					
Striped bass	20	19.25 (6.47)	0.3000 (0.1025)		
White perch	136	18.23 (2.22)	0.2577 (0.0387)	0.085	0.7702
Outside release					
Striped bass	317	>211	0.9842 (0.0070)		
White perch	1,394	>211	0.9871 (0.0031)	0.348	0.5550

modified Schnabel calculation (Table 92). Based on the study period impingement rate, an estimated 3,649 white perch and striped bass from this population were impinged. If all fish residing in the embayment were vulnerable to impingement, approximately 1.6% of this population was impinged over the 9-d study period, a daily impingement rate of 0.18%.

Gill-Net Studies

Twenty-five species were collected by gill net from inside and outside the barrier net; white perch constituted 79% and striped bass 4% of the total catch. Ninety percent and 84% of the catch were collected in 2.54-cm (or smaller) mesh panels at stations outside and inside the barrier net, respectively. The mean gill-net catch rate (38.2 fish/h) outside the barrier was significantly greater ($P \leq 0.05$) than the mean catch rate inside the barrier (7.6 fish/h; Table 93).

Deployment Modifications

Barrier-net deployment techniques and net construction have been modified over the evaluation period. The initial modification was the addition of an air bubbler along the entire length of the net to minimize ice buildup, which had caused the net to lift from the bottom. A debris boom was added to stop large floating debris from tearing or being caught in the net and pulling it down. Construction of the net in panels (currently seven), laced together at the time of deployment, facilitates repair and cleaning because individual panels can be removed. Large cement weights used during the initial deployment years required heavy equipment and divers during deployment and maintenance. Replacement of these weights with buoyed Danforth anchors in conjunction with the use of panels permits net maintenance and deployment without divers.

TABLE 92.—Percentage of yearling and older striped bass and white perch impinged at the Bowline Point plant during a 9-d mark–recapture experiment. A barrier net was deployed in front of the water intake structure, and another blocked the embayment from which intake water was drawn. Numbers in parentheses are 95% confidence intervals about the population means.

Population estimation procedure	Sampling date[a]	Cumulative number marked	Number recaptured[b]	Number captured[c]	Population size	% of population impinged
Modified Petersen	6 Apr	968	4	891	172,870	
	7 Apr	1,068	10	1,107	107,677	
	8 Apr	1,300	6	1,004	186,786	
	12 Apr	1,683	6	1,116	268,872	
	13 Apr	1,667	6	1,408	335,745	
Mean					214,390 (133,342–295,438)	1.7
Modified Schnabel	6–13 Apr				229,555 (164,324–332,251)	1.6

[a]Collection and marking of fish began on 5 April and ended on 8 April.
[b]Thirteen additional recaptures were not used for population estimates, including six recaptured on the first day (5 April) and five collected in impingement samples for the period 9–11 April when no embayment trawls were conducted.
[c]Captured in trawl and impingement collections.

TABLE 93.—Gill-net catches inside and outside a barrier net located around the cooling water intake at the Bowline Point plant.

Date net was set	Surface (S) or bottom (B) deployment	Catch/h inside[a] (gill nets deployed)	Catch/h outside[a] (gill nets deployed)
14 Mar 1979	B	9 (1)	18.3 (3)
28 Mar 1979	B	29 (1)	32.3 (3)
11 Apr 1979	B	0 (1)	122.7 (3)
25 Apr 1979	B	9 (1)	52.0 (3)
22 May 1980	B	1 (1)	13.0 (3)
19 Nov 1980	B	3 (1)	11.0 (3)
17 Dec 1980	B	0 (1)	8.7 (3)
3 Apr 1981	B	8 (1)	104.0 (4)
28 Apr 1981	B	42 (1)	79.5 (4)
12 May 1981	B	11 (1)	54.8 (4)
31 Mar 1982	B	5.0 (2)	60.8 (4)
3 May 1982	B	7.5 (2)	37.2 (4)
25 May 1982	B	5.5 (2)	23.8 (4)
31 Mar 1982	S	0.5 (2)	3.5 (4)
3 May 1982	S	5.0 (2)	20.0 (4)
25 May 1982	S	4.5 (2)	24.2 (4)

[a] Mean catch when more than one gill net was deployed.

Several survey methods rather than one technique were selected to evaluate the effectiveness of the barrier net. Each method had its limitations, including possible marking mortality bias in the mark–recapture program, influence of trends in riverwide populations, and an open embayment population during the Edwards and Hutchison (1980) marking studies and gill-net studies reported in this paper. The combination of methods, however, demonstrated that the barrier net reduced the rate at which fish were drawn into the Bowline Point water intake.

Acknowledgments

We are grateful to the many staff members at Lawler, Matusky & Skelly Engineers and at Orange and Rockland Utilities, Incorporated, who assisted in the design, deployment, and evaluation of the barrier net. Special thanks go to Ronald J. Klauda, Lawrence W. Barnthouse, and Douglas S. Vaughan for their critical reviews and suggestions for this manuscript and to Alan W. Wells for analysis of the survival data and review of the overall program results. The studies were conducted under contract with Orange and Rockland Utilities, Incorporated, and jointly funded by Central Hudson Gas & Electric Corporation, Consolidated Edison Company of New York, Incorporated, New York Power Authority, Niagara Mohawk Power Corporation, and Orange and Rockland Utilities, Incorporated.

References

Barnthouse, L. W., J. Boreman, T. L. Englert, W. L. Kirk, and E. G. Horn. 1988 Hudson River settlement agreement: technical rationale and cost considerations. American Fisheries Society Monograph 4:267–273.

Battelle. 1983. 1980 and 1981 year class report for the Hudson River estuary monitoring program. Report to Consolidated Edison Company of New York.

Boreman, J., and R. Klauda. 1988. Distributions of early life stages of striped bass in the Hudson River estuary, 1974–1979. American Fisheries Society Monograph 4:53–58.

Cutler, S. J., and F. Ederer. 1958. Maximum utilization of the life table method in analyzing survival. Journal of Chronic Diseases 8:699–713.

Edwards, S. J., and J. B. Hutchison, Jr. 1980. Effectiveness of a barrier net in reducing white perch (*Morone americana*) and striped bass (*Morone saxatilis*) impingement. Environmental Science and Technology 14:210–213.

Hutchison, J. B., Jr. 1988. Technical descriptions of Hudson River electricity generating stations. American Fisheries Society Monograph 4:113–120.

Klauda, R., J. McLaren, R. E. Schmidt, and W. P. Dey. 1988. Life history of white perch in the Hudson River estuary. American Fisheries Society Monograph 4:69–88.

Miller, R. G., Jr. 1981. Survival analysis. Wiley, New York.

Ricker, W. E. 1975. Computation and interpretation of biological statistics of fish populations. Fisheries Research Board of Canada Bulletin 191.

Siegel, S. 1956. Non-parametric statistics for the behavioral sciences. McGraw-Hill, New York.

American Fisheries Society Monograph 4:286–291, 1988

Feasibility of Supplementary Stocking of Age-0 Striped Bass in the Hudson River

JAMES B. MCLAREN,[1] JOHN R. YOUNG,[2] THOMAS B. HOFF,[3]
IRVIN R. SAVIDGE,[4] AND WILLIAM L. KIRK[2]

Texas Instruments Incorporated, Ecological Services Group
Buchanan, New York 10511, USA

Abstract.—A study was performed to determine the survival, movements, growth, and sexual maturation of hatchery-reared striped bass released into the Hudson River. During 1973 through 1975, 318,585 age-0 striped bass derived from Hudson River brood stock were marked with fin clips and magnetized nose tags. Stocked fish were distributed into river regions approximately in proportion to the expected distribution of resident wild young of the year. Recapture efforts from 1973 through 1980 yielded 1,925 returns of age-0 to age-VI fish. The 6–9-month postrelease survival of hatchery-reared fish was significantly better than that of wild fish for the 1975 stocking, and did not differ significantly from wild fish survival for the 1973 and 1974 stockings. Recaptured fish that had been at large up to 6 years indicated the growth and sexual maturation of stocked fish were similar to those of wild fish.

Artificial propagation of striped bass from Hudson River brood fish was initially proposed as a mitigative measure to offset losses resulting from operation of the proposed pumped-storage hydroelectric plant at Cornwall, New York. The feasibility of a hatchery and the possible detrimental effects of stocking on the native striped bass population were debated during hearings before the Federal Power Commission in 1966. A study to determine the feasibility of this approach, including the establishment of a hatchery along the river, was stipulated as a condition of the license issued by the Federal Power Commission in 1970 for the proposed pumped-storage plant. This study began in 1973 at a pilot hatchery in Verplanck, New York, 5 km south of Peekskill (see Figure 3). Artificial propagation of striped bass from the Verplanck hatchery continues as a condition of the case settlement regarding the Cornwall pumped-storage plant and other Hudson River generating stations.

Fishery managers have recently been considering artificial propagation as a means of supplementing or reestablishing striped bass stocks in coastal rivers and estuaries (ASMFC 1987). States where programs have been conducted include Georgia (Hornsby and Hall 1981), Florida (Wooley and Crateau 1983), South Carolina (Harrell 1978, 1979, 1980, 1981), and Mississippi (McIlwain 1981). There have been few attempts to estimate the survival of released fish or to quantify the ultimate contribution of striped bass of hatchery origin to a naturally reproducing population.

Three years of intensive striped bass stocking in the Hudson River (1973, 1974, and 1975), and the ensuing results through 1980 will be discussed. The primary objective at the beginning of the study was to determine the poststocking survival of age-0 striped bass released into the Hudson River. Specific objectives were to estimate immediate mortality attributable to handling, to compare the survival of stocked fish to that of wild striped bass of the same year class during the 3–9 months after stocking, and to compare growth and sexual maturation of hatchery-reared and wild striped bass returning to the Hudson River during spawning runs.

Methods

Spawning and rearing.—Striped bass were reared during 1973, 1974, and 1975. Parental fish (16–36 females and 19–35 males) were selected each year from among wild fish captured from the Hudson River, primarily by gill nets, during the spawning season (May and June). Spawning and

---

[1]Present address: Beak Consultants Incorporated, 12072 Main Road, Akron, New York 14001, USA.

[2]Present address: Consolidated Edison Company of New York, 4 Irving Place, New York, New York 10003, USA.

[3]Present address: Mid-Atlantic Fishery Management Council, Room 2115, Federal Building, 300 South New Street, Dover, Delaware 19901, USA.

[4]Present address: Department of Chemistry, University of Colorado, Box 215, Boulder, Colorado, 80309, USA.

hatching techniques (TI 1977a) were adapted from those developed by Stevens (1966) and subsequently refined by Bayless (1972) and Bonn et al. (1976).

Most brood females received intramuscular injections of chorionic gonadotropin to hasten ovulation. Males were injected only if milt production was limited or if repeated use was anticipated. The progress of ovarian development was monitored during early stages by catheterization and microscopic examination (Bayless 1972) and during later stages by abdominal palpation. Eggs and milt were manually stripped. Fertilized eggs were incubated in MacDonald jars for 34–101 h until hatching. Hatching larvae were carried by water flow from jars into 75-L aquaria with screened drains, where they were held 1–5 d until shipment. Larvae were shipped by air in plastic bags (13,000–23,000 larvae; 0.3–0.7 g/L) to facilities in Florida and Oklahoma in 1973 and North Carolina in 1974 and 1975; survival was 90–99%. After up to a week of acclimation in troughs, the larvae were stocked into rearing ponds or tanks. When the fish were 75–150 mm long, they were returned by air or tank truck (60 g/L; 98–99% survival) to New York for stocking.

Marking and stocking.—Hatchery-reared juveniles received two types of marks to permit future identification. In addition to clips of either dorsal fin or both pelvic fins, which were specific to release time, wire nose tags were inserted into the nasal cartilage and magnetized. In 1975, the wire tags were color-coded for release sites to provide information on postrelease movement patterns. Because of a lack of adequate holding facilities at Verplanck, fish shipped by air in 1973 and 1974 were marked at the rearing facilities by project crews, and were stocked upon arrival at Verplanck. Fish shipped by truck in late fall 1974 and throughout 1975 were marked after arrival at Verplanck by the same crews. Fish were anesthetized in 50 mg quinaldine/L prior to marking and held in solutions containing NaCl (5.1 g/L) and Combiotic (10 mg/L; Combiotic, produced by Pfizer, Incorporated, is a broad-spectrum antibiotic containing penicillin-G and dihydrostreptomycin).

Stocking was done at night in shallow water less than 4 m deep. Fish release was geographically apportioned according to the expected distribution of wild young of the year at the time of stocking (Table 94). During 1973, catch-per-effort data collected with 30-m seines in late August and early September 1973 were used to represent the

TABLE 94.—Numbers of 75–150-mm-long hatchery-reared striped bass stocked in the Hudson River by geographical area during 1973, 1974, and 1975.

Year	River kilometer[a]	Number stocked
1973	82–100	12,666
	61–76	6,637
	42–45	9,371
1974	90–98	3,058
	76–89	6,301
	63–74	21,648
	55–61	37,429
	39–53	33,088
1975	90–98	12,802
	76–89	12,732
	63–74	42,066
	55–61	69,352
	39–53	51,435

[a]River kilometers are measured from the Battery at the tip of Manhattan Island (river kilometer 0) to Troy Dam (km 243).

distribution of wild young of the year. During 1974 and 1975, catch-per-effort data from August and October of each previous year (1973 and 1974, respectively) were used. The numbers of hatchery fish stocked increased from 28,674 in 1973 to 101,524 in 1974 to 188,387 in 1975.

Wild age-0 striped bass were marked during the autumns of 1973–1975 as part of a mark–recapture program to estimate population size in the Hudson River (numbers and distribution of marked wild fish are presented by Young et al. 1988, this volume). They were caught by 30-m seines and 0.9 × 0.9 × 1.8-m box traps, and were released near the capture site. Wild fish did not receive nose tags but were marked by clipping a combination of two fins that differed from fin clips assigned to hatchery fish.

Short-term survival.—Marked and unmarked hatchery fish were held for 14 d at the Verplanck hatchery to determine the immediate effects of handling and marking. Subsequent recapture rates of marked hatchery fish and marked wild fish were compared over 6–9-month periods after release. Both wild and hatchery fish were recaptured in collections from seines, bottom trawls, epibenthic sleds, box traps, and power plant screens as part of the mark–recapture program described by Young et al. (1988). Survival of hatchery-reared juveniles relative to that of wild juveniles finclipped and released during the same time interval was evaluated by the change-in-ratio technique (Paulik and Robson 1969):

$$S_h/S_w = (R_{h2}/R_{h1})/(R_{w2}/R_{w1});$$

S_h and S_w are the survivals of hatchery (h) and of wild (w) fish; R_{h1}, R_{h2}, R_{w1}, and R_{w2} are the

TABLE 95.—Mortality of marked and control age-0 striped bass during 14-d postmarking observation periods, 1973–1975.

Year	Month(s)	Marked fish[a]		Control fish	
		Percent mortality	N	Percent mortality	N
1973	Sep	3.7	108	1.0	101
	Oct	2.0	100	7.0	100
1974	Sep–Oct	4.5	200	0.5	200
	Nov	6.0	50	0.0	63
1975	Sep	5.0	100	3.0	100
	Sep–Oct	5.0	100	1.0	100
	Oct	9.0	100	15.0	100
	Nov	34.0[b]	100	2.0	100
All		8.2	858	3.5	864

[a]Marked with both nose tag and fin clip.
[b]Only test in which mortality of marked fish differed significantly from that of unmarked fish ($\chi^2 = 34.7$; $P \leq 0.05$).

numbers of hatchery and wild fish recaptured in time intervals (1) and (2), respectively. The statistical significance of the survival ratios was determined by chi-square tests for differences between two proportions (R_{h2}/R_{h1} versus R_{w2}/R_{w1}).

Long-term growth and maturity.—The recovery effort for hatchery-reared fish continued through June 1980. All striped bass that were caught during the river-wide surveys of juvenile and adult striped bass from 1973 through 1980 (Hoff et al. 1988, this volume; Young et al. 1988) were examined for fin clips that would signify hatchery fish. Striped bass still showing evidence of a hatchery fin clip were passed through a magnetizer and a detector. If a nose tag seemed to be present, it was confirmed by dissection or X-ray examination.

During each year from 1976 to 1980, a portion of the spawning run was collected and processed in the laboratory for studies on sex ratio, age of maturity, and fecundity. All fish of sizes compatible with those of stocked 1973, 1974, or 1975 year classes were examined for nose tags whether fins appeared clipped or not. Gonads were excised and weighed, and the ratio of body weight to gonad weight was used to evaluate maturity (Hoff et al. 1988).

Results

Short-Term Survival

Handling and marking mortality of age-0 hatchery striped bass averaged only 8% over 14 d, of which 5% could be attributed to fin clips and insertion of nose tags (Table 95). Only in November did mortality of marked and control fish differ significantly ($P \leq 0.05$). Most deaths in all tests occurred within the first week of the 2-week trials.

Hatchery fish survived as well as wild fish over the 1974–1975 winter and significantly better than wild fish over the 1975–1976 winter (Table 96). Wild fish appeared to have survived the 1973–1974 winter better than did hatchery fish, but the recapture ratios were not significantly

TABLE 96.—Short-term relative survival estimates for hatchery-reared and wild striped bass in the Hudson River estuary, from time of stocking to the following spring. Asterisk (*) denotes significant χ^2 test ($P \leq 0.05$).

Period of stocking	Recapture period[a]	Number of recaptured fish		Survival ratio hatchery:wild	χ^2
		Hatchery-reared	Wild		
Sep–Oct 1973	Nov–Dec 1973	9	22	0.303	1.933
	Jan–Jun 1974	3	20		
Sep–Oct 1974	Nov 1974	43	5	1.179	0.072
	Jan–Jun 1975	71	7		
Sep 1975	Oct–Nov 1975	351	106	4.832	5.528*
	Jan–Jun 1976	32	2		
Oct 1975	Nov 1975	115	49	2.272	4.337*
	Jan–Jun 1976	48	9		

[a]Includes only fish recaptured at least 2 weeks after completion of marking to allow time for dispersal.

TABLE 97.—Summary of recaptures, through 30 June 1980, of hatchery-reared striped bass stocked in the Hudson River in 1973, 1974, and 1975. Percentages of recaptured fish retaining nose tags are in parentheses.

Stocking		Number of fish recaptured (% with nose tags) in								
Year	Number	Total	1973	1974	1975	1976	1977	1978	1979	1980
1973	28,674	49	46 (17)	3 (33)	0	0	0	0	0	0
1974	101,524	656		164 (82)	422 (82)	6 (63)	6 (67)	21 (52)	12 (33)	5 (20)
1975	188,387	1,220			925 (75)	259 (62)	10 (60)	11 (82)	13 (62)	2 (0)
Total	318,585	1,925	46	167	1,367	265	16	32	25	7

different ($P > 0.05$) for the two groups. Two factors that may have affected the survival comparisons were the loss of marks and dispersal of marked fish.

Fin regeneration was observed but, for the recovered fish, was not complete enough to make the mark undetectable. Fin regeneration may have been slight because fish were marked late in the growth season and recovered prior to the resumption of rapid growth the following year. Nose-tag retention during the recovery period ranged from 17 to 82%, the lowest retention rates being observed for the fall 1973 releases (Table 97).

Hatchery fish exhibited more extensive dispersal within the Hudson River during the marking season than wild fish (Figure 86). Whereas hatchery fish dispersed throughout the river after

release, movement of marked wild fish was principally downriver from the initial area of capture.

Long-Term Growth and Maturity

Hatchery-reared striped bass were recaptured up to 6 years after they were stocked in the Hudson River, and many reached maturity. As of 30 June 1980, 1,925 hatchery-reared fish had been recaptured from the Hudson River and adjacent waters (Table 97). The number of recaptures declined sharply as the hatchery fish reached 2 years of age. If hatchery fish behaved like wild fish (McLaren et al. 1981), this decline was likely due to emigration from the river into more saline waters surrounding western Long Island and northern New Jersey. No fish from the 1973 stocking were recaptured after 1974, possibly because few fish were released that year.

Hatchery fish participated in the spawning run into the Hudson River during late March through June. Mature hatchery males began to appear in the 1977 catch at age III (Table 98). Mature hatchery females did not appear until 1979 at age V. Although sample sizes for hatchery fish were small, the onset of their maturity appeared to be similar to that of wild fish (Table 98).

Hatchery and wild striped bass of the same age and sex did not significantly differ in length or weight after reaching 2 years of age (Figure 87). As expected (Hoff et al. 1988), mean total lengths and weights of females were generally larger than those of males of the same age. Detection of size differences was hampered, however, by small sample sizes for hatchery-reared fish.

Discussion

Hudson River striped bass can be artificially propagated, by a combination of intensive and extensive culture methods, from local brood stock. Furthermore, stocked fish appear to survive as well as wild age-0 fish. The relative survival of two groups was assessed by changes in the ratio of their recaptures through time. For that

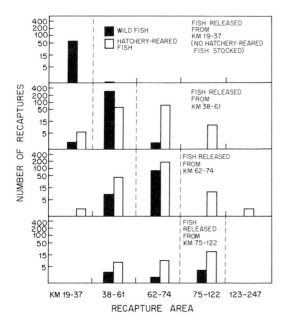

FIGURE 86.—Recovery distribution of hatchery-reared and wild fin-clipped striped bass released and recaptured during fall 1975, Hudson River estuary.

TABLE 98.—Percentage of hatchery-reared and wild striped bass mature by age for the 1974 and 1975 year classes in the Hudson River; fish were recaptured between 1977 and June 1980. Samples sizes are in parentheses.

Sex	Year class	Origin	II	III	IV	V	VI
Male	1974	Hatchery		83 (6)	88 (16)	100 (6)	67 (3)
		Wild		35 (34)	88 (82)	83 (66)	85 (46)
	1975	Hatchery	0 (3)	25 (4)	20 (5)	(0)	
		Wild	12 (25)	41 (37)	76 (71)	80 (46)	
Female	1974	Hatchery		(0)	0 (2)	40 (5)	50 (2)
		Wild		0 (27)	2 (59)	24 (50)	64 (39)
	1975	Hatchery	0 (3)	0 (7)	0 (7)	0 (2)	
		Wild	0 (27)	0 (43)	0 (51)	15 (53)	

inference to be valid, the relative vulnerabilities of hatchery and wild fish to capture must not change between recapture time intervals. Relative vulnerability may change because of differences in mark recognition, dispersal patterns (resulting in changes in the relative exposures to sampling gear), or growth (resulting in changes in gear avoidance capability).

Varying regeneration rates among fins could have caused differences in mark recognition for hatchery and wild fish. Most hatchery fish had

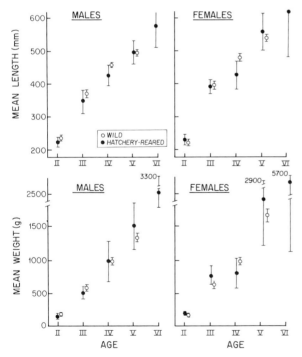

FIGURE 87.—Mean total lengths and weights of hatchery-reared and wild striped bass recaptured during March–June, 1976–1978, Hudson River estuary. Bars depict one standard error.

their first or second dorsal fin clipped; mutilation of these fins frequently was more recognizable than that of other fins after regeneration had occurred. This could have biased estimates toward greater survival of hatchery fish than wild fish. Marked wild fish may not have been recognized or may have been mistaken for hatchery fish if they were marked with a fin-clip combination involving either of the dorsal fins (the other fin clip in the combination would go unrecognized). A continual increase in the percentage of fish having hatchery-type fin clips but lacking nose tags, as observed after fish were at large 1 year (Table 97), may have been due as much to misidentification of wild fish as hatchery fish as to tag shedding.

Larger size generally is considered to confer better gear avoidance capability; thus a size-related bias would cause the relative survival of the larger hatchery fish to be underestimated. Mean total lengths at time of release during the 3 years were greater for hatchery fish than for wild fish, ranging from 98 to 115 mm versus 80 to 86 mm in September and early October, and from 106 to 143 mm versus 83 to 108 mm in October and November. The difference in sizes of hatchery and wild fish at the time of stocking may also have resulted in better survival of hatchery-reared fish. The larger size of hatchery fish could confer an advantage in escape from predators, decrease the number of potential predators, increase the number of potential prey, or increase the energy stores for overwintering. Despite taking precautions to stock hatchery fish according to the distribution of wild fish, we cannot eliminate the possibility that dispersal patterns could have affected relative survival estimates to some degree. Hatchery fish exhibited more extensive dispersal than did wild fish. If this more extensive dispersal resulted in a larger proportion of hatchery fish remaining in the Hudson River over the winter and being available for recapture during the following spring, this could have inflated the relative survival estimates for these fish. However, the occurrence of hatchery-reared striped bass below the George Washington Bridge (river kilometer 19) in spring 1975 (TI 1977b) indicates that many of the hatchery fish do exhibit emigration behavior typical of wild fish.

The evidence does indicate, however, that stocked age-0 fish survived at least as well as wild age-0 fish through the winter and spring following stocking, and that managers may expect the stocked fish to survive to maturity. The actual return rates of hatchery-reared striped bass as spawning adults, and therefore the true benefits of an enhancement program for the Hudson River, remain to be determined.

Acknowledgments

We appreciate the financial support of the Consolidated Edison Company of New York. Many individuals were responsible for the completion of this study, but we especially thank Bruce Friedmann, Dick Pugh, Barry Smith, Lauren and Cheryl Watson, and Phil Welsh for their skilled operation of the Verplanck hatchery. The following hatcheries and organizations cooperated with the extensive rearing of fingerlings: Edenton National Fish Hatchery, North Carolina; Marine Protein Corporation, Florida; Oklahoma Department of Wildlife Conservation; and Welaka National Fish Hatchery, Florida. We would like to pay a special tribute to the late Jack D. Bayless for his early contribution to the study.

References[5]

ASMFC (Atlantic States Marine Fisheries Commission). 1987. Draft ASMFC striped bass management plan. Martin Marietta Environmental Systems, Columbia, Maryland.

Bayless, J. B. 1972. Artificial propagation and hybridization of striped bass, *Morone saxatilis* (Walbaum). South Carolina Wildlife Resources Department, Columbia.

Bonn, E. W., W. M. Bailey, J. D. Bayless, K. E. Erickson, and R. E. Stevens, editors. 1976. Guidelines for striped bass culture. American Fisheries Society, Southern Division, Striped Bass Committee, Bethesda, Maryland.

Harrell, R. M. 1978. Enhancement of striped bass population in Santee River. South Carolina Wildlife and Marine Resources Division, Job Progress Report AFS-8-1, Columbia.

Harrell, R. M. 1979. Enhancement of striped bass population in Santee River. South Carolina Wildlife and Marine Resources Department, Job Progress Report AFS-8-2, Columbia.

Harrell, R. M. 1980. Enhancement of striped bass population in Santee River. South Carolina Wildlife and Marine Resources Department, Job Progress Report AFS-8-3, Columbia.

Harrell, R. M. 1981. Enhancement of striped bass population in Santee River. South Carolina Wildlife and Marine Resources Department, Job Progress Report AFS-8-4, Columbia.

Hoff, T. B., J. B. McLaren, and J. C. Cooper. 1988. Stock characteristics of Hudson River striped bass. American Fisheries Society Monograph 4:59–68.

Hornsby, J. H., and C. S. Hall. 1981. Impact of supplemental stocking of striped bass fingerlings in the Ogeechee River. Georgia Department of Natural Resources, Atlanta.

McIlwain, T. D. 1981. Striped bass, restoration program—Mississippi Gulf Coast. Gulf Coast Laboratory, Annual Progress Report, Project AFCS-2, Ocean Springs, Mississippi.

McLaren, J. B., J. C. Cooper, T. B. Hoff, and V. Lander. 1981. Movements of Hudson River striped bass. Transactions of the American Fisheries Society 110:158–167.

Paulik, G. L., and D. S. Robson. 1969. Statistical calculations for change-in-ratio estimators of population parameters. Journal of Wildlife Management 33:1–27.

Stevens, D. R. 1966. A report on the operation of Moncks Corner striped bass hatchery, 1961–1965. South Carolina Wildlife Resources Department, Columbia.

TI (Texas Instruments). 1977a. Feasibility of culturing and stocking Hudson River striped bass. An overview, 1973–1975. Report to Consolidated Edison Company of New York.

TI (Texas Instruments). 1977b. 1974 year class report for the multiplant study of the Hudson River estuary, volume 3. Lower estuary study. Report to Consolidated Edison Company of New York.

Wooley, C. M., and E. J. Crateau. 1983. Biology, population estimates, and movement of native and introduced striped bass. Apalachicola River, Florida. North American Journal of Fisheries Management 3:383–394.

Young, J. R., R. J. Klauda, and W. P. Dey. 1988. Population estimates for juvenile striped bass and white perch in the Hudson River estuary. American Fisheries Society Monograph 4:89–101.

---

[5]See Table 1 for sources of legal documents and unpublished reports pertaining to the Hudson River.

American Fisheries Society Monograph 4:293, 1988

SECTION 5: CLOSING PERSPECTIVES

Introduction

LAWRENCE W. BARNTHOUSE

Environmental Sciences Division, Oak Ridge National Laboratory[1]
Post Office Box 2008, Oak Ridge, Tennessee 37831-6036, USA

The preceding sections of this monograph have presented the scientific results of the Hudson River studies and discussed their use in the settlement negotiations. The monograph would be incomplete, however, without something more. The structure chosen for the monograph (i.e., peer-reviewed scientific papers emphasizing research findings) tends to obscure the fact that most of the work appeared originally as documentation for license applications or as testimony in hearings. The Hudson River studies cannot be properly understood or evaluated without consideration of their role in the judicial proceedings through which concrete environmental decisions were derived from abstract environmental laws. The papers in this section discuss the relationship of the Hudson River studies to the regulatory process.

The first two papers treat the roles of science and scientists in judicial proceedings, as viewed by Thomas J. Yost, the administrative law judge who presided over the U.S. Environmental Protection Agency (EPA) hearings, and by G. S. Peter Bergen, the lead counsel for the Hudson

River utility companies. These authors also present sharply divergent interpretations of the key sections of the Clean Water Act; their discussions illustrate the contribution of legal ambiguity to the complexity and duration of the EPA hearings.

The third paper, authored by two scientists (Sigurd Christensen and Ronald Klauda), discusses the relationship between scientists and lawyers from the perspective of the scientists. These authors worked side by side with EPA (Christensen) and utility (Klauda) lawyers throughout the EPA hearings. They give a first-hand account of their experiences and provide suggestions to other scientists for minimizing the inevitable conflicts and problems that arise when scientists become involved in judicial proceedings.

The final two papers present evaluations of (1) the successes and failures of the Hudson River studies and (2) the implications of the work for future environmental assessment studies. They do not represent a consensus of all of the monograph authors, although most of the authors probably would support most of conclusions. Both papers reflect perspectives of direct participants in the studies and in the hearings they supported; readers with other perspectives are invited to make their own independent evaluations.

[1]Operated by Martin Marietta Energy Systems, Incorporated, under contract DE-AC05-84OR21400 with the U.S. Department of Energy.

American Fisheries Society Monograph 4:294–301, 1988

Science in the Courtroom

Thomas B. Yost[1]

U.S. Environmental Protection Agency
345 Courtland Street, Atlanta, Georgia 30365, USA

Abstract.—The purpose of a trial is to resolve a dispute, not to answer scientific questions. It is important that expert witnesses offer their informed opinions about scientific evidence, but resolution of the dispute will not necessarily bring resolution of the scientific issues involved. In the Hudson River case, the dispute over the need to construct cooling towers at power plants was settled even though the key scientific issue—the biological response of fish populations to power-plant-induced mortality—was not decided. The length and complexity of the Hudson River case would have been markedly reduced had consultants to the utilities and to the regulatory agency cooperated more fully from the beginning in data acquisition and analysis instead of debating these matters reactively during the hearing. It helps, not hinders, judicial procedure if areas of technical agreement are maximized before a hearing occurs, and frequent scientific interaction between parties to a dispute is recommended for future regulatory cases of this type.

Most scientists and technical people of my acquaintance are apparently mystified about the purpose of the trial and how the entire trial process works. The vast majority of the people in this country have never been in court and probably never will be; consequently, their exposure to the court process consists of television and movie versions that, in all too many cases, differ substantially from the real world of jurisprudence. The function of a trial is to resolve controversies involving factual differences within a framework of a rigidly structured process that involves rules of evidence, established procedures, and a great deal of arcane verbiage. In the normal trial, whether it be civil or criminal, the evidence is presented to a jury whose function it is to determine the facts. The law applicable to these facts is presented to the jury by the court in the form of instructions. It is the jury's function to determine the facts of the case and then apply these facts to the law as handed down to them by the court. In administrative procedures, there is no jury and, therefore, the presiding judge must function both as the decider of the facts and the interpreter of the applicable law.

In all cases involving National Pollutant Discharge Elimination System (NPDES) permits that the U.S. Environmental Protection Agency (EPA) hears, because of the technical nature of the issues, all testimony is required to be prefiled. By that I mean that all parties must prepare their direct testimony in writing; following an exchange of that testimony by the parties, rebuttal testimony is then prepared which is, likewise, filed with the court and exchanged between the parties prior to the hearing. The purpose of this procedure is twofold. It allows the parties (1) to adequately prepare rebuttal testimony, and (2) to take the necessary time along with their expert witnesses to carefully examine the testimony, data, and other materials submitted by the other party. This procedure also facilitates the potential for settlement of the case, or, in the alternative, stipulation of certain facts that, based on the testimony of all the parties, do not appear to be in controversy. Many factual issues and, in some cases, conclusions related thereto can be agreed upon by the parties prior to the hearing and, thus, save everyone's time and money in the actual trial.

The trial then consists primarily of the cross-examination of the witnesses sponsored by each party in an attempt to discredit, refute, or otherwise reduce the weight or impact of that testimony for the record. Typically, cross-examining counsel will have at his or her side one or more experts on the subject of a particular witness' testimony to assist in asking the questions and to provide follow-up questions, depending on the answers given by the witness being examined.

At the end of the trial, following the presentation of all the evidence, the parties then prepare briefs that support their particular view of the case and propose to the judge findings of fact and legal conclusions, which they urge the court to

[1]Judge Yost, an Administrative Law Judge with the Environmental Protection Agency, presided at the Hudson River case.

accept.[2] Following the receipt by the court of the transcript and the proposed findings and briefs submitted by the parties, a decision is issued which must address all the factual issues and legal issues presented by the case.

Role of the Scientist as Expert Witness

Throughout this dissertation, I have used the phrase "expert witness," which has a special meaning in the world of law. The sole difference between an expert witness and a lay witness is that an expert witness is permitted to express an opinion on a subject within her or his area of expertise, whereas a lay witness is prohibited from doing so. An example would be a personal injury case involving an automobile accident. Various lay witnesses will be called upon to testify and all they are permitted to testify to is what they actually saw: the facts, in other words. In that same trial, however, there will appear expert witnesses, usually medical witnesses, who are permitted to not only describe what they observed as to the injuries sustained by the plaintiff, but also express an opinion as to their cause and the prognosis for permanent injury or disability resulting from such injuries. In a case like the Hudson River case, lay witnesses might testify as to such factual matters as the number of samples taken, where taken, the number of aquatic species captured, a description of their variety and size, etc. The expert witness will then use the factual data presented by the lay witness to form opinions as to what all these facts and figures mean, and to express an opinion as to the effects of continued cropping of a particular species by whatever source. Although the role of the expert witness is relatively straightforward, many scientists seem to have a great deal of confusion over their role in the trial.

Scientists, even those who have some familiarity with the trial process, view both attorneys and the judicial process with some antagonism, skepticism, and fear. These emotions seem, at least in my opinion, to stem from the fear most scientists have of being ridiculed before their peers. In this regard, scientists and politicians have much in common in that their good reputations are their most valued possessions. If a scientist loses his creditability in a public forum, his future in his chosen field of expertise is likely to be irreparably damaged. Although I have not attended many scientific meetings, I am of the opinion that, even in cases where they disagree with one another, scientists are much more reserved and circumspect in expressing their disagreements than attorneys are in exposing what they perceive to be erroneous opinions expressed by scientific experts in the courtroom. Lawyers are, after all, advocates for their respective clients and, as such, are tenacious and aggressive in exposing what they perceive to be erroneous and, in some cases, insincere testimony.

The lawyer, on the other hand, views the world of science with as much uncertainty and confusion as the scientist views the world of law. Lawyers tend to want to get to the bottom line of a controversy or issue and determine what the ultimate answer is. Scientists tend to be reluctant to express opinions or to come to conclusions on a complex scientific issue because they never feel they have enough data at their disposal to express an opinion about what is likely to occur in the future given an array of scientific data. It is this reluctance on the part of the scientist to express an opinion that is most frustrating to the attorney. It is likely that this dichotomy is based upon the different ways in which lawyers and scientists view a problem. The law only requires that a scientist have an honestly held belief that one occurrence is more likely to happen than another, based solely upon a preponderance of the evidence, in order to express an opinion. A preponderance of the evidence means simply a majority—that is to say, 51%. In normal scientific practice, on the other hand, a confidence level in the high 90 percentages is required. It is this difference in standards of proof that causes the majority of the problems between the attorney and the scientist. To put it in a nutshell, the scientist never has enough data and the attorney can never get enough answers.

Nothing herein said should be construed to suggest that the law wishes expert witnesses to make wild or unsupported guesses as to what is likely to occur in the future. All the law expects of the expert is that he or she express an opinion based on his or her best judgement as to which of two alternative results is more likely to occur. In making such a choice, there is nothing either scientifically wrong or legally improper about an opinion based on this type of analysis. The scien-

---

[2]The bizzare rules of practice in existence during the Hudson River case did not authorize the presiding judge to issue the decision, but rather vested that responsibility on the Regional Administrator of EPA. The rule has since been amended to reflect the process described here.

tist seems to forget that it is training, knowledge, and experience, in combination, that is on display; even in the absence of unlimited data, the expert should not be reluctant to express an opinion based on her or his judgement of whatever evidence is available. There may be incidences where sufficient data do not exist to make any kind of decision at all. If this circumstance arises, the scientist should advise counsel and refuse to testify or, alternatively, testify that, in his or her judgement, there are insufficient data upon which a conscientious, trained scientist could make a decision either way. In the Hudson case, many government and private sector witnesses took the position that, even in view of the wealth of data available, no scientifically supportable prediction could be made concerning the fate of the various species of fish in the Hudson River. Many attorneys would view this position as a "cop out." I express no opinion on that subject, except to say that this type of response causes great frustration among attorneys in dealing with scientists in a trial situation.

Role of the Lawyer

Much of the apprehension experienced by expert witnesses who have not testified before can be allayed through proper preparation by attorneys. Contrary to popular belief, there is nothing improper about coaching a witness. It is improper for an attorney to direct the witness's testimony and to urge the witness to testify in a manner inconsistent with honestly held beliefs. This is not coaching, but rather subornation of perjury, which is a far different matter. A skillful attorney will prepare a witness for trial by initially working with her or him on testimony to make sure that it does not contain inconsistencies or opinions unsupported by the evidence. The attorney will instruct the witness as to what kinds of questions may be asked by opposing counsel and advise the witness to respond honestly and concisely to the questions asked, neither arguing with counsel or volunteering answers to questions not asked. Counsel must assure the witness of protection from harassment and abuse on the part of opposing counsel, and must let the witness know that the attorney views their appearance as a joint effort and that the witness will not be left alone on the witness stand to fend off the sometimes aggressive cross-examination of opposing counsel. The witness should be instructed to carefully review all of the articles and papers that the witness has written or will rely upon for support

during testimony. The witness can be assured that both opposing counsel and the opposing expert advisors have carefully read every paper written by the witness on the subject of the testimony, as well as all papers and articles cited in the testimony. The attorney sponsoring the witness should advise the witness of the weaknesses of the testimony to the extent that they exist and advise the witness how to deal with questions on these areas. Counsel will realize that any body of expert testimony, no matter how soundly based, has some areas over which there may be serious disagreement. In addition to the protection that an expert witness will receive from her or his own counsel, the presiding judge also has a role to play in this area and should not hesitate to interrupt cross-examination when, in the court's judgement, such cross-examination has gone beyond the normal limits of proper questioning. If an expert witness's testimony is honestly prepared and well documented, he or she should have nothing to fear from the trial process inasmuch as truth is the best defense to any cross-examination.

Why It Did Not Work on the Hudson

To prepare an effective trial strategy, one must first examine the applicable statutes and regulations to determine precisely what the issues are and what elements of proof the statutes require. The Hudson River case, like most power plant cases, was held pursuant to §316 of the Clean Water Act (see Appendix). Section 316 is divided into three subsections—(a), (b), and (c). Subsection (a) has to do with thermal discharges from a power plant. The statute states that whenever the owner or operator of any such source, after an opportunity for public hearing, can demonstrate to the satisfaction of the EPA Administrator that any effluent limitation proposed for the control of the thermal component of any *discharge* from such source will require effluent limitations *more stringent* than necessary to assure the protection and propagation of a balanced indigenous population of shellfish, fish, and wildlife in or on a body of water into which the discharge is made, the Administrator may impose an effluent limitation with respect to such thermal component that will assure the protection, propagation, etc. Counsel for the utilities in the Hudson River case, throughout the course of the hearing, continued to cite the requirements for relief set forth in subsection (a), to the effect that the utilities' sampling and other studies indicated that there was, in fact, a balanced population of fish and shellfish in the Hudson River and, therefore, the utilities had met the burden of proof imposed upon them by the statute and were entitled to relief from the requirements that they build cooling towers at the various power plants involved. However, the Hudson River case was not being held pursuant to subsection (a), but rather subsection (b), which has to do with *intake* structures. Subsection (b) makes no mention of the protection of a balanced population of shellfish or fish, but rather requires the location, design, construction, and capacity of cooling water intake structures reflect the best technology available for minimizing adverse environmental impact. Therefore, utility counsel's continued reference to the protection of a balanced population of fish was entirely irrelevant to the proceeding in the Hudson case because that was not the standard to which the statute held the utilities.

Therefore, the utilities' task in the Hudson case was to demonstrate that cooling towers did not necessarily represent the best technology available for minimizing adverse environmental impact. In the context of that task, the utilities were required to put into evidence data and testimony concerning the ability of various alternative designs of intake structures to prevent impingement and entrainment of aquatic organisms, particularly fishes. Most of the alternative intake technologies are primarily designed to minimize impingement; the utilities, therefore, attempted to prove that the mortality experienced by the various fish species passing through the plant was less than 100%.

A necessary adjunct to the utilities' strategy was to attempt to prove that the through-plant mortality they postulated would not adversely affect the fish populations at risk. To demonstrate this position, they utilized the concept of *compensation*. Although all scientists recognize that the mechanics of compensation are at work in the biological world, not everyone was willing to accept that it works in precisely the fashion propounded by the utilities' scientists. Needless to say, the vigorous examination of the utilities' predictions in this area consumed most of the participants' time and effort, and captured the interest of most of the scientists involved. Nevertheless, the case was settled before the scientific validity of the utilities' position was resolved.

Another problem that contributed to the length of the Hudson case was that every attorney employed by all parties wished to participate in the hearing process—in many cases for no other

reason than to gain trial experience. Although I do not have at my fingertips an exact count of the number of attorneys that appeared in the case and actively participated in cross-examination, my recollection is that it was in excess of 20. Although I recognize that the complexity of the case and the variety of issues and disciplines involved required that several attorneys concentrate on certain areas of the testimony so that they could intelligently cross-examine the witnesses, I do not feel that it was necessary that all of the attorneys who ultimately participated in the case actually needed to be involved. Many attorneys engaged in rather useless and repetitious cross-examination, concerning themselves with minutia (i.e., grammar, punctuation, and editorial matters) that are not really the proper subject of good cross-examination. Because the court was not privy to the strategy adopted by any of the parties in the Hudson case, it was reluctant to limit or interfere with seemingly, irrelevant cross-examination, taking the position that the attorneys, hopefully, had some point in mind to which the seemingly endless cross-examination was leading. In some cases, however, when it became apparent to the court that the cross-examination being conducted was not leading anywhere or helping to bring forth facts that would be helpful to the ultimate decision maker, such cross-examination was curtailed.

How It Can Work Better Next Time

A major underlying problem in the Hudson case was that all of the data were collected and presented by only one party to the controversy. The government scientists were in the position of having to analyze and critique data presented by others rather than being able to collect and interpret data of their own. Unfortunately, this situation is the norm, rather than the exception, in that neither the government nor the private-sector environmentalists have the financial resources to collect data of sufficient quantity and quality to build a case of their own. It occurs to me that a better procedure would be for the utilities' consultants and the government's consultants to sit down together, analyze the problems associated with a particular course of action, mutually agree on the best techniques, and study protocol to be used in order to obtain data that both sides agree would be useful in attempting to provide answers to the questions at issue.

My observations and experience with Environmental Protection Agency personnel seem to indicate that agency people are reluctant to participate in the design of a study for fear that the utilities might later argue that the agency's participation in the planning process amounted to an endorsement by the agency of the entire study and the results derived therefrom. I feel that this perception on the part of the agency is both shortsighted and legally unsound. Simply because the agency participates in the design of a particular study does not prevent the agency from thereafter criticizing the results derived from such study or the conclusions reached by the utilities based on their analysis of the data produced. Rather, such a cooperative effort would, in my judgement, save vast amounts of time and money on both sides of the case. In the Hudson case, such a procedure would have, no doubt, resulted in a more streamlined and coordinated data collection effort and would have resulted in the generation of far fewer data and reports than the procedure actually used. At least a year's worth of agency and consultant time and effort was expended on trying to analyze the utilities' data to understand how the utilities processed the data, and to understand the inner workings of the various mathematical models used by the utilities to support the conclusions of their experts. Whether or not such a joint effort would have produced a body of better data is, of course, speculative, but at least both sides of the case would have understood the scope and aims of whatever study strategy was ultimately developed.

Additional time could also have been saved by the simultaneous transmission of all data collected to the agency as well as to the utilities, thus resulting in a parellel effort and a continual interplay of ideas and analyses leading to the production of the final report.

As the participants in the hearing will recall, a great deal of time was spent on cross-examination of utilities' experts on just exactly how the data were handled and manipulated. If there had been a joint effort, this type of cross-examination would have been unnecessary. If, at some point in the procedure, agency experts and their consultants felt that the utilities' consultants were improperly analyzing the data, such advice could have been immediately transmitted and an attempt to reconcile the two viewpoints could have been accomplished immediately instead of later at the trial.

Early in the process, I suggest that attorneys for both sides be involved to make sure that the study plan recommended by the consultants for both sides of the issue will provide data appropriate to

answer the questions that the statutes required to be answered. This is not to suggest that the attorneys for the parties would have to be involved in the day-to-day monitoring of progress; rather, their primary function would be to assure the scientists that the data they proposed to collect will be relevant to the issues actually existing between the parties. If, in the context of this ongoing joint effort, legal issues arise between the parties, the attorneys could be called in to discuss among themselves the legal implications. The utilities' experts and the agency's experts would be simultaneously analyzing the data that are produced. To the extent different results occur, these differences would be addressed at the hearing. To the extent the agency consultants' results agree with those of the utilities, these areas of agreement could be stipulated, thus reducing to a considerable degree the length and complexity of the hearing. If the procedure suggested here had been used for the Hudson River case, I believe that the length and complexity of the case could have been reduced by approximately 80%, resulting in a sizeable saving of time and money on behalf of both the utilities and the agency.

Why the Major Scientific Issues Were Not Resolved

It has been estimated that the utilities spent about $35 million in the preparation of their case. The scientists involved could not accept that the expenditure of this vast sum of money did not produce firm answers to the various biological questions posed by the proceedings. This posture results from an apparently widely held belief among nonlawyers that the purpose of a trial is to answer scientific questions. Such is not the case. The purpose of a trial is to resolve disputes. In the Hudson River case, the issue was whether or not the utilities would be required to build cooling towers at their several power plants. If, in the course of answering that question, some scientific "truths" emerged, that was merely a beneficial side effect.

Had the case proceeded to its conclusion, it is quite likely that the various scientific theories presented would have been thoroughly examined and accepted or rejected. Because the case was settled, this examination was not done. It is not in the nature of settlements to provide the kind of answers the scientists desire, but rather to provide a mechanism for resolving the primary dispute between the parties, which, in the Hudson case, was the issue of cooling towers.

Perhaps the ongoing study of the Hudson River, provided for by the settlement document, will provide some of the answers that the trial did not.

Comments on G. S. Peter Bergen's Paper

In response to Bergen's assertion (1988, this volume) that my interpretation of the meaning of §§316(a) and (b) is a "recipe for disaster," let me say that his notion that the two sections should be read in pari materia is true only to the extent that both sections have to do with power plants and, where enacted, with protection of the aquatic environment. To suggest that §316(b) should be interpreted to permit impingement and entrainment up to some nebulous level of mortality just short of disaster[3] is not warranted by the language of the statute. The requirements of §§316(a) and (b) are entirely different; in my judgement, §316(b) imposes the more stringent requirements. Section 316(b) clearly requires that the cooling water intake structures *shall* reflect the *"best technology available to minimize* adverse environmental impact" (emphasis mine). "Webster's Ninth New Collegiate Dictionary" (Merriam-Webster, Springfield, Massachusetts) defines *minimize* as "to reduce to a minimum," and *minimum* as "the least quantity assignable, admissible, or possible." Another definition, which sounds suspiciously like some testimony in the hearing, advises a *minimum* to be "the smallest value assumed by a continuous function defined on a closed interval." I understand the first definition better, and I see no reference to any margin for cost–benefit analysis of including a *larger quantity* after allowing for some unknown latitude of degradation.

Some commentators have suggested that Bergen, the utilities' lead counsel, and I have philosophical differences concerning the correct interpretation of sections §§316(a) and (b) of the Clean Water Act. I do not believe that our differences are as much philosophical as they are positional. Bergen represents paying clients as an advocate. I represent no one. I am merely a referee, or at least that was my role at the time of the Hudson River hearing.

A reading of the Clean Water Act will disclose that §316 confers special treatment upon the electric utility industry, a gift not made to any other

discrete segment of our business community.[4] The inclusion of that section in the law is a tribute to the effectiveness of the electric utilities' lobbying power.

Originally, the section only included subsection (a), which provided the utilities with a way to avoid cooling towers if they could convince EPA that they were not needed. That section only applied to the *thermal discharge* from a power plant. Concerned citizens, environmentalists, marine and aquatic biologists, and other scientists alerted Congress to the dangers presented to the aquatic community by the mortality induced by impingement and entrainment. Thus, in a last-minute amendment, Congress added subsection (b), which was directed at the *intake* portion of the power plant's generating processes. In some cases, the subsection (b) problem could be solved by relocating the intake structure away from sensitive biological areas. This was the solution used in the Seabrook nuclear power plant case in New Hampshire. In that case, the utility moved the intake structure from a sensitive estuary to a point well offshore in the Atlantic Ocean. Similarly, some utilities have attempted to solve their subsection (a) (thermal discharge) problems by piping or otherwise directing their discharge to an area which either contained few desirable aquatic species or to larger and colder water bodies that could absorb the thermal component more efficiently.

Neither of these options was available to the Hudson River utilities. As noted above, the NPDES permits issued to the Hudson utilities would have required cooling towers under both subsections (a) and (b). The waiver or exemption provided by subsection (a) was not at issue in the Hudson cases. In the Hudson cases, the utilities presented three primary theories to support their position that the intake aspect of their plants would not violate §316(b). One involved the notion of compensation, that is, that impingement and entrainment mortality did not really threaten the fish species of concern because cropping somehow benefits those fish spared by reducing the population size. The second theory was that impingement mortality could be significantly reduced by installing state-of-the-art traveling and angled screens on the intakes in combination with some form of fish return system. The third prong

---

[3]Also read "imbalance," whatever that means. I suppose that the South American anchovetta fishery could be called imbalanced. Others would hold that the fishery has crashed.

[4]Although it is true that other industries operating their own power stations did benefit from this section, the electric utility industry was its primary beneficiary.

of their defense was unique to the Hudson case. It involved the notion that through-plant mortality of entrained eggs, larvae, and postlarval organisms was not total, but, rather, relatively insignificant. With the exception of defense number two, the validity of the utilities' theories were never proven, primarily because the case was ultimately settled and secondarily because the scientific basis for them was not well supported by the available data.

As suggested above, the utility's defenses were interesting but not particularly relevant to the demands posed by the statute. The statute requires use of the *best* available technology. There is no doubt that cooling towers met that test. In my opinion, the utilities could only have avoided installing cooling towers by demonstrating that the use of some other technology would equally minimize adverse environmental impacts. That, in my judgement, is a very hard burden of proof.

An examination of Bergen's "recipe"[5] would require that the agency or the utility identify an unknown and possibly unknowable level of population imbalance, allow thermal pollution to impose mortality up to that point, then look at the mortality imposed by impingement and entrainment, and let it go once again up to "imbalance." One would, then, I suppose, have to introduce some factor that would account for the cumulative or possibly synergistic effects of these two separate sources of mortality, add a margin of safety, and then vote yea or nay on cooling towers. Oh, would that science were that exact!

I agree with Bergen's analysis of the situation as it relates to what the existing permits required, and agree that EPA's position in the case was unclear and poorly articulated. It is clear, however, that the case was tried under §316(b) and that the agency had not yet made a decision on the thermal aspects of the case, because the utilities were still providing data and information regarding a possible waiver of the §316(a) requirements. Had the agency denied the utilities' request for a waiver under §316(a), that issue would also had to have been tried in another hearing or a continuation of the ongoing one. I had planned to retire with this case still in progress.

---

[5]Combine:
    1 large grain of salt with
    2 cups of self-interest.
Stir vigorously with tame scientist until slightly thick. Adjust seasoning to taste.

Reference

Bergen, G. S. P. 1988. The Hudson River cooling tower proceeding: interface between science and law. American Fisheries Society Monograph 4:302–306.

Appendix: Section 316 of the Clean Water Act

Within the Clean Water Act, more formally known as the "Federal Water Pollution Control Act Amendments of 1972" (Public Law 92-500), Section 316 deals with "Thermal Discharges." The text of this section follows; bracketed phrases are editorial additions.

Sec. 316. (a) With respect to any point source otherwise subject to the provisions of section 301 ["Effluent Limitations"] or section 306 ["National Standards of Performance"] of this Act, whenever the owner or operator of any such source, after opportunity for public hearing, can demonstrate to the satisfaction of the [Environmental Protection Agency] Administrator (or, if appropriate, the State [in which the source is located]) that any effluent limitation proposed for the control of the thermal component of any discharge from such source will require effluent limitations more stringent than necessary to assure the projection [sic] and propagation of a balanced, indigenous population of shellfish, fish, and wildlife in and on the body of water into which the discharge is to be made, the Administrator (or, if appropriate, the State) may impose an effluent limitation under such sections for such plant, with respect to the thermal component of such discharge (taking into account the interaction of such thermal component with other pollutants), that will assure the protection and propagation of a balanced, indigenous population of shellfish, fish, and wildlife in and on that body of water.

(b) Any standard established pursuant to section 301 or section 306 of this Act and applicable to a point source shall require that the location, design, construction, and capacity of cooling water intake structures reflect the best technology available for minimizing adverse environmental impact.

(c) Notwithstanding any other provision of this Act, any point source of a discharge having a thermal component, the modification of which point source is commenced after the date of enactment of the Federal Pollution Control Act Amendments of 1972 and which, as modified, meets effluent limitations established under section 301 or, if more stringent, effluent limitations established under section 303 ["Water Quality Standards and Implementation Plans"] and which

effluent limitations will assure protection and propagation of a balanced, indigenous population of shellfish, fish, and wildlife in or on the water into which the discharge is made, shall not be subject to any more stringent effluent limitation with respect to the thermal component of its discharge during a ten year period beginning on the date of completion of such modification or during the period of depreciation or amortization of such facility for the purpose of section 167 or 169 (or both) of the Internal Revenue Code of 1954, whichever ends first.

American Fisheries Society Monograph 4:302–306, 1988

The Hudson River Cooling Tower Proceeding:
Interface between Science and Law

G. S. Peter Bergen[1]

LeBoeuf, Lamb, Leiby & MacRae, 520 Madison Avenue, New York, New York 10022, USA

Abstract.—As the Hudson River power plant case proceeded, the regulatory ground shifted under the utility companies. At first, the U.S. Environmental Protection Agency (EPA) contended that the utilities should build expensive closed-cycle cooling towers at three plants to minimize the plants' discharge of heated effluents to the river. When the formal hearing began, however, EPA claimed that cooling towers were needed to minimize the number of organisms impinged at and entrained through the plants. These two issues—thermal discharges and cooling water intakes—fall under separate subsections of the 1972 federal Clean Water Act; whether the two subsections should be considered independently of one another, as the EPA asserted, or in combination to address the total power plant impact on adult fish populations, as the utilities believed, was a matter of contention. The Hudson River proceeding became a policy dispute over what the appropriate standard of environmental conduct should be, instead of a determination of whether a standard had been met or not. Such policy issues, which arise when legal precedent has yet to be developed for new laws like the Clean Water Act, are better addressed by a rule-making proceeding than by the adjudicatory hearing format used in the Hudson case. A rule-making proceeding would have markedly shortened the Hudson deliberations, probably without substantive change in the final settlement, and is recommended for future cases in which ambiguity in legislation or the lack of precedent has left policy matters unresolved.

The issue in the Hudson River cooling tower proceeding was whether or not natural-draft cooling towers should be installed at some or all of six large existing electric generating units on the Hudson River north of New York City. The six units are located at three generating stations in which five utilities have various ownership interests.

Background: Permits and Hearings

The cooling tower proceeding began formally in 1975, when the U.S. Environmental Protection Agency (EPA) issued proposed wastewater discharge permits for each generating station. The agency's proposed National Pollutant Discharge Elimination System (NPDES) permits were issued under §402 of the Clean Water Act[2] and, by their terms, limited the allowable rates of heat discharge from each of the units to less than 5% of their existing heat discharge rates. These proposed heat discharge limitations could only be achieved if six closed-cycle cooling towers were installed, one at each generating unit. The utilities objected to the heat discharge limitations because of their cost and energy requirements. They requested hearings in

1975. Also raised in each of the four requests for a hearing were certain objections to other "nonthermal" conditions in the proposed permits.

Each utility met separately with EPA's regional staff to identify disputed issues and to attempt to resolve their differences. By mid-1976, each utility had resolved virtually all disputed issues concerning the proposed permits, except for the proposed heat discharge limitations. The EPA staff advised that the installation of cooling towers was "non-negotiable." Due to this impasse, and in accordance with established administrative procedure, EPA ordered that hearings be held for the purpose of developing a record upon which the Regional Administrator of EPA could decide the disputed thermal issues. Administrative Law Judge Thomas B. Yost was appointed to preside at the hearings, and to submit the record to the Regional Administrator, who would make a decision.

At the first prehearing conference, held in February 1977, the four separate permit proceedings were consolidated into a single proceeding. The EPA staff advised at that time that the issue with which it was concerned was not primarily heat discharge rates, but instead whether or not closed-cycle cooling should be installed to reduce the volume of river water withdrawn and thereby minimize entrainment of fish eggs, larvae, and other aquatic life through the power plant condensers. As of 1977, policy

---

[1]Mr. Bergen, who specializes in environmental law, was lead counsel for the utilities in the Environmental Protection Agency's Hudson River cooling tower proceeding.

[2]Public Law 95-500 (October 18, 1972), 33 U.S.C. §1251 et seq.

makers and consulting scientists concerned with cooling water discharges had made a conservative assumption that entrainment mortality was on the order of 100% (this assumption was later modified, based on subsequent studies). Thus, EPA staff had concluded that closed-cycle cooling was needed to protect fish regardless of whether or not the effects of thermal discharges on aquatic life were significant. The EPA staff would not concede, however, that the discharges of heated water from the generating units to the river had no adverse impact on fish and aquatic life in the ambient river water. Rather, the staff's position was that closed-cycle cooling needed to be installed to reduce entrainment pursuant to §316(b) of the Clean Water Act. If the staff failed in that effort, it planned to contend later that subsequent hearings should be held to determine whether cooling towers should be imposed pursuant to §316(a) of the Act, which relates to discharges of heat. The inability, and I believe reluctance, on the part of EPA, at the early stages of the proceeding, to define precisely the factual issues to be adjudicated is substantially responsible for the protracted and rather directionless record that was compiled.

The utilities were instructed by the Administrative Law Judge at the initial prehearing conference in February 1977 to file all the evidence and data that they then had available. These data were the result of years of prior sampling and analysis of the river fishery, the results of entrainment and impingement studies, and economic and engineering analyses. What were all the facts and data introduced as evidence supposed to show? In any trial or administrative proceeding, a proponent (the party with the burden of going forward with evidence) attempts to demonstrate with facts that an established standard of conduct (the law) has been met or has been exceeded as the case may be. For example, to show that an auto driver was drunk, a prosecutor will attempt to show that the driver's blood alcohol concentration (a fact) exceeded a specified concentration (a legal standard). The evidence introduced by the utilities in the Hudson River hearing was necessary to show that applicable statutory standards were achieved, thus obviating the necessity to build cooling towers. However, the Hudson River case had two major departures from the norm of more ordinary adjudication. First, the applicable standard of conduct was unclear. Second, the burden and nature of proof needed to show that the standard had been met was unknown. These problems arose both because the Clean Water Act was new and no precedent was available, and because the statutory standards as

written by the U.S. Congress were imprecise. In writing the Clean Water Act's legislative standards for thermal discharges and intake structure design, Congress left gaps and interstices to be filled through administrative processes.

I will, therefore, next outline the applicable standard, §316 of the Clean Water Act; I will then discuss how the administrative process dealt with the Act in the Hudson River case. I will also give my thoughts on how that process might be improved in future cases.

Section 316 of the Clean Water Act

Section 316 of the Clean Water Act, entitled "Thermal Discharges," sets out standards applicable to discharges of heat (subsection "a") and applicable to the design of intake structures (subsection "b"). The text of §316 is reproduced in the appendix to Yost (1988, this volume).

Section 316(a) provides (in substance) that thermal effluent limitations established by EPA's effluent limitation guidelines may be relaxed (allowing more heat to be discharged) provided the discharger can show that the increased discharge of heat will assure maintenance of a "balanced, indigenous population" of fish in the river.

Section 316(b) of the Act provides (in substance) that the design of cooling water intake structures shall reflect the "best technology available to minimize adverse environmental impact."

It is my interpretation that these two parts of §316 were intended by Congress to be read together, and not as independent, unrelated subsections. Judge Yost (1988) asserts, however, that "the Hudson River case was not being held pursuant to subsection (a), but rather subsection (b)," and that, accordingly, the utilities' evidence was "entirely irrelevant." He further asserts that the "utilities' task in the Hudson case was to demonstrate that cooling towers did not necessarily represent the best technology available for minimizing adverse environmental impact." I believe, on the other hand, that to contend that §316(a) was not relevant ignores the reality that the very permits at issue in the proceeding only restricted the rate of heat discharge. They did not mention cooling towers. They contained no effluent limitations with regard to entrainment or impingement.

The interrelationship between §316(a) and §316(b) is quite important.

Relationship between §§316(a) and 316(b)

My emphatic disagreement with Judge Yost is that I believe he incorrectly reads §316(b) in isolation. I have written elsewhere (Bergen 1978) that subsections 316(a) and 316(b) need to be read together, in pari materia. In drafting §316(a), Congress said, in effect, that waters would be allowed to receive discharges of heat, but short of the point of creating an "imbalance" in the communities of fish and aquatic life in the receiving water body. This reflects, I submit, a Congressional intention that the capital and operating costs and aesthetic effects of large cooling towers should not be imposed upon electric consumers except where needed to prevent "unbalanced" fish populations in the receiving waters. Costs are significant: in the Hudson River case, six cooling towers would impose a capital cost on the order of $358 million (in 1982 dollars).[3]

Congress provided in §316(a) that heat could be discharged to receiving waters so long as aquatic life was not heated to a point of imbalance, and that, under §316(b), impingement and entrainment impacts on the same aquatic populations would be permitted, but only to the same extent—a point short of an "imbalance."

Yost (1988) contends that a point short of an "imbalance" cannot be determined, at least by any precise scientific formula. It is true that Congress rarely legislates formulas. It legislates policy, and leaves the derivation of formulae to the administrative agencies, such as EPA. In §316(a), Congress set a verbal standard of "balanced indigenous population," requiring that judgments be made by EPA as to the definition of "balanced" populations. Yost's argument that the verbal standards of §316 are unhelpful is irrelevant—unless made directly to Congress.

If Yost was correct that only §316(b) applied to the Hudson River case, and that §316(b) mandated the installation of towers, then Congress must have legislated in 1972 that cooling towers should be retrofitted at every existing intake structure in the country. This suggests, in turn, that §316(a) would have no meaning whatsoever, because towers would be needed regardless of §316(a). Basic tenets of statutory interpretation preclude such a result. Cooling towers have not been installed on every existing power plant, so it is evident that Yost's interpretation is not the prevailing one.

Therefore, I submit respectfully that Yost's interpretation of §316 is incorrect as a matter of law—and would be a recipe for disaster as a matter of public policy, if adopted, because it

---

[3]Opinion 63 of EPA's General Counsel admits that costs are a relevant fact under §316.

would require tower installation whether or not towers would be needed to protect the fish.

Burden of Proof

As noted above, EPA staff initially drafted the proposed NPDES permits to include effluent limitations that, if put into effect, would have reduced the rate of heat input by over 95%. However, although the staff proposed these conditions, the burden of proving that some alternative permit condition, or no permit condition, was more appropriate fell upon the utilities. Note that although a government prosecutor in a criminal case must prove the guilt of a defendant "beyond a reasonable doubt," one who applies to a government agency for a permit normally bears the burden of proving that the standard of conduct necessary to warrant the permit's issuance is met.

Yost (1988) suggests that the utilities should have had the burden of proving that "cooling towers did not necessarily represent the best technology available" under §316(b). Such a standard of proof, if imposed, would be impossible to meet and would, therefore, be meaningless. This point should not be lost on scientists, who are prone to advise lawyers that it is impossible to prove a negative. Under Yost's formula, utilities would need to prove that once-through cooling was a better technology than closed-cycle cooling for protecting aquatic life without regard to economics, aesthetics, or other factors. According to Yost, if fewer fish eggs and larvae would be killed by installation of closed-cycle cooling than with once-through cooling, towers should be built. Even if a "balanced population" were maintained, and regardless of cost, towers would be needed under the Yost formula. If the Yost formula was correct, there clearly was no need for the years of hearings that he conducted and allowed to be carried on. If the Yost formula was the correct one, it should have been articulated clearly at the outset of the hearing, and the proofs should have been limited to that narrower standard.

Compensation

A great bulk of the testimony prepared for the hearings concerned whether the data presented could support the existence of compensation in Hudson River striped bass, white perch, and other fish populations. Opinions of some utility scientists contended that there was ample evidence that compensation exists, and scientists for EPA contended that the evidence was not sufficient to justify a conclusion that compensation exists in a sufficient degree to warrant continued once-through cooling of power plants along the river. This entire controversy was hotly debated by the scientists, some of whom assured their respective legal counsel early on that this was *the* key issue in the case. On reflection, I think there was too much emotion expended and emphasis placed on compensation, an issue that created more heat than light as far as the case was concerned.

As the record developed, the question (from my perspective at least) seemed to evolve into whether or not the impact of power plant operations on the striped bass population as a whole, due to the loss of eggs and larvae by entrainment and impingement, was relatively insignificant by comparison to the taking of adults by fishing. A very serious question arose, I think, as to whether the destruction of adult striped bass by fishing is going to wipe out the striped bass population even if power plant entrainment and impingement mortality is totally eliminated. Put directly, if all the power plants were shut down, but the fishing continued, would the striped bass population be wiped out anyway by the fishermen? The evidence seemed to point in that direction. As of 1986, ironically, striped bass fishing had been completely prohibited, ostensibly because of PCB contamination.

If the hearings had continued, there would have been increased attention given by the utilities to the age structure and cyclical reproductive character of the striped bass and other estuarine species inhabiting the Hudson. Failure to address the population dynamics of striped bass earlier in the proceedings may have been an important omission. I believe it was. For example, striped bass are multiple-age, broadcast spawners. Spawning success depends, in large measure, upon variable factors of temperature and salinity in the river during and after the spawning season. It seems essential that substantial numbers of adults in each of the various reproductive year classes be maintained to assure spawning success in those years when river temperature and salinity are most suited to development of young fish. The importance of these factors may be so great that impacts of entrainment and impingement may be negligible by comparison, insofar as preservation and maintenance of the existing adult stock is concerned. Assuming, as I do, that the overall Congressional objective in §316 of the Clean Water Act is to maintain healthy adult stocks of indigenous fish species, I believe that more attention to the Hudson species' population dynamics,

and less to the narrower compensation question, would have been very desirable.

Policy Objective of §316(b)

This raises a broader question, one of policy and philosophy: What is the objective of the legislative standard of §§316(a) and 316(b)? To preserve individual eggs and larvae? To preserve and maintain adult fish stocks?

Perhaps Congress, or its legislative draftsmen, never consciously considered these questions. On the other hand, one could quite rationally assert that Congress intended to protect adult fish populations, and was not concerned about individual eggs and larvae. Given that adults, each of which is capable of producing millions of eggs and young, are allowed to be killed for sport, power-plant-induced mortality of a small portion of the eggs and larvae should not, of itself, be socially unacceptable.

The hearing's focus on the compensation issue may have diverted attention from analysis of the population dynamics of the adult fishery stocks and the factors that threaten or enhance the stocks, including the impacts of fishing on adult year classes and of power plant mortality on the development of young of the year. The adversary format of the proceeding may have detracted from a more considered analysis of the population dynamics issue.

This type of analysis would have attempted to match the statutory standards of §316. Does ambient river temperature remain low enough to assure maintenance of a balanced community of aquatic life? Do the existing intake structures result in "adverse" environmental impact that results in an "imbalance" of the aquatic community?

The analytic approach suggested above differs from EPA's legal theory of the case: that cooling towers were the "best technology" to minimize entrainment and impingement—period! Under EPA staff's and Judge Yost's legal theory, it would be irrelevant to consider estimates of entrainment and impingement impacts on the stability of the adult population, and the costs of installing cooling towers would also be irrelevant. But Congress did not seem to intend that a pure technology-based standard be applied under §316. In creating §316 of the Clean Water Act, Congress, I believe, wanted to protect fish stocks, but to avoid, when possible, capital expenditures and energy usage. Section 316's subsections (a) and (b) mean that if the adult fish stock will not be impaired significantly, then towers should not be built.

In the absence of a precedent-setting interpretation of §§316(a) and 316(b), however, the parties needed to address all possible legal approaches in their evidentiary presentations, which undoubtedly prolonged the proceeding. Moreover, the continuance of river surveys, and availability of new data and new analyses added to the time needed for the hearings, and might have meant they would never end. Accordingly, resolution of the case through settlement made good sense.

Legal versus Policy Dispute

As I said above, I think that the extensive evidence and controversy over the existence or nonexistence of compensation may have been less important than some of the participants actually believed at the time. However, many of the scientists who were witnesses in the case seem, to this day, to be greatly disappointed that the proceeding was terminated by settlement, leaving unresolved a final judgment on the compensation issue. Indeed, they believed that somehow a new scientific truth would arise from the findings of fact and conclusions of the EPA's Regional Administrator in the case. It should be made clear, on the other hand, that administrative and courtroom proceedings are not intended to resolve scientific disputes, such as the extent to which compensatory mechanisms may offset entrainment mortality in the Hudson River striped bass population. The question in the Hudson River case was more mundane: whether cooling towers would or would not be required to be built.

Adversary proceedings are not suited to resolve scientific issues, which require the ability to replicate experiments. On the other hand, expert opinion testimony is invaluable in developing records for use by judges in deciding factual disputes, provided the issues are specific and narrowly focused against established legal standards.

The Hudson River proceeding amounted to a policy dispute. The real issue was "What should the proper standard of conduct be?" not whether the standard had been met or had been exceeded. From the utilities' perspective at least, resolution of the issue needed to consider whether the cost of electricity generation should be increased by an additional $358 million of capital expense and additional operating expense, in order to reduce, to some unknown extent, impingement and entrainment of fish eggs and larvae in the Hudson River. Resolution of this policy matter compelled that an estimate be made of the seriousness of the entrainment and impingement, and that such an estimate be weighed against the social and economic costs to the public. Although some might urge that the loss of a single fish egg or larva might entail a priceless social cost, others might contend that even total eradication of the fish population would not be of any importance whatsoever. Perhaps most of us would contend that the correct balance lies somewhere in between. The policy-maker's task should be to choose the appropriate point of balance, given all of the relevant facts. In essence, this was achieved in the settlement process.

One of the most important questions raised by the history of the Hudson River proceeding, it seems to me, is whether or not an adjudicatory hearing format is appropriate for resolving this type of policy question. The adjudicatory format's utility is greatly diminished when factual issues are not clearly subdivided and specified in relation to a legal standard before the testimony is prepared. The alternative to the adjudicatory format, the rule-making proceeding, is traditionally and effectively used to establish and revise policy issues—and to do so subject to judicial review (the standard for review of which is whether the result of the rule-making is "rational" or "arbitrary and capricious"). An appropriately fashioned rule-making proceeding, with some "quasi cross-examination," may well have resolved the Hudson River cooling tower matter in much less than 3 months—and with about the same outcome as that which actually occurred in the settlement. Indeed, the settlement amounted to a negotiated rule-making proceeding, a type of proceeding that has gained in acceptance in complex EPA policy matters in the mid-1980s.

Despite the frustration of the scientists involved, especially because the case closed by settlement rather than a decision on the merits, their participation aided and encouraged the settlement which was reached. The legal system worked because the dispute was resolved. As a by-product, scientific thinking and knowledge about the Hudson River ecosystem was provoked. I hope the scientists' observations will help to develop and maintain reasonable Hudson River use and conservation policies in future years.

References

Bergen, G. S. P. 1978. Thermal discharges and power-plant intakes: section 316 in perspective. Natural Resources Lawyer 9:305–325.

Yost, T. B. 1988. Science in the courtroom. American Fisheries Society Monograph 4:294–301.

American Fisheries Society Monograph 4:307–315, 1988

Two Scientists in the Courtroom:
What They Didn't Teach Us in Graduate School

SIGURD W. CHRISTENSEN

Environmental Sciences Division, Oak Ridge National Laboratory
Post Office Box 2008, Oak Ridge, Tennessee 37831-6036, USA

RONALD J. KLAUDA[1]

Texas Instruments Incorporated
Ecological Services Group, Buchanan, New York 10511, USA

Abstract.—Scientists and lawyers come to an adjudicatory hearing with different purposes: scientists wish to address objective truths; lawyers want to win the case. It is not easy for the two groups to accommodate one another, even when scientists and lawyers are on the same team. Scientists must educate lawyers on the technical issues involved, and they must learn from lawyers a judicial procedure that is outside their normal experience. Scientists may control the technical content of their testimony, but lawyers control the presentation of testimony and the way a case develops. Cross-examination by skilled counsel is one of the most unnerving features of a trial for scientists new to the courtroom. New scientific knowledge or truth rarely emerges during a trial; the most scientists can hope for is to preserve their objectivity and professional dignity. In the Hudson River power case, scientists found their normally reasonable differences becoming polarized, strident, and disrespectful under the pressures of trial procedures, and the regulatory hearing never achieved a resolution of the issues. During out-of-court preparations for a settlement, in contrast, opposing scientists worked constructively together and compromises were readily reached. An adversarial judicial procedure seems an unsuitable means of resolving issues that are both highly controversial and highly technical.

The Hudson River power case meant different things to the different parties involved. To the utilities, participation was the only alternative to building and operating expensive cooling towers. To the U.S. Environmental Protection Agency (EPA), it was viewed as a national test case for a requirement that utilities retrofit cooling towers. The EPA was convinced that if such retrofitting could not be required on the Hudson River, it probably could not be required elsewhere. To the lawyers involved, the case was a confrontation on a massive scale compared with the personnel, preparation, and effort usually accorded an environmental impact decision. And to the scientists involved, it was the most technical and controversial impact confrontation yet encountered.

The data, information, and analyses generated by hundreds of persons employed full time for the better part of a decade (see Appendix to this volume) were funneled into the hearings process via scores of scientists who prepared written testimony. This paper is intended to share some of

our insights regarding scientists' interactions with lawyers in the courtroom, for the benefit of those who missed the experience but wonder what it was like or who anticipate someday being in a similar situation. Specifically, we reflect on our experiences and describe what scientists can expect during their involvement in a controversial adjudicatory hearing like the Hudson River power case. We also suggest strategies that can minimize conflict between lawyers and scientists, members of two distinct disciplines that must develop an effective working relationship in the courtroom.

We caution readers not to expect an exhaustive treatise on how the interaction between science and law in the United States could be improved. Rigorous discussion of this topic would require an entire monograph by itself and extends well beyond the goal of our paper. Our perspectives on the case are certainly not the only scientists' views of courtroom events, and we have not attempted to summarize other views. Rather, we present the observations and insights of two scientists who actively participated in the EPA's Hudson River power case hearings from opposite sides of the courtroom aisle. We worked with different lawyers who used an array of courtroom

---

[1]Present address: The Johns Hopkins University, Applied Physics Laboratory, Environmental Sciences Group, Shady Side, Maryland 20764, USA.

strategies in an effort to win their case. We participated in the review and cross-examination of each other's testimony. What we present here is a blend of our independent and shared experiences.

Basis of Technical Controversies

The Hudson River power case did not begin as a technical controversy. Rather, a dispute between proponents and opponents of the proposed Cornwall pumped-storage project at Storm King Mountain intensified in the mid-1960s, moved into the courtroom, and identified fisheries issues as one area of disagreement. Scientific studies and legal proceedings followed each other in a complicated sequence (Christensen et al. 1981), and cause-and-effect relationships are difficult to discern. The various legal actions shaped the types of scientific studies conducted and ensured a large-scale research effort to try to resolve the controversial issues.

Depending on the legislation under which a hearing is conducted, entire issues of potential importance might be excluded from study and debate. For example, the question of the contribution of the Hudson River striped bass stock to the Atlantic coastal fishery was a major issue of contention during the mid-1970s in U.S. Atomic Energy Commission and U.S. Nuclear Regulatory Commission hearings on Indian Point units 2 and 3 concerning the National Environmental Policy Act of 1969 (U.S. Congress 1970). The stock contribution issue was not considered in EPA's Hudson River power case hearings, because the agency was concerned with the Federal Water Pollution Control Act Amendments (FWPCA) of 1972 (U.S. Congress 1972). However, because of the working of FWPCA, a new concept, maintenance of a "balanced indigenous community," was introduced into the EPA hearings and supplemented other issues relating to fisheries impacts.

Although stimulated, shaped, and bounded by legislation and court rulings, the scientific issues in a particular case are logically determined by scientists, not lawyers, judges, or legislators. Scientists should decide at the outset if attention is to be concentrated on potential effects on primary producers, zooplankton, fish, or ecosystems. In the Hudson River power case, issue definition was more a matter of reaffirming the importance of historically identified issues than of deliberate decision making. As long as importance, relevance, and tractability could be argued, any of these issues could have become the focus. The focus on fish, particularly striped bass, emerged in

the mid-1960s (HRPC, undated) and only partly for scientific reasons. The scientific basis of this emphasis was reaffirmed by C. P. Goodyear, the ecologist assigned by Oak Ridge National Laboratory on behalf of the Atomic Energy Commission to analyze the effects of Indian Point unit 2 on the aquatic environment (USAEC 1972), and by others since then (e.g., Christensen et al. 1976). Similarly, fish entrainment and impingement were identified as major concerns early in the controversy. Had thermal or chemical discharges associated with power plant operations been of greater concern on the Hudson, the emphasis of the case would likely have shifted to embrace those issues.

One principle is clear: scientists and lawyers must work together but play different roles. Scientists should identify the key environmental issues. Lawyers and interested parties working with the scientists should identify the appropriate judicial body and enabling legislation to resolve those issues. This principle was inconsistently followed during the Hudson River power case for several understandable but counterproductive reasons.

Disagreements among Scientific Experts

Mazur (1981) took "the existence of a dispute between experts as a defining characteristic of technical controversies." The Hudson River power case stemmed from disagreements among scientific experts extending back at least to the mid-1960s. Those disagreements were aired in a series of adjudicatory hearings leading up to the Hudson River power case. In these hearings, scientific experts with divergent opinions on the environmental impacts testified in a role known to lawyers as the "expert witness." Whereas courtroom witnesses are generally permitted to testify only about matters of fact, expert witnesses are also permitted to state opinions in their areas of expertise. This practice is not only fortunate but also necessary, because much of science relies on analyses based on professional judgments (e.g., about which set of assumptions underlying the various alternative ways of analyzing a given set of data is most consistent with the data and the real world).

Scientists bring to the courtroom their individual views of the world, shaped by background, experience, and intelligence. No two scientists will see the same problem in exactly the same way; hence, different scientists working with the same universe of data can arrive at different conclusions. This phenomenon occurs in part because scientists have some latitude in the process of "doing science." They may work with different subsets of the available data. Their personal backgrounds and judgments may lead to different ways of weighting and analyzing the data. Their interpretations of the results are likely to include some influence from their own values.

To approach the "truth" in science, however, a scientific judgment must be free of nonscientific and potentially biasing considerations that push objectivity toward advocacy. Scientists are trained to avoid any tendency toward biased judgments. Our experiences in the Hudson River power case, however, indicate that political and economic forces can purposely or inadvertently encourage bias. The legal arena seems to accentuate these forces, tends to polarize the scientific participants, and promotes a superficial exaggeration of differences between opposing sides. We observed that, in the courtroom, scientists working on one side of the case usually espoused similar views and that these views were very different from those espoused by scientists working for the opposing side. The adversarial nature of the adjudicatory hearing seemed to conceal the true range of technical opinions held by the scientists on both sides of the case.

The apparent polarization of technical positions, coupled with the adversarial courtroom setting and the influence of lawyers, actively inhibited candid interactions between members of the opposing sides, particularly between the respective scientists. When informal communication diminished, levels of suspicion bordering on mutual distrust crept into the courtroom. Often ill-founded questions about the other side's basic attitudes became disturbingly commonplace topics of conversation among the participants. For example, do scientists who believe that nature is resilient and persistent tend to select industry for the opportunity to practice their beliefs? Do scientists who believe that nature is fragile and delicately balanced gravitate toward conservation organizations? Do those who wish to be totally unconstrained in their analyses choose government? Does industry select scientists whose views are consistent with the industry's attitude toward environmental issues? Do scientists ever succumb to subtle peer pressure and respond to the prevailing climate of opinion where they work? Are some scientists so motivated by financial reward and job security that they are willing to construct impressive-sounding arguments in which they may not believe just to help their clients and "feather their nests"? Are some sci-

entists such strong preservationists that they oppose any uses of ecological resources that generate profits for industry?

The answers to such questions from the scientists involved in the Hudson River power case would likely be as varied as the individual scientists themselves. Regardless of the justifiability of these suspicions, mere concern that such predilections would influence some scientists' judgments was a real and often counterproductive factor that influenced attitudes, interactions, and lines of cross-examination. We believe that all scientists favor unbiased analyses, but our experiences suggest that the courtroom setting discourages complete openness and objectivity. Scientists can consider the advice of Dunn and Kiersch (1976) who wrote that "Your responsibility is to be loyal to your client within the framework of ethical practice, which places truth above all considerations. Advise your client of the nature of your findings. . . . Such information is not to be divulged to others unless you are legally required to do so." But such constrained discussions of the technical issues among the scientists participating in a hearing will likely retard the resolution of many environmental controversies. Many scientists may wish for a "science court" approach (USDOC et al. 1977). It is also important to consider that scientists who begin to function as "hired guns" are no longer scientists but advocates. If their roles as advocates can be demonstrated, their credibility in the expert witness role will be severely undermined.

The differences in technical opinions among the scientists involved in the Hudson River power case ranged from mild to strong depending on the technical issue and individual scientist. The issue of biological compensation evoked more of a confrontational "Hatfields and McCoys" atmosphere than any other single technical issue. Many fascinating anecdotes could be told. For example, there was the utility panel witness who came to be admired for his candid responses to questions and thereby became the specific target of many questions. There was the EPA consultant who, after listening to a particularly eloquent utility witness draw an analogy between the inevitability of compensation and the inevitability of the sun's rising the next morning, exclaimed "Oh, my [deity]!" in a stentorian stage whisper and stalked from the hearings room in disgust, never to return. There was the witness who, recalling a key conversation between one day and the next, avoided by a matter of minutes an accusation of perjury the next morning.

The scientists involved in the case clearly differed in personality and in desire to play expert-witness roles. Many appeared intimidated by a courtroom appearance and sought minimal participation. Others accepted the challenge and stimulation of an adversarial milieu, shouldered their responsibilities, and tried to clearly communicate complex technical points to a heterogeneous audience. A few scientists viewed the expert-witness role as formal recognition of their professional stature, a battle of wits with lawyers and other scientists, and an open invitation to simultaneously pontificate their favorite theories and display their oratorical skills. These personality differences heightened rather than diminished the level of controversy.

Because the focus of this monograph is on the technical lessons learned from the Hudson River power case, the adversarial tone that permeated the testimony and the hearings record has been intentionally minimized. Still, to communicate accurately all facets of this case, we must acknowledge that (especially during cross-examination) tempers sometimes flared, the atmosphere frequently resembled that of a pitched battle, and paranoia was not unknown.

This situation may be typical of adjudicatory proceedings. Scientists involved as expert witnesses, especially for the first time, should realize that they will face opposing lawyers, advised by opposing scientists, whose strategy may be to attempt to convince them to reverse their positions or, failing that, to retract their testimony. If all else fails, the opposition will settle for discrediting the other side's experts or reducing them to lumps of quivering jelly. A scientist's defense is to be sure that any damage to their testimony involves only window dressing and not the basic structure. The best way to accomplish this is to adhere to good scientific principles, to anticipate the courtroom atmosphere, and to prepare for cross-examination accordingly.

Enter the Lawyers

In technical analyses (e.g., environmental impact statements) or controversies destined to end in hearings, lawyers eventually become involved and the participating scientists' roles are never again the same. No lawyers attended the first few technical meetings held by the EPA's Region II staff to discuss the Hudson River power case. There, EPA scientific staff and consultants evalu-ated the information and analyses and candidly defined technical areas in need of further work. Then, during a part of one such meeting, an EPA lawyer explained the situation from his perspective. He ascertained that there were significant issues on which the gathered scientists disagreed with the utilities' analyses enough to justify a serious legal case. Every succeeding meeting was strongly influenced by lawyers, until eventually they took charge.

A scientist's primary goals are to seek scientific truth and to communicate conclusions to others. A lawyer's primary goal is to win the case. From a lawyer's point of view, data, analyses, and conclusions that increase his or her likelihood of winning are desirable; those that decrease the chances of winning are anathema. If scientists are slow to grasp this perspective at the outset, they learn it quickly enough. Hence, a tension between scientists and their own lawyers is created. Scientists can best maintain balance and integrity in this process by not getting carried away with the enthusiasm of a "team spirit"; by remaining the detached, impartial, objective scientists they were trained to be in graduate school; and by striving to look at all relevant sides of the issues.

Scientists are in control of the technical aspects of the case. They have the lead in defining the key technical issues and in stating their positions as best they know them to their own lawyers. Scientists' positions may change as new information or results become available, and they should prepare the lawyers for this eventuality. In turn, lawyers have their own ethics, and they know that they cannot win by attempting to put words in scientists' mouths—unless the scientists let them, which is a risky proposition indeed. The key principle for scientists here is to remain objective and to be alert for pressures that may compromise objectivity.

Educating the Lawyers

Few of the lawyers we encountered had taken more than a course or two in college biology, yet they were intimately involved in the progress of a major biological controversy, the Hudson River power case, right up to the final negotiated settlement (see Section 4 of this monograph). Lawyers, not the scientists, were in charge of the conduct of the case. In cross-examination, the lawyers normally asked the questions of the scientific experts and chose or modified the next question depending on the response. It was essential for the lawyers to understand the technical issues involved, in at least as much depth as the advanced layman could understand them.

We observed that the lawyers involved in the Hudson River power case differed in their respective approaches to the learning process. A few lawyers wanted to master and become thoroughly familiar with the basic scientific principles underlying important technical topics. When successful, these lawyers interacted very effectively with scientists; together, they occasionally conceived elegant lines of questioning that penetrated directly to the heart of the key topics, and they were not easily deflected to softer and less important points by alert expert witnesses. Possessing a clear comprehension of the science being debated in the case also permitted cross-examining lawyers to operate more independently from their scientific advisors and occasionally step away from the prepared questions to "shoot from the hip" with less fear of drawing damaging return fire. Other lawyers seemed content with a more superficial grasp of the technical issues, so they leaned heavily on their scientific advisors.

Lawyers tend to be overworked, just like scientists, and they may put off the inevitable learning exercise as long as possible. If so, the scientist may need to take the first step. On the EPA side of the Hudson River case, the scientists applied early pressure to get the lawyers more involved with the technical details. Soon this was unnecessary; the EPA lawyers organized two "retreats" with the scientists, each lasting several days, for the joint purposes of being educated in the scientific issues and preparing cross-examination questions. Technical experts gave lectures on their respective topics and practiced being cross-examined by the lawyers. Many such educational sessions were also held between the utility lawyers and their consultant scientists.

Although these mutual learning processes were arduous, we observed that most lawyers learned quickly. At this preparatory stage, scientists can maximize the integrity of the future hearing process by giving the lawyers a balanced view of the technical issues and an appreciation of potentially vulnerable areas in their case. There will always be vulnerable areas, because the practice of science involves judgments and few judgments are indisputable. The ability of scientists, at this stage, to see (and minimize) those vulnerable areas will enable them to do their best scientific job, to avoid misleading their lawyers about these areas, and to anticipate eventual cross-examination of their own testimony.

Establishing the Ground Rules

Usually at a prehearing conference, but sometimes as a matter of policy, key decisions will be made by the judge or hearing board and the lawyers for the parties involved. Some of these decisions will shape the future course of events. One of the most critical legal questions that affected all participants in the Hudson River power case was: Who had the burden of proof? At one extreme, the utilities would be required to prove that power plant operation would not cause substantial harm, or else they would have to retrofit expensive cooling towers. At the other extreme, the EPA would be required to demonstrate that the power plants would cause substantial harm, or else accept continued once-through cooling operations at the three major power plants (see Hutchison 1988, this volume). Any scientist familiar with the Hudson River situation or other environmental impact assessments will realize that either requirement would have been impossible to meet. The lawyers should have been (and, in some cases, were) told such things early and repeatedly. Their limited scientific backgrounds could not have prepared them for the degree of uncertainty inherent in biological systems and data.

In the Hudson River power case, the actual burden of proof was not entirely clear-cut. It seemed likely at the outset that the EPA Administrator, who would render a decision without ever having been present at the hearings, would sift the evidence in the written record and seek demonstration that the costs of retrofitting cooling towers would not be wholly disproportionate to the benefits gained from cessation of once-through cooling operations. Determining not only the question of burden of proof but also the scope of the technical issues involved was a frustrating problem for the scientists, but even more so for the lawyers and the judge (Bergen 1988; Yost 1988, both this volume).

Another critical question concerned the order of proceedings. Should direct testimony be filed simultaneously or sequentially? What about rebuttal testimony? At what point does cross-examination occur, and in what order? Such questions became critical, especially when one party (EPA) had to rely on data collected by the other party (the utilities).

Preparing Written Testimony

The written testimony is the foundation of the legal case. In the end, the lawyers for each side (perhaps with scientists' help) draft proposed "findings of fact" that they hope the judge or hearing board will accept as the legal "truth." These findings of fact are based primarily on the written record. The lawyers also draft proposed "conclusions of law," acceptance of which would constitute winning the case. Just as lawyers want to win, scientists want their analyses to be accepted in adjudicatory cases, in the same manner as they want their papers to be published in journals. Although peer review of journal articles can be thorough, it rarely erects the kind of gauntlet that written testimony must run in a hotly contested case.

Our experience in the Hudson River power case suggests that preparing testimony can be an unusually demanding process. One EPA consultant averaged 80 hours a week for 16 consecutive weeks completing analyses, preparing and revising his testimony, and reviewing others' testimony. His colleagues were not far behind. The utility consultants who prepared testimony for the case were part of a large research team. Hence, testimony preparation for them extended beyond writing and revising the documents. They also spent many hours poring over detailed field and laboratory procedures, quality control records, and data entry protocols to comprehend fully and be able to explain clearly how the data forming the basis for their testimony were collected, processed, checked for errors, and stored in computer files prior to analysis. This level of detailed understanding was required if they were to survive the barrage of cross-examination and emerge with their written testimony mostly intact.

Scientists must explain to the lawyers that certain technical statements simply cannot be made. What we were able to conclude from the analyses presented in written testimony was usually much less than what the lawyers wanted us to say. The lawyers grumbled a bit but eventually accepted the inevitable. Whatever the prevailing legal attitude, scientists must maintain control of the technical content of their testimony and should resist the lawyers, if necessary, to preserve this control. They are the ones who must defend their testimony under cross-examination. Lawyers have veto power, however. If they judge that the net effect of a piece of testimony is negative, they can refuse to file it.

Scientists should be very receptive to comments from lawyers about organization, style, and level of detail and explanation needed in their written testimony. The testimony is, after all, being written for a different audience than usual, ranging from laymen to experts. The lawyers understand the needs of this audience better than most scientists.

Cross-Examining the Opposition

After scientists have prepared and filed written testimony, and the opposing side has had time to examine it in detail, they are cross-examined on their work by the opposing side (as well as by other parties to the hearing). This can be a difficult, frustrating, and time-consuming process. In the Hudson River case, cross-examination on biological compensation in striped bass populations and related topics alone occupied approximately 25 days of hearings time spread over a period of 6 months, and it generated more than 5,000 transcript pages.

Cross-examination is usually a negative process. The lawyers conducting it wish to do as much damage to the opposing case as they can. They try to achieve this by attacking the testimony, the expert witness, or both. It is important that scientists defending their testimony not take such attacks personally. The lawyers are just doing their job.

The rules of the game require that lawyers, not scientists, conduct the cross-examination. The lawyers can be advised by scientists whispering in their ear, but too much advising of the cross-examiner is usually discouraged by the presiding judge because it wastes time. Scientists, after discussion with lawyers, conceive most of the lines of cross-examination and write out sequences of questions. The next step is rehearsing the lawyers. Scientists can play the role of the opposition and try to confuse their own lawyers with the kinds of answers that may be given in the actual hearings. During cross-examination, a good lawyer with a sound technical understanding of the issues can really come into her or his own. The process is, however, truly grueling. Lawyers and scientists worked days, nights, and weekends, often on travel status, in preparation for the Hudson River power case hearings. It was extremely frustrating at times. Many scientists began to wish for a forum allowing scientist-to-scientist, back-and-forth exchanges, perhaps modeled after the "science court" concept.

As a rule, lawyers never ask a question to which they do not already know the answer they expect or want. Therefore, cross-examination is rarely intended to get "new" information from the opposition into the record. During the EPA's cross-examination of the utilities' panel of experts on biological compensation, however, some constructive results were achieved. Some questions asked of the panel were intended to establish points of agreement about assumptions underlying their techniques. Establishing points of agreement among panel members enabled subsequent EPA testimony (Christensen et al. 1979) to reference the hearings transcripts on these points, so that a justification did not have to be developed for each point. This strategy saved considerable time in preparing the testimony, and it would likely have shortened subsequent cross-examination had the case not been first settled out of court.

Being Cross-Examined

Significant gains are not to be expected when the written testimony of an expert witness is cross-examined. The major objective is to lose nothing or, failing that, to lose as little as possible. A good performance, in terms of the written transcript record, may also favorably impress the eventual decision maker and provide helpful material for proposed findings of fact.

Scientists can best prepare for cross-examination by being thoroughly familiar with their testimony and by identifying parts of it that the opponents will likely attack. Scientists should also reread their own previous publications in technical journals and the popular literature. Lawyers routinely dredge up old papers that a scientist wrote many years ago, extract statements that appear to contradict the testimony being cross-examined, and base often-embarrassing questions on these statements. Scientists are, of course, permitted to have changed their views over time, but they should ponder their answers to such questions before the testimony is filed. Their lawyers or peers can rehearse them in dealing with difficult questions.

Scientists testifying in an adjudicatory hearing must tell the truth—being under oath is a compelling although hopefully an unnecessary motivation—and they should take time, if needed, to think through their answers before beginning to talk. Otherwise, the written record may be nearly unreadable and embarrassing to the persons who spoke. Contrary to what we wanted to believe when subsequently reading the transcripts, courtroom stenographers rarely were responsible for garbled material. If scientists do their jobs well, they can be confident about their work and their testimony, and this part of the hearings can actually be almost enjoyable.

Living with the Decision

Adjudicatory hearings have as their basic purpose the rendering of a decision. The adversary method on which they depend is "ascientific not only in its procedure, but in its greater commitment to victory than to truth" (Hammond and Adelman 1976). Normally, one side wins and the other side loses. From the scientists' point of view, being on the winning side brings little professional credit. Being on the losing side, however, is likely to be hard on the scientists' egos, if only because "findings of fact" will be written indicating that their analyses were not as good as their opponents'. Scientists may also realize it is quite conceivable that political considerations could outweigh scientific merit. A concern voiced by some scientists involved in the Hudson River power case was that the decision might actually be made on political rather than scientific grounds, but that it might be made to appear justified on scientific grounds regardless of the scientific validity of these grounds. If such a cynical view were correct, it would imply a serious misuse of science.

The negotiated, out-of-court settlement of the Hudson River power case rendered these concerns moot. Several of the formerly opposing scientists had the opportunity to work together within an agreed-upon technical framework to produce analyses needed as the basis for compromises (Englert et al. 1988, this volume). Few adjudicatory proceedings end so amiably. For many participating scientists, however, their only satisfaction came from knowing they prepared and defended valid testimony, maintained their objectivity, and were a credit to their profession.

Epilogue

Opportunity for the utility and government scientists to candidly and informally discuss the controversial technical issues in the Hudson River power case (e.g., biological compensation in the striped bass population) was sharply curtailed whenever hearings were underway. The overriding legal goal—to win the case—polarized the views of opposing-side scientists on many issues and apparently masked the true range of opinions held by the utility and government scientists. The gap between the espoused views of opposing scientists tended to widen as cross-examination intensified and the lawyers repeatedly "aimed for the jugular vein." The courtroom atmosphere of this adversary hearing was clearly not the appropriate setting for the diverse group of scientists to challenge, critique, consider, and compromise on the difficult biological issues.

From our vantage point as scientists, the legal approach to resolving technical disagreements during the EPA hearing seemed laboriously ineffective. The case was settled out of court in December 1980, so perhaps the adversary hearings process, as applied to the Hudson River power case, was killed before maturity. Perhaps the hearing could have neatly resolved the major issues and led to an equitable decision before the end of the 20th century. We doubt it.

The negotiated settlement agreement was hammered into shape by a subgroup of the scientists who participated in the adversary hearing. However, the candid, settlement-related discussions of complex technical issues were out of court, off the record, and controlled by the scientists. Polarization decreased over time, allowing the true range of scientific views among all participants to emerge and be openly debated. The strengths and weaknesses of all views were freely critiqued, and compromises on most issues were quickly reached.

Our experiences demonstrated that scientists and lawyers can educate one another and meld into reasonably effective courtroom teams. We are not convinced, however, that such scientist–lawyer teams could have equitably resolved the Hudson River power case issues within the courtroom setting, even if the EPA hearing had not ended prematurely.

The out-of-court settlement agreement achieved what the adversary hearings process did not—compromise among scientists on those technical issues that contained much uncertainty. Timing contributed to the success of the settlement agreement process: the scientists came into the settlement process predisposed to reaching a compromise in part because of their frustration with the questionable technical progress and the slow pace of the interminable hearings. Nevertheless, we are convinced that the lack of restrictions upon open scientific discussion of critical technical issues contributed more significantly than timing to the success of the settlement negotiation process, in contrast with failure of the EPA hearing process to resolve the Hudson River power case. In our view, the constraints on discussion in the courtroom make the adversarial process an inherently unsuitable means of resolving issues that are both highly controversial and highly technical.

Acknowledgments

L. W. Barnthouse, C. W. Gehrs, S. G. Hildebrand, D. S. Vaughan, and N. D. Vaughan critically reviewed the manuscript and provided helpful suggestions. S. W. Christensen's contribution was supported in part by the Office of Nuclear Reactor Regulation, U.S. Nuclear Regulatory Commission, under interagency agreement (IAG) DOE 40-544-75 with the U.S. Department of Energy (DOE); in part by the Office of Health and Environmental Research, DOE, under contract DE-AC05-84OR21400 with Martin Marietta Energy Systems, Incorporated; and in part by the Office of Research and Development, U.S. Environmental Protection Agency, under IAG DOE 40-740-78 (EPA 79-D-X0533). This is publication 2963 of the Environmental Sciences Division, Oak Ridge National Laboratory.

References[2]

Bergen, G. S. P. 1988. The Hudson River cooling tower proceeding: interface between science and law. American Fisheries Society Monograph 4: 302–306.

Christensen, S. W., C. P. Goodyear, and B. L. Kirk. 1979. An analysis of the validity of the utilities' stock-recruitment curve-fitting exercise. Testimony (Exhibit EPA-208) before the U.S. Environmental Protection Agency, Region II, in the matter of Adjudicatory Hearing Docket C/II-WP-77-01.

Christensen, S. W., W. Van Winkle, and J. S. Mattice. 1976. Defining and determining the significance of impacts: concepts and methods. Pages 191–219 in R. K. Sharma, J. D. Buffington, and J. T. McFadden, editors. The biological significance of environmental impacts. U.S. Nuclear Regulatory Commission, NRCONF-002, Washington, D.C.

Christensen, S. W., W. Van Winkle, L. W. Barnthouse, and D. S. Vaughan. 1981. Science and the law: confluence and conflict on the Hudson River. Environmental Impact Assessment Review 2:63–88.

Dunn, J. R., and G. A. Kiersch. 1976. The professional fisheries scientist as an expert witness. Fisheries (Bethesda) 1(6):2–4, 44–46.

Englert, T. L., J. Boreman, and H. Y. Chen. 1988. Plant flow reductions and outages as mitigative measures. American Fisheries Society Monograph 4:274–279.

Hammond, K. R., and L. Adelman. 1976. Science, values, and human judgement. Science (Washington, D.C.) 194:389–396.

HRPC (Hudson River Policy Committee). Undated [1968]. Hudson River fisheries investigations (1965–1968). Report to Consolidated Edison Company of New York.

Hutchison, J. B., Jr. 1988. Technical descriptions of

---

[2]See Table 1 for sources of legal documents and unpublished reports pertaining to the Hudson River.

Hudson River electricity generating stations. American Fisheries Society Monograph 4:113–120.

Mazur, A. 1981. The dynamics of technical controversy. Communications Press, Washington, D.C.

USAEC (U.S. Atomic Energy Commission). 1972. Final environmental statement related to operation of Indian Point nuclear generating plant, unit no. 2, volumes 1 and 2. Docket 50–247.

U.S. Congress. 1970. National Environmental Policy Act of 1969. (PL 91-190, January 1, 1970.) Pages 852 et seq in U.S. statutes at large 83. U.S. Government Printing Office, Washington, D.C.

U.S. Congress. 1972. Federal Water Pollution Control Act Amendments of 1972. (PL 92-500, October 18, 1972.) Pages 816 et seq in U.S. statutes at large 86. U.S. Government Printing Office, Washington, D.C.

USDOC (U.S. Department of Commerce), National Science Foundation, and American Association for the Advancement of Science. 1977. Proceedings of the colloquium on the science court (September 19–21, 1976, Leesburg, Virginia). USDOC, Commerce Technical Advisory Board, Washington, D.C.

Yost, T. B. 1988. Science in the courtroom. American Fisheries Society Monograph 4:294–301.

American Fisheries Society Monograph 4:316–328, 1988

What We Learned about the Hudson River:
Journey toward an Elusive Destination

RONALD J. KLAUDA[1]

Texas Instruments Incorporated, Ecological Services Group, Buchanan, New York, 10511, USA

LAWRENCE W. BARNTHOUSE AND DOUGLAS S. VAUGHAN[2]

Environmental Sciences Division, Oak Ridge National Laboratory
Post Office Box 2008, Oak Ridge, Tennessee 37831-6036, USA

Abstract.—The major findings of the Hudson River power plant studies, a landmark environmental impact assessment conducted during the 1960s and 1970s, are summarized in this paper. The studies focused on the effects of power plant operations on finfish populations and contributed knowledge of (a) the biology of Hudson River fishes, (b) sampling design and collection procedures, (c) measurement of entrainment and impingement mortality, (d) quantitative estimates of year-class strength reductions attributable to entrainment and impingement, (e) mitigation techniques, and (f) resolution of technical controversy in the face of uncertainty and an ineffective litigative process. Many insights contributed by the studies became the technical tools used to forge a negotiated settlement agreement in December 1980. The agreement ended the long-standing dispute among the utilities, regulatory agencies, and environmental groups regarding plans for a pumped-storage facility at Storm King Mountain and the need for cooling towers at several operating steam electric stations. These insights also provide a framework for research and monitoring tasks that have continued. In spite of its failure to achieve all study objectives, the Hudson River studies found acceptable solutions to the major environmental problems posed by power plants along the river.

The signing of the out-of-court settlement agreement on 19 December 1980 was a significant event in a long series of environmental studies and legal proceedings that is collectively labeled the Hudson River power plant case. The negotiated agreement suspended dispute over the environmental consequences of power plant operations along the river. The agreement truncated almost 20 years of environmental impact assessment studies even though several scientific objectives were not met. Since the settlement agreement, the question of whether the Hudson River studies were collectively a success or a failure has been debated in terms both of the studies' applied objectives and of their contributions to basic aquatic science. In this paper, we examine the question from the viewpoint of an "optimistic realist" (Van Winkle 1977). In the next paper (Barnthouse et al. 1988b), the question is addressed from the perspective of a "realistic pessimist."

Goals and Objectives: The Elusive Destination Defined

Most of the Hudson River studies were financed by the utilities, whose goal was to obtain state and federal permits to construct and operate electrical generating stations along the river. In testimony before the Atomic Safety and Licensing Board, McFadden and Woodbury (1973) described the major objectives of the utilities' biological studies program: (a) to determine the biological significance for the Hudson River ecosystem of chemical and thermal power plant effluents and of entrainment and impingement of fishes in plant cooling-water flows; (b) to determine the acute and chronic effects of temperature on life stage survival and migratory behavior of key fish species; and (c) to develop and test measures for minimizing adverse biological effects of power plant operations. These objectives bounded the research efforts of all scientific groups involved in the Hudson River power plant studies through December 1980, and qualified the studies for classification as an environmental impact assessment, broadly defined by Larkin (1984) as "a process aimed at guarding the public interest in the proper use of resources in the aggregate."

As the studies rapidly evolved during the early 1970s, the major objectives were refined and

[1]Present address: The Johns Hopkins University, Applied Physics Laboratory, Environmental Sciences Group, Shady Side, Maryland 20764, USA.

[2]Present address: National Marine Fisheries Service, Southeast Fisheries Center, Beaufort Laboratory, Beaufort, North Carolina 28516, USA.

redefined. From the outset, the emphasis was on fish populations and not ecosystem dynamics, even though the phrase "significance for the Hudson River ecosystem" in McFadden and Woodbury (1973) could be interpreted otherwise. Further, by middecade, efforts had moved away from the determination of chemical and thermal effects of effluents and were sharply focused on entrainment and impingement effects on selected finfish species (Klauda et al. 1988c).

The Hudson River studies comprised a reactive research program whose course was controlled more by societal controversy than by basic tenets of scientific inquiry. Larkin (1984) recognized the inherent conflicts among parties to environmental assessments. Among the parties are administrators, who desire publicly visible research; (b) project managers, who desire mission-oriented results; and (c) scientists, who are motivated by their peers to produce credible research with no constraints on publication and few constraints on objectives. Although constraints on publication did not exist within the Hudson River program, other conflicts did, and they shaped many research tasks.

The focus on fish populations adopted and maintained during the Hudson River studies was typical of environmental assessments in the 1960s and 1970s. By definition, environmental assessment research is predominantly mission oriented. Larkin (1984) pointed out that the objectives of impact assessments need not contribute to scientific knowledge so much as they should foster public knowledge about the threatened resource and the degree to which the resource is at risk. When the level of controversy surrounding an environmental assessment becomes an important stimulus for the arrangement of issue priorities, curiosity-oriented research is likely to be neglected whereas mission-oriented and publicly visible research are promoted. Therefore, even if a broader perspective on power plant effects had been adopted for the Hudson River and major study objectives had expanded from selected finfish populations to questions of multispecies interactions, we are certain that litigative forces would have steadily narrowed the window of investigation and steered the study back to striped bass and a few other fishes.

The Hudson River studies demonstrated that, like resource management, environmental assessment is fraught with uncertainties. Uncertainties slowed progress toward study goals and objectives, but the long series of research programs

yielded a wealth of insights about this major estuarine system, several important finfish resources, and approaches to environmental impact assessment. When the power plant case participates moved from the stalemate that had developed by the late 1970s (Christensen and Klauda 1988, this volume) and began to seek solutions (Barnthouse et al. 1988a, this volume), many of these insights became technical tools used to forge the compromise agreement.

Accomplishments: Major Milestones along the Journey

The Hudson River studies contributed importantly to six general topics: (a) biology of finfishes, (b) sampling design and collection procedures, (c) measurement of entrainment and impingement mortality, (d) quantitative estimates of reductions in year-class strength attributable to entrainment and impingement, (e) mitigation techniques, and (f) resolution of environmental controversy in the face of uncertainty and an ineffective litigative process. These contributions were invaluable during the negotiated settlement of the power plant case and are now available to aid future environmental assessments.

Fish Biology

To date, 140 diadromous, resident, and marine fish species have been found in the Hudson River, making this fauna one of the most diverse fish groups in Atlantic coast rivers (Beebe and Savidge 1988, this volume); several species were recorded for the first time in the period since the mid 1960s. The finfish community is composed of resident native species that repopulated the river after the last glacial episode, several introduced species that gained access to the Hudson River through canals, and marine species that regularly or incidentally enter the river from coastal areas. The proximity of the Hudson River to the Gulf Stream in the Atlantic Ocean enhances the probability that tropical and pelagic fishes will occasionally stray into the estuary. Beebe and Savidge (1988) concluded that human activities on the Hudson River have been accompanied by an increase in fish species diversity. Extensive surveys from 1965 through 1980 demonstrated that despite a long period of pollution and other anthropogenic stresses such as entrainment and impingement at power plants, the Hudson estuary (the lower 243 km of tidally influenced river below the Troy Dam) sustains a diverse and abundant fish fauna.

The distributions and abundances of migrant and resident finfish species appear to be directly and indirectly regulated, to a large degree, by freshwater inputs to the estuary (Gladden et al. 1988, this volume). Freshwater flows import organic carbon, enhance phytoplankton and zooplankton standing crops, and alter the spatiotemporal distributions of river salinity and temperature, all of which affect fish habitats. The Hudson River studies showed that spatial, temporal, and trophic partitioning occurs among the dominant fish species. Residents occupy primarily the shore zone and shoal-bottom habitats, where they feed on benthic and epibenthic invertebrates. Migrants occur in all major habitats but predominate in deeper pelagic regions. Compared to resident species, migrants feed more frequently on the higher trophic levels (zooplankton, macroinvertebrates, fish) and exhibit larger annual variations in abundance. Residents occupy relatively stable, low-risk habitats and gain persistence at the expense of abundance. Migrants occupy more dynamic areas of the river where risks are high but the energetic benefits of short-term resource exploitation can be great (Gladden et al. 1988).

Less than a dozen finfish species became primary or secondary foci of the studies. The quantity of biological information compiled on each species was roughly proportional to the species' "importance" based on economic value, population status, exposure to the major power plants, and role in the ecological food web.

Striped bass, white perch, and Atlantic tomcod were designated primary focal species because of their relatively high degree of exposure to the major operating or proposed power plants. Striped bass, American shad, Atlantic sturgeon, and, to a lesser extent, bluefish and weakfish were focal species because of their importance to existing or potential recreational and commercial fisheries. Blueback herring, alewife, and bay anchovy were focal species because of their importance as prey organisms (Dew and Hecht 1976; Gardinier and Hoff 1982), and also because of their high degree of exposure to power plants during at least part of their life cycles.

Sturgeons.—Atlantic and shortnose sturgeons became secondary focal species in the late 1970s because the latter had been classified as an endangered species by the U.S. Government. Interest in Atlantic sturgeon in the Hudson River intensified along with concern for shortnose sturgeon. Catches of young and adult sturgeons during the studies were mostly incidental to sampling programs directed at other focal species, particularly adult striped bass. Compilation and reporting of sturgeon catches (Hoff et al. 1977; Hoff and Klauda 1979) coincided with the activities of the Shortnose Sturgeon Recovery Team (SSRT), coordinated by the National Marine Fisheries Service and the U.S. Fish and Wildlife Service. The SSRT was assembled to evaluate the status of shortnose sturgeon stocks along the east coast of North America (Dadswell et al. 1984).

Data collected during the Hudson River studies contributed to the SSRT's evaluation of the shortnose sturgeon stock, and also stimulated investigations on both sturgeon species in the late 1970s. These studies were sponsored by the New York Department of Environmental Conservation, the Boyce Thompson Institute, the Oceanic Society, and others (Dovel 1979). The investigators learned that the adult shortnose sturgeon population in 1979-1980 numbered at least 13,000 (possibly as many as 30,000), making it the largest or second largest population in eastern North America (Dadswell et al. 1984). New insights were gained about sturgeon spawning areas, population age structure, growth rates, migratory patterns, and management potential for both species (Dovel and Berggren 1983; Hoff et al. 1988a; Young et al. 1988a).

Other fishes were designated focal species in the Hudson River studies because their populations were exposed and vulnerable to entrainment and impingement at power plants, either year round or seasonally, during some periods in their life cycles. Detailed information was obtained on the spatiotemporal distributions of early life stages of these species, which allowed assessment of their vulnerability to power plants.

Striped bass and Atlantic tomcod.—Migratory species such as striped bass and Atlantic tomcod that spawn close to one or more power plants (Boreman and Klauda 1988, this volume; Klauda et al. 1988b; McLaren et al. 1988b, this volume) were expected to be at particularly high risk from power plant operations. The narrowness of the tidal Hudson River (Cooper et al. 1988, this volume) and the small number of major embayments and tributary confluences in the mesohaline and lower oligohaline zones (0.3–5 ‰ salinity) meant that many immigrating adults and their emigrating eggs, larvae, and juveniles were forced into areas of cooling-water withdrawals. Despite this vulnerability, the Hudson River population of striped bass remained stable during the 1970s and showed no clear evidence of overexploitation from sport and commercial fisheries or power plant operations. The population produced moderate to strong year classes throughout the 1970s (Klauda et al. 1980; Boreman and Austin 1985) and into the 1980s (Martin Marietta Environmental Systems 1986; R. E. Brandt, New York Department of environmental Conservation, personal communication). The 1983 year class was one of the strongest on record. In contrast, other Atlantic and Pacific coast striped bass populations declined during this period (Boreman and Austin 1985; Stevens et al. 1985).

Annual distributions of young striped bass differed little in the Hudson River during the last half of the 1970s (Boreman and Klauda 1988). The low variability in annual abundance and distribution of young hindered definitive conclusions about a relationship between early life stage distribution (an index to the degree of exposure to major power plants) and year-class success. This dilemma also meant that little could be discerned about a stock–recruitment relationship for the species and population-level effects of power plant operations. The task was further complicated by annual variations in several biotic and abiotic factors (Dey 1981). More than 3,700 MW of electrical generating capacity came on-line along the Hudson River between 1974 and 1976. The commercial fishery for striped bass within the Hudson River was closed in early 1976 because fish became contaminated with polychlorinated biphenyls (PCBs: McLaren et al. 1988a; Sloan and Armstrong 1988). Striped bass from the Hudson River continued to be exploited outside the river, where they mingle with other Atlantic coastal stocks (McLaren et al. 1981) and make up 10–50% of the migrants from Maine to North Carolina (Berggren and Lieberman 1978; Van Winkle et al. 1988, this volume).

The major studies of adult striped bass in the Hudson River began in 1976, extended through 1979, and coincided with the commercial fishing moratorium in the river (Hoff et al. 1988b, this volume). The spawning population at that time comprised at least 12 age-groups and included females approaching 20 years old. Age-3, -4, and -5 fish usually were most abundant, but the predominant age-group varied with annual fluctuations in year-class strength (Young and Hoff 1988). The spring abundance of striped bass larger than 500 mm (total length), ages 4–6, varied from about 100,000 in 1976 to over 250,000 in 1979.

Hoff et al. (1988b) observed annual variations in maturity and growth in the striped bass population between 1976 and 1979, but could detect no obvious link between power plant operations and changes in any stock characteristics. They pointed out, as did Vaughan and Van Winkle (1982), that fish population responses to increased generating capacity may take decades to occur, let alone be detected, and recommended that large-scale studies of striped bass be repeated every 5 years. We doubt that even 15 years of data would be sufficient to detect and explain trends in abundance of striped bass or any fish species whose life span extends to 20 years or more and whose population has a mean generation time of 4–5 years.

In contrast to striped bass, the migratory Atlantic tomcod population in the Hudson River has a short mean generation time of about 1 year and is composed of individuals that rarely live beyond age 2. Age-1 females (11–13 months old) exhibit relatively low fecundity of about 10,000 to 17,000 eggs compared to about 300,000 to 800,000 eggs per female striped bass in the more abundant age-groups (Hoff et al. 1988b). Age-1 Atlantic tomcod made up over 90% of the spawning population from 1975 to 1980 (McLaren et al. 1988b). Atlantic tomcod spawn during the winter months in proximity to some of the major power plants (Klauda et al. 1988b). Their demersal eggs and larvae are not very vulnerable to entrainment, but young juveniles become so as they move downriver to cooler and more saline waters.

The Hudson River studies contributed almost all our knowledge of Atlantic tomcod biology in the Hudson River (Dew and Hecht 1976; Nittel 1976; Grabe 1978, 1980; Watson 1987; Klauda et al. 1988b; McLaren et al. 1988b). During the 1970s, the adult population during the spawning period ranged from about 2.5 million to just over 10 million fish (McLaren et al. 1988b). No trends in annual abundance were discernible. The life history of the Hudson River population may represent the southern end of a temperature-related latitudinal gradient of growth, age, and maturity characteristics.

The 25% incidence of hepatic hepatomas detected in the Hudson River Atlantic tomcod population during the late 1970s (Smith et al. 1979), accompanied by tissue contamination with PCBs (Klauda et al. 1981), is an adverse condition that could diminish reproductive success. Summer temperatures in the Hudson River may also inhibit feeding, retard growth, and diminish survival of Atlantic tomcod in some years (Grabe 1978).

The population may have declined during the early 1980s (EA 1983; Normandeau Associates 1984a, 1984b) but changes in the sampling programs do not allow a firm conclusion about this.

White perch.—The white perch is a common estuarine resident of the Hudson River that was designated a primary focal species because a large portion of its widely distributed population is exposed to the major power plants (Klauda et al. 1988a, this volume). Like its congener, striped bass, annual production of white perch juveniles in the Hudson River showed no discernible trend during the 1970s, but white perch abundance indices appeared to be more variable (Young et al. 1988b, this volume). Since 1980, several abundance indices suggest that juvenile white perch production has been declining (Martin Marietta Environmental Systems 1986; R. E. Brandt, New York Department of Environmental Conservation, personal communication), but the accuracy of these indices are being evaluated.

In contrast to striped bass, fishing mortality of white perch in the Hudson River during the 1970s was low, probably less than 5% (Klauda et al. 1988a). Compared to other estuarine populations at similar latitudes along the east coast of the United States, Hudson River white perch in the mid-1970s were smaller at age, suggesting the population was overcrowded. White perch are exposed to all operating Hudson River power plants, but their widely dispersed distribution and use of freshwater spawning habitats, which include tributaries and embayments, appear to minimize their realized vulnerability to entrainment at the major power plants clustered in the brackish reaches of the tidal river. Juvenile and older white perch are more vulnerable to impingement, however, particularly during the winter.

Alosids.—Migratory alosid species that spawn upstream in the limnetic zone (typically <0.3‰ salinity) of the Hudson River (e.g., American shad, blueback herring, alewife) are also relatively invulnerable, as eggs and larvae, to entrainment at the major power plants (Schmidt et al. 1988). The vulnerability of juveniles to power plants increases during their fall emigration to the ocean. Many juvenile alosids were impinged on cooling water intake screens in September and October during the 1970s (TI 1981). Despite this impact, the current stocks of migratory alosids in the Hudson River appear to be abundant and lightly exploited by sport and commercial fisheries (Brandt 1987). For example, annual abundance indices for juvenile American shad increased about sixfold from 1980 to 1986.

The biological information compiled during the 1960s and 1970s in the Hudson River established a series of population bench marks for three primary focal fish species and yielded important insights into the status of several others. None of these species exhibited exaggerated annual variations or discernible multiyear trends in distribution, abundance, and other key biological variables such as growth or fecundity. Therefore, power plant operations could not be convincingly implicated as a major source of mortality that was clearly distinguishable from other abiotic factors. Perhaps the power plants had no effect or perhaps 10–15 years of intensive studies were not long enough for any effects of power plants on these fish populations to be manifested. The Hudson River studies of the 1960s and 1970s were unable to resolve this controversy (Barnthouse et al. 1988b).

The role of power plants as a mortality factor in the population dynamics of Hudson River fishes emerged from the 1970s as an open challenge to scientists of the 1980s and probably decades beyond. Fortunately, several core monitoring efforts have been continued by the utilities into the 1980s, in addition to new programs aimed at entrainment and impingement mitigation. These recent studies are establishing an updated series of population bench marks for focal fish species. Striped bass, American shad, blueback herring, and alewife apparently produced average to above-average year classes between 1980 and 1987. During this same period, the white perch and Atlantic tomcod populations showed evidence of decline. Convincing answers to the intriguing question of power plant effects will require a lengthy series of carefully designed studies that must be based on knowledge accumulated during the 1960s, 1970s, and early 1980s.

Sampling Methodology

During the Hudson River studies, several finfish species and life stages were sampled during all seasons, day and night, in a 225-km segment of the tidal and unimpounded Hudson River. Intensive field surveys were carried out adjacent to the sites of six operating and several proposed power plants. Study objectives required that the sampling programs provide data appropriate for determination of several important biological variables for the focal species (e.g., abundance, mortality, age, growth, sex, ratio, age at maturity, fecundity, food habits), for quantification of fish mortality associated with entrainment and impingement, for calculation of direct impacts on selected fish populations associated with power plant operations (including entrainment factors and impingement mortality rates: Vaughan 1988, this volume), and for evaluation of mitigation strategies designed to reduce adverse power plant effects. To maximize data credibility, the Hudson River studies incorporated conventional sampling methods, wherever possible, that were widely accepted by the scientific community. Many standard methods were also modified to increase their effectiveness. New sampling methods were developed and tested.

Biases introduced by sampling designs and gear were addressed during the Hudson River studies. Several gear-performance studies were designed to measure and compare the catch efficiencies of plankton nets, trawls, and seines and to improve collection techniques (Klauda et al. 1988c). Variable-mesh gill nets, several haul seines, and a large midwater trawl were used collectively in the adult striped bass program to interpret gear selectivity biases and to decrease collection-induced mortality (TI 1981; Hoff et al. 1988b).

In-plant sampling of entrained organisms required development and implementation of sophisticated techniques designed to minimize collection-induced mortality. Collecting fish eggs and larvae that are enroute to or have been carried through a power plant condenser system with the cooling water is a difficult process that can damage these fragile organisms and hinder accurate estimates of entrainment-induced mortality. Collection efficiency studies were carried out during the Hudson River studies, in conjunction with in-plant sampling, to find methods that would improve the estimates of fish entrainment mortality. One major development was the larval table (McGroddy and Wyman 1977). This collection device improved the measurement of in-plant entrainment mortality by minimizing collection-induced mortality and consequently increased the accuracy and precision of mortality estimates (Vaughan and Kumar 1982). Development of the larval table was an important step in the progression of knowledge that led initially to a convergence of utility and regulatory agency estimates of conditional entrainment mortalities and ultimately to the settlement agreement (Boreman and Goodyear 1988; Englert and Boreman 1988, both this volume).

The loss of impinged fish from power plant traveling screens during impingement monitoring and the extent of reimpingement were evaluated at the major power plants throughout the Hudson River studies (Klauda et al. 1987c). Estimates of collection efficiency and reimpingement rates were used to adjust impingement counts and more accurately quantify impingement mortality (TI 1980; Barnthouse 1982). The survival of impinged fish was also quantified (Muessig et al. 1988a, this volume).

Fish impingement sampling at the Hudson River power plants was time consuming and costly. Throughout the studies of the 1960s and 1970s, impingement sampling at the Indian Point power plant was done each day. Tests of alternative sampling designs, augmented by computer-simulated sampling, allowed collection effort and processing time to be reduced by 70% with no important loss of precision in the impingement estimates (Mattson et al. 1988, this volume).

Some Hudson River study objectives were not reached because state-of-the-art limitations precluded successful use of available sampling gear and analytical methods or hampered the development of more effective approaches. Although studied intensively, the problems associated with estimates of population size for migratory or highly mobile fish species in the open-ended Hudson River estuary were never completely resolved during the power plant studies. Young et al. (1988b) discuss the efforts aimed at estimating annual abundances of juvenile striped bass and white perch. For striped bass, population size estimation based on density-extrapolation methods (with adjustments for diel distributions and gear efficiencies) and mark–recapture methods generally yielded similar and acceptable results. For juvenile white perch, however, estimates of population size calculated by density-extrapolation methods were well below the mark–recapture estimates. Selection of appropriate adjustments to account for capture-gear efficiencies could explain some but not all of the discrepancies between methods.

Impact Estimation

The Hudson River studies showed that the chemical, thermal, and physical stresses imposed on entrained organisms did not necessarily cause 100% mortality. For example, more than half of the striped bass entrained at Indian Point could be expected to survive passage through the power plant (Muessig et al. 1988b, this volume). These insights contributed to the development of strategies for power plant operations designed to manipulate cooling water flow rates and excess temperatures to minimize entrainment mortality (Schubel et al. 1979; Steen and Schubel 1986).

The studies brought increased understanding of mortality estimation errors; improvements in the sampling gear used to collect fish eggs, larvae, and juveniles in the river and at the major power plants (McGroddy and Wyman 1977); insights into the relative importance of entrainment stressors on through-plant mortality (Poje et al. 1981); and development of better models for estimating the thermal-stress component of entrainment mortality (Jinks et al. 1978; Kellogg et al. 1984; Kellogg and Jinks 1985). These advances, in turn, led to more reliable inputs to the models used to calculate conditional entrainment mortality rates, defined as the fractional reduction in year-class strength due to entrainment at the power plants if other sources of mortality were density independent.

The series of models developed during the Hudson River studies and used to evaluate entrainment impacts evolved from simplistic to complex and then back toward simplistic (Christensen and Englert 1988, this volume). Eventually, all parties to the power plant case agreed that empirical models were more practical and defensible than complex mechanistic models for estimating entrainment impacts on age-0 fishes. An empirically derived age-, time-, and space-variant equation was the final model used to estimate entrainment mortality on seven focal fish species (striped bass, white perch, Atlantic tomcod, American shad, blueback herring, alewife, and bay anchovy) at five operating power plants (Bowline Point, Lovett, Indian Point, Roseton, and Danskammer). This entrainment equation was referred to as the empirical transport model (ETM; Boreman et al. 1981). Estimates of conditional entrainment mortality based on historical and projected once-through cooling operations of five power plants ranged from 5–7% for Atlantic tomcod to 35–79% for bay anchovy (Boreman and Goodyear 1988). Post–yolk-sac larvae and entrainable juveniles were the life stages most vulnerable to entrainment.

Estimates of conditional impingement mortality were less than 5% for Atlantic tomcod, American shad, alewife, and blueback herring, and less than 10% for striped bass. For these species, impingement is not likely to be a biologically important source of mortality unless combined with other more serious stressors (Barnthouse and Van Winkle 1988, this volume). Only for white perch were the calculated losses due to impingement high enough, 10–60%, to be considered important. Barnthouse and Winkle (1988) concluded that the understanding of white perch population dynamics and the species' interactions with other components of the Hudson River ecosystem was insufficient in the 1970s for predicting the long-term effects of these impingement-related losses on the white perch population.

The Hudson River studies were also relatively unsuccessful in extrapolating conditional entrainment and impingement mortality rates to obtain reliable estimates of long-term effects on other focal fish species populations (Barnthouse et al. 1984, 1988b). Pursuit of quantitative estimates of biological compensation (which represents the ability of a fish population to offset, either in whole or in part, reductions in numbers caused by an increase in mortality due to entrainment and impingement) was a key element of the projection of long-term effects. Various attempts were made to apply stock–recruitment theory to this problem. These exercises added fuel to the fires of technical controversy and ultimately consumed countless hours of vigorous debate inside and outside the courtroom (Christensen and Klauda 1988).

The utility consultants were criticized for their application of stock–recruitment theory to the question of fish population-level effects of power plant operations. For example, Christensen and Goodyear (1988, this volume) question the validity of methods used to fit stock–recruitment models to the time series of commercial catch and effort records for the Hudson River striped bass population. Fletcher and Deriso (1988, this volume) question the usefulness of stock–recruitment models for power plant impact assessment on the grounds that fisheries managers had found these models to be unreliable. It should be remembered, however, that the general view of stock–recruitment modeling in the mid-1970s was relatively positive. The utilities' efforts were stimulated by a substantial published literature on the development and application of stock–recruitment models (e.g., Cushing and Harris 1973; Garrod 1977). Widespread recognition of the limitations of this approach due to (a) statistical problems caused by measurement errors in catch and effort data (Ludwig and Walters 1981; Walters and Ludwig 1981) and (b) the complexity of regulatory processes in fish populations (Beverton et al. 1984; Rothschild 1986; Walters 1986) did not occur until after the settlement of the Hudson River power case in late 1980. Hudson River study experiences contributed to this general reassessment of stock–recruitment models and their value in ecological assessment.

Goodyear (1988, this volume) recognized the difficulty of accurately forecasting the effects of power plants on the striped bass stock in the Hudson River. He suggests basing the management of power plant impacts on estimates of the lifetime reproductive potential of a 1-year-old female fish (an indicator of the compensatory capacity of the population) rather than on estimates of long-term yield or population size. By Goodyear's approach, reductions in reproductive potential in the striped bass stock caused by entrainment and impingement could be offset by reductions in other sources of fish mortality, such as fishing, so that the reproductive potential of a typical fish could be maintained at the pre–power-plant level.

The commercial fishery for striped bass in the river was closed in 1976 because of PCB contamination; more recently, harvest restrictions have been applied to adjacent coastal waters as part of the management strategies for east coast stocks. Both strictures have reduced adult mortality, so Goodyear's approach to managing power plant impacts on Hudson River striped bass has been at least qualitatively implemented. The sustained production of juveniles during the late 1970s and into the 1980s (Boreman and Austin 1985) may be related, in part, to reduced fishing mortality on the adults. Evaluation of this approach is continuing. A similar analysis of reproductive potential in the Chesapeake Bay striped bass stock (Boreman and Goodyear 1984) contributed to the recent decisions to close the recreational and commercial fisheries for the species in the Maryland portion of Chesapeake Bay and to restrict harvests in Virginia waters.

Mitigation

Minimizing adverse biological effects of power plant operations was pursued vigorously during the Hudson River studies (Klauda et al. 1987c). These efforts, coupled with mitigative measures spelled out in the 1980 settlement agreement, fostered development and implementation of several procedures designed to reduce entrainment and impingement mortality and also to supplement the system's annual production of striped bass. Mitigation has been and continues to be a major utility research goal in the 1980s (Barnthouse et al. 1988a).

Information collected during the 1970s on spatiotemporal distributions of entrainable life stages of focal finfish species and on temporal patterns of entrainment at the three major power plants (Bowline Point, Indian Point, and Roseton) was the basis for a detailed scheme of cooling-water flow reductions and scheduled outages implemented at these plants in 1981. These mitigative measures are based on the premise that properly timed reductions in cooling-water withdrawals would substantially reduce the number of organisms entrained (Barnthouse et al. 1988a). The ETM was used to develop a set of flow reductions and outages at each power plant that would collectively reduce entrainment losses of striped bass, white perch, and Atlantic tomcod; the program also would allow the utilities to meet electric power demands and to receive credit for unscheduled outages (Englert et al. 1988, this volume).

Studies aimed at reducing fish impingement at Hudson River power plants began in the early 1960s and have continued since. Research efforts sought to characterize the impingement process and elucidate the suite of plant operation schemes and environmental factors that affected impingement rates (TI 1980; Klauda et al. 1988c). Muessig et al. (1988a) discuss the important finding that continual rotation of conventional vertical traveling screens significantly increased postimpingement survival of several focal species, especially striped bass, white perch, and Atlantic tomcod. Other intake screen modifications at Indian Point, Roseton, and other power plants were tested during the 1960s, 1970s, and early 1980s. To date, no major modifications have been installed, but negotiations are under way pursuant to installing advanced conventional traveling screens at Indian Point (Barnthouse et al. 1988a).

A barrier net deployed from October through May in a 49-hectare embayment at the Bowline Point power plant intake proved effective during the 1970s in reducing fish impingement by over 91% (Hutchison and Matousek 1988, this volume). This fish protection device was designed specifically for Bowline's unique intake configuration (Hutchison 1988, this volume). Because the net was so successful, its use has continued into the 1980s, as stipulated by the settlement agreement (Barnthouse et al. 1988a).

Artificial propagation of striped bass was proposed by the utilities in the late 1960s as a mitigative measure to offset projected losses from operation of a planned pumped-storage hydro-electric station along the river at Cornwall (Hutchison 1988). Construction plans for this facility were abandoned in 1980 but, based on research conducted by Texas Instruments from 1973 to 1975 (McLaren et al. 1988c, this volume), the utilities agreed, as part of the settlement agreement, to construct and operate a hatchery designed to produce 600,000 fingerling striped bass each year for supplemental stocking in the Hudson River (Barnthouse et al. 1988a). Research had demonstrated that the 6- to 9-month postrelease survival of stocked juveniles was as good as or better than the survival of wild juveniles. As of 30 June 1980, 1,925 hatchery-reared fish (0.6% of the juveniles that had been marked and released during 1973–1975), ranging in age from 0 to 6 years, had been recaptured in the Hudson River and adjacent waters.

Resolution of Technical Issues in the Legal Arena

From our perspective as scientists, the process of legally resolving controversial impact issues associated with the Hudson River power plant case was a major learning experience. Many scientists initially perceived the hearings process as a forum of lively debate on biological compensation mechanisms, predictive-effects models, and other intriguing technical issues. But most of us soon assimilated the message crisply summarized by Judge Yost (1988, this volume): "The purpose of a trial is to resolve disputes. In the Hudson River case, the issue was whether or not the utilities would be required to build cooling towers at their several power plants. If, in the course of answering that question, some scientific 'truths' emerged, that was merely a beneficial side effect."

The case stumbled along in the courtroom with little progress, then moved quickly to resolution soon after informal and candid negotiations began. Science and law did not completely mix during the power plant case, but an emulsion of sorts was eventually created after much stirring and shaking. This science–law emulsion, however, was not particularly effective in the courtroom. The adjudicatory process included a microscopic examination of the data, models, and major interpretations (voluminous discourses now entombed in a mountain of testimony and transcripts), but it resolved few technical controversies.

Most parties to the Hudson River power plant case, legal and scientific, now agree that the adjudicatory hearing format prevented develop-ment and persistence of a cooperative spirit between government and utility scientists (Bergen 1988, this volume; Christensen and Klauda 1988; Yost 1988). Only after the hearings were abandoned in favor of a negotiated settlement process were scientists on both sides of the case free to candidly discuss, question, and modify their views on the major technical issues. This search for a common ground on which all parties felt comfortable was sufficiently successful outside the courtroom that a technically based agreement was finally negotiated in late 1980 (Barnthouse et al. 1988a).

Epilogue

This paper highlights the technical contributions of the Hudson River power plant studies carried out during the 1960s and 1970s; a greater breadth of knowledge gained during this landmark environmental assessment is exposed by other papers in this volume and elsewhere in the scientific literature. Ongoing research programs sponsored by the utilities, the Hudson River Foundation for Science and Environmental Research, the New York Department of Environmental Conservation, other governmental agencies, and universities are contributing new data to the information base. New data are expensive to collect, so maximum usage of extant data should be encouraged. We hope the rich lode of information still buried in numerous data files will be mined for even more knowledge of the structure and function of the Hudson River ecosystem. To promote such activities, the Hudson River Foundation is sponsoring the preparation of a user's guide to the major ecological data sets collected during the 1970s (Keiser and Klauda 1985). If this retrospective monograph on the power plant case stimulates continuing exploration of existing Hudson River data bases, another of our major goals will be achieved.

The successes and failures of the power plant case make up a valuable learning experience that has contributed significantly to the youthful field of environmental impact assessment. The concept of societal impacts on environmental resources is so widely discussed in the public media, scientific literature, and bureaucratic circles that we can easily forget that impact assessments are recent phenomena. Prior to the early 1960s, the level of public concern for the environment was low. Most governmental agencies did not institutionalize environmental impact assessments until the early 1970s (Hecky et al. 1984). By then, the

Hudson River power plant studies were into a second decade. In their infancy, environmental assessment projects were typically carried out by young scientists who were relatively inexperienced, refreshingly optimistic, and unabashedly enthusiastic, but occasionally naive. The Hudson River studies were no exception.

Society has increasingly turned to science during the latter half of the 20th century for advice on managing important natural resources. This need has drawn many scientists out of their laboratories and thrust them into the limelight of public attention. These often self-appointed spokesmen were expected to provide unequivocal solutions to perceived problems that could be quickly implemented. Predictably, scientists have labored diligently to assist the decision makers to be truly rational, if the decision makers choose to be so. But rarely have workable solutions met the schedule of public expectations. Scientific solutions of environmental issues are commonly uncertain, complicated, and more costly than anticipated. The research necessary to provide adequate solutions almost always takes longer and costs more than society is willing to accept; thus actions must be based on a partial understanding of the impacts, and scientific disputes will predictably arise (Egerton 1985).

Perhaps the clearest message from the Hudson River power plant case is a reminder that the elucidation of cause-and-effect relationships in fish population dynamics is a difficult task. For almost 20 years, controversy surrounding the questions of power plant effects on several fish species raged. Unequivocal conclusions were uncommon. Nevertheless, the studies found acceptable answers to many of the long-standing impact questions in spite of the failure to achieve all study objectives. As optimistic realists, we conclude, therefore, that the Hudson River power plant studies should be labeled a success.

Acknowledgments

R. L. Kendall and N. D. Vaughan reviewed the manuscript and offered helpful comments. Our views were stimulated and sharpened by lively conversations with many colleagues over the years, especially with J. Boreman, S. W. Christensen, J. C. Cooper, M. J. Dadswell, C. P. Goodyear, E. G. Horn, W. L. Kirk, L. C. Kohlenstein, C. J. Lauer, J. T. McFadden, T. T. Polgar, E. M. Portner, C. L. Smith, and W. Van Winkle.

References[3]

Barnthouse, L. W. 1982. An analysis of factors that influence impingement estimates at Hudson River power plants. Page I-1 to I-49 in L. W. Barnthouse and seven coeditors. The impact of entrainment and impingement on fish populations in the Hudson River estuary, volume 2. Oak Ridge National Laboratory, ORNL/NURGL/TM-385/V2, Oak Ridge, Tennessee.

Barnthouse, L. W., J. Boreman, S. W. Christensen, C. P. Goodyear, W. Van Winkle, and D. S. Vaughan. 1984. Population biology in the courtroom: the Hudson River controversy. BioScience 34:14–19.

Barnthouse, L. W., J. Boreman, T. L. Englert, W. L. Kirk, and E. G. Horn. 1988a. Hudson River settlement agreement: technical rationale and cost considerations: American Fisheries Society Monograph 4:267–273.

Barnthouse, L. W., R. J. Klauda, and D. S. Vaughan. 1988b. What we didn't learn about the Hudson River, why, and what it means for environmental assessment in the future. American Fisheries Society Monograph 4:329–335.

Barnthouse, L. W., and W. Van Winkle. 1988. Analysis of impingement impacts on Hudson River fish populations. American Fisheries Society Monograph 4: 182–190.

Beebe, C. A., and I. R. Savidge. 1988. Historical perspective on fish species composition and distribution in the Hudson River estuary. American Fisheries Society Monograph 4:25–36.

Bergen, G. S. P. 1988. The Hudson River cooling tower proceeding: interface between science and law. American Fisheries Society Monograph 4:302–306.

Berggren, T. J., and J. T. Lieberman. 1978. Relative contribution of Hudson, Chesapeake and Roanoke striped bass, Morone saxatilis, stocks to the Atlantic coast fishery. U.S. National Marine Fisheries Service Fishery Bulletin 76:335–345.

Beverton, R. J. H., and eleven coauthors. 1984. Dynamics of single species: group report. Pages 13–58 in R. M. May, editor. Exploitation of marine communities. Springer-Verlag, Berlin.

Boreman, J., and H. M. Austin. 1985. Production and harvest of anadromous striped bass stocks along the Atlantic coast. Transactions of the American Fisheries Society 114:3–7.

Boreman, J., and C. P. Goodyear. 1984. Effects of fishing on the reproductive capacity of striped bass in Chesapeake Bay, Maryland. National Marine Fisheries Service, Northeast Fisheries Center, Woods Hole Laboratory Reference Document 84–29, Woods Hole, Massachusetts.

Boreman, J., and C. P. Goodyear. 1988. Estimates of entrainment mortality for striped bass and other fish species inhabiting the Hudson River estuary. American Fisheries Society Monograph 4:152–160.

Boreman, J., C. P. Goodyear, and S.W. Christensen.

---

[3]See Table 1 for sources of legal documents and unpublished reports pertaining to the Hudson River.

1981. An empirical methodology for estimating entrainment losses at power plants sited on estuaries. Transactions of the American Fisheries Society 110:255–262.

Boreman, J., and R. J. Klauda. 1988. Distributions of early life stages of striped bass in the Hudson River estuary, 1974–1979. American Fisheries Society Monograph 4:53–58.

Brandt, R. E. 1987. Current status of American shad, alewife, and blueback herring in New York waters. Report to Atlantic States Marine Fisheries Commission, Washington, D.C.

Christensen, S. W., and T. L. Englert. 1988. Historical development of entrainment models for Hudson River striped bass. American Fisheries Society Monograph 4:133–142.

Christensen, S. W., and C. P. Goodyear. 1988. Testing the validity of stock–recruitment curve fits. American Fisheries Society Monograph 4:219–231.

Christensen, S. W., and R. J. Klauda. 1988. Two scientists in the courtroom: what they didn't teach us in graduate school. American Fisheries Society Monograph 4:307–315.

Cooper, J. C., F. R. Cantelmo, and C. E. Newton. 1988 Overview of the Hudson River estuary. American Fisheries Society Monograph 4:11–24.

Cushing, D. H., and J. K. G. Harris. 1973. Stock and recruitment and the problem of density-dependence. Rapports et Procès-Verbaux des Réunions, Conseil International pour l'Exploration de la Mer 164:142–155.

Dadswell, M. J., B. D. Taubert, T. S. Squiers, D. Marchette, and J. Buckley. 1984. Synopsis of biological data on shortnose sturgeon, *Acipenser brevirostrum* 1818. NOAA (National Oceanic and Atmospheric Administration) Technical Report NMFS (National Marine Fisheries Service) 14.

Dew, C. B., and J. H. Hecht. 1976. Ecology and population dynamics of Atlantic tomcod (*Microgadus tomcod*) in the Hudson River estuary. Paper 25 *in* Hudson River ecology, 4th symposium. Hudson River Environmental Society, New Paltz, New York.

Dey, W. P. 1981. Mortality and growth of young-of-the-year striped bass in the Hudson River estuary. Transactions of the American Fisheries Society 110:151–157.

Dovel, W. L. 1979. The biology and management of shortnose and Atlantic sturgeon of the Hudson River. New York Department of Environmental Conservation, Albany.

Dovel, W. L., and T. J. Berggren. 1983. Atlantic sturgeon of the Hudson estuary, New York. New York Fish and Game Journal 30:140–172.

EA (Ecological Analysts). 1983. Spawning characteristics of the Hudson River Atlantic tomcod population during the 1980–81 and 1981–82 spawning seasons. Report to Consolidated Edison Company of New York.

Egerton, F. N. 1985. Overfishing or pollution? Case history of a controversy on the Great Lakes. Great Lakes Fishery Commission Technical Report 41.

Englert, T. L., and J. Boreman. 1988. Historical review of entrainment impact estimates and the factors influencing them. American Fisheries Society Monograph 4:143–151.

Englert, T. L., J. Boreman, and H. Y. Chen. 1988. Plant flow reductions and outages as mitigative measures. American Fisheries Society Monograph 4:274–279.

Fletcher, R. I., and R. B. Deriso. 1988. Fishing in dangerous waters: remarks on a controversial appeal to spawner–recruit theory for long-term impact assessment. American Fisheries Society Monograph 4:232–244.

Gardinier, M. N., and T. B. Hoff. 1982. Diet of striped bass in the Hudson River estuary. New York Fish and Game Journal 29:152–165.

Garrod, D. J. 1977. The North Atlantic cod. Pages 216–242 *in* J. A. Gulland, editor. Fish population dynamics. Wiley, New York.

Gladden, J. B., F. R. Cantelmo, J. M. Croom, and R. Shapot. 1988. Evaluation of the Hudson River ecosystem in relation to the dynamics of fish populations. American Fisheries Society Monograph 4:37–52.

Goodyear, C. P. 1988. Implications of power plant mortality for management of the Hudson River striped bass fishery. American Fisheries Society Monograph 4:245–254.

Grabe, S. A. 1978. Food and feeding habits of juvenile Atlantic tomcod, *Microgadus tomcod,* from Haverstraw Bay, Hudson River. U.S. National Marine Fisheries Service Fishery Bulletin 76:89–94.

Grabe, S. A. 1980. Food of age 1 and 2 Atlantic tomcod, *Microgadus tomcod,* from Haverstraw Bay, Hudson River, New York. U.S. National Marine Fisheries Service Fishery Bulletin 77:1003–1006.

Hecky, R. E., R. W. Newbury, R. A. Bodaly, K. Patalas, and D. M. Rosenberg. 1984. Environmental impact prediction and assessment: the Southern Indian Lake experience. Canadian Journal of Fisheries and Aquatic Sciences 41:720–732.

Hoff, T. B., and R. J. Klauda. 1979. Data on shortnose sturgeon (*Acipenser brevirostrum*) collected incidentally from 1969 through June 1979 in sampling programs conducted for the Hudson River ecology study. Report to U.S. National Marine Fisheries Service and U.S. Fish and Wildlife Service, Shortnose Sturgeon Recovery Team, Danvers, Massachusetts.

Hoff, T. B., R. J. Klauda, and B. S. Belding. 1977. Incidental catch and distribution of shortnose sturgeon and Atlantic sturgeon in the Hudson River estuary, 1969 through 1977. Report to the U.S. National Marine Fisheries Service and U.S. Fish and Wildlife Service, Shortnose Sturgeon Recovery Team, Danvers, Massachusetts.

Hoff, T. B., R. J. Klauda, and J. R. Young. 1988a. Contribution to the biology of shortnose sturgeon in the Hudson River estuary. Pages 171–189 *in* Smith (1988).

Hoff, T. B., J. B. McLaren, and J. C. Cooper. 1988b. Stock characteristics of Hudson River striped bass. American Fisheries Society Monograph 4:59–68.

Hutchison, J. B., Jr. 1988. Technical descriptions of Hudson River electricity generating stations. American Fisheries Society Monograph 4:113–120.

Hutchison, J. B., Jr., and J. A. Matousek. 1988. Evaluation of a barrier net used to mitigate fish impingement at a Hudson River plant intake. American Fisheries Society Monograph 4:280–285.

Jinks, S. M., T. Cannon, D. Latimer, L. Claflin, and G. Lauer. 1978. An approach for the analysis of striped bass entrainment survival at the Hudson River power plants. Pages 343–350 in L. D. Jensen, editor. Proceedings of the fourth national workshop on entrainment and impingement. Ecological Analysts Communications, Sparks, Maryland.

Keiser, R. K., Jr., and R. J. Klauda. 1985. User's guide to the Hudson River data base. Volume 1—Overview. Report to Hudson River Foundation for Science and Environmental Research, New York.

Kellogg, R. L., and S. M. Jinks. 1985. Short-term thermal tolerance of 10 species of Hudson River ichthyoplankton. New York Fish and Game Journal 32:41–52.

Kellogg, R. L., R. J. Ligotino, and S. M. Jinks. 1984. Thermal mortality prediction equation for entrainable striped bass. Transactions of the American Fisheries Society 113:794–802.

Klauda, R. J., W. P. Dey, T. B. Hoff, J. B. McLaren, and Q. E. Ross. 1980. Biology of Hudson juvenile striped bass. Marine Recreational Fisheries 5:101–123.

Klauda, R. J., J. B. McLaren, R. E. Schmidt, and W. P. Dey. 1988a. Life history of white perch in the Hudson River estuary. American Fisheries Society Symposium 4:69–88.

Klauda, R. J., R. E. Moos, and R. E. Schmidt. 1988b. Life history of Atlantic tomcod, *Microgadus tomcod*, in the Hudson River estuary with emphasis on spatio-temporal distribution and movements. Pages 219–251 in Smith (1988).

Klauda, R. J., P. H. Muessig, and J. A. Matousek. 1988c. Fisheries data sets compiled by utilities-sponsored research in the Hudson River estuary. Pages 7–85 in Smith (1988).

Klauda, R. J., T. H. Peck, and G. K. Rice. 1981. Accumulation of polychlorinated biphenyls in Atlantic tomcod (*Microgadus tomcod*) collected from the Hudson River estuary, New York. Bulletin of Environmental Contamination and Toxicology 27:829–835.

Larkin, P. A. 1984. A commentary on environmental impact assessment for large projects affecting lakes and streams. Canadian Journal of Fisheries and Aquatic Sciences 41:1121–1127.

Ludwig, D., and C. J. Walters. 1981. Measurement errors and uncertainty in parameter estimates for stock and recruitment. Canadian Journal of Fisheries and Aquatic sciences 38:711–720.

Martin Marietta Environmental Systems. 1986. 1984 year class report for the Hudson River estuary monitoring program. Volume 1—Text. Report to Consolidated Edison Company of New York.

Mattson, M. T., J. B. Waxman, and D. A. Watson. 1988. Reliability of impingement sampling designs: an example from the Indian Point station. American Fisheries Society Monograph 4:161–169.

McFadden, J. T., and H. G. Woodbury. 1973. Indian Point studies to determine the environmental effects of once-through vs. closed-cycle cooling at Indian Point unit no. 2. Testimony (February 5, 1973) before the U.S. Atomic Energy Commission in the matter of Consolidated Edison Company of New York, Inc. (Indian Point station, unit no. 2). Docket 50–247.

McGroddy, R. M., and R. L. Wyman. 1977. Efficiency of nets and a new device for sampling live fish larvae. Journal of the Fisheries Research Board of Canada 34:571–574.

McLaren, J. B., J. C. Cooper, T. B. Hoff, and V. Lander. 1981. Movements of Hudson River striped bass. Transactions of the American Fisheries Society 110:158–167.

McLaren, J. B., R. J. Klauda, T. B. Hoff, and M. N. Gardinier. 1988a. Striped bass commercial fishery of the Hudson River. Pages 89–123 in Smith (1988).

McLaren, J. B., T. H. Peck, W. P. Dey, and M. Gardinier. 1988b. Biology of Atlantic tomcod in the Hudson River estuary. American Fisheries Society Monograph 4:102–112.

McLaren, J. B., J. R. Young, T. B. Hoff, I. R. Savidge, and W. L. Kirk. 1988c. Feasibility of supplementary stocking of age-0 striped bass in the Hudson River. American Fisheries Society Monograph 4:286–291.

Muessig, P. H., J. B. Hutchison Jr., L. R. King, R. J. Ligotino, and M. Daley. 1988a. Survival of fishes after impingement on traveling screens at Hudson River power plants. American Fisheries Society Monograph 4:170–181.

Muessig, P. H., J. R. Young, D. S. Vaughan, and B. A. Smith. 1988b. Advances in field and analytical methods for estimating entrainment mortality factors. American Fisheries Society Monograph 4:124–132.

Nittel, M. 1976. Food habits of Atlantic tomcod (*Microgadus tomcod*) in the Hudson River. Paper 26 in Hudson River ecology, 4th symposium. Hudson River Environmental Society, New Paltz, New York.

Normandeau Associates. 1984a. Abundance and stock characteristics of the Atlantic tomcod (*Microgadus tomcod*) spawning population in the Hudson River, winter 1982–83. Report to Consolidated Edison Company of New York.

Normandeau Associates. 1984b. Abundance and stock characteristics of the Atlantic tomcod (*Microgadus tomcod*) spawning population in the Hudson River, winter 1983–84. Report to Consolidated Edison Company of New York.

Poje, G. V., S. A. Riordan, and J. M. O'Connor. 1981. Power plant entrainment simulation utilizing a condenser tube simulator. New York University Medical Center, NUREG/CR-2091 RE, Tuxedo.

Rothschild, B. J. 1986. Dynamics of marine fish populations. Harvard University Press, Cambridge, Massachusetts.

Schmidt, R. E., R. J. Klauda, and J. M. Bartels. 1988. Distribution and movements of the early life stages of three *Alosa* spp. in the Hudson River estuary, with comments on mechanisms that reduce interspecific competition. Pages 193–215 *in* Smith (1988).

Schubel, J. R., H. H. Carter, and J. M. O'Connor. 1979. Effects of increasing ΔT on power plant entrainment mortality at Indian Point, New York. State University of New York, Marine Sciences Research Center, Special Report 19, Stony Brook.

Sloan, R. J., and R. W. Armstrong. 1988. PCB patterns in Hudson River fish: II. Migrant and marine species. Pages 325–350 *in* Smith (1988).

Smith, C. E., T. H. Peck, R. J. Klauda, and J. B. McLaren. 1979. Hepatomas in Atlantic tomcod *Microgadus tomcod* (Walbaum) collected in the Hudson River estuary in New York. Journal of Fish Diseases 2:313–319.

Smith, C. L., editor. 1988. Fisheries research in the Hudson River. State University of New York Press, Albany.

Steen, A. E., and J. R. Schubel. 1986. An application of a strategy to reduce entrainment mortality. Journal of Environmental Management 23:215–228.

Stevens, D. E., D. W. Kohlhorst, L. W. Miller, and D. W. Kelley. 1985. The decline of striped bass in the Sacramento–San Joaquin estuary, California. Transactions of the American Fisheries Society 114:12–30.

TI (Texas Instruments). 1980. Hudson River ecological study in the area of Indian Point, 1978 annual report. Report to Consolidated Edison Company of New York.

TI (Texas Instruments). 1981. 1979 year class report for the multiplant impact study of the Hudson River estuary. Report to Consolidated Edison Company of New York.

Van Winkle, W. 1977. Conclusions and recommendations for assessing the population level effects of power plant exploitation: the optimist, the pessimist, and the realist. Pages 365–372 *in* W. Van Winkle, editor. Proceedings of the conference on assessing the effects of power-plant-induced mortality on fish populations. Pergamon Press, Elmsford, New York.

Van Winkle, W., K. D. Kumar, and D. S. Vaughan. 1988. Relative contributions of Hudson River and Chesapeake Bay striped bass stocks to the Atlantic coastal population. American Fisheries Society Monograph 4:255–266.

Vaughan, D. S. 1988. Introduction [to entrainment and impingement impacts]. American Fisheries Society Monograph 4:121–123.

Vaughan, D. S., and K. D. Kumar. 1982. Entrainment mortality of ichthyoplankton: detectability and precision of estimates. Environmental Management 6:155–162.

Vaughan, D. S., and W. Van Winkle. 1982. Corrected analysis of the ability to detect reductions in year-class strength of the Hudson River white perch (*Morone saxatilis*) population. Canadian Journal of Fisheries and Aquatic Sciences 39:782–785.

Walters, C. J. 1986. Adaptive management of renewable resources. Macmillan, New York.

Walters, C. J., and D. Ludwig. 1981. Effects of measurement errors on the assessment of stock–recruitment relationships. Canadian Journal of Fisheries and Aquatic Sciences 38:704–710.

Watson, L. C. 1987. Spawning and hatching Atlantic tomcod. Progressive Fish-Culturist 49:69–71.

Yost, T. B. 1988. Science in the courtroom. American Fisheries Society Monograph 4:294–301.

Young, J. R., and T. B. Hoff. 1988. Age-specific variation in reproductive effort in female Hudson River striped bass. Pages 124–133 *in* Smith (1988).

Young, J. R., T. B. Hoff, W. P. Dey, and J. G. Hoff. 1988a. Management recommendations for a Hudson River Atlantic sturgeon fishery based on an age-structured population model. Pages 353–365 *in* Smith (1988).

Young, J. R., R. J. Klauda, and W. P. Dey. 1988b. Population estimates for juvenile striped bass and white perch in the Hudson River estuary. American Fisheries Society Monograph 4:89–101.

American Fisheries Society Monograph 4:329–335, 1988

What We Didn't Learn about the Hudson River, Why, and What It Means for Environmental Assessment

LAWRENCE W. BARNTHOUSE

Environmental Sciences Division, Oak Ridge National Laboratory
Post Office Box 2008, Oak Ridge, Tennessee 37831-6036, USA

RONALD J. KLAUDA[1]

Texas Instruments Incorporated
Ecological Services Group, Buchanan, New York 10511, USA

DOUGLAS S. VAUGHAN[2]

Environmental Sciences Division, Oak Ridge National Laboratory

Abstract.—Many of the major objectives of utility- and agency-sponsored Hudson River research programs were not achieved. Among these were identification and quantification of regulatory mechanisms and discovery of factors controlling year-class strength in striped bass and other important fish populations. Questions about community- and ecosystem-level effects were not seriously addressed. Because of these limitations, an unambiguous assessment of the effects of power plants on the long-term production and persistence of Hudson River fish populations was not possible. In this paper we argue that the failure to reach a scientifically defensible "bottom line" was largely due to (1) institutional constraints on the design and conduct of assessment studies, (2) the complexity and spatiotemporal variability of estuarine ecosystems, and (3) the inadequacy of existing population and ecosystem theory. We conclude that, for the foreseeable future, estimates of short-term impacts on populations will continue to be the most useful indices of power plant effects. Long-term monitoring and basic research on ecological processes in estuaries, funded and managed independently of the regulatory process, are essential to improving future environmental impact assessments.

Klauda et al. (1988a, this volume) discussed the major objectives of the utilities' Hudson River biological studies program. These objectives encompassed the general objectives of all of the scientific groups involved in assessing the impacts of electric power generation on Hudson River biota. Substantial success was achieved in describing the natural history of important fish populations in the Hudson, in measuring the thermal and mechanical mortality of entrained and impinged organisms, in estimating the reductions in year-class abundance attributable to power plant operation, and in evaluating techniques for mitigating entrainment and impingement losses.

However, the studies failed to identify and quantify regulatory mechanisms and factors controlling year-class strength in striped bass and other fish populations. Without quantification of these processes, it was impossible to develop credible estimates of the long-term effects of power plants on the abundance, production, or persistence of Hudson River fish populations. Community- and ecosystem-level effects of power plants were not seriously addressed. Understanding of these effects would have permitted an assessment of the effects of power plants on the production, diversity, and trophic structure of the estuary as a whole. Because of these limitations, a comprehensive assessment of the effects of power plants on the Hudson River ecosystem was not possible.

In this paper we explore the reasons for the failure of the Hudson River biological studies to provide definitive conclusions regarding the ultimate impacts of power plants on the Hudson River ecosystem. We argue that the primary reasons involve institutional constraints and scientific limitations that would be encountered in any large-scale assessment study. These constraints and limitations should be clearly recognized by

---

[1]Present address: The Johns Hopkins University, Applied Physics Laboratory, Environmental Services Group, Shady Side, Maryland 20764, USA.

[2]Present address: National Marine Fisheries Service, Southeast Fisheries Center, Beaufort Laboratory, Beaufort, North Carolina 28516, USA.

industries and regulatory agencies contemplating such studies.

Institutional Constraints

Institutional constraints that, in our experience, are not generally understood within the scientific community at large profoundly affect all phases of environmental impact studies, from definition of objectives through interpretation of results. Environmental impact studies are components of a complex decision-making process that includes both political and legal elements. As clearly pointed out by Yost (1988, this volume) and Bergen (1988, this volume), the purpose of such studies is to aid decision makers in implementing environmental law. The objectives must, therefore, meet specific statutory requirements such as the language of the Federal Water Pollution Control Act Amendments of 1972 (Public Law 92-500) regarding "a balanced, indigenous population" and "best technology available." Study objectives must also respond to the specific concerns of private citizens and institutions whose interests may be threatened by a proposed project.

We doubt that a perfectly rational study of the dynamics of the Hudson River ecosystem would have devoted most of the available resources to investigating the life history, distribution, and abundance of the striped bass population. However, after interested citizens persuaded the federal courts that the Federal Power Commission was obliged to give this particular species careful consideration before issuing a license for the Cornwall pumped-storage facility (Scenic Hudson Preservation Conference versus Federal Power Commission, U.S. Court of Appeals, Second Circuit, December 29, 1965), an emphasis on striped bass became unavoidable. White perch (Klauda et al. 1988b, this volume), Atlantic tomcod (McLaren et al. 1988, this volume), and other species discussed in this monograph were added to the study program at the insistence of the Nuclear Regulatory Commission (NRC) and the U.S. Environmental Protection Agency (EPA). Analyses of the contribution of the Hudson River striped bass stock to the Atlantic coastal fishery (Berggren and Lieberman 1978; Van Winkle et al. 1988, this volume) were specifically intended to address the cost–benefit provisions of the National Environmental Policy Act (NEPA; PL 91-190). Attempts to quantify population regulation by striped bass (Lawler 1988; Savidge et al. 1988, both this volume) were deemed necessary by the utility companies to convince the federal agencies

that the construction of the Cornwall pumped-storage facility and the operation of thermal power plants with once-through cooling would be consistent with federal law.

It must also be recognized that the results of environmental impact studies are judged not according to the scientific process of peer review and publication but according to the judicial process of testimony, cross-examination, and rebuttal. Any scientist who has testified at hearings will agree that the judicial system can be extraordinarily efficient at ferreting out sloppy reasoning or outright dishonesty (Christensen and Klauda 1988, this volume). However, this system also encourages scientists to develop one-sided arguments, discourages open communication among groups of scientists employed by different sides, and makes it dangerous to admit to changing one's mind. These factors probably explain the thousands of pages of hearings transcripts devoted to biological compensation and stock–recruitment modeling.

Even if scientists had complete control over assessment studies and did not have to deal with lawyers and judges, the adequacy of the results would still be constrained by limitations on the time and resources available for collecting data and performing assessments. Determining the biological significance of entrainment and impingement, including understanding the life histories and regulatory mechanisms of numerous fish populations and how they are affected by environmental fluctuations, would be the work of many lifetimes. The biological significance of fishing is still a major subject of research even though many exploited populations have been studied for decades. Yet, decades were not available to biologists studying the Hudson. The Atomic Safety and Licensing Appeals Board, for example, allowed the utilities only 3 years to come forward with new evidence to support their contention that cooling towers were not needed at Indian Point (Atomic Safety and Licensing Appeal Board 1974). Although it might appear that utility and agency biologists had a leisurely 10–15 years to study the river, a great deal of this time was spent preparing for hearings held in 1972–1973, 1974, 1976, and 1977–1980.

A further compounding difficulty is institutional inertia on the part of both the agencies and the project sponsors. In an ideal world, scientists should be free to modify research programs to exploit new techniques, pursue promising areas of investigation, and abandon efforts that have proven unsuccessful. However, in the real world of assessment, major program components often reflect either a consensus between regulatory agencies and project sponsors or on-the-record commitments. Program modifications sometimes require approval by federal or state agencies. Under these circumstances, both the sponsors and the regulatory authorities are reluctant to modify an established program. The only major changes possible are add-ons, which increase the cost of the program and are, therefore, discouraged.

Environmental Variability, Spatial Heterogeneity, and Sampling Error

Aside from institutional problems, it should be clear from the papers in Section 2 of this monograph that estuarine ecosystems are highly heterogeneous, variable, and difficult to study. Cooper et al. (1988, this volume) documented the high variability of the salinity regime in the middle Hudson River estuary and related it to the high variability in freshwater discharge. The spatial distribution of salinity, in turn, strongly influences the distributions of many fish populations. Variations in flow and salinity are believed to influence the survival of early life stages of striped bass (Ulanowicz and Polgar 1980; Hoff et al. 1988, this volume). Variations in temperature, also documented by Cooper et al. (1988), were found to strongly affect the survival of larval striped bass (Hoff et al. 1988) and juvenile Atlantic tomcod (McLaren et al. 1988).

Year-to-year variability in environmental conditions and, consequently, in year-class abundance make it extremely difficult to detect changes in year-class abundance that might be related to power plant operation. Using a variety of indices of white perch year-class abundance, Van Winkle et al. (1981) and Vaughan and Van Winkle (1982) showed that approximately 20 years of monitoring would be required to detect a 50% decline in average year-class abundance, given 5 years of preimpact base-line data. Klauda et al. (1988b) noted that variations in year-class abundance are lower for white perch than for many other Hudson River fish species, including striped bass. In comparison to white perch, power-plant-induced reductions in these other species would be even more difficult to detect. The spatial distribution of the life stages of interest are complex and temporally variable, as shown by Boreman and Klauda (1988, this volume). Primarily striped bass data are discussed in this monograph, but distributions for other species described by

Boreman (1979) are equally variable. Given this variability, abundance estimates obtained from sampling at one or a few standard stations are inevitably subject to large errors; even estimates based on riverwide sampling are subject to a variety of uncertainties (Young et al. 1988, this volume).

The survival of young-of-the-year fish probably varies, depending on the week they are spawned, and may also vary among regions (Boreman and Klauda 1988). The efficiencies of the gear used to sample fish vary with age and size of the fish, as well as with the weather at the time sampling is conducted. For all of these reasons, even the most intensive sampling efforts can produce only approximate estimates of the population parameters of interest in power plant impact assessment.

Inadequacy of Existing Population and Ecosystem Theory

In the early 1970s, when the first models of the Hudson River striped bass population were being developed, it was widely believed that, given enough of the right kind of data, the future behavior of fish populations could be reliably predicted. Fisheries research journals were filled with papers, such as those cited by Fletcher and Deriso (1988, this volume), describing the theoretical development of stock–recruitment models and presenting fits of the models to data. Enthusiasm for modeling was not limited to fisheries scientists. Entomologists and wildlife biologists were also developing models intended for management applications (Watt 1968). Ambitious efforts aimed at modeling the behavior of complex aquatic and terrestrial ecosystems were under way under the auspices of the International Biological Programme (e.g., Park et al. 1974; Kremer and Nixon 1978). Given the optimistic spirit of the times, it should not be surprising that utility and agency biologists were confident that they could develop models useful for forecasting and managing the impacts of entrainment and impingement on Hudson River fish populations.

In the intervening years, the validity of the approaches employed in the Hudson River modeling studies and other similar studies has been seriously challenged. Fletcher and Deriso (1988) showed that fisheries managers find stock–recruitment models to have little or no value in managing exploited stocks. This is true even for the various Pacific salmon species, despite abundant data and an enormous number of modeling studies. More complex models, including the striped bass entrainment models and life cycle models reviewed by Swartzman et al. (1978), have been no more successful. The reasons for the failure of these models in practice are now becoming clear.

The problem of quantifying population regulation, biological compensation, or renewal has been called by Fletcher and Deriso (1988) "the central problem of fishery science." All long-term predictions about the future behavior of populations are critically sensitive to assumptions made about regulatory mechanisms. For example, a 20% annual reduction in striped bass year-class abundance could cause either a trivial or a catastrophic reduction in the abundance of the adult stock, depending on the ability of the population to compensate for this mortality through increased survival or fecundity of the remaining fish. The perceived need to identify these mechanisms, or at least to quantify their effects, stimulated the controversial modeling studies described in Section 3 of this monograph (Christensen and Goodyear 1988; Lawler 1988; Savidge et al. 1988), as well as previously published reviews and modeling studies by Christensen et al. (1977), De-Angelis et al. (1977), and McFadden (1977).

The great variability of year-class abundance and the large uncertainties in estimates of population parameters have been noted above. Obviously, variability and uncertainty reduce the reliability of model predictions. However, when parameter estimates for stock–recruitment models or other nonlinear models are obtained by fitting the models to time-series data, the consequences of variability and uncertainty have been shown to be disastrous. Independently, Walters and Ludwig (1981) and Christensen and Goodyear (1988) showed that when the Ricker stock–recruitment model is fitted to a time series of catch–effort data subject to random variation, measurement errors, or both, the parameter estimates obtained are biased. For realistic levels of variability, the fitted models tend to significantly overestimate the capacity of heavily exploited populations to sustain additional mortality. Both Walters and Ludwig (1981) and Christensen and Goodyear (1988) concluded that basing management decisions on such model-fitting exercises can easily lead to overexploitation and population collapse.

It appears that, because of variability and uncertainty inherent in estimates of catch size and year-class abundance, compensation in fish populations cannot be quantified by fitting models to catch–effort data. The alternative to this approach is to identify and quantify the specific biological mechanisms responsible for population regulation. These may be either intrinsic to the population being studied (e.g., competition for spawning sites) or may involve interactions with competitors, predators, and prey species. Few, if any, of the scientists on either side of the compensation controversy disagree with the general notion espoused by Tanner (1966) and Royama (1977) that density-dependent regulatory processes operate, at least occasionally, in animal populations. However, there is no consensus regarding the specific processes that may, or may not, be operating in estuarine populations such as the Hudson River striped bass population. On what spatial and temporal scales do compensatory and depensatory mortality occur? Do competitive interactions influence the relative abundance of fish species that inhabit the same regions and feed on the same organisms? No satisfactory answers to such questions were provided by the Hudson River studies.

The effort devoted to basic biological questions such as these was small in comparison to that devoted to quantifying the direct effects of entrainment and impingement on young-of-the-year fish. However, we doubt that any significant success could have been achieved within the roughly 5 years available for intensive research on the Hudson, regardless of the effort expended. Not only fisheries biologists, but population biologists in general have struggled for decades with the problem of identifying and quantifying the role of biotic interactions in determining the distribution and abundance of organisms. The current controversy over the role of interspecific competition in ecological communities (Connor and Simberloff 1979; Connell 1980; Roughgarden 1983; Strong et al. 1984) demonstrates that there is no consensus among ecologists regarding the importance of biotic interactions, the evidence needed to demonstrate their importance, or the methods appropriate for obtaining such evidence. We believe that until a credible scientific theory of population regulation is developed, it will not be possible to develop credible models of the long-term dynamics of fish populations.

Implications for Impact Assessment Studies

Clearly, many intuitively appealing approaches to predicting and measuring impacts of human activity on estuarine fish populations cannot be implemented because they (1) are incompatible with the institutional aspects of the assessment process, (2) require measurements that are impossible to perform, or (3) assume an understanding of estuarine ecology that does not currently exist. In particular, attempts at ecosystem-level quantification, whether based on the concepts of resistance–resilience (Holling 1973), energy flow (Odum 1971; Kemp et al. 1977), or mechanistic simulation modeling (Kremer and Nixon 1978) are unlikely to succeed, given the current decision-making process and degree of understanding of estuarine ecosystems. The idea of studying the whole ecosystem rather than a few of its component populations is intellectually appealing to all ecologists; however, we must agree with Van Winkle (1981) that, at present, populations are the most appropriate foci for assessment studies.

The Hudson River experience also demonstrates that, even if attention is restricted to fish populations, the inadequacy of existing understanding of regulatory processes precludes credible prediction of the long-term biological significance of power-plant-induced mortality. Moreover, regardless of the means used to predict impacts, only very large impacts are likely to be detected through routine plant-specific environmental monitoring programs. Given that as long as two decades might be required to demonstrate a 50% average reduction in abundance of a typical fish population (Van Winkle et al. 1981; Vaughan and Van Winkle 1982), decisions regarding the need for mitigation should not be made contingent on observing a decline in abundance of indicator species.

The Hudson River experience compels us to conclude that estimates of short-term impacts on fish populations (e.g., estimated reductions in year-class abundance) provide the best compromise between what we would like to know and what is possible to learn. When liberally supplemented with expert judgment, these estimates can provide a rational if imperfect basis for decision making. As described by Barnthouse et al. (1988, this volume), utility and agency scientists were able to use short-term impact estimates to identify species of concern, to compare the impacts of power plants and fishing on a common scale, and to evaluate the relative merits of alternatives for reducing power-plant-induced mortality. Identical or similar methods have been successfully employed in many other entrainment–impingement impact assessments (e.g., Christensen et al. 1975; USNRC 1979; Vaughan 1981).

Given the expense of even limited population-level assessment studies, efficient allocation of resources is essential. We concur with Yost's (1988) recommendations that project sponsors and regulatory agencies should work together to develop a mutually agreeable study plan, and the data collected should immediately be made available to all parties. The alternative to cooperation is a time-consuming and expensive adversarial process in which the sponsors design and conduct the studies without any input from the agencies and, after the sponsors present their findings, the agencies spend months studying project reports in an attempt to understand how the studies were performed and whether the data support the sponsor's findings. Even if direct sponsor–agency cooperation is legally or politically impossible, biologists could substantially improve the assessment process by developing objective criteria for determining the need for population-level studies. Barnthouse and Van Winkle (1988, this volume) and Klauda et al. (1988a) discussed the biological phenomena and plant-siting and design characteristics that influence the vulnerability of Hudson River fish populations to entrainment and impingement. A more general evaluation of power plant impact studies performed at a variety of sites, covering a variety of ecosystem types and plant designs, could be used as the basis for more general criteria. To ensure its credibility, any such study should be a joint industry–government effort.

The long-term need to understand and manage the many sources of direct and indirect mortality that affect estuarine fish populations, including not only power plants but also pollution and habitat modification, can be met only through long-term monitoring and basic ecological research. Long-term monitoring serves two purposes. First, it provides a means of testing predictions concerning qualitative and quantitative impacts. Assessments of future power plants would be greatly improved if good information were available on the impacts of existing plants. Second, the likelihood that adverse changes could be detected before they become irreversible is much greater if a long time series of comparable data is available. Despite their potential value, comprehensive postoperational monitoring programs are rare and, to our knowledge, no attempts have been made to synthesize the scattered data that are reported to state and federal regulatory agencies.

Research into the biotic and abiotic processes governing the production and persistence of fish populations is necessary both to discriminate between anthropogenic impacts and natural environmental variability and to make long-term predictions feasible. The entire ecosystem, not simply the fish populations themselves, should be the subject of this research. A complete understanding of these processes may never be achieved, but even incremental progress will improve future assessments.

It is unrealistic to expect the above kinds of studies to be conducted routinely as part of the licensing of power plants or other industrial facilities. Institutional constraints described in this paper, which are unlikely to change, limit most assessment studies to narrowly defined, short-term objectives. Long-term monitoring and research require funding and management independent of the regulatory process. We believe that such programs are critical for improving the scientific basis of environmental protection and resource management decisions.

The Hudson River utility companies have set an excellent example for other industries through their continued monitoring of year-class strength in selected fish populations (Barnthouse et al. 1988). However, the continuity of this effort is guaranteed only for the 10-year duration of the Hudson River settlement agreement, which will expire in 1990. Because the value of the data collected increases with every year the program continues, it would be highly desirable to insulate it from the 10-year permit cycle. The Hudson River Foundation established in the settlement (Barnthouse et al. 1988) provides an excellent example of the value of an independent funding institution. Besides sponsoring basic estuarine research, the Foundation has been active in public education and in fostering communication among scientists studying a variety of environmental problems affecting the Hudson River and other estuarine ecosystems. The activities of the Hudson River Foundation, coupled with a strong long-term monitoring program, provide the best hope that the problems encountered by contributors to this monograph will be less constraining to the next generation of assessment scientists.

Acknowledgments

We thank C. C. Coutant, W. Van Winkle, and S. G. Hildebrand for their thoughtful comments on this manuscript. Research was sponsored by the Office of Health and Environmental Research, U.S. Department of Energy, under contract DE-AC05-84OR21400 with Martin Marietta Energy Systems, Incorporated. This is publication 3018 of the Environmental Sciences Division, Oak Ridge National Laboratory.

References[3]

Atomic Safety and Licensing Appeals Board (of the U.S. Atomic Energy Commission). 1974. Decision in the matter of Consolidated Edison Company of New York, Inc. (Indian Point station, unit no. 2). Docket 50-247.

Barnthouse, L. W., J. Boreman, T. L. Englert, W. L. Kirk, and E. G. Horn. 1988. Hudson River settlement agreement: technical rationale and cost considerations. American Fisheries Society Monograph 4:267–273.

Barnthouse, L. W., and W. Van Winkle. 1988. Analysis of impingement impacts on Hudson River fish populations. American Fisheries Society Monograph 4:182–190.

Bergen, G. S. P. 1988. The Hudson River cooling tower proceeding: interface between science and law. American Fisheries Society Monograph 4:302–306.

Berggren, T. J., and J. T. Lieberman. 1978. Relative contribution of Hudson, Chesapeake, and Roanoke striped bass, *Morone saxatilis*, stocks to the Atlantic Coast fishery. U.S. National Marine Fisheries Service Fishery Bulletin 76:335–345.

Boreman, J. 1979. Life histories of seven fish species that inhabit the Hudson River estuary. Exhibit EPA-198, U.S. Environmental Protection Agency (Region II) Adjudicatory Hearing No. C/II-WP-77-01.

Boreman, J., and R. J. Klauda. 1988. Distributions of entrainable life stages of striped bass in the Hudson River estuary, 1974–1979. American Fisheries Society Monograph 4:53–58.

Christensen, S. W., D. L. DeAngelis, and A. G. Clark. 1977. Development of a stock–progeny model for assessing power plant effects on fish populations. Pages 196–226 in W. Van Winkle, editor. Assessing the effects of power-plant-induced mortality on fish populations. Pergamon, New York.

Christensen, S. W., and C. P. Goodyear. 1988. Testing the validity of stock–recruitment curve fits. American Fisheries Society Monograph 4:219–231.

Christensen, S. W., and R. J. Klauda. 1988. Two scientists in the courtroom: what they didn't teach us in graduate school. American Fisheries Society Monograph 4:307–315.

Christensen, S. W., W. Van Winkle, and P. C. Cota. 1975. Effect of Summit power station on striped bass populations. Testimony (March 1975) before the U.S. Atomic Energy Commission in the matter of Delmarva Power and Light Co., and Philadelphia Electric Co. (Summit power station, units 1 and 2). Dockets 50-450 and 50-451.

Connell, J. 1980. Diversity and the coevolution of competitors, or the ghost of competition past. Oikos 35:131–138.

Connor, E., and D. Simberloff. 1979. The assembly of species communities: chance or competition? Ecology 60:1132–1140.

---

[3]See Table 1 for sources of legal documents and unpublished reports pertaining to the Hudson River.

Cooper, J. C., F. R. Cantelmo, and C. E. Newton. 1988. An overview of the Hudson River estuary. American Fisheries Society Monograph 4:11–24.

DeAngelis, D. L., S. W. Christensen, and A. G. Clark. 1977. Responses of a fish population model to young-of-the-year mortality. Journal of the Fisheries Research Board of Canada 34:2124–2132.

Fletcher, R. I., and R. B. Deriso. 1988. Fishing in dangerous waters: remarks on a controversial appeal to spawner–recruit theory for long-term impact assessment. American Fisheries Society Monograph 4:232–244.

Hoff, T. B., J. B. McLaren, and J. C. Cooper. 1988. Stock characteristics of Hudson River striped bass. American Fisheries Society Monograph 4:59–68.

Holling, C. S. 1973. Resilience and stability of ecological systems. Annual Review of Ecology and Systematics 4:1–23.

Kemp, W. M., and six coauthors. 1977. Energy cost–benefit analysis applied to power plants near Crystal River, Florida. Pages 507–543 in C. A. S. Hall and J. Day, editors. Ecosystem modeling in theory and practice. Wiley-Interscience, New York.

Klauda, R. J., L. W. Barnthouse, and D. S. Vaughan. 1988a. What we learned about the Hudson River: journey toward an elusive destination. American Fisheries Society Monograph 4:316–328.

Klauda, R. J., J. B. McLaren, R. E. Schmidt, and W. P. Dey. 1988b. Life history of the white perch in the Hudson River estuary. American Fisheries Society Monograph 4:69–88.

Kremer, J. N., and S. W. Nixon. 1978. A coastal marine ecosystem. Springer-Verlag, Berlin.

Lawler, J. P. 1988. Some considerations in applying stock–recruitment models to multiple-age spawning populations. American Fisheries Society Monograph 4:204–218.

McFadden, J. T. 1977. An argument supporting the reality of compensation in fish populations and a plea to let them exercise it. Pages 153–183 in W. Van Winkle, editor. Assessing the effects of power-plant-induced mortality on fish populations. Pergamon, New York.

McLaren, J. B., T. H. Peck, W. P. Dey, and M. Gardinier. 1988. Biology of Atlantic tomcod in the Hudson River estuary. American Fisheries Society Monograph 4:102–112.

Odum, H. T. 1971. Environment, power, and society. Wiley-Interscience, New York.

Park, R. A., and 24 coauthors. 1974. A generalized model for simulating lake ecosystems. Simulation 21:33–50.

Roughgarden, J. 1983. Competition and theory in community ecology. American Naturalist 122:583–601.

Royama, T. 1977. Population persistence and density-dependence. Ecological Monographs 47:1–35.

Savidge, I. R., J. B. Gladden, K. P. Campbell, and J. S. Ziesenis. 1988. Development and sensitivity analysis of impact assessment equations based on stock–recruitment theory. American Fisheries Society Monograph 4:191–203.

Strong, D. R., D. Simberloff, L. G. Abele, and A. B. Thistle. 1984. Ecological communities: conceptual issues and the evidence. Princeton University Press, Princeton, New Jersey.

Swartzman, G. L., R. B. Deriso, and C. Cowan. 1978. Comparison of simulation models used in assessing the effects of power-plant-induced mortality on fish populations. University of Washington, College of Fisheries, Center for Quantitative Science, UW-NRL-10, Seattle.

Tanner, J. T. 1966. Effects of population density on growth rates of animal populations. Ecology 47:733–737.

Ulanowicz, R. E., and T. T. Polgar. 1980. Influence of anadromous spawning behavior and optimal environmental conditions upon striped bass (Morone saxatilis) year-class success. Canadian Journal of Fisheries and Aquatic Sciences 37:143–154.

USNRC (U.S. Nuclear Regulatory Commission). 1979. Draft environmental statement related to construction of New England power units 1 and 2. Dockets STN 50-568 and STN 50-569. NUREG-0529.

Van Winkle, W. 1981. Population-level assessments should be emphasized over community/ecosystem-level assessments. Pages 63–66 in L. Jensen, editor. Issues associated with impact assessment. EA Communications, Sparks, Maryland.

Van Winkle, W., K. D. Kumar, and D. S. Vaughan. 1988. Relative contributions of Hudson River and Chesapeake Bay striped bass stocks to the Atlantic coastal population. American Fisheries Society Monograph 4:255–266.

Van Winkle, W., D. S. Vaughan, L. W. Barnthouse, and B. L. Kirk. 1981. An analysis of the ability to detect reductions in year-class strength of the Hudson River white perch (Morone americana) population. Canadian Journal of Fisheries and Aquatic Sciences 38:627–632.

Vaughan, D. S. 1981. An age structure model of yellow perch in western Lake Erie. Pages 189–216 in D. G. Chapman and V. F. Galluci, editors. Quantitative population dynamics. International Cooperative Publishing House, Fairland, Maryland.

Vaughan, D. S., and W. Van Winkle. 1982. Corrected analysis of the ability to detect reductions in year-class strength of the Hudson River white perch (Morone americana) population. Canadian Journal of Fisheries and Aquatic Sciences 39:782–785.

Walters, C. J., and D. Ludwig. 1981. Effects of measurement errors on the assessment of stock–recruitment relationships. Canadian Journal of Fisheries and Aquatic Sciences 38:704–710.

Watt, K. E. F. 1968. Ecology and resource management. McGraw-Hill, New York.

Yost, T. B. 1988. Science in the courtroom. American Fisheries Society Monograph 4:294–301.

Young, J. R., R. J. Klauda, and W. P. Dey. 1988. Population estimates for juvenile striped bass and white perch in the Hudson River estuary. American Fisheries Society Monograph 4:89–101.

List of Technical Personnel Associated with the Hudson River Power Plant Case, 1964–1980

Inside the cover page of a Texas Instruments document titled "First Annual Report for the Multiplant Impact Study of the Hudson River Estuary" and dated July 1975, the following statement can be found:

This report is dedicated to the memory of DEAN R. SPAULDING who lost his life while seining at Con Hook on 30 June 1975 and to all the men and women who have labored under often hazardous conditions to obtain the data used in this report. Their courage, dedication, and integrity is the foundation upon which this project rests.

This quote serves as a reminder that a great many people other than authors represented in this monograph made significant contributions to the Hudson River power plant case. To acknowledge these individuals, we have attempted to list in this appendix those who contributed technically (i.e., in data collection, data analysis, technical consultation, or expert testimony) on behalf of any of the parties in the case. Because the monograph is a scientific document, we did not attempt to list individuals whose primary contributions were legal, administrative, or clerical. Aside from our own records and memories, we consulted the personnel departments of the Hudson River utility companies and wrote to the two major environmental groups (the Scenic Hudson Preservation Conference and the Hudson River Fishermen's Association) that played major roles in the case. Despite these efforts, we have undoubtedly omitted a few individuals whose names should have appeared. We sincerely apologize for these omissions.

Members and Consultants of Major Environmental Groups

E. Alexander	J. Clark	A. Glowka	E. Raney
J. Alexander	J. Cronin	M. Hair	R. Sandler
R. Boyle	R. Dagon	D. Perrone	E. Sworge
A. Butzel			

Employees and Consultants of U.S. and State Regulatory Agencies

L. Barnthouse	C. Coutant	G. Kinser	D. Robson
F. Blossum	M. Dadswell	B. Kirk	F. Rohlf
J. Boreman	D. DeAngelis	K. Kumar	B. Schubel
W. Bossert	R. Deriso	D. Lee	D. Seymour
R. Brandt	W. Dovel	S. Levin	R. Sharp
B. Brown	C. Edwards	R. Levins	A. Sherk
G. Cada	A. Eraslan	J. McCann	C. Shuster
J. Cannon	T. Fikslin	W. Muller	M. Sissenwine
F. Carlson	D. Flemer	T. O'Hare	L. Slobodkin
E. Carpenter	R. Fletcher	F. Ossiander	J. Sohn
H. Chadwick	J. Golumbek	B. Pastalove	S. Storey
S. Christensen	P. Goodyear	D. Policansky	W. Van Winkle
J. Clark	C. Hall	E. Radle	D. Vaughan
B. Cohen	R. Henshaw	P. Rago	L. Wareham
	E. Horn	W. Ricker	

Employees and Consultants of Hudson River Utilities

M. Aaronson	G. Abe	L. Abrams	S. Ackerman
B. Abarmo	K. Abood	G. Accarino	D. Akusis

M. Alavanja	D. Balint	M. Berquist	J. Branco
B. Albanese	T. Ballinger	A. Bertolino	R. Brandt
J. Alber	J. Bannerman	D. Bertone	C. Brannen
L. Albroneda	J. Barnes	C. Bessey	J. Braun
A. Alderman	R. Barnett	R. Bessey	J. Bray
R. Alevras	S. Barry	B. Beverly	A. Brebbia
H. Alex	E. Bartnett	G. Beverly	J. Brennan
R. Alley	F. Barton	J. Bible	R. Breyer
R. Allgauer	J. Bascietto	E. Bidoski	D. Bridge
S. Altensan	M. Baslow	D. Bigelow	B. Brienza
M. Altizer	D. Bath	M. Billy	R. Brier
T. Alworth	J. Battle	N. Bilyov	L. Brigman
L. Amanna	L. Bauer	P. Binns	E. Brissing
A. Amato	D. Baumann	D. Black	H. Brock
A. Amatruda	P. Baumann	M. Black	J. Brooks, Jr.
M. Ambrosi	C. Baummer	S. Blackwood	D. Brown
J. Amrhein	B. Baxter	M. Bladel	F. Brown
S. Amrhein	J. Bayless	E. Blair	C. Browne
B. Andersen	E. Bazarian	J. Blake	W. Brtalik
D. Anderson	R. Beatty	T. Blank	R. Brulotte
K. Anderson	R. Beazley	R. Blatt	A. Brundage
R. Anderson	C. Beckler	J. Block	A. Buatti
S. Angelone	C. Beebe	J. Blossum	G. Buchanan
R. Anselmo	D. Begbie	D. Blumel	S. Budney
K. Antonsen	K. Begbie	B. Bodian	T. Bull
G. Apicella	J. Beiler	N. Boehm	G. Bumpus
T. Arcadi	C. Beisser	S. Boettcher	S. Buneo
P. Arline	D. Beisser	S. Bohan	B. Burgan
R. Armao	D. Belcher	C. Bohr	C. Burge
K. Armistead	B. Belding	C. Bolz	J. Burghardt
R. Arndt	N. Belinko	J. Bomersbach	E. Burgher
L. Arnstein	B. Bell	W. Bonaventure	M. Burke
B. Arthur	F. Bell	W. Bonczek	P. Burr
R. Arthur	J. Bell	M. Bonomo	L. Bush
L. Arvidson	M. Bell	O. Boody	D. Butel
J. Asp	D. Bellamy	D. Booth	E. Butler
R. Atherton	P. Belyea	M. Booth	K. Butler
S. Atkins	S. Bender	Y. Boozier	M. Butler
D. Auclair	J. Beni	D. Borgatti	R. Byer
A. Auld	D. Benjamin	R. Bosco	A. Byers
D. Avery	J. Bennett	E. Botka	L. Bynum
F. Aydin	S. Bennett	B. Boulware	A. Byrd
	J. Benson	D. Bow	T. Byrne
B. Babcock	J. Benton	W. Bowden	
G. Back	T. Benton	K. Bowman	D. Cabe
L. Badalucco	G. Beretta	D. Bowser	S. Cabrera
K. Badlam	P. Berg	D. Boy	S. Cackowski
D. Bagley	E. Berger	G. Boyhen	W. Cahill
R. Bagley	R. Berger	D. Boykins	L. Cahson
D. Bailey	C. Berggren	P. Bozoin	W. Calabrese
C. Baker	T. Berggren	D. Bradshaw	C. Caldwell
H. Baker	A. Berkowitz	D. Brady	D. Caldwell
E. Baldocchi	R. Bernard	L. Brady	N. Caldwell
J. Baldwin	S. Bernard	R. Bragg	R. Caldwell

R. Califano	F. Chin	W. Cooper	M. Daley
R. Callahan	R. Christie	A. Cordisco	R. Daniel
T. Callina	B. Christopher	G. Corley	S. Dansler
K. Campbell	J. Chu	R. Corn	M. Darder
C. Cann	L. Chu	M. Cornell	C. Darrah
P. Cann	P. Chu	M. Corneymyer	J. Davanzo
T. Cann	S. Church	G. Correale	B. David
W. Cann	L. Ciaccio	E. Corrigan	C. Davidowsky
T. Cannon	J. Cianci	T. Cosper	D. Davies
F. Cantelmo	S. Cieplik	L. Costa	P. Davino
G. Capuano	A. Ciesluk	R. Costa	D. Davis
P. Capuano	E. Cirenza	R. Costanza	R. Davis
D. Caputo	P. Clapp	J. Costello	D. Dawkins
C. Carayas	B. Clark	P. Cota	D. Dawson
J. Cardenas	E. Clark	A. Councill	M. Day
C. Cardiges	S. Clark	J. Cove	D. Dean
S. Carey	C. Clarke	G. Cowherd	J. DeBellis
F. Carlson	R. Clarke	S. Cowton	W. Decken
C. Carman	T. Class	S. Coyle	N. Decker
D. Carminucci	P. Claudio	C. Cozart	P. Decker
K. Carney	M. Clayborne	K. Craig	J. DeGeorge
J. Carnright	R. Clearwater	R. Crana	M. Degregorio
R. Carreau	E. Clock	F. Crawford	J. Delgreco
A. Carriveau	J. Clock	W. Crawford	R. Dellechiai
B. Carroll	S. Close	R. Creeden	D. Deloatch
D. Carroll	B. Cobey	D. Crestin	J. DelPup
J. Carroll	J. Cochran	J. Crittenden	S. DelPup
M. Carroll	W. Coffin	M. Croce	L. DeLuca
M. Carter	P. Cogburn	J. Croom	R. Deluca
E. Carulli	E. Cohen	M. Crouse	B. Demaio
R. Caruso	K. Cohen	J. Crowley	M. DeMarco
D. Carver	B. Cole	W. Crum	D. DeMaria
S. Casne	H. Cole	F. Cuccia	E. Demers
L. Cassa	W. Cole	T. Culbertson	J. Demetriou
K. Cassidy	S. Colman	J. Cullen	D. Denisi
S. Cassidy	F. Colon	T. Cullen	I. Denney
J. Castrogiovanni	J. Condello	E. Cummerow	R. Dennis
V. Catalmo	D. Condon	R. Cummins	M. Denove
A. Cathey	M. Conenello	P. Cuneo	R. DePace
D. Catlin	S. Congdon	P. Cunningham	J. Derevjanik
W. Catlin	R. Congedo	K. Curry	A. DeRoberts
P. Cattania	L. Conklin	M. Curry	V. DeThomasis
J. Caufield	S. Conley	L. Curtis	G. Devens
C. Cavicchio	J. Connors	M. Curtis	J. Devine
C. Cawein	P. Conoyer	D. Cushing	A. Devita
K. Cea	M. Considine	D. Cutter	R. DeVries
B. Cella	J. Consoli	N. Cutwright	B. Dew
R. Cerasoli	L. Conte		W. Dey
J. Chailer	G. Cook	C. D'Agostino	N. Diaz
T. Chambers	R. Cook	R. D'Amica	R. Dicapua
P. Chandra	E. Cooke	E. Dafgard	M. Dichiaro
M. Charpentier	M. Cooke	H. Dagaev	T. Dickson
A. Cheifetz	J. Cooper	T. Dahlem	L. Diehl
H. Chen	S. Cooper	C. Daley	K. Dillard

P. Dillon	J. Eldridge	E. Fiske	R. Furlong
B. Dingee	K. Eldridge	P. Fitch	
R. Dingee	P. Eldridge	P. Fithian	E. Gaillard
E. Dino	D. Elliot	J. Fitzgerald	J. Gallagher
L. Ditommaso	L. Ellis	B. Fitzpatrick	P. Gallagher
D. Dizefalo	R. Ellis	L. Flach	W. Galloway
J. Dobi	R. Ellsworth	J. Flanagan	M. Gamerman
C. Dobin	R. Ellwood	L. Flanagan	M. Gardinier
R. Doering	E. Elmiger	P. Flaschenberg	T. Gardinier
P. Doniger	P. Englehardt	R. Flaxman	T. Garland
J. Donlan	T. Englert	S. Fleckenstein	M. Garlinghouse
M. Dorfman	G. Entzminger	P. Fleming	F. Garrabrant
G. Dorosz	W. Epps	N. Flieger	T. Garrett
B. Dorris	T. Ermenville	L. Floersheimer	D. Garretto
M. Doshi	L. Ertell	C. Flynn	B. Garrison
V. Doshi	D. Evans	E. Fodrill	D. Garrison
J. Dougher	W. Ewald	J. Foley	P. Gaynor
W. Dougherty		F. Ford	T. Geary
T. Dowler	B. Fairbanks	J. Ford	C. Gebo
B. Dowling	P. Fairbanks	R. Formica	A. Geiger
J. Downing	M. Fairman	E. Fortune	S. Geiger
D. Driever	B. Faist	J. Foster	K. Gelman
P. Driscoll	S. Fanning	L. Foster	M. Gelman
L. Drost	R. Fares	R. Foster	D. Gelok
G. Drucker	J. Farmer	S. Foster	M. Gentile
W. Drummer	D. Farnam	D. Fotopoulos	P. Gentile
R. Dube	I. Farnam	J. Fowler	C. George
E. Dubolsky	T. Farnam	G. Fox	S. George
D. Dubose	P. Farrell	T. Fox	D. Gerbitz
D. Duchin	E. Farrington	J. Fraine	L. Gerry
D. Duden	J. Fava	E. Franct	F. Gertsen
N. Dumser	A. Feldman	R. Fraser	D. Gessler
W. Duncan	J. Feldman	S. Frederick	R. Gessler
B. Dunn	J. Feldsine	T. Freeland	J. Gettier
D. Dunning	J. Fellenzer	P. Fremgen	W. Gewant
C. Dureyea	J. Feltman	D. Fremont	J. Giacchetti
L. Durfey	R. Fenner	M. Fremont	S. Giacobello
D. Durland	L. Feraca	M. French	J. Gibson
C. Durrett	L. Ferenz	S. French	A. Gieger
P. Duskin	M. Ferenz	R. Frick	J. Gift
F. Dutcher	M. Ferguson	J. Friedman	V. Gigante
	R. Ferguson	S. Friedman	E. Gilbert
T. Earle	M. Feriel	B. Friedmann	D. Ginder
J. Early	G. Ferrante	E. Frost	T. Ginn
R. Edson	L. Ferrante	R. Frost	F. Giordano
B. Edwards	S. Ferrer	R. Fry	J. Gladden
S. Edwards	T. Festa	B. Fulton	B. Gleason
J. Egan	C. Figorito	N. Funicelli	M. Glynn
D. Ehorn	W. Fil	L. Furgal	T. Goertemiller
L. Ehrsam	R. Finkeldie	C. Furlong	R. Goff
D. Eignor	M. Fischer	K. Furlong	S. Goff
M. Eisenbud	G. Fisher	M. Furlong	D. Goggin
E. Ekman	J. Fisher	N. Furlong	W. Going
F. El-Shamy	S. Fisher	P. Furlong	R. Goldsmith

R. Goldstein	T. Hanna	T. Hilliard	M. Iagnocco
G. Goldwyn	J. Hannon	W. Hinelstein	J. Illovsky
B. Golemboski	B. Hanover	G. Hitchcock	S. Infante
J. Golomb	C. Hansen	D. Hjorth	P. Inman
V. Gonella	J. Hansen	P. Ho	G. Irsich
M. Goodale	R. Hansen	F. Hoell	R. Irvine
S. Goodbred	M. Hapeman	D. Hoerr	J. Isaacs
T. Goode	R. Haras	J. Hoff	P. Isaacson
G. Goodman	A. Hardesty	T. Hoff	D. Israel
S. Gordon	P. Hark	C. Hoffacker	C. Italiano
S. Gottehrer	S. Harney	H. Hoffer	D. Ivanick
C. Gove	A. Harris	R. Hoffman	M. Ivanick
S. Grabe	H. Harris	F. Hogan	
V. Grabe	M. Harris	T. Hogan	D. Jackson
J. Grace	P. Hart	W. Hohmann	G. Jacob
P. Graham	E. Hartgrove	D. Hollenbeck	I. Jacobs
R. Graham	J. Harvey	J. Hollister	V. Jamison
D. Grannas	F. Haug	D. Holly	J. Jannarone
D. Grass	N. Havens	J. Holsapple	E. Jappen
G. Gray	G. Havranek	P. Holt	V. Jeanty
T. Greco	S. Hayes	M. Holtzner	M. Jenkens
S. Greene	S. Hayford	M. Holzer	B. Jennings
J. Greenwell	T. Hayhurst	L. Hondo	T. Jennings
R. Greer	J. Hecht	W. Honeycutt	L. Jensen
K. Gref	G. Hecker	C. Honeywell	H. Jett
J. Grim	R. Heffner	P. Hontz	J. Jhaveri
M. Grimm	A. Heindl	L. Hoodes	S. Jinks
P. Grimm	A. Heller	S. Hope	J. Joa
S. Grimm	M. Heller	H. Hopgood	L. Joachim
J. Gross	R. Heller	T. Horst	C. Johanny
D. Grosse	J. Hellmann	R. Horstman	B. Johnson
K. Grosso	J. Hemingway	R. Horton	C. Johnson
T. Grove	G. Hempstead	D. Howe	J. Johnson
H. Groves	P. Hempton	S. Howe	R. Johnson
G. Gruntowicz	M. Henderson	D. Howell, Jr.	S. Johnstone
H. Guigli	R. Henning	W. Howell	G. Jolicoeur
G. Gundersen	W. Henry	G. Howells	N. Jolly
S. Gunsalus	D. Hernandez	J. Huang	D. Jones
R. Gurgui	K. Hernandez	K. Hubbard	E. Jones
T. Gustainis	B. Herring	S. Hubley	G. Jones
	D. Herring	H. Hudson	L. Jones
J. Haas	W. Herring	L. Hudson	R. Jones
C. Habel	F. Hershkowitz	S. Hufnagel	S. Jones
L. Haffen	R. Hertz	T. Huggins	K. Juba
M. Hahn	T. Hess	R. Hulit	R. Juby
D. Haith	L. Hibbitts	J. Hulme	S. Julie
M. Hajny	J. Hickam	E. Humphries	A. Jung
J. Hall	J. Hickman	S. Huns	C. Juritsch
H. Hallenbeck	J. Higgins	L. Hunt	R. Juritsch
C. Hamilton	R. Hildreth	B. Hunter	
J. Hamilton	E. Hilinski	M. Hurd	K. Kabat
B. Hamm	M. Hilinski	J. Huryn	V. Kaczinski
S. Hammalian	J. Hill	J. Hutchison, Jr.	R. Kakerbeck
T. Handsel	J. Hillegas	J. Hutlock	A. Kane

D. Kane	A. Klein	T. Lanotte	D. Lord
V. Kane	C. Klein	N. Lansing	L. Lovas
R. Kao	D. Klibinoff	G. Lanza	K. Lovin
D. Kaplan	S. Kling	J. Lanzilotti	D. Lowell
F. Kaplan	M. Klotz	A. Larocca	C. Lu
R. Kapp	D. Kmckinstrie	J. Larsen	S. Lucas
R. Karfiol	M. Knafou	L. Larsen	C. Luce
A. Karpel	D. Kneip	R. Larsen	D. Luce
S. Kartiganer	S. Kneip	K. Lasher	D. Luciano
D. Kassell	V. Kobos	M. Laspina	R. Ludt
B. Katz	Z. Kobos	L. Lasure	T. Lujan
S. Kaulfers	K. Koenig	W. Latenser	M. Lulu
P. Kayko	M. Koksvik	D. Latimer	C. Lundy
H. Kazemi	J. Kolb	G. Lauer	R. Lutts
M. Keefe	K. Konopka	J. Lawler	R. Lutz
R. Keegan	K. Konrad	K. Lawler	B. Lyddon
R. Keene	K. Kopera	T. Lawler	F. Lyles
P. Keesser	F. Korensky	G. Lawley	A. Lyman
P. Keibel	G. Kornheisl	J. Leavitt	D. Lynch
M. Keicher	J. Koskella	N. Lebo	R. Lyons
R. Keiser	R. Koski	R. Lee	B. Lytle
W. Kelleher	R. Kossman	E. Leff	
D. Keller	D. Koster	C. LeGuen	E. MacHaffie
K. Kelly	R. Koster	M. Lehr	D. MacIver
N. Kelly	C. Kozlow	N. Leleau	D. Mackey
J. Kenealy	S. Kramer	M. Lemakis	K. Mackey
B. Kenton	G. Kreamer	D. LeMarie	T. Macur
J. Kepler	W. Krempel	R. Lennon	R. Madow
R. Keppel	W. Kretser	R. Lentz	M. Maduakolam
M. Kersaint	R. Krol	J. Leslie	M. Maher
P. Kiang	R. Krueger	D. Lettieri	J. Mahoney
M. Kihm	K. Krychear	T. Levine	K. Mahoney
T. Kilcer	L. Krychear	F. Lewandowski	P. Mahoney
J. Kilduff	D. Kuhn	N. Lewi	E. Maikish
W. Kilgore	J. Kunowski	W. Lewis	S. Majeski
R. Killam	M. Kunzmann	L. Libordi	J. Major
D. Kimball	M. Kuperman	M. Lickers	P. Maksden
D. Kimble	C. Kutas	V. Lickers	J. Maldonado
R. Kimmel		M. Ligotino	F. Mallick
J. King	M. Labanowski	R. Ligotino	I. Mallick
L. King	B. Labrenz	A. Lin	M. Malloy
T. King	E. Lacouara	B. Lindstrom	C. Maloy
W. King	E. Ladny	B. Lippincott	J. Maniscalco
E. Kinslow	R. LaForge	L. Liso	R. Manlove
W. Kinslow, Jr.	C. Laidlaw	D. Lispi	D. Mann
S. Kiraly	R. Lamar	B. Little	R. Mann
K. Kirby	D. Lamparelli	J. Little	N. Manter
S. Kirby	R. Land	L. Livsey	J. Manuso
S. Kiremidjian	R. Landa	M. Loftus	T. Marasco
W. Kirk	V. Lander	C. Logan	K. Marcellus
T. Kivisalu	W. Landy	D. Logan	D. Maret
S. Kizer	K. Lane	S. Lopez	L. Marino
K. Klauda	K. Lang	R. Lopopollo	T. Mark
R. Klauda	R. Lange	R. Lorberbaum	C. Marshall

J. Marshall	D. Meddaugh	M. Moran	R. Nielson
W. Marshall	W. Meeks	S. Moriates	R. Nisbet
C. Marsin	A. Megdanis	D. Moro	Y. Nixon
R. Marsin	G. Mehta	C. Moroney	D. Nodhturft
G. Marston	J. Meister	S. Morphew	T. Noonan
J. Martin	R. Meistrich	A. Morris	C. Normann
P. Martin	J. Meldrin	C. Morris	L. Norris
R. Martino	F. Melio	T. Morris	R. Norris
R. Marxreiter	L. Meloro	J. Morrison	J. Nutant
W. Mason	C. Meltz	R. Morrison	
A. Masotti	C. Menzie	V. Morrison	M. O'Brien
M. Master	D. Merriman	B. Morrissey	A. O'Connell
M. Mateo	K. Mersery	W. Morton	C. O'Connell
J. Matousek	D. Mesec	C. Moser-Saylor	R. O'Connell
R. Matsis	D. Metcalf	D. Moss	A. O'Connor
J. Matthews	G. Metti	H. Moutal	D. O'Connor
P. Matthi	G. Meyer	P. Mowery	J. O'Connor
M. Mattson	E. Meyers	S. Moy	K. O'Connor
M. May	J. Mikolajczyk	B. Muchmore	S. O'Connor
D. Mayercek	S. Milazzo	P. Muchmore	R. O'Donnell
D. Mayhew	A. Miller	J. Mudge	M. O'Handley
K. McCall	D. Miller	P. Muessig	B. O'Neill
A. McCammon	G. Miller	J. Mularadelis	M. O'Prey
E. McCann	J. Miller	G. Mulford	E. Oakley
J. McCann	K. Miller	H. Mullane	G. Oakley
L. McCarthy	M. Miller	H. Mulligan	M. Oakley
R. McCarthy	N. Miller	J. Mundy	S. Oakley
G. McCormick	W. Miller	I. Murawaka	P. Occhiogrosso
H. McCoy	L. Milliger	L. Murdie	T. Occhiogrosso
M. McDonald	K. Milliken	A. Murphy	M. Ochoa
S. McDonald	E. Mills	K. Murphy	S. Ochs
J. McElhenney	J. Miner	M. Murphy	N. Oglesby
K. McElroy	R. Miniter	T. Murphy	A. Oguntuase
N. McEwen	A. Minthorn	J. Murray	D. Olympia
J. McFadden	F. Mintun	K. Murray	K. Omland
R. McFerran	A. Mitchell	Y. Mussalli	W. Ondler
D. McGilvray	M. Miyasaki	F. Muthig	B. Ood
M. McGinn	P. Moccio	J. Myers	J. Opiekum
K. McGrath	D. Moerman		J. Oprandy
J. McGregor	J. Moerman	A. Nachman	K. Orimertl
T. McGrief	F. Moffo	V. Naclerio	P. Orrick
P. McGroddy	J. Mohr	M. Naierman	G. Ortiz
C. McGuire	M. Moment	S. Naplachowski	B. Ortquist
E. McGurgan	M. Monjeau	C. Nash	M. Ortquist
B. McKenna	S. Monroe	W. Neff	T. Orvosh
D. McKenzie	H. Monsees	P. Neito	D. Ostrye
W. McKeon	D. Monteleone	A. Neitupski	M. Otter
G. McKillop	A. Montemurro	L. Nekvapil	H. Ottey
P. McKillop	C. Montgomery	D. Neumann	Y. Ottoviano
J. McLaren	R. Mood	B. Newman	C. Ouellette
B. McLaughlin	L. Moon	K. Newmann	D. Outlaw
A. McLeod	T. Moore	C. Newton	K. Owens
K. McLeod	R. Moos	G. Newton	
L. McQuiston	R. Morales	O. Nichols	E. Paccione

A. Packard
E. Page
C. Palen
D. Palochko
J. Panarelli
P. Panchak
M. Pappas
P. Parekh
A. Parisi
C. Parker
K. Parker
F. Parkinson
R. Parks
L. Partee
S. Pasternack
T. Peck
R. Peplow
J. Perazzo
A. Perlmutter
E. Perry
L. Perry
C. Peterson
S. Peterson
B. Petrosky
D. Pettinato
F. Phelan
T. Philbin
C. Phillips
J. Phillips
D. Piccorelli
N. Piggery
J. Pike
K. Pilon
N. Pinerio
D. Piniat
B. Pinney
D. Pizzuto
J. Platt
R. Pomerantz
B. Ponwith
J. Porpiglia
H. Porter, Jr.
A. Potter
D. Powers
C. Pratt
M. Predmore
G. Prescott
S. Pristash
F. Pritchard
J. Privitera
P. Probst
M. Prokopchak
R. Pugh

N. Quackenbush
N. Quenzer
A. Quill
K. Quinn
R. Quinn
T. Quirk

J. Raccuglla
J. Racino
P. Raczynski
M. Randolph
E. Raney
M. Raney
E. Rankin
A. Rapco
M. Rappaport
S. Ray
J. Raymo
J. Read
G. Reed
J. Reichle
L. Reichman
R. Reider
C. Reilly
R. Reincker
B. Reineke
P. Reis
E. Reisenhauer
J. Reisinger
C. Reitz
O. Renfroe
L. Resnick
G. Reusch
C. Reynold
E. Reynolds
F. Reynolds
L. Rezendes
A. Rhein
R. Rhoads
A. Riccardi
D. Ricci
M. Ricci
V. Ricci
J. Richards
P. Richards
W. Richardson
J. Richburg
J. Richichi
A. Richner
D. Ries
J. Rigano
J. Riley
K. Riner
M. Riner

P. Ringlehan
T. Rippilon
E. Ritchie
C. Rittenhouse
S. Rives
R. Roberts
S. Roberts
T. Robinson
M. Robison
R. Rockefeller
S. Rodgers
G. Rodriguez
L. Rodriquez
S. Roeseke
B. Rogers
D. Rogers
R. Rogers
J. Rommel
L. Roosa
S. Roosa
R. Rosa
T. Rosario
A. Rosato
I. Rose
M. Rose
R. Rose
J. Rosenthal
M. Rosenthal
D. Ross
G. Ross
M. Ross
N. Ross
Q. Ross
G. Roth
R. Rothwell
A. Rouse
R. Rowland
A. Roy
R. Rubel
M. Rubin
B. Ruble
R. Ruby
W. Rue
A. Ruffel
G. Ruffino
R. Rulifson
D. Russell
P. Russell
C. Russo
T. Rutledge
J. Ryan
D. Rydzak

E. Sacca

K. Salamon
J. Salerno
A. Salkin
J. Saltes
C. Salvo
R. Sampson
J. Samuels
J. Sands
R. Sandy
S. Santiago
R. Santmire
M. Santora
D. Sarvbbi
M. Sarvis
J. Sassano
L. Saunders
I. Savidge
K. Sawnson
F. Scala
P. Scala
D. Scanlan
M. Scarpa
J. Schaeffer
D. Schaezler
G. Scheinin
J. Scherer
J. Scheurick
C. Schilling
P. Schilling
P. Schindo
R. Schirripa
R. Schlicht
G. Schlosser
E. Schmidt
R. Schmidt
D. Schmitt
M. Schmitt
B. Schneider
J. Schneider
R. Schnell
C. Schnier
C. Schoder
L. Schoonmaker
R. Schreiber
J. Schultz
G. Schumacher
D. Schupack
A. Schupp
D. Schussler
G. Schuster
B. Schwartz
L. Scifo
A. Scivolette
B. Scofield

J. Scott
L. Scotton
C. Scotzko
F. Scribner
L. Scrofani
C. Sedivec
J. Seebach
G. Seegert
R. Seela
J. Sefcik
P. Segraves
N. Seifert
E. Seifts
L. Seltzer
P. Seplow
J. Serven
S. Seth
R. Setterlund
T. Severino
J. Seymour
S. Seymour
L. Shackelford
U. Shah
S. Shaner
J. Shannon
R. Shapot
S. Shapot
R. Sharma
L. Shaternik
B. Shepard
D. Shepard
P. Shepard
M. Sherer
S. Shiboski
T. Shiel
N. Shih
J. Shirk
J. Shorey
J. Shulman
E. Shumbris
J. Shute
M. Siciliano
C. Sicina
C. Sidor
P. Siebold
M. Sieman
B. Siers
T. Sigmon
T. Silvertein
J. Sime
J. Simons
T. Simpson
D. Singlemann
S. Sininsky

R. Sinos
J. Skelly
K. Skelly
B. Skolnik
P. Slaight
J. Slater
K. Sloat
B. Small
C. Small, III
A. Smart
D. Smat
B. Smith
C. Smith
D. Smith
E. Smith
M. Smith
P. Smith
R. Smith
S. Smith
V. Smith
H. Smothers
J. Snedeker
W. Snow
T. Snyder
W. Soderberg
Z. Soderberg
K. Soeder
S. Solner
D. Solon
P. Soloway
E. Soper
D. Soracco
C. Soranno
R. Sosnowski
C. Soutar
P. Souza
P. Spana
D. Spaulding
W. Speers
S. Spenser
C. Squicciarino
M. Sramek
M. St. Peter
P. St. Pierre
C. Staats
D. Stack
B. Stamp
S. Stanely
A. Staples
A. Steen
B. Stein
K. Stephens
W. Stepien
E. Stern

F. Stevens
R. Stevens
W. Stevens
P. Stewart
R. Stimpfle
R. Stira
D. Stiriz
C. Stokes
R. Stokes
P. Storm
S. Storms
I. Stover
C. Stowell
K. Straut
D. Strickert
B. Stroppel
M. Stroppel
F. Stroup
D. Strout
L. Strout
J. Strube
R. Sugatt
D. Sugerman
A. Sullivan
B. Sullivan
J. Sung
I. Sussman
N. Suway
W. Suway
C. Swartz
E. Sweitzer
T. Swetz
W. Sydor
D. Symonds
R. Szechtman
S. Szedlmayer
F. Szeli
E. Szymanowicz

M. Tabery
B. Tabor
E. Tacopina
S. Tacopina
E. Taft
A. Tangredi
L. Taormino
W. Tarplee
M. Tartanian
M. Tatro
G. Tauber
D. Taylor
H. Taylor
J. Taylor
M. Taylor

C. Teece
L. Teitelbaum
G. Templeton
R. Tephabock
B. Tervenski
T. Theisen
M. Thomas
J. Thompson, Jr.
V. Thompson
R. Tiedeman
F. Tiene
R. Tierney
R. Till
R. Tillman
S. Tilmont
D. Tolderlund
G. Tomaselli
R. Toole
G. Torpy
O. Torres
B. Toscano
J. Trainor
J. Trapani
K. Trazino
G. Treacy
J. Trizzino
R. Trottier
J. Tucker
J. Turetsky
B. Turnbaugh
G. Turnbaugh
L. Tuttle
R. Tuttle

S. Ugolino
M. Uhlfelder
B. Ullrich
F. Unangst
N. Unger
S. Unseld
R. Uy

D. Vacek
P. Valente
J. Valentine
C. Van Cleave
S. Van Dijk
R. Van Hoesen
T. Vanderbeek
I. Vanderpot
M. Vantosh
J. Vanwick
D. Vanwyck
J. Vanwyck

J. Vecchione
H. Velasquez
A. Venuto
D. Versteeg
S. Vibert
M. Victor
J. Vigliotti
S. Vignola
J. Villeto
M. Visalli
D. Voigts
J. Volkman
H. VonBorstel
S. Voor

R. Wahler
B. Waisala
W. Waisala
B. Wald
D. Wallace
S. Wallenstein
T. Waller
W. Waller
C. Walsh
M. Walsh
P. Walsh
R. Walsh
M. Walters
C. Wang
J. Wang
N. Wasserman
C. Watson
D. Watson
L. Watson
M. Watson
L. Watters

J. Waxman
A. Weber
W. Wegner
L. Wehr
M. Weinstein
S. Weisman
J. Weiss
S. Weiss
B. Weissert
J. Weitzel
R. Wells
T. Wells
P. Welsch
B. Wendler
M. Weng
C. Werner
J. Werner
W. Werth
E. West
M. Westerfield
D. Westin
K. Wheelan
S. Wheelan
C. Wheeler
J. Wheeler
R. Wheeler
K. White
L. White
R. White
S. White
L. Whitley
E. Whitney
L. Whitney
D. Whittaker
T. Whitted
J. Whitworth

K. Wich
B. Wickes
C. Wideberg
R. Wieder
A. Wiedow
H. Wiegard
K. Wiessner
C. Wigand
K. Wigderson
M. Wilcken
I. Wilenitz
D. Wiley
H. Wiley
D. Wilkes
B. Williams
D. Williams
F. Williams
J. Williams
K. Williams
R. Williams
D. Wilson
L. Wilson
R. Wilson
E. Winser
K. Winters
D. Witman
J. Wodarski
M. Wodka
V. Wong
W. Wood
B. Woodard
H. Woodbury
L. Woodstock
H. Woodward
R. Woodyear

J. Wright
R. Wyman

C. Yacopino
P. Yaffie
D. Yamasaki
R. Yannuzzi
C. Yass
M. Yeaple
B. Yedvobnick
H. Yeh
J. Yost
L. Yost
J. Young

H. Zajd, Jr.
J. Zambrano
H. Zeliger
H. Zeller
K. Zenobia
T. Zent
C. Ziegler
J. Ziesenis
D. Zimmer
J. Zimmerman
H. Zion
F. Zito
Z. Zo
M. Zoller
J. Zoukis
S. Zrake
K. Zseleczky
L. Zubarik
J. Zuck
P. Zweiacker

Author Index

Asterisks (*) denote senior authors; C denotes comment